P9-APJ-901

INTEGRATION OF FUNDAMENTAL POLYMER SCIENCE AND TECHNOLOGY

The proceedings of an international meeting on polymer science and technology held at Rolduc Abbey, Limburg, The Netherlands, 14–18 April 1985

INTEGRATION OF FUNDAMENTAL POLYMER SCIENCE AND TECHNOLOGY

Edited by

L. A. KLEINTJENS

DSM Central Research, Geleen, The Netherlands

and

P. J. LEMSTRA

Eindhoven University of Technology, The Netherlands

ELSEVIER APPLIED SCIENCE PUBLISHERS
LONDON and NEW YORK

ELSEVIER APPLIED SCIENCE PUBLISHERS LTD
Crown House, Linton Road, Barking, Essex IG11 8JU, England

Sole Distributor in the USA and Canada
ELSEVIER SCIENCE PUBLISHING CO., INC.
52 Vanderbilt Avenue, New York, NY 10017, USA

WITH 43 TABLES AND 359 ILLUSTRATIONS

British Library Cataloguing in Publication Data

Integration of fundamental polymer science and
technology.
1. Polymers and polymerization
I. Kleintjens, L. A. II. Lemstra, P. J.
547.7 QD381

Library of Congress Cataloging in Publication Data

Integration of fundamental polymer science and technology.

Bibliography: p.
Includes index.
1. Polymers and polymerization—Congresses.
I. Kleintjens, L. A. II. Lemstra, P. A.
TP1081.156 1986 668.9 85-27465

ISBN 0-85334-416-7

Printed in Great Britain by Galliard (Printers) Ltd, Great Yarmouth

FOREWORD

'Integration of Fundamental Polymer Science and Technology' is a theme that admits of countless variations. It is admirably exemplified by the scientific work of R. Koningsveld and C. G. Vonk, in whose honour this meeting was organized. The interplay between 'pure' and 'applied' is of course not confined to any particular subdiscipline of chemistry or physics (witness the name IUPAC and IUPAP) but is perhaps rarely so intimate and inevitable as in the macromolecular area. The historical sequence may vary: when the first synthetic dye was prepared by Perkin, considerable knowledge of the molecular structure was also at hand; but polymeric materials, both natural and synthetic, had achieved a fair practical technology long before their macromolecular character was appreciated or established.

Such historical records have sometimes led to differences of opinion as to whether the pure or the applied arm should deserve the first place of honour. The Harvard physiologist Henderson, as quoted in Walter Moore's *Physical Chemistry*, averred that 'Science owes more to the steam engine than the steam engine owes to Science'. On the other hand, few would dispute the proposition that nuclear power production could scarcely have preceded the laboratory observations of Hahn and Strassmann on uranium fission. Whatever history may suggest, an effective and continuous working relationship must recognize the essential contributions, if not always the completely smooth meshing, of both extremes. An outsider gains the impression that this principle must have long been firmly recognized at DSM in order to have enabled Koningsveld and Vonk to make the broadly impressive contributions that are here recognized. The vagaries of the market-place have led in many industrial establishments to painful fluctuations in fundamental research effort, generally to no useful ultimate purpose. The papers assembled in this volume could, in our view, achieve no more desirable goal than to create strong impressions that might contribute to the stabilization of long-range research programmes.

(As a frivolous aside, we may recall that a similar question, regarding the relative importance of words and music, has long plagued opera-lovers and has inspired some pithy replies. When asked whether text or musical score was more important to his light operas with Sullivan, W. S. Gilbert answered by the single word 'Yes!' Actually, the science/technology dualism itself has been musically addressed by the composer Koningsveld in the second movement, 'Contrasts: Polypenteneamer and Phantom Networks', of his *Polymer Music*, Op. 2.)

The programme of the Rolduc meeting was divided into a number of areas, several of which (Thermodynamics, Blends, Morphology) embrace directly the major research interests of the honorees. The other major sections (Diffusion and Barriers, Processing and Rheology, Chain Dynamics, Characterization and Solution Behaviour, New Developments) round out a full picture of modern polymer research and its technological spinoffs. Readers of this volume, as well as attendants at the Rolduc gathering, owe thanks to DSM for its sponsorship, and above all a debt of profound gratitude to Drs P. Lemstra and L. Kleintjens and their helpers for organizing the meeting and for their herculean efforts during its actual course. They have set an example which one hopes will inspire many followers.

WALTER H. STOCKMAYER
Dartmouth College
Hanover, New Hampshire, USA

PREFACE

In April 1985, Ron Koningsveld and Chris Vonk of DSM Central Research retired after a lifetime devoted to the transfer of fundamental polymer science to technology. For his work on the integration of academic and industrial research, the University of Bristol awarded the degree of Doctor of Science *honoris causa* to Ron Koningsveld, announced during the conference and presented on 9 July. The simultaneous retirement stimulated the idea of organizing an international discussion meeting on the theme 'Integration of Fundamental Polymer Science and Technology'.

The aim of the meeting was to stimulate discussions between both academic and industrial polymer scientists and processing engineers. About 250 participants took part in the conference, which was held in Rolduc Abbey, a well-preserved medieval monument in the rural province of Limburg, The Netherlands.

In close consultation with the advisory committee, Sir Geoffrey Allen (London), Sir Charles Frank (Bristol), Professor G. Smets (Louvain), Professor W. Beek (Rotterdam) and Professor G. Menges (Aachen), the organizing committee, encouraged by honorary chairman A. H. de Rooij (Research Director of DSM), composed a programme consisting of the following topics: Thermodynamics/(Co)polymer Blends; Diffusion/Barrier Properties; Chain Mobility/Morphology; Processing; Polymer Networks; and Speciality Polymers. Each topic was introduced by invited academic or industrial experts, after which contributed papers and posters were presented.

All active participants were asked to submit a paper for this book. The editors are not to be held responsible for the scientific contents or views which are expressed herein, but we are convinced that the reader will enjoy some of the controversial opinions presented in this volume.

We thank all contributors to this volume and we hope that the 'Rolduc Polymer Meeting' will become an established event with the aim of the transfer of polymer science to technology.

L. A. K.
P. J. L.

CONTENTS

Part II: Characterization/Solution Behaviour

Part III: Blends

Part IV: Networks

Part V: Diffusion/Barrier Properties

Part VI: Chain Dynamics

Part VII: Processing/Rheology

Part VIII: Structure and Morphology

Part IX: New Developments

Part I

THERMODYNAMICS

Thermodynamics and Engineering Needs

R. Koningsveld

DSM, Research & Patents, 6160 MD Geleen, Netherlands

Synopsis

Selected molecular models for small-molecule and macromolecular systems are discussed from the standpoint that the only theoretical approaches worth-while considering are those capable of reliable predictions. This condition restricts the discussion to models accounting for disparity in numbers of nearest neigbour contacts, for free volume, and for non-uniform segment density in systems, dilute in macromolecular constituents.

Introduction

To date free energy expressions, that can reliably predict thermodynamic equilibrium properties, are still lacking greatly. Zudkevitch, lucidly showing that the quality of plant and equipment design is determined by the reliability of the data used, has pointed out that basic thermodynamic information belongs to such data[1,2]. In addition, improving engineering skill, in particular control of temperature-pressure regimes in large-scale equipment, calls for ever more precise descriptions and estimations of equilibrium data. Further, one has to deal with increasing complexity of the systems in hand, regarding the number of components and their chemical structure. All this, and the urge for reduction of investments pose a challenge to theory to come up with truly quantitative descriptions of known data and the ability to reliably predict information required[1,2].

The periods since Van der Waals for small-molecule systems on the one hand, and Flory-Huggins-Staverman[3-7] for macromolecular system on the other, have seen many attempts at improving the unsatisfactory situation stated above. Relatively simple models have shown to be amenable to adaptation and to come closer to the actual situation. However, quantitative descriptions still call for the introduction of empirical parameters, the molecular origin of which can often only be guessed.

Accepting this inelegance, one can come within reach of the required objective. Molecular parameters of primary importance prove to be:

1) the disparity in size and shape between the various molecules and chain segments in the nearest neighbour contacts they can establish,

2) the equilibrium free volume and its dependence on temperature, pressure and concentration,

3) as far as macromolecular systems are concerned, the connectivity within polymer chains, i.a. necessarily leading to highly dilute systems showing inhomogeneous segment densities[8-10].

This paper attempts to summarize the situation, seen from the author's self-opiniated point of view. One aspect of this view is that small-molecules and polymeric substances are not all that different, except for point 3) above. Hence, the examples presented below will alternatively refer to small or large molecules in fluid phase equilibria, as the discussion may require.

2. Lattice Models

2.1. Rigid lattice

The regular lattice stil is a useful tool to formulate how conceivable molecular characteristics work out in the (Gibbs) free energy expression. Using the model in this way one often finds the strict lattice to become an abstraction[11] but this does not detract from its usefulness.

For systems containing molecules of various sizes, the Flory-Huggins-Staverman[3-7] equation is appropriate:

$$\Delta G/NRT = \phi_o \ln\phi_o + (\phi_1/m_1)\ln\phi_1 + g\phi_o\phi_1 \tag{1}$$

where ΔG= the Gibbs free energy of mixing, N= total number of sites, m_1= number of sites occupied by chain molecules 1, ϕ_o,ϕ_1= volume fractions of solvent o and solute 1, and RT has its usual meaning.

The simplest interpretation of the interaction function g defines it in terms of the interchange energy (see e.g. ref. 9) but experimental data suggest interpretation as a free energy to be preferable[12]:

$$g = g_o + g_1/T \qquad (2)$$

Shultz and Flory analysed miscibility gaps in several solvent-polymer systems for a series of polymer chain lengths[13]. They identified the maximum separation temperature T_c with the critical point for which eqs (1) and (2) yield

$$g_1/T_c = 0.5 - g_o + (m_1^{-\frac{1}{2}} + 0.5/m_1) \qquad (3)$$

A 'Shultz-Flory' plot of 1/Tc vs the expression between brackets thus provides the parameters g_o and g_1. The fact that it proved impossible to describe the miscibility gaps in any detail can be ascribed to a) the systems not having been binary ones and b) the g-function normally to depend on concentration[14].

Another often heard objection against the Flory-Huggins equation ((1) and (2)) is its alleged incapability of describing upper and lower critical miscibility in the same system. Such criticism is not justified[14]. Once the semi-empirical adaptation by the term g_o has been accepted, it should also be accepted that there are fundamental classical thermodynamic relations to be obeyed by any model, viz.:

$$\Delta H = \int \Delta c_p dT; \quad \Delta S = \int (\Delta c_p/T) dT \qquad (4)$$

where ΔH, ΔS and Δc_p are the enthalpy, entropy and specific heat changes upon mixing. The specific heat at constant pressure c_p of a liquid is known to depend on temperature. In addition, Δc_p must be expected to vary with concentration, for instance

$$\Delta c_p = (c_o + c_1 T)\phi_o \phi_1 \qquad (5)$$

Eqs (4) and (5), together with (1), can now be used to define the interaction function g. One finds

$$g = g_o + g_1/T + g_2 T + g_3 \ln T \qquad (6)$$

where g_o and g_1 arise from integration constants (eq (4)) and $g_2 = - c_1/2NR$; $g_3 = - c_o/NR$. Use of the Flory-Huggins equation in the form of eqs (1) an (2), like the critics to, unrealistically ignores the temperature dependence of Δc_p. The latter was formulated theoretically by Delmas et al[15] who found

Prigogine's model[16] to supply expressions for g_1 and g_2.
The prime reason for g to depend on concentration is to be sought in the disparity in size and shape between solvent molecules and polymer segments. Staverman[17], working out earlier suggestions by Langmuir[18] and Butler[19], assumed the number of nearest-neighbour contacts a molecule or segment can make to be proportional to its accessible surface. Thus he obtained an expression for ΔH which, in terms of g reads

$$g = \{B(T)/(1 - \gamma_1\phi_1)\} \tag{7}$$

where B(T) summarizes the temperature dependence and σ_1/σ_o = the ratio of the surface areas of polymer segments and solvent molecules ($\gamma_1 = 1-\sigma_1/\sigma_o$).
Figure 1 illustrates the viability of Staverman's approach. It is interesting to note that the curve in fig. 1b was calculated with Bondi's[22] σ_1/σ_o value for p-xylene and benzene of 1.35, and passes right through the experimental points.

2.2. Lattice gas

A phenomenological treatment similar to that for g(T) can be applied to assess how, with the framework of eq (1), g might depend on pressure. Such a treatment has proved useful in describing cloud-point pressures in the system ethylene/polyethylene[23]. It also allowed extrapolation into pressure and temperature regions not covered by the experimental data used in the description.
An alternative route is the extension of the rigid lattice by the addition of vacant sites which contribute to the volume of the system but not to its mass. This lattice-gas approach has a long history[24-29]. It was recently developed so as to supply fairly complete quantitative descriptions of fluid phase behaviour in single-component systems and mixtures, either or not polymeric, polar as well as non-polar in nature[30-34].
The Helmholtz free energy of 'mixing' vacant and occupied sites (single substance) reads

$$\Delta F/NRT = \phi_o \ln\phi_o + (\phi_1/m_1)\ln\phi_1 + g\phi_o\phi_1 \tag{8}$$

where ϕ_o is now the volume fraction of empty sites and ϕ_1 that of the molecules, each occupying m_1 sites. The density ρ_1 of the substance defines ϕ_o by

$$\rho_1 = (1-\phi_o)M_1/m_1 v_o \tag{9}$$

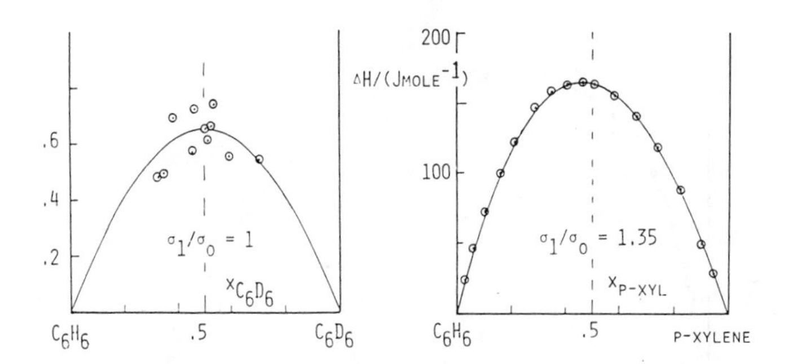

Figure 1. Enthalpies of mixing (ΔH) for benzene with perdeutereous benzene (left, ref. 20) and with p-xylene (right, ref. 21). The curves were calculated according to Staverman[17)] with molecular surface-ratios of 1 (left) and 1.35 (right).

where v_o is the molar volume of the vacant sites, and M_1 is the substance's molar mass.

So far a quantitative application of eqs (8) and (9) necessitated the g function to be defined as

$$g = \alpha + (\beta_o + \beta_1/T)/(1 - \gamma_1\phi_1) \tag{10}$$

where β_1 is related to the contribution to the internal energy per contact of occupied sites, but α and β_o are empirical parameters, having unpredictable values depending on the value chosen for v_o. Also here, introduction of the surface area σ_1/σ_o ($=1-\gamma_1$) proved necessary, figure 2 shows its effect for the vapour-liquid coexistence curve of carbon dioxide.

The molecular origin of the empirical parameters α and β_o can be appreciated by a simple and not rigorous consideration of the numbers of nearest neighbours (lattice coordination numbers z_{ij}) in a binary liquid (vacant sites can be treated as a component in this approach). If the two component molecules differ in size but not too much in shape there will at least be three coordination numbers z_{oo} (assumed equal to z_{11}), z_{o1} and z_{1o}, defining the numbers of j-contacts of molecules i. If all z_{ij} are equal, the number of arrangements on the lattice is given by the well-known expression

$$\Omega = N!/(N_o!N_1!) \tag{11}$$

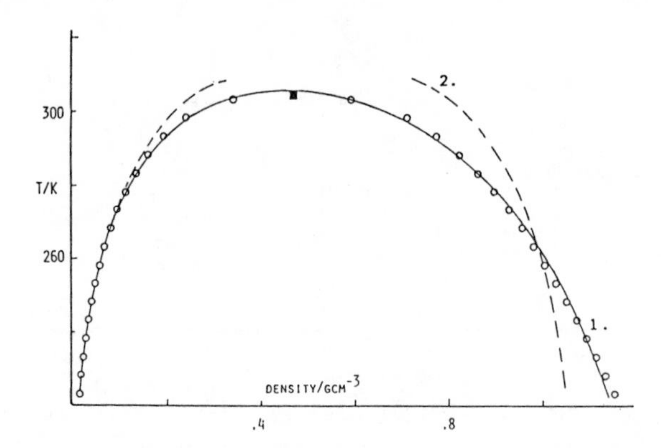

Figure 2. Fit of vapour-liquid coexistence data (o) of carbon dioxide to the MFLG expressions (8)-(10) with σ_1/σ_o = 0.993 (1) and 1 (2). Critical point: ■

where N_o and N_1 are the numbers of molecules o and 1 and $N = N_o + N_1$. If $z_{oo} \neq z_{o1} \neq z_{1o}$ eq (11) must be corrected because over- and underestimations referring to z_{o1} and z_{1o}. Following a procedure suggested by Huggins[35] and Silberberg[36] one might write

$$\Omega = \{N!/(N_o!N_1!)\}(z_{o1}/\bar{z})^{P_{o1}}(z_{1o}/\bar{z})^{P_{1o}} \qquad (12)$$

where $P_{o1} \equiv P_{1o}$ is the number of unequal contact pairs, and $\bar{z}$ is an average number depending on concentration (or density).

Arbitrarily writing $(\sigma_o\phi_o + \sigma_1\phi_1)$ for $\bar{z}$ and the regular-solution approximation for P_{o1}, one obtains

$$\alpha = 2z_{11}(\ln Q)/Q; \quad \beta_o = -z_{11}\ln(z_{o1}z_{1o}/z_{oo}^2) \qquad (13)$$

where $Q = 1-\gamma_1\phi_1$.

It is seen that eq (13) supplies an after-the-fact explanation for the empirical parameters α and β_o in terms of disparity of contact numbers, though in a qualitative sense only. It should be noted that Staverman recently developed a rigorous treatment of contact statistics[37].

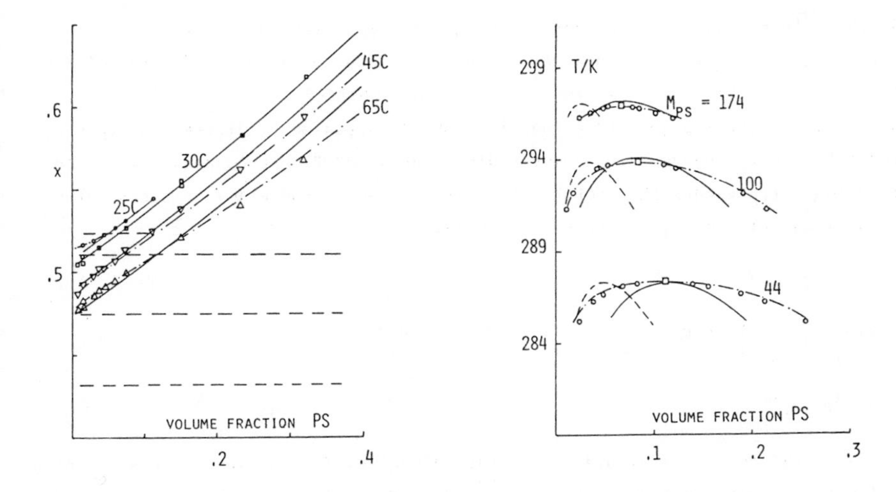

Figure 3. The system cyclohexane-polystyrene. Descriptions by the Shultz-Flory method (---), eq (7) with σ_1/σ_0 = 0.89 (——) and Nies' equation (16) (-.-.-)
Left: Excess chemical potential of cyclohexane for M_{ps} = 164 kg/mole (data by Scholte[41]).
Right: Near-binary binodals (data by Fujita et al[42]) for indicated molar-mass values. Critical points: □ .

3. Examples

3.1. Cyclohexane-Polystyrene

Nies et al[38-40] have carried out a rigid-lattice analyse of various sets of data on the system cyclohexane-polystyrene, a selection is presented in figure 3.

The Shultz-Flory analysis leads to a poor prediction of both excess solvent chemical potentials and binodals (data by Scholte[41] and Fujita et al[7]), respectively). Introduction of eq (7) yields a major improvement and shifts the calculated curves into the correct region. On both ends of the concentration scale, however, discrepancies remain.

On the dilute end, the agreement can be greatly improved upon introduction of the bridging function P into g, as suggested by Stockmayer et al[10]. The dilute range is characterized by polymer coils being separated by regions of pure solvent, and the appropriate g function must be expected to differ from that referring to the homogeneous segment distribution at higher concentration. The function P represents the probability that a given volume element in the solution is not within any of the coil domains:

$$\ln P \propto -(\phi_1/\phi_1^*) \tag{14}$$

where ϕ_1^* is the overlap concentration. The interaction function can now be written as[10]

$$g = g^* P + g^c \tag{15}$$

where g* refers to the dilute concentration regime. It was found necessary to allow for a molar mass and a temperature dependence in g* in order to obtain a correct description[10,38]. Dilute-solution theory[43] may be used to specify expressions for $g^*(m_1)$. The term g^c represents the uniform segment density range and is given by eq (7) or (10).
So far the new parameters introduced have a clear physical origin[10]. To improve the description for concentrated solutions the empirical parameter α must also be assumed to depend on chain length as well as temperature and the final result

$$g = g^*(m_1,T)P + \alpha(m_1,T) + B(T)/(1-\gamma_1\phi_1) \tag{16}$$

can be fitted to the data as shown in figure 3 (details in ref. 38). The σ_1/σ_0 values arising from Nies' curve fitting procedures range from 0.65 to 0.89, depending on the complexity of the g-function tested. Bondi's scheme[22] yields 0.87 and confirms the data-fit which suggests the cyclohexane surface area to be larger than that of styrene units.
A justification of the curve-fitting procedure was given by Nies in the form of predictions of data not having been used in the estimation of the parameters. Perfect agreement between calculated and experimental curves was obtained for Krigbaums' osmotic pressure measurements and Fujita et al's ternary binodals[38-40]. Simpler versions than eq (16) for the g-function will indeed give a less satisfactory description of the fitted data, and fail in accuracy of prediction. Eq (16) has six adaptable parameters in addition to those that have a physical meaning. The various contributions to the small resulting value of ΔG differ in sign and magnitude and, obviously, none of them may be ignored.

As to a possible molecular origin of the α (m_1, T) term, Nies points out that a) Staverman's calculation of the effect the chain's bending back on itself has on the entropy of mixing can at least partly account for $\alpha(m_1,T)$[37,45],
b) free volume effects have been left out of consideration and might also be partly responsible.

The mere fact that the entropy of mixing was found to be markedly chain-length dependent may have far-reaching consequences for the theory of dilute polymer solutions[8,43]. Summarizing those one might write

$$A_2 \propto (\tfrac{1}{2} - \chi_1)h(z) \tag{17}$$

where A_2 = the osmotic second virial coefficient and h(z) is the excluded-volume function. The latter is thought to fully account for the chain-length dependence of A_2 while the interaction parameter χ, for the concentrated regime is assumed to be independent of chain length.
The present findings do not support this latter assumption. In view of the possible molecular origins of the $\alpha(m_1)$ function one might expect the phenomenon to occur in other systems as well. As a consequence, the excluded-volume function had better not be tested with eq (17) but rather with

$$A_2 \propto \{\tfrac{1}{2} - \chi_1(m_1)\}h(z) \tag{18}$$

3.2. Polyvinylmethylether-Polystyrene

Recent measurements of miscibility gaps in PVME-PS mixtures[46] supply an interesting application of Staverman's surface-ratio concept. Figure 4 summarizes the results which include an estimation of the region of spinodal decomposition and, thereby, of the location of the critical points. The latter are remarkable in that they occur at the minimum of the cloud point curves. The polystyrene samples had narrow molar mass distributions, but the PVME was characterized by relatively wide spread in chain lengths. Polymolecularity tends to shift the critical point away from the extreme of a cloud point curve, unless there is a counterbalancing effect. We found the difference in surface area between the VME and S repeat units to supply enough concentration dependence in the interaction function to qualitatively account for the remarkable phase behaviour.
The spinodal curves I in figure 4 calculated with σ_S/σ_{VME} = 1.8, and with a reasonable value for the polydispersity index of PVME, show an acceptable

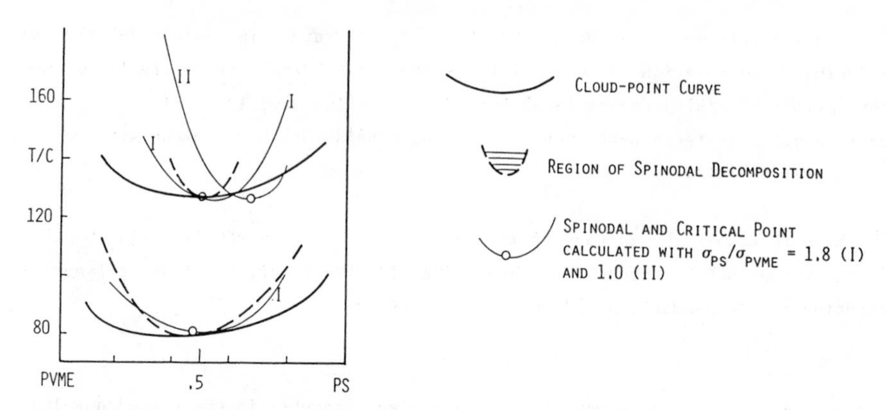

Figure 4. The system polyvinylmethylether-polystyrene PVME: M_w = 75 kg/mole; PS: M_w = 17 kg/mole and M_w = 54 kg/mole (upper and lower curves, resp.). Data by Voigt-Martin et al[46].

agreement with the experimental findings. Bondi's procedure yields 1.4 and confirms our assumption that styrene units have a larger surface than the VME segments. The curve II calculated with a surface ratio of 1.0 obviously cannot deal with the data.

3.3. Helium-4

The MFLG eqs (8)-(10) provide an excellent description of PVT data He-4, even in the very simple version where $\alpha = 0$ and $\beta_o = 0$[47]. Figure 5 illustrates the description thus obtained with $v_o = 5$ cm^3/mole, $\sigma_1/\sigma_o = 1.8$.
The parameters were fitted to the coexistence data covering an extremely small temperature range, and yet the predicted isotherms are surprisingly accurate up to 450 K ($T_c = 5.2$ K). An enlargement of the coexistence region (fig. 6) indicates that, close to the critical point, the two branches are slightly steeper than the data, and a 'non-classic' exponent[49] is not to be expected from this description. However, at higher υ_o values and with a density-dependent α, it is quite possible to obtain the usual scaling behaviour close to the critical point. Figure 7 illustrates that, contrary to current opinion[49], an analytic equation of state like the MFLG expression is quite capable of supplying linear double-log plots as required by scaling laws. What such laws cannot do, is provide engineers with accurate density values far from the critical state.

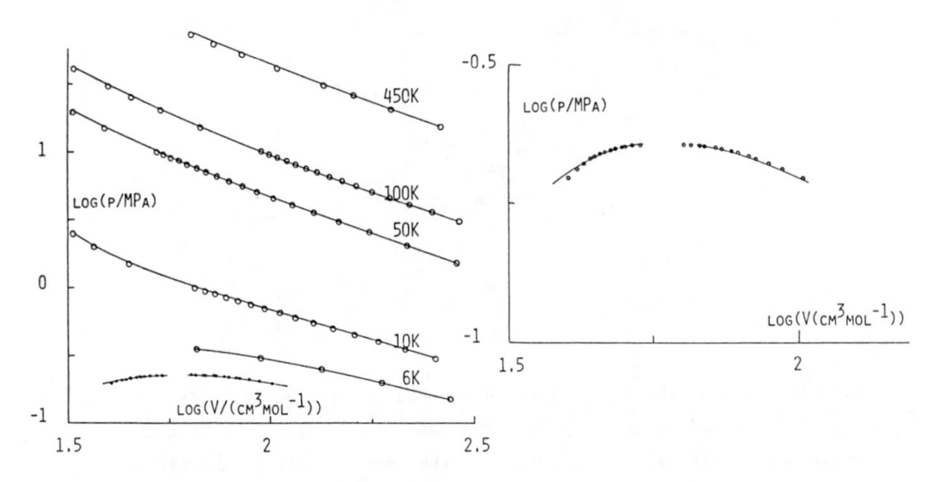

Figure 5: Mean-field lattice-gas description of He-4 with $\alpha = 0$, $\beta_o = 0$, $\upsilon_o = 5$ cm^3/mole, $\sigma_1/\sigma_o = 1.8$[47]).
Isotherms (o) and coexisting phases (•).

Figure 6: Enlargement of the coexistence region in fig. 5.

3.4. Noble-Gas Mixtures

The MFLG model also serves a useful purpose in the treatment of gas-gas demixing in binary mixtures around the critical point of the heavier component. Keller et al[50]) showed that the widely varying critical curves of binary mixtures of noble gases could be well described, with values of the parameters m_1 and σ_1/σ_0 that make physical sense, if only qualitatively (figure 8).

3.5. Block-Copolymer Systems

Roe and Zin[51]) recently published remarkable phase diagrams showing the influence the addition of a homopolymer has on the microphase separation in a block copolymer, sharing one of its repeat units with the homopolymer. Figure 9 illustrates that the temperature of domain formation goes up if polystyrene is added to a styrene-butadiene blockcopolymer, and goes down if polybutadiene is added.

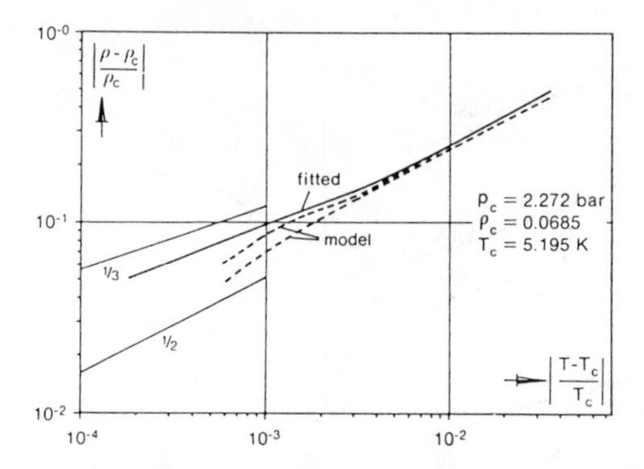

Figure 7: Reduced density, vs reduced temperature for He-4. 'Model' curves: data by Roach[48]) fitted to MFLG eqs (8)-(10) with $\alpha = \alpha(\varphi_1)$. 'Fitted' curves: coexistence densities by MFLG fitted to the usual 'scaling' quantities.

It is an interesting albeit for polymers rather unrealistic consequence of free-volume models that they include a critical state and coexistence between phases differing in density. The MFLG model, for instance, predicts such a critical state in a homopolymer at impossibly low polymer density. However, if a copolymer $P_{\alpha\beta}$ is considered, and the two surface ratios differ from unity, quite acceptable densities could be involved. For instance, with σ_α/σ_0 = 0.4 and σ_β/σ_0 = 1.6 one calculates p_c = 2000 atm., T_c = 450 K and has reasonable values for coexisting densities at slightly lower pressure and temperature. Addition of homopolymer $P_{\alpha\alpha}$ or $P_{\beta\beta}$ would affect the stability of the system roughly in the manner Roe and Zin found. Yet, the calculation did not include domain formation and might therefore be considered to be not very meaningful.
However, if domain formation is taken into account by the addition to ΔF of the simplest expression available, i.e. that by Bianchi et al[52]), we find the equilibrium pressure and temperature to be shifted towards ambient conditions. We used reasonable values for domains size and interfacial energy in Bianchi's equation to obtain this result.
This example indicates that
a) it is obviously necessary to consider the effect microphase separation has in the free energy,
b) it is inadequate to neglect details of the bulk free energy and treat is as a simple Flory-Huggins case.

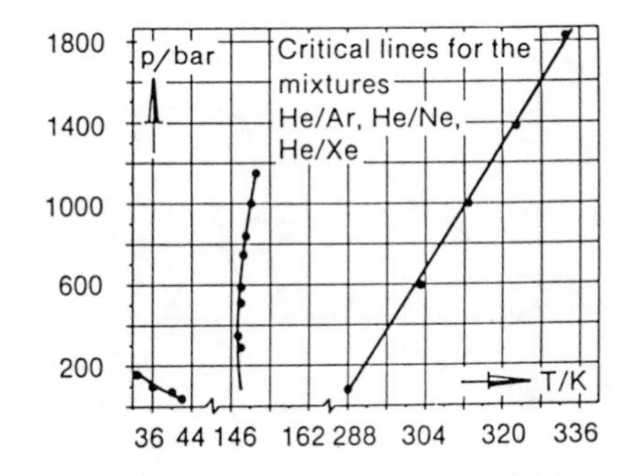

MFLG parameters for noble gases

	m_1	σ_1/σ_o
He	1	1.891
Ar	1.2	2.346
Kr	1.45	2.550
Xe	1.7	2.631

Figure 8: MFLG description of binary noble gas mixtures (——), experimental (•)[50].

3.6. Kanig's Treatment of the Glass Transition

Many years ago Kanig[26] proposed a treatment of the glass transition which is an application of the mean-field lattice-gas model. In the present terminology, Kanig set $\alpha = 0$, $\beta_o = 0$ and he assumed the surfaces of occupied sites and holes to be proportional to the contribution of each to the total free volume. The theory requires a criterion to judge whether a melt has reached its glass temperature. Kanig's criterion is the 'hole' free volume to equal about two thirds of the total free volume. The description of the glass transition in copolymers and plasticized polymers, so obtained, is remarkably good and seems to suffer neglect in the literature.

Conclusions

Two obvious reasons exist why predictions (extrapolations) frequently come out so poorly. First, the total free energy is the usually small sum of a number of terms that may each be large and inaccurate. Yet, fluid phase behaviour is determined by details in the resulting free energy function. Consequently, the often applied practice of neglecting small contributions altogether is to be avoided.

A second problem is posed by the combination rules required for most of the molecular and empirical parameters in a model, when mixtures are concerned. Mostly, such rules represent more or less arbitrary assumptions that have to survive threefold differentiation if critical states are to be dealt with. Small wonder that predictions are often very unsuccesful.

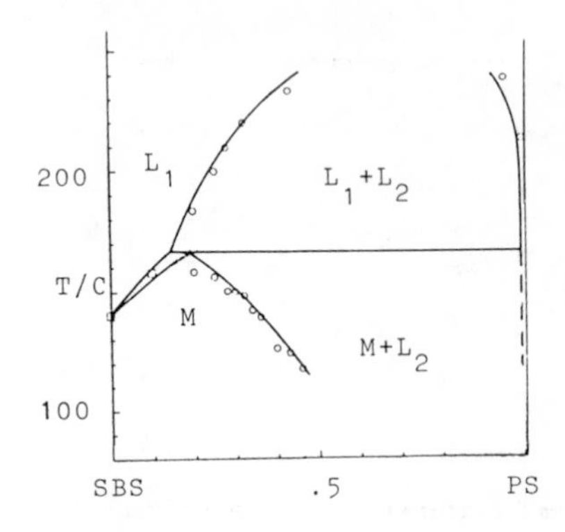

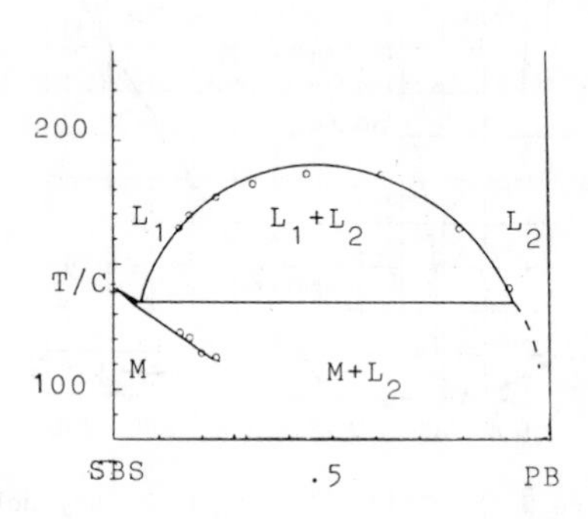

Figure 9: Phase relations in mixtures of a styrene-butadiene blockcopolymer (SBS) with polystyrene (PS) (left)) and with polybutadiene (PB) (right).
L_1 and L_2 refer to homogeneous liquid phases, M to a microphase-separated state. Data by Roe and Zin[51]).

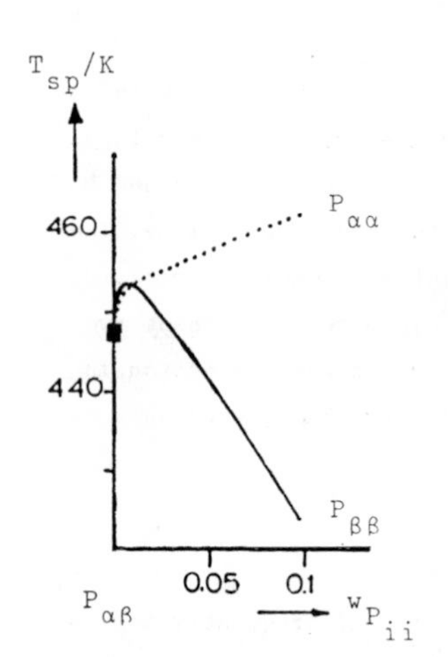

Figure 10: MFLG description of a copolymer $P_{\alpha\beta}$ showing two phases differing slightly in density (■). The stability is affected by the addition of homopolymer $P_{\alpha\alpha}$ ($\sigma_\alpha/\sigma_o < 1$; ....) or $P_{\beta\beta}$ ($\sigma_\beta/\sigma_o >$ 1; ——); $w_{P_{ii}}$ = volume fraction of homopolymer i.

Molecular models will probably invariably be too crude to avoid the treatment of their parameters in a qualitatitve sense. They can at most be expected to come up with a useful mathematical framework, amenable to empirical adaptation.

Acknowledgement

The author is much indebted to Mrs. A.M. Leblans-Vinck and to Drs. L.A. Kleintjens and E. Nies for helpful discussions and assistance in the preparation of the manuscript.

References

1. Zudkevitch D., Proc. 5th Conference on Mixtures of Non-Electrolytes and Intermolecular Interactions, 1983, p. 69; Kongress- und Tagungsber. Martin-Luther-Univ., Halle-Wittenberg, GFR, 1984.
2. Zudkevitch D., Encyclop. Chem. Proc. Design, Vol. 14, 431 (1982)
3. Flory P.J., J. Chem. Phys., Vol. 9, 660 (1941); Vol. 10, 51 (1942).
4. Huggins M.L., J. Chem. Phys., Vol. 9, 440 (1941).
5. Huggins M.L., Ann. N.Y. Acad. Sci., Vol. 43, 1 (1942)
6. Staverman A.J. and Van Santen, Rec. Trav. Chim., Vol. 60, 76 (1941).
7. Staverman A.J., Rec. Trav. Chim., Vol. 60, 640 (1941).
8. Flory P.J., J. Chem. Phys., Vol. 17, 1347 (1949).
9. Flory P.J., 'Principles of Polymer Chemistry', Cornell, 1953.
10. Koningsveld R., Stockmayer W.H., Kennedy J.W. and Kleintjens L.A., Macromolecules, Vol. 7, 73 (1974).
11. Rowlinson J.S., 'Liquids and Liquid Mixtures', Butterworth, 1959.
12. Guggenheim E.A., 'Mixtures', Oxford, 1952.
13. Shultz A.R. and Flory P.J., J. Am. Chem. Soc., Vol. 74, 4760 (1952).
14. Koningsveld R., Adv. Coll. Interf. Sci., Vol. 2, 151 (1968).
15. Delmas G., Patterson D. and Somcynski T., J. Polym. Sci., Vol. 57, 79 (1962).
16. Prigogine I., 'The Molecular Theory of Solutions', North Holland, Publ. Co., 1957.
17. Staverman A.J., Rec. Trav. Chim., Vol. 56, 885 (1937); PhD Thesis, Leiden 1938.
18. Langmuir I., Colloid Symposium Monograph, Vol. 3, 48 (1925).
19. Butler J.A.V., J. Chem. Soc., Vol. 19, 681 (1933).
20. Lal M. and Swinton F.L., Physica, Vol. 40, 446 (1968).
21. Singh J., Pflug H.D. and Benson G.C., J. Phys. Chem., Vol. 72, 1939 (1968).
22. Bondi A., J. Phys. Chem., Vol. 68, 441 (1964).
23. Koningsveld R., Diepen G.A.M. and Chermin H.A.G., Rec. Trav. Chim., Vol. 85, 504 (1966).
24. Frenkel J., 'Kinetic Theory of Liquids', Dover, 1947.
25. Cernuschi F. and Eyring H., J. Chem. Phys. Vol. 7, 547 (1939).
26. Kanig G., Kolloid Z & Z. Polym., Vol. 190, 1 (1963).
27. Trappeniers N.J., Schouten J.A. and Ten Seldam, C.A., Chem. Phys. Lett., Vol. 5, 541 (1970).

28. Schouten J.A., Ten Seldam C.A. and Trappeniers N.J., Physica, Vol. 73, 556 (1974).
29. Sanchez I.C. and Lacombe R.H., J. Phys. Chem., Vol. 80, 2352, 2568 (1976).
30. Kleintjens L.A., PhD Thesis, Essex UK, 1979.
31. Kleintjens L.A. and Koningsveld R., Colloid & Polym. Sci., Vol. 258, 711 (1980).
32. Kleintjens L.A. and Koningsveld R., Sep. Sci. J. Technol. Vol. 17, 215 (1982).
33. Kleintjens L.A., Fluid Phase Eq., Vol. 10, 183 (1983).
34. Leblans-Vinck A.M., Koningsveld R., Kleintjens L.A. and Diepen G.A.M., Fluid
35. Huggins, M.L., J. Phys. Colloid Chem., Vol. 52, 248 (1948).
36. Silberberg A., J. Chem. Phys., Vol. 48, 2835 (1968).
37. Staverman A.J., This Volume.
38. Nies E., PhD Thesis, Antwerp, 1983.
39. Nies E., Koningsveld R. and Kleintjens L.A., ref. 1, p. 208.
40. Nies E., Koningsveld R. and Kleintjens L.A., 29th International Symposium on Macromolecules, Bucharest, September 1983; Plenary and Invited Lectures, Part 2, p. 121.
41. Scholte Th.G., J. Polym. Sci. A2, Vol. 8, 841 (1970), Vol. 9, 1553 (1971); Europ. Polym. J., Vol. 6, 1063 (1970).
42. Hashizume I., Teramoto A. and Fujita H., J. Polym. Sci., Pol. Phys. Ed., Vol. 19, 1405 (1981).
43. Yamakawa H., 'Modern Theory of Polymer Solutions', Harper & Row, 1971.
44. Krigbaum W.R., J. Am. Chem. Soc., Vol. 76, 3758 (1954).
45. Staverman A.J., Rec. Trav. Chim., Vol. 69, 163 (1950).
46. Voigt-Martin I.G., Leister K.H., Rosenau R. and Koningsveld R, Submitted to J. Polym. Sci.
47. Leblans-Vinck A.M., to be published.
48. Roach P.R., Phys. Rev., Vol. 170, 213 (1968).
49. See e.g.: McGlashan M.L., 'Chemical Thermodynamics', Acad. Press., 1979, p. 133.
50. Keller P., Kleintjens L.A. and Koningsveld R., IUPAC Conference on Chemical Thermodynamics, London, 1982.
51. Roe R.J. and Zin W.C., Macromolecules, Vol. 17, 189 (1984).
52. Bianchi U., Pedemonte E. and Turturro A., Polym. Lett., Vol. 7, 785 (1969).

STATISTICS OF SURFACE CONTACT DISTRIBUTIONS

A.J. Staverman

University of Leiden, Gorlaeus Laboratories, P.O.Box 9502,
2300 RA Leiden, The Netherlands

SYNOPSIS

While many authors have proposed thermodynamic equations for liquid mixtures containing surface fractions besides volume fractions, also the most sophisticated treatment of Guggenheim is not applicable to solutions of folded macromolecules with intramolecular contacts. In this paper a complete and consistent treatment is proposed, based upon Guggenheim' equations on one hand and on a model of molecules with intramolecular contacts proposed by the author in 1950 on the other hand. The treatment results in equations for the free energy and the heat of mixing as functions of ϕ, the volume fraction of the polymer. It turns out, that the free energy of mixing depends on two parameters, both representing surface ratio's, γ and γ_m. The latter measures the surface of the macromolecule when it is free of intramolecular contacts, while γ is the real surface ratio and depends on ϕ.
Suggestions for the experimental determination of these parameters are given. The generally employed erroneous definition of "athermal solutions" is replaced by two correct definitions of athermal and random solutions respectively.

1. INTRODUCTION

The basic equation for the description of thermodynamic properties of liquid solutions and mixtures of macromolecules is the equation of Flory and Huggins for ΔG, the free energy of mixing. For a binary mixture of a solvent, component 0, and a polymer, component p, this equation can be written

$$\Delta G/kT = N_0 \ln \phi_0 + N_p \ln \phi_p + \chi \phi_p N_0 = N_0 \ln \phi_0 + N_p \ln \phi_p + \chi \phi_0 r N_p$$

N_i is the number of molecules of component i

ϕ_i is the volume fraction of component i
χ is the vanLaar-Hildebrand interaction constant[1]
r is v_p/v_0, the ratio of the molecular volumes.
For a multicomponent mixture the equation reads

$$\Delta G/kT = \Sigma N_i \ln \phi_i + \sum_{j>i} \chi_{ij}\phi_j \cdot \Sigma r_i N_i$$

with $\Sigma r_i N_i = V/v_0$, V is the volume of the mixture and $r_i = v_i/v_0$. The first term of the r.h.s. of these equations represents the combinatorial contribution to the free energy, the contribution of the number of ways of arranging hard particles with the size and shape of the molecules in the given volumes of the mixture and the pure components respectively. The second term represents the difference in free energy of interaction between unlike and like molecules respectively.
The only molecular property appearing in these equations is the molecular volume. This is due to the assumption, that the probability of interaction between two molecules is proportional to the volumes of these molecules. In liquid mixtures this assumption is justifiable only in case the molecules are of nearly equal size and shape. For general mixtures a much more reasonable assumption is that the probability of intermolecular interaction is proportional to the number of nearest neighbours. This, however, implies the introduction of an additional molecular property, the number of nearest neighbours, or, in other words "the surface" or "the number of contact points" of a molecule. In the following we will use these two terms indiscriminately. Hildebrand (ref.1, p.133) uses the term "the effective volume fraction".
The introduction of molecular surfaces in the expression for the probability of intermolecular interaction not only affects the interaction term in the Flory Huggins equation, but also the combinatorial term, and thus profoundly transforms the theory of thermodynamics of macromolecular mixtures.

Surface statistics of mixtures has a long history:

1925 Langmuir[2], "The Principle of independent Surface Action"

1937 Staverman[3] Demonstrates that results of ΔH_m-measurements (Hirobe) are not described by volume fractions, but by "Surface Fractions"; the vanLaar-Hildebrand equation is ruled out for simple liquid mixtures.

1948-1958, also 1970-1976, Huggins[4], Polymer solutions

1952 Guggenheim[5]), "Mixtures". Based upon a lattice model

1. "Regular Solutions": Mixtures of molecules "sufficient alike in size and shape to be interchangeable on a lattice".
2. Solutions of macromolecules. Segments of macromolecules are considered interchangeable with solvent molecules. Calculation of Q_c restricted to "open macromolecules".

1964-1967 Flory[6]), "Equation of State" for liquid mixtures.

1967-1985 Koningsveld, Kleintjens, Chermin, Onclin, Nies a.o. Phaserelations in polymer solutions and mixtures.

Guggenheims treatment is the most general and sophisticated. In the following we will extend, generalize and correct Guggenheims treatment on the following points.

1. Abolish the lattice model.
2. Correct Guggenheims expressions for ΔH_m .
3. Derive Guggenheims correct expression for ΔF_e from correct starting point.
4. Generalize Guggenheims expressions for ΔF_r^c for macromolecules to macromolecules containing intramolecular rings.
5. Replace the erroneous definition of "athermal mixtures" by two correct definitions.
6. Envisage variations of the surface of macromolecules with concentration.

2. DEFINITIONS AND ASSUMPTIONS

The contribution of repulsive forces to the free energy is called F_r^c (c for "combinatorial"); the contribution of attractive forces is called F_e, the "excess free energy".

$$F = F_r^c + F_e \tag{1}$$

with

$$F_r^c = kT \ln Q_c \tag{2}$$

Q_c is the number of configurations of "hard" molecules in the available volume. Mixtures are considered at zero pressure.

$$F = G, \ \Delta F = \Delta G \ , \ H = U, \quad \Delta H = \Delta U \tag{3}$$

volume changes on mixing,if any,are attributed to nearest neighbour interactions; their effect on F is accounted for in F_e. Q_c and F_r^c depend on the ratio's only, of volumes and surfaces. These are assumed to be independent of composition.

The free energy of mixing of a mixture of N_0 and N_1 molecules is

$$\Delta F(N_0,N_1) = F(N_0,N_1) - F(N_0,0) - F(0,N_1) \tag{4}$$

$$\Delta F = \Delta F_r^c + \Delta F_e \tag{5}$$

ΔF_r^c calculated from Q_c, volume statistics.

ΔF_e is calculated from surface statistics.

3. EXPRESSING ΔF_e AND ΔH_m IN SURFACE FRACTIONS

Consider N_0 and N_1 molecule with volumes v_0 and v_1 and surfaces s_0 and s_1. Two quantities characterize ΔF_e : w and γ.

$$w = w_{10} - \tfrac{1}{2}(w_{00} + w_{11}) \tag{6}$$

note that

$$w = u - T\,\eta \ = u + T\,\frac{dw}{dT} \tag{7}$$

w is a contribution to the free energy, u to the heat.

Definitions:

$$r \equiv v_1/v_0 \tag{8}$$
$$t \equiv s_1/s_0 \tag{9}$$
$$\gamma \equiv 1 - (s_1 v_0)/(s_0 v_1) \tag{10}$$

ΔF_e is completely given by γ and w, but $\Delta F = \Delta F_r^c + \Delta F_e$
ΔH_m is completely given by γ and u, while $\Delta H_m = \Delta H_e$
In a "random mixture" the relation between ΔH_m and the volume fractions ϕ and $(1-\phi)$ is

$$\Delta H_m^* = \phi(1-\phi)\,\frac{(1-\gamma)V}{(1-\gamma\phi)v_0}\,s_0 u \tag{11}$$

Compare with "vanLaar-Hildebrand":

$$\Delta H_m = \phi(1-\phi)\,Vg \tag{11a}$$

where g is the interaction parameter
The star indicates: "random mixture".
Derivation of (11): if X is the number of 0-1 contacts:

$$\Delta H_m = Xu \tag{12}$$

In a random mixture

$$X = \sigma_0\sigma_1 S = X^* \tag{13}$$

with σ_1 , the surface fraction

$$\sigma_i = N_i s_i / \Sigma_i N_i s_i = N_i s_i / S \tag{14}$$

Koningsveld and collaborators found experimentally strong deviations from vanLaar-Hildebrand with constant g. Considerable improvement resulted from replacing (11a) by (15) with $\gamma \sim 0.24$:

$$\Delta F_e = \Delta F_e^* = X^* w = \phi(1-\phi)\,\frac{(1-\gamma)V}{(1-\gamma\phi)v_0}\,s_0 w \tag{15}$$

However replacing (11a) by (15) did not lead to quantitative agreement between theory and experiment. Deviations are expected from two sources: ΔF_r^c and deviations from randomness.

4 EFFECTS OF NON-RANDOMNESS

Deviations from randomness depend on y

$$y = w/kT \qquad \text{(not } ws_0/kT\text{)} \tag{16}$$

If w = 0, mixture is "random", not "athermal"

If u = 0, mixture is "athermal" (ΔH_m=0, Hildebrand), not necessarily random.

For non-random mixtures Guggenheim proposes

$$X^2 = (N_0s_0-X)(N_1s_1-X).\exp(-2w/kT) \tag{17}$$

This yields:

$$X = 2X^*/(\ \beta+1\ ) \tag{18}$$

with

$$\beta = [1 + 4\sigma_0\sigma_1\exp(2w/kT)]^{\frac{1}{2}} \tag{19}$$

Expanding versus y

$$X = X^*(1 - \frac{1}{\sigma_0\sigma_1}\sum_{i=2}\frac{\ell_i}{i!}y^{i-1}) \tag{20}$$

and also:

$$\Delta H_m = \Delta H_m^*(1 - \frac{1}{\sigma_0\sigma_1}\sum_{i=2}\frac{\ell_i}{i!}y^{i-1}) \tag{21}$$

$$\ell_2 = (\sigma_0\sigma_1)^2;\ \ell_3 = (1-2\sigma_1)^2 \text{ a.s.o.}$$

<u>N.B. In the limits</u> $\sigma_0\to 0$ and $\sigma_1\to 0$ <u>, all mixtures are random.</u>

If $$\Delta H_m^0 \equiv \lim_{\phi_0\to 0}\frac{\Delta H_m}{\phi_0} \text{ and } \Delta H^1{}_m \equiv \lim_{\phi_1\to 0}\frac{\Delta H_m}{\phi_1} \tag{22}$$

then

$$\Delta H_m^1/\Delta H_m^0 = \frac{s_1 v_0^u}{v_1 s_0^u} = 1 - \gamma \text{ provided } \gamma \text{ independent of } \phi \tag{23}$$

The limits are taken at constant V/v_0 of the mixture.

$$\text{If} \quad \gamma = \gamma(\phi) \text{ then } \Delta H_m^1/\Delta H_m^0 = \frac{s_1(0)\ v_0}{s_1\ v_1} = 1-\gamma(0) \tag{23a}$$

<u>Effect of non-randomness on ΔF .</u>

Not only is

$$\Delta F = \Delta F_r^c + \Delta F_e \text{ also } \Delta F_e = \Delta F_e^x + \Delta F_e^c \tag{24}$$

because non-randomness gives rise to a combinatorial entropy.

Do not confuse ΔF_e^c with ΔF_{rep}^c. $\Delta F_e^x = Xw$ (25)

Extending Guggenheims calculations we find:

$$\Delta F_e^x = \Delta F_i^* [1 - \frac{1}{\sigma_0 \sigma_1} \sum_{i=2} \frac{i\ \ell_i}{i!} y^{i-1}] \tag{26}$$

$$\Delta F_e^c = \Delta F_i^* [1 + \frac{1}{\sigma_0 \sigma_1} \sum_{i=2} \frac{(i-1)\ell_i}{i!} y^{i-1}] \tag{27}$$

+

$$\Delta F_e = \Delta F_i^* [1 - \frac{1}{\sigma_0 \sigma_1} \sum_{i=2} \frac{\ell_i}{i!} y^{i-1}] \tag{28}$$

In ΔF_e, the effect of non-randomness in ΔF_e^x is compensated partly (the ith term for a part $(i-1)/i$) by ΔF_e^c .

In polymer solutions the random approximation will probably be satisfactory in most cases. Guggenheim (MIXTURES p.65) shows that in (non-polymer) regular solutions the series in (20) is convergent above the critical temperature T_c. For polymer solutions T_c is relatively higher, so above T_c the series will converge rapidly, whereas below T_c the mixture will separate in two phases with small values of the coefficients ℓ_2, ℓ_3 .....

Nevertheless in a complete treatment of polymer mixtures the possible effect of non-randomness on F_e must be considered. Much more important is the contribution ΔF_r^c to ΔF_e.

5. THE COMBINATORIAL FREE ENERGY, ΔF^c_{rep}, FOR MACROMOLECULAR MIXTURES

Guggenheim calculates ΔF^c_r from a lattice model as follows: Consider a domain on the lattice that can accomodate a macromolecule. Then calculate a quantity α defined by

$$\alpha \equiv \frac{\text{probability of occupation by a single macromolecule}}{\text{probability of occupation by r solvent molecules}} \tag{29}$$

or $$\alpha = f(\overline{rP})/f(rS) \tag{29a}$$

Guggenheim shows that

$$\frac{\Delta F^c_r}{kT} = \frac{N_0+rN_p}{r}\left[\int_0^\phi \ln\alpha\, d\phi - \phi\int_0^1 \ln\alpha\, d\phi\right] \tag{30}$$

$f(\overline{rP})$ is proportional to $\phi_p=\phi$ while $f(rS)= \phi_0^r = (1-\phi)^r$. The crudest approximation of α is, therefore:

$$\alpha = K\,\phi/(1-\phi)^r \tag{31}$$

This leads to Flory's equation for a random ("athermal") solution

$$F^c_r = kT(N_0\ln\phi_0 + N_p\ln\phi_p) \tag{32}$$

However, Guggenheim, Huggins and Miller have observed, that (31) is a crude approximation. In the calculation of f(rS) the probability for the first solvent mol= $(1-\phi)$, but for the r-1 following molecules it is larger, because we know it is a neighbour of a solvent molecule. This leads to

$$f(rS) = \phi_0^r \cdot (\sigma_0/\phi_0)^{r-1} \tag{33}$$

σ_0 can be expressed in γ and also in Guggenheims q:

$$\gamma = 1 - \frac{q}{r} \tag{34}$$

and

$$\sigma_0 = \frac{1-\phi}{1-\gamma\phi} \tag{35}$$

Guggenheim derives from (33):

$$\Delta F_r^c = kT[N_0 \ln \phi_0 + N_p \ln \phi_p + \frac{r-1}{r-q} \{(N_0+qN_p)\ln \frac{N_0+rN_p}{N_0+qN_p} - q N_p \ln \frac{r}{q}\}] \quad (36)$$

but this relation is valid for "open macromolecules" only, macromolecules without intramolecular rings. For open molecules on a lattice with lattice constant z:

$$q^0 z = rz - 2z + 2 \text{ or } \frac{r-q^0}{r-1} = 2/z \text{ (open molecules)} \quad (37)$$

Putting r-1 = r, this can be written:

$$1 - \frac{q^0}{r} = \gamma^0 \quad \text{(open molecules)} \quad (38)$$

The restriction to open molecules is severe. It excludes bad solvents at all concentrations, non-dilute solutions in good solvents and solutions near critical conditions.

Why is validity of (36) so restricted?

The total surface is: $S = (N_0+qN_p)s_0$ (39)

The total volume is : $V = (N_0+rN_p)v_0$ (40)

For mixtures: $S = (N_0+ \Sigma_i q_i N_i)s_0$ (41)

Eq.(36) can be written:

$$\frac{\Delta F}{kT} = N_0 \ln \phi_0 + N_p \ln \phi_p - \frac{r-1}{r-q} [\frac{S}{s_o} \{\ln \frac{S}{V} - \ln \frac{s_0}{v_0}\} + B N_p] \quad (42)$$

For mixtures of macromolecules this equation remains valid, if

$$\frac{r-1}{r-q} = \frac{2}{z} \quad \text{(open molecules)} \quad (43)$$

holds for all macromolecules. If (43) does not hold, the equation for the free energy for mixtures is inconsistent.[5)8)9)10)]

In 1950 Staverman[8)] proposed a solution:

Replace (33): $f(rS) = \phi_0(\sigma_0/\phi_0)^{r-1}$

by: $f(rS) = \phi_0 (\sigma_0/\phi_0)^{\ell}$ (44)

where ℓ is the number of intramolecular bonds, chemical and physical together. In

(33) r-1 is the number of chemical bonds (number of bonds between singly connected segments). For more-than-singly-connected segments eq.(44) is more justified than eq.(33). Indeed, if a factor σ_0/ϕ_0 is added for a segment known to be adjacent to a segment considered earlier, then it is consistent to add 2 factors for a segment known to be adjacent to 2 previously considered segments. This is the justification of eq.(44).

If $\gamma=(1-\frac{q}{r})$ is the fraction of contacts occupied in intramolecular bonds and $\gamma_m=(1-q_m/r)$ is the fraction occupied by chemical bonds then

$$\ell/r = \gamma/\gamma_m = \frac{r-q}{r-q_m} \tag{45}$$

q_m is the maximum value of q (singly connected segments)

γ_m is the minimum value of γ.

Substitution of (44) in (30) and (31) gives:

$$\frac{\Delta F}{kT} = N_0 \ln \phi_0 + N_p \ln \phi_p - \frac{1}{\gamma_m} [\frac{S}{s_o} \{\ln \frac{S}{V} - \ln \frac{s_0}{v_0}\} + B N_p] \tag{46}$$

This equation is consistent for mixtures of macromolecules with different γ-values, provided that γ_m has the same value for all macromolecules. Since γ_m depends on the extent of screening of segments by connected segments, γ_m may be expected to be constant in a series of homologous macromolecules. Experimentally γ_m could be determined from heats of mixing on oligomers of sufficiently small M to be ring-free.

The effect of replacing (36) by (46) on the free energy is shown by comparing the virial coefficients, A_2, for random solutions (w=0).

$$A_2 = \tfrac{1}{2} \quad \text{(Flory, } w = 0) \tag{47}$$

$$A_2 = \tfrac{1}{2}(1-\frac{2}{z}) = \tfrac{1}{2}(1-\gamma_m) \quad \text{(Huggins-Guggenheim } w=0) \tag{48}$$

$$A_2 = \tfrac{1}{2}(1-\gamma^2/\gamma_m) \quad \text{(eq.46, } w=0) \tag{49}$$

Tompa[10] has observed that eq.(43) breaks down for large ℓ because it will lead eventually to values of f(rS)>1, which is physically impossible. However, this occurs only at large values of ℓ (ℓ=2r). For $\gamma/\gamma_m<2$ eq.(43) appears "a very reasonable approximation" (Tompa p.91).

F_c depends on <u>two</u> surface parameters: γ_m and γ. $1-\gamma_m$ is the surface-to-volume-

ratio of singly connected segments, expected to be independent of ϕ: in the lattice model $\gamma_m=2/z$ (fraction of occupied surface in string of singly connected sites).
$1-\gamma$ is the surface-to-volume-ratio of the macromolecules: γ depends on w(T), M and ϕ (except at T= Θ). The dependence of $F\gamma$ on T, M and ϕ is closely related to similar dependencies of R_g (radius of gyration). The assumption $\gamma=\gamma(\phi)$ has a sound physical basis in contrast to the assumption $\chi=\chi(\phi)$ of the vanLaar-parameter.

6. DETERMINATION OF γ AND γ_m. CONCENTRATION DEPENDENCE OF γ.

The conclusion of sec.5 is, that a consistent description of thermodynamic properties of polymer mixtures requires the introduction and determination of two surface parameters, γ and γ_m; γ_m is the surface ratio of singly connected segments (of volume v_0), wheras γ is this surface ratio not counting the surfaces in intramolecular contacts. Experimental determination of γ can be achieved by measuring the limiting values of ΔH_m, ΔH_m^o and ΔH_m^1, and calculating γ from eq.(23). Once γ is known, $\Delta H_m(\phi)$ can be calculated at all concentrations with (11) provided that γ is constant. Since macromolecules tend to shrink with increase of ϕ, except at T= Θ, γ will decrease. Thus differences between theoretical and experimental values of $\Delta H_m(\phi)$ can be explained from variations of $\gamma=\gamma(\phi)$. On one hand this provides freedom for adaption of γ to experimental results, on the other hand the relation between γ and ϕ is an interesting complement of the well-known relation between the radius of gyration and ϕ. The assumption of a concentration dependent surface ratio has a better physical basis than the assumption of a concentration dependent vanLaar-Hildebrand parameter.

For the experimental determination of γ_m measurements of ΔH_m or ΔF of solutions of ring-free oligomers or even monomers may be suggested.

7. THE BRIDGING FUNCTION

The equations 11,15,21,28,46 and 49 are all based upon the assumption of a statistically uniform segment distribution. However, in very dilute solutions of macromolecules the segment distribution is definitely not uniform. In order to describe the effect of non-uniformity on the function $\Delta F(\phi)$, Koningsveld and Stockmayer a.o.[11] introduced a bridging function of the form

$$P(\phi) = \exp-(\phi/\phi^*) \tag{50}$$

Here ϕ^* is a critical concentration, characterising the transition from dilute to semi-dilute solutions as defined by de Gennes[12]. This critical concentration decreases with increasing molecular weight, M. In Θ-solutions $\phi^* \sim M^{-\frac{1}{2}}$ as proposed by Koningsveld c.s., while for random ("athermal") solutions de Gennes (ref.12, p.77) gives $\phi^* \sim M^{-4/5}$. The effect of non-uniformity on ΔH_m and ΔF_e in surface statistics is simple. Again $\Delta H_m(\phi)$ is more informative than ΔF ; from measurements of $\Delta H_m(\phi)$ an experimental bridging function can be derived immediately. Indeed, if X_{ij} is the number of contacts i-j, then it follows from (12):

$$\Delta H_m = X_{10}u = (N_1 s_1 - 2X_{11})u \tag{51}$$

The effect of the non-uniformity of the segment distribution is, that in dilute solution, with $\phi<\phi^*$, the number of 1-1-contacts is reduced: $X^n_{11}<X^u_{11}$. Here indices u and n indicate "uniform" and "non-uniform" respectively. Thus we can define a bridging fuction $f(\phi)$ by

$$X^n_{11} = f(\phi)\, X^u_{11} \tag{52}$$

This function must be very small for small ϕ and approach unity for $\phi \to \phi^*$. A possible expression for $f(\phi)$ could be

$$f(\phi) = 1 - P(\phi) \tag{53}$$

with $P(\phi)$ the bridging function of Koningsveld and Stockmayer. Since ϕ^u_{11} is quadratic in N_1, its value does not contribute to the limit $\lim_{N_1 \to 0} (\Delta H_m/N_1)$ and this limit is not affected if X_{11} is multiplied with a function increasing with N_1. Since $f(\phi)$ must increase with ϕ, non-uniformity will not affect the value of ΔH^1_m in (22) nor the value of $\gamma(0)$ derived from it with (33). Once γ is known the theoretical value of $H^u_m(\phi)$ can be calculated with (11) and this quantity can be compared with experimental values of $H^n_m(\phi)$ in dilute solutions. The bridging function follows then simply from

$$f(\phi) = \frac{\phi \Delta H^1_m - \Delta H^n_m(\phi)}{\phi \Delta H^1_m - \Delta H^u_m(\phi)} \tag{54}$$

A complication could arise if γ depends on ϕ. However, the effect of non-uniformity is perceptible only if $\phi<\phi^*$, where the macromolecules are mainly isolated from each other, so, accordingly γ is expected to be constant in this region.

The simplicity of the effect of non-uniformity is limited to the effect on ΔH_m and ΔF_e and is due to the fact, that for these quantities the restriction to nearest neighbour interactions appears acceptable. Non-uniformity will also affect ΔF_r^c and this effect will be manifest in an experimental value of the virial coefficient, A_2, differing from the theoretical value of eq.(49), calculated for a uniform segment distribution. Once experimental values of ϕ^* and $f(\phi)$ are known from measurements of ΔH_m, the effect of non-uniformity on ΔF_r^c can be calculated along the lines of sec.5. The effect on ΔF_e is considered separately and added in ΔF. It should be noted, that, although the effect of non-uniformity on ΔH_m and ΔF_e is small and restricted to a small region $\phi<\phi^*$ without affecting the limiting values for $\phi \to 0$, nonetheless the initial slope of the plot of $\Delta H_m/\phi$ versus ϕ may be affected strongly. In case (53) is valid, $f(\phi)$ is proportional to ϕ for small ϕ, and, since X_{11}^u in quadratic in ϕ it follows from (52), that X_{11}^n is proportional to ϕ^3. In this case the development of $H_m(\phi)/\phi$ has no linear term in ϕ. Eq.(54) affords an interesting and sensitive test of the form of $f(\phi)$.

8. CONCLUSIONS

The proposed treatment is admittedly crude. It lacks the finesse of equation of state theories (Prigogine, Flory). However, its crudeness may well correspond adequately with the chaotic conditions in liquid mixtures, and with the pretence to describe merely effects of variation of composition, not to predict values of w, u, ΔV or γ. As a first approximation the quantities r, γ, γ_m, w and u are assumed to be independent of composition. Effects are considered at zero pressure; change of volume does not affect the free energy: $\Delta F=\Delta G$, $\Delta H=\Delta U$. Possible variations of volume with composition are assumed to originate from nearest neighbour interactions and are, therefore, absorbed in w and u. In case of variations of volume on mixing, thermodynamic functions are considered not at constant volume but at constant constant $V/v_o = N_0 + rN_p$, with r assumed to be constant.

REFERENCES

1. Hildebrand, J.H., The Solubility of Non-electrolytes, N.Y. 1950.
2. Langmuir, I., Colloid Symposium Monograph, vol. 3, 48, 1925.
3. Staverman, A.J., Rec.Trav.Chim., vol. 56, 885, 1937.
4. Huggins, M.L., J.Phys.Coll.Chem., vol.52, 248, 1948;
   Physical Chemistry of High Polymers, N.Y., 1958;
   J.Phys.Chem., vol.74, 371, 1970;
   J.Phys.Chem., vol.75, 1255, 1971;
   J.Phys.Chem., vol.80, 1317, 1976.
5. Guggenheim, E.A., Mixtures, Oxford, 1952.
6. Flory, P.J., J.Am.Chem.Soc., vol. 87, 1833, 1965;
   J.Am.Chem.Soc., vol. 89, 6814, 1967.
   Trans.Faraday Soc., vol. 67, 2251, 2270, 2275, 1971.
7. Koningsveld, R., Thesis, Leiden 1967.
   Koningsveld, R., Kleintjens, L.A., Macromolecules, vol. 4, 637, 1971.
   Koningsveld, R., Ber.Bunsengesellschaft, vol. 81(10), 960, 1977.
   Chermin, H.A.G., Thesis, Essex, 1971.
   Onclin, M., Thesis, Antwerpen, 1980.
   Nies, E., Thesis, Antwerpen, 1983.
8. Staverman, A.J., Rec.Trav.Chim., vol. 69, 163, 1950
9. Miller, A.R., Proc.Camb.Phil.Soc., vol.43, 422, 1947
10. Tompa, H., Polymer Solutions, London 1956.
11. Koningsveld, R., Stockmayer, W.H., Kennedy, W., Kleintjens, L.A., Macromol., vol. 7, 73, 1974.
12. de Gennes, P.G., Scaling Concepts in Polymer Physics, London 1979.

NOTATION

A_2	Virial coefficient of the osmotic pressure
$F=G$	Free energy
F_r^c	kT ln Q_c combinatorial free energy
F_e	Excess free energy
ΔF_e^*	X^*w (random solution)
$f(rS)$, $f(\overline{rP})$	see eq.29a
$f(\phi)$	"Bridging function" eq.52
$H=U$	Energy, Heat
ΔH_m	Heat of mixing

$\Delta H_m^* = X^* u$	(random solution)
k	Boltzmanns' constant
ℓ	Total number of intramolecular bonds between segments (eq.44)
N_i	Number of molecules of component i
P	"Bridging function" eq.50
Q_c	Partition function of <u>hard</u> molecules
q/r	$1- \gamma=(s_1N_0)/(s_0N_1)$
$q^0/r=q_m/r=1- \gamma^0$	for "open molecules"
r	$v_1/v_0=v_p/v_0$
s_i	"surface" (number of contacts) of molecules i
$S = \Sigma N_i s_i$	Total surface
T	Kelvin Temperature
u_{ij}	energy of contact i en j
u	$u_{10}-\frac{1}{2}(u_{00}+u_{11})$
v_i	volume of molecule i
$V = \Sigma N_i v_i$	Total volume
w_{ij}	Free energy of contact i-j
$w =$	$w_{10}-\frac{1}{2}(w_{00}+w_{11})$
$X=X_{ij}$	Number of contacts i-j
X^*	Number of contacts in random solution
$y =$	w/kT
z	Lattice constant in lattice model
α	see eq.29
β	Measure of randomness (eq.18,19)
γ	$1-(s_1v_0)/(s_0v_1)$
$\gamma^0=\gamma_m$	Value of γ for open molecules (singly connected segments)
ϕ	Volume fraction of polymer
ϕ_i	Volume fraction of component χ
ϕ^*	Critical concentration (eq.50)
σ_i	Surface fraction of component i
χ	vanLaar-Hildebrand interaction constant

POLYMER MELT AND GLASS: THERMODYNAMIC AND DYNAMIC ASPECTS

ROBERT SIMHA

Department of Macromolecular Science, Case Western Reserve University, Cleveland, Ohio 44106, U. S. A.

SYNOPSIS

We discuss the following topics: 1. The scaled pressure-volume-temperature surface of the equilibrium melt. 2. The temperature and pressure dependence of the melt viscosity: Scaling and the relation to the thermodynamic scaling parameters. 3. The glass: Quasi-equilibrium aspects. 4. Molecular kinetics of physical aging processes in the glass.

A connecting link between the four subjects is indicated by a particular free volume function, its dependence on the state variables and, in the glass, formation history and aging time.

1 INTRODUCTION

The equilibrium properties of homogeneous dense fluids and their mixtures have been an object of intensive experimental and theoretical research, following van der Waals' work. His theory, moreover, provided an impetus to the exploration of relationships between equilibrium and transport. For practical and fundamental reasons, a particular consideration was the connection between temperature and pressure coefficients of viscosity and diffusion on the one hand, and volume, i.e. the equation of state, on the other. Physical intuition suggested relations in terms of unoccupied or "free" volume quantities as an expression of molecular mobility required for the transport process. Van der Waals' equation of state contains such a quantity, and understandably became the basis for an early correlation for low molecular weight fluids.[1]

In dense assemblies of flexible chain molecules, additional features in the dynamics of molecular motions appear, resulting from internal motions. The changes of the physical properties in the glass transition process and the highly retarded approach to the equilibrium state in polymer glasses are issues of particular interest. The configurational thermodynamics of the melt and its mixtures for low and high molecular weights, of course, retains its importance.

We pursue these topics in terms of a particular model of the amorphous polymer assembly. It was originally devised for the configurational thermodynamics of the equilibrium melt and has been subsequently introduced with appropriate modifications in the other problem areas mentioned. These matters form the subject of this paper.

2 THE LIQUID EQUILIBRIUM STATE

2.1 The Model

The basic feature is a lattice, with sites occupied by polymer segments of one, and several types in mixtures, or otherwise empty. In the absence of such vacancies this reduces to the well known cell model of Lennard-Jones and Devonshire[2] and its modification for chain molecular fluids by Prigogine, Trappeniers and Mathot.[3] The partition function contains three contributions,[4,5] namely a lattice energy, a statistical mechanical free volume term, and a combinatory term. The latter arises from the mixing of molecules and holes and is the primary element in the quantitative success of the theoretical equation of state of polymer and oligomer melts.[5]

2.2 The Thermodynamic Functions

A summary in implicit form brings out the essential features. Let $h = 1 - y$ be the fraction of unoccupied sites. The Helmholtz free energy has the form

$$F = F[V, T \; ; \; y(V, T)] \tag{1}$$

The y-function satisfies the minimum condition on the free energy, i.e.

$$(\partial F/\partial y)_{V,T} = 0 \tag{2}$$

which yields the hole fraction as a function of volume V and temperature T. Equations 1 and 2 provide the temperature and volume derivatives of F. The equation of state is

$$-P = (\partial F/\partial V)_T = (\partial F/\partial V)_{T,y} + (\partial F/\partial y)_{V,T} \times (\partial y/\partial V)_T \tag{3}$$

At equilibrium, the second term on the right hand side vanishes. Note that the hole fraction $h(V, T)$ can be regarded as an <u>excess</u> free volume fraction over that existing in the absence of holes, $h = 0$. The theory, eqs. 1 and 2, has been generalized to multicomponent fluid systems.[6]

2.3 The Characteristic Parameters: Scaling

Two quantities characterizing the chain molecule appear. One is the chain-length, s, the other, a dynamic quantity, is $3c$, the number of effectively external degrees of freedom.[3] For a rigid, spherically symmetrical unit, $c = 1$. The characteristic parameter is the "flexibility" ratio c/s, which will be of the order of unity for a flexible chain. The introduction of the c-factor simply bypasses the difficult question of the perturbation of internal degrees of freedom, such as bond rotations and sidechain librations, by the intermolecular interactions in the dense medium.[5] Strictly, one should expect c to be a temperature and pressure dependent quantity. However, extensive experimental data (see below) have demonstrated no discernible change. For liquid argon we find that there also c may be treated as a constant between 100 and 200 K and pressure up to 6 kbar.[7] However, with $s = 1$, a value of 1.8 for $3c$ must be adopted, rather than the expected 3. Furthermore, the ethylene data in the liquid and compressed gas region do not lead to conclusive results.[8] Since it is not feasible to derive specific values for a given molecular structure, we adopt a universal assignment, namely $3c/(s+3) = 1$, or for a polymer $3c = s$. The particular choice is irrelevant for the quantitative success or failure of the theory. However, it determines the definition of the segment relative to the actual repeat unit as a distinguising element between different polymers. For an n-mer with molar mass M_{rep} of the unit and M_o of the segment, we have

$$sM_o = nM_{rep} \; ; \; 3c/n = (3c/s)(s/n) = s/n$$

and it remains to obtain M_o in order to define the segment and derive the flexibility ratio $3c/n$ based on the monomer unit. Defining moreover an intersegmental attraction energy ε^* and repulsion volume v^*, the following scaling quantities for volume, pressure and temperature respectively result:

$$v^* = V^*M_o/N_A \; ; \; P^* = [s(z-2)+2]\varepsilon^*/(sv^*) \; ; \; T^* = [s(z-2)+2]\varepsilon^*/(ck) \qquad (4)$$

Here V^* is the specific repulsion volume and z the coordination number of the lattice. Equation 4 then yields the relation:

$$(P^*V^*/T^*)M_o = R(c/s) \qquad (5)$$

If the theory is quantitative, the scaling quantities will be reasonably constant. The assumed universality of c/s then results in a principle of corresponding states, and the superposition of the theoretical reduced $\tilde{P}\tilde{V}\tilde{T}$ on the measured PVT surface yields the scaling parameters and hence M_o.

2.4 Experimental Test

The literature is by now replete with experimental data on polymer melts; as examples see refs. 9 and 10. Equations 1 and 2, and the principle of corresponding states have been extensively confirmed. The results for a wide variety of polymer structures show the anticipated parallelism between structural complexity of the repeat unit and magnitude of the ratio $3c/n$. Although absolute magnitudes of the c-factor can not be taken literally, as shown earlier by the example of argon,[7] the relative ordering is in accord with expectation on physical grounds.

3 VISCOSITY AND EQUATION OF STATE

Batschinski[1] was the first to investigate a correlation between the Newtonian viscosity η and the van der Waals "free volume" $v - b$ of the form

$$\eta \propto (v - b)^{-1}$$

In more recent times and in the polymer area, Doolittle's relation [11]

$$\ln \eta = A + B/f \tag{6}$$

has found wide application. Here f represents the free volume fraction $(V - V_o)/V$, with V_o an occupied volume. According to these expressions η should depend on temperature and pressure only implicitly through the dependence on volume. Various efforts at improvement of what turns out to be a deficiency, appear in the literature.

One may inquire into the possible role of our free volume function h, which depends explicitly on two rather than a single variable of state, and which can be scaled. This question has been recently addressed by Utracki.[12] A functional relationship

$$\ln \eta = \phi[Y_s(\tilde{P}, \tilde{T})] \; ; \; Y_s = 1/(1 - y) \tag{7}$$

obtains, with ϕ in general not a linear function. Both oligomers and polymers have been investigated, for which the equation of state had been analyzed and the Y_s - function therefore was known. As an example, Fig. 1 shows the result for n-pentadecane, encompassing a temperature range exceeding 100 degrees and with a maximum pressure of ca. 3 kbar. Clearly, the correlation is valid and moreover, the temperature and pressure dependent thermodynamic and transport properties are scaled by the identical parameters. Hence viscosities are predictable, based on the evaluation of the equation of state and limited information on the viscosity, e.g., low pressure data. The same conclusions obtain for other members of the n-paraffin series, investigated up to C_{18}.[12] For high polymer melts, superposition of temperature and pressure data is again accomplished, and for some

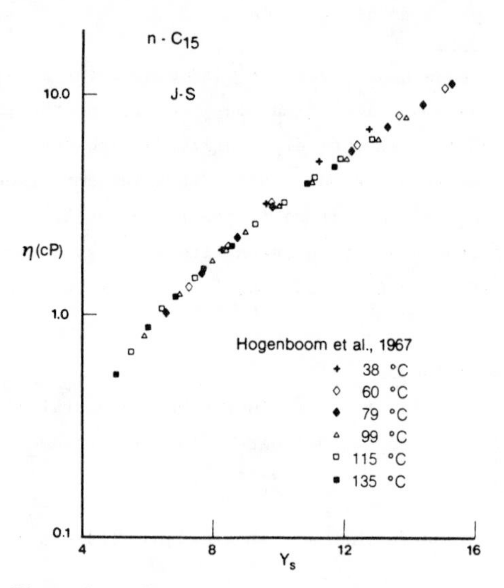

Fig. 1 - Viscosity of liquid n-pentadecane as a function of Y_s, eq. 7, indicating scaling characteristics. Pressures $\leq$ 3.2 kbar.

polymers, viz. poly(dimethyl siloxane) or polypropylene, with the scaling parameters defined by the equation of state.[12] For polystyrene, polyethylenes of various types, or poly(methyl methacrylate), the temperature scale is maintained; low pressure viscosities follow the analogous pattern of the n-paraffins. However, the scaling of viscosity at elevated pressures requires a different pressure scale than the equation of state, i.e. a change from P^* to $2P^*$. Thus no universal distinction between polymer vs. oligomer emerges, and it is not possible to consider the differences as a molecular weight effect. However throughout, a scaled relationship between viscosity and excess free volume is established, which encompasses temperature and pressure effects. The required increment in the pressure scale and reduction of the reduced pressure $\tilde{P}$ imply an increase in the free volume fraction and, by virtue of eq. 5, a decrease in the segment size, utilized in the flow process. One may expect that fluctuations of the free volume rather than average values will be pertinent, but this remains to be investigated.

4 THE GLASSY STATE: QUASI-EQUILIBRIUM

What modifications in the equation of state theory are required by the non-equilibrium character of the glass? The important element is the influence of formation history, such as formation pressure and rate of cooling. Given these, one can define a quasi-equilibrium state, when with sufficiently rapid experimentation, time-dependent relaxation in the drive to equilibrium can be disregarded. The notion of an equation of state for specified formation conditions can then be maintained,[13] and one may inquire into the behavior of the free volume function. Clearly, the equilibrium condition, eq. 2, must be abandoned. On the other hand, the assumption that this function degenerates into a constant at the glass transition was very early shown to be invalid, the more so, the higher the transition temperature,[14] as one may expect.

In what appears to be the absence of a general, even if history dependent condition on y, we retain eq. 1 as pertinent for a mixture of molecules and vacancies and appeal to the PVT experiment to provide the y-function in this equation. In eq. 3, the two free energy derivatives on the right hand side are known as a function of V, T and y. This equation is then combined with the pressure data to obtain a partial differential equation for y.[15] The result is shown in Fig. 2 for two glasses of poly(vinyl acetate) formed by cooling the liquid at a

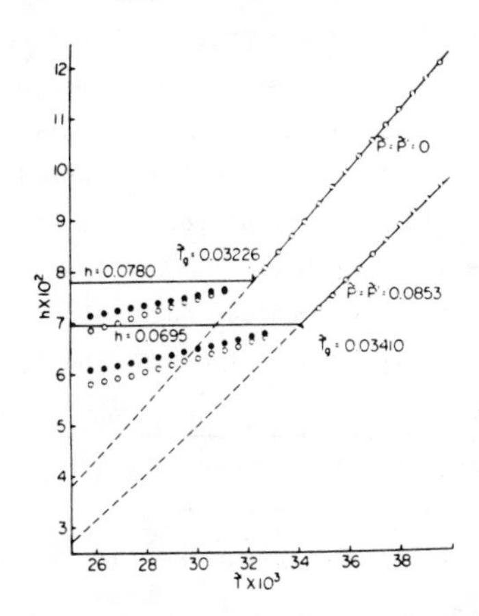

Fig. 2 - Free volume fraction h as a function of temperature for two glasses of poly(vinyl acetate). For explanation see text.

fixed rate and pressures $P' = 1$ and 800 bar respectively, which are then maintained in the glass. The free volume fraction $h = 1 - y$ is plotted as a function of temperature in the melt and in the glass. For $T > T_g$, the coincidence between measured (i.e. extracted from eq. 1) and computed (from eq. 2) h attests to the theory's quantitative performance in the liquid. For $T < T_g$, the full circles are the result of the procedure described and they indicate the extent of freeze-in by their intermediate position between the horizontal lines and the computed extension of the liquid lines into the glassy domain. The open circles are the result of a simplification, which replaces a highly non-linear differential by an algebraic equation. That is, one adopts the theoretical pressure equation in the form $P = P(V, T ; y)$ and treats y as a disposable quantity, to be extracted from experiment.[13,16]

The internal consistency of the above approach may be examined by predictions of other quantities which are functions of the hole fraction. The mean square thermal density fluctuations are such a quantity. The computations, employing the results for h in Fig. 2 for the atmospheric pressure glass,[17] are indeed in accord with experiment.[18]

5 DYNAMICS OF RELAXATION PROCESSES IN THE GLASS

Viscous flow may be viewed as a relaxational process. Another type of relaxational process, namely physical aging, is considered now. The aging properties in question may be extensive thermodynamic functions, e.g. volume or enthalpy; on the other hand they could be for example mechanical properties. Theoretical interpretations of the aging process should serve the double purpose of deriving long-time predictions from short-time experimentation, and of providing some insight into the underlying molecular dynamics in the dense medium. Two points of view and hence methodologies may be adopted. One aims at connections between the relaxation patterns of different quantities and predictions of one, based on the measurement of another. The other attempts a kinetic formulation for a particular quantity. We consider both aspects and select volume as the primary relaxing quantity, since detailed experimental results are available.

5.1 Visco-Elastic Relaxation and Free Volume: Melt

The temperature dependence of the visco-elastic relaxation spectrum in the melt is given by a frequency independent shift factor a_T. It has the form obtained empirically and rationalized through Doolittle's eq. 6,[19] viz.

$$\ell n\, a_T = B[f^{-1}(T) - f^{-1}(T_r)] \tag{6'}$$

with T_r a reference temperature. To proceed, an assumption about the temperature dependence of the free volume fraction f is necessary. The h-function on the other hand, is explicitly determined in the melt by eq. 2 and one may investigate its relation to $\ell n\, a_T$. For poly(methyl methacrylate)[20] and poly(vinyl acetate),[21] a linear relation ensues, at least over the limited temperature range available. Note that no disposable parameter in eq. 6' with $f \equiv h$ remains.

5.2 Aging System

The major consequence of the aging process is a shift of the visco-elastic spectrum on the log-time axis with no change in shape.[22] A time-dependent shift factor $a_T(t)$ can then be introduced and the question of its relation to a time-dependent $h(t)$ be raised.[20] To obtain the latter, the procedure of Section 4 is extended. That is, we assume that the aging volume passes through a series of consecutive quasi-equilibrium states in which stress relaxation proceeds with comparative rapidity. Hence $a_T(t)$ can be related to the volume $V(t)$ at a given temperature and pressure. Application to mechanical and dilatometric data for poly(methyl methacrylate)[23] and poly(vinyl acetate)[24,25] indeed demonstrates the predictability of the temperature shift $a_T(t)$ from data on volume relaxation.[20,21]

5.3 Kinetics of Volume Relaxation: Diffusion Model

The foregoing discussion suggests free volume relaxation as the central issue. That is, we make the transformation $h(t \to V(t)$, whereas in 5.2 the reverse transformation was applied. The basic ingredient is the existence of free volume gradients, rapidly eliminated by molecular motions which establish dynamic equilibrium above T_g. In the glass this process is slow and governed by the diffusion rate of holes.[26] The diffusion coefficient D is related to the <u>local</u> free volume by a Doolittle-type equation, viz.

$$\ell nD = \ell nD_r - B(h^{-1} - h_r^{-1})$$

A characteristic, for simplicity spherical volume of radius ℓ is considered and the total free volume h obtained by integration over this volume. The time variable t is scaled by the ratio $\tau = \ell^2/D_r$, where ℓ is a kind of mean free length for vacancy diffusion. In Fig. 3 the circles indicate the free volume of

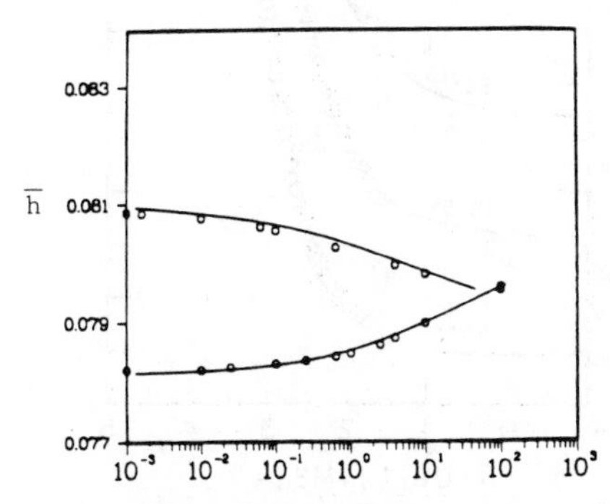

Fig. 3 - Volume recovery of poly(vinyl acetate) as a function of time, following temperature jumps. Lines theory. For explanation see text.

poly(vinyl acetate), as extracted from experiment[25] for two temperature jumps from 38 and 32°C respectively to 35°C. Superposition of one of the computed curves yields $\tau = 7800$ s and defines without further adjustment the location of the second.

5.4 Kinetics of Volume Relaxation : Stochastic Model

Here the time evolution of the h-function is ascribed to free volume dependent conformational changes,[27] rather than to a diffusion process. Again, a characteristic volume, containing a fixed number of chain units is considered.[28] This number is to account for the dependence of the mutual rotation of a pair on the cooperation of neighbors in the dense medium. The region contains an ensemble of free volume states and we seek their distribution function $w_i(t_o)$ as determined by transition probabilities $P_{ij}(t - t_o)$ from state i to j in a time interval $t - t_o$. A system of rate equations for the P_{ij} ensues. The rate parameters are expressed as a function of free volume states h_i and the average free volume of z adjacent domains serving as sources or sinks for free volume changes. We note that this approach operates not only with the values of h, but also their fluctuations $< \delta h^2 >$, derived from the free energy F(V, T, y), eq. 1. The solutions provide two important functions, namely the distribution of relaxation frequencies and the time-dependent distributions of free volume in the aging system. Some results are illustrated in the following figures. Figure 4

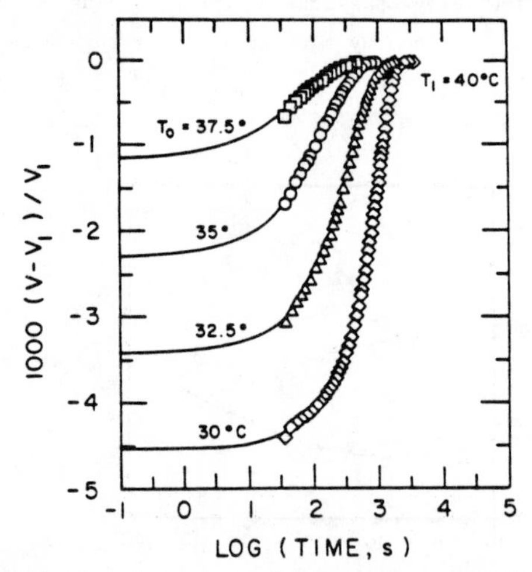

Fig. 4 - Relative departure of volume from equilibrium value as a function of time for poly(vinyl acetate), following temperature jumps from T_o to T_1. Lines, theory.

shows the measured[25] and computed[28] departures of the volume from the equilibrium value of poly(vinyl acetate) as a function of time for a series of temperature steps from the temperature levels indicated to the final temperature T_1 . The corresponding distribution of free volume for the largest temperature jump is seen in Fig. 5. Note the contrasting rates of approach to the equilibrium distribution by large and small free volume fractions. Isobaric annealing experi-

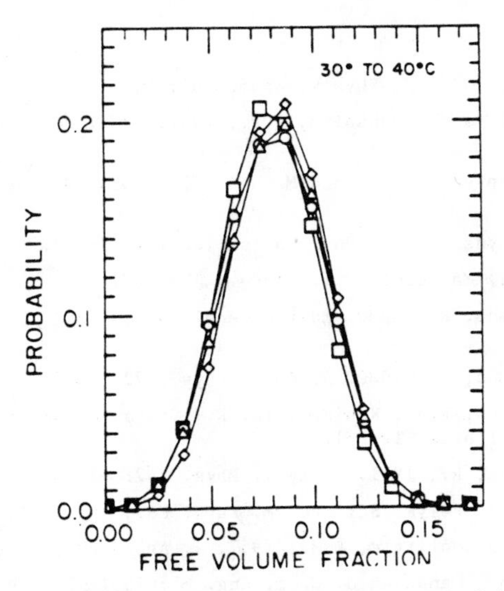

Fig. 5 - Distribution of free volume fraction h following a temperature jump from 30 to 40°C at 4 stages: initial, 0 and intermediate, final.

ments at 1 and 1000 bar have been carried out by Rehage and Goldbach for polystyrene.[29] Again, satisfactory agreement between experimental and computed volume recovery results with suitably altered parameter values. These are then employed to calculate isothermal pressure-jump curves.[30] A much smaller spread in the data than seen in the temperature-jump experiments is in qualitative accord with observation.[29] Finally, utilizing the relation between h and the thermal density fluctuations, we find that these relax much more slowly than volume or enthalpy in poly(vinyl acetate).[31]

In the light of results obtained for the free volume, its thermal fluctuations and size distribution, we draw attention to some recent spectroscopic[32] and kinetic[33] studies in aging glasses. They involve probes and labels and suggest the pertinence of free volume considerations. The results invite future systematic investigations into the role of the h-function as a connecting link between different relaxing properties, and as a measure of intermolecular effects in the dynamics of molecular motions.

ACKNOWLEDGEMENT. It is a pleasure indeed to present on this occasion some ideas which profited from discussions with Ronald Koningsveld. We thank the National Science Foundation for support through Grant DMR 84-08341 Polymers Program.

REFERENCES

1. Batschinski, A.J., 1913. Z. Physik. Chem., 84: 643.
2. Lennard-Jones, J.E. and Devonshire, A.F., 1937, Proc. Roy. Soc. (London), A163: 53.
3. Prigogine, I., Trappeniers, N. and Mathot, V., 1953. Discuss. Faraday Soc., 15: 93.
4. Simha, R. and Somcynsky, T., 1969. Macromolecules, 2:342.
5. Simha, R., 1980. J. Macromol. Sci. -Phys., B18: 377.
6. Jain, R.K. and Simha, R., 1980. Macromolecules, 13: 1501 ; 1984. Macromolecules, 17: 911.
7. Jain, R.K. and Simha, R., 1980. J. Chem. Phys., 72: 4909.
8. Nies, E., Kleintjens, L.A., Koningsveld, R., Simha R. and Jain, R.K., 1983. Fluid Phase Equilibria, 12: 11.
9. Quach, A. and Simha, R., 1971. J. Appl. Phys., 42: 4592.
10. Zoller, P., 1978, J. Polym. Sci. Polym. Phys. Ed., 16: 1491.
11. Doolittle, A.K. and Doolittle, D.B., 1957. J. Appl. Phys., 28: 901.
12. Utracki, L.A., 1983. Canadian J. Chem. Eng. 61: 7531 ; 1983. Polymer Preprints., 24 No. 2: 113.
13. McKinney, J.E. and Simha, R., 1974. Macromolecules, 7: 894.
14. Somcynsky, T. and Simha, R., 1971. J. Appl. Phys., 42: 4545.
15. McKinney, J.E. and Simha, R., 1976. Macromolecules, 9: 430.
16. Simha, R. and Wilson, P.S., 1973. Macromolecules, 6: 908.
17. Balik, C.M., Jamieson, A.M. and Simha, R., 1982. Colloid Polym. Sci., 260: 477.
18. Tribone, J., Jamieson, A.M. and Simha, R., 1984. J. Polym. Sci. : Polym. Symp., 71: 231.
19. Williams, M.L., Landel, R.F. and Ferry, J.D., 1955. J. Am. Chem. Soc., 77: 3701.
20. Curro, J.G., Lagasse, R.R. and Simha, R., 1981. J. Appl. Phys., 52: 5892.
21. Lagasse, R.R. and Curro, J.G., 1982. Macromolecules, 15: 1559.
22. Struik, L.C.A., 1978. 'Physical Aging in Amorphous Polymers and Other Materials', Elsevier.
23. Cizmecioglu, M., Fedors, R.F., Hong, S.D. and Moacanin, J., 1981. Polym. Eng. Sci., 21: 940.

24. Kovacs, A.J., Stratton, R.A., Ferry, J.D., 1963. J. Phys. Chem., 67: 152.

25. Kovacs, A.J., 1963. Adv. Polym. Sci., 3: 394.

26. Curro, J.G., Lagasse, R.R. and Simha, R., 1982. Macromolecules, 15: 1621.

27. Robertson, R.E., 1981. Ann. N. Y. Acad. Sci., 371: 21.

28. Robertson, R.E., Simha, R. and Curro, J.G., 1984. Macromolecules, 17: 911.

29. Rehage, G. and Goldbach, G. 1966. Ber. Bunsenges. Phys. Chem., 70: 1144.

30. Robertson, R.E., Simha, R. and Curro, J.G., in preparation.

31. Jain, S.C. and Simha, R., 1982. Macromolecules, 15: 1522.

32. Tsay, F.-D., Hong, S.D., Moacanin, J. and Gupta, A., 1982. J. Polym. Sci. Polym. Phys. Ed., 20: 763.

33. Sung, C.S.P., Gould, I.R. and Turro, N.J., 1984. Macromolecules, 17: 1447.

A FRESH LOOK AT SOLUTIONS OF POLYMER MIXTURES

Karel Šolc

Michigan Molecular Institute, Midland, Michigan 48640

SYNOPSIS

Analysis of phase equilibria in ternary systems of the type solvent - polymer(1) - polymer(2) is facilitated by introduction of two new parameters replacing traditional separation factors σ_i, namely $\eta = (\sigma_1 \sigma_2)^{1/2}$ and $\xi = (\sigma_2/\sigma_1)^{1/2}$. The projection of the latter one into the complex plane of the reduced interaction term creates a one-dimensional global space of all ternary systems, ordered according to the tendency of their polymeric components to separate into two different phases. Surprisingly, ξ changes very little over the entire span of the cloud-point curve of a given polymer mixture. The geometrical average η is a measure of the distance of a cloud-point from the critical point, just as σ was for quasibinary systems. It can be utilized to derive criteria for existence of multiple critical points. Conditions for single, double and triple critical points are established. The approximate plait point criterion derived years ago by Scott is shown to arise, with a minor modification, as an exact solution for a special case.

1 INTRODUCTION

It has been shown that the separation factor σ of the Flory-Huggins theory[1] is perhaps intrinsically the most important parameter of phase separation in quasibinary solutions of polydiperse polymers. It may be defined by the partition of a polymer species i with chain length r_i between the incipient (*) and bulk ( ) phases,

$$\phi^*_{r_i}/\phi_{r_i} = \exp(\sigma r_i) \tag{1}$$

where ϕ_{r_i} denotes the volume fraction of the species i. It is the only phase equilibrium variable that (unlike ϕ or temperature T) determines uniquely a point of the cloud-point curve (CPC), even in the presence of cusps. Its introduction and manipulation made possible, e.g., (i) to formulate a single equation for the cloud point;[2] (ii) to calculate the cloud-point curve for polymers with logarithmic-normal distribution of molecular weight where other methods failed;[3] (iii) to derive criteria for single and multiple critical points,[4,5] thereby providing background for understanding the development of multiphase equilibria.[6]

In this paper we demonstrate that an analogous parameter can be defined for true ternary systems consisting of a solvent (0) and two chemically different polymer species (1 and 2), and explore some of the benefits this formalism offers.

2 PARAMETERS η AND ξ

Straightforward generalization of the Flory-Huggins theory yields two separation factors, one for each polymer, that are tied to other variables as

$$\sigma_i = r_i^{-1}\ln(\phi_i^*/\phi_i) = \Delta A + (g_j-g_i-g_x)\Delta(\phi_0\phi_j) - g_i\Delta\phi_0{}^2 - g_x\Delta\phi_j{}^2 \qquad (2)$$

$$i,j=1,2 \; ; \; i\neq j$$

where g_1 and g_2 are the interaction parameters of the two polymers with the solvent, g_x is the parameter characterizing the interaction between the two polymers (related to one polymer segment, i.e., $g_x \equiv \chi_{23}/r_2$ of Flory[1]), Δ denotes the difference between the two phases, $\Delta X \equiv X^* - X$, and $A \equiv \sum_0^2\phi_k/r_k$.

The existence of two parameters, σ_1 and σ_2, however, is unpleasant and inconvenient since the incipient phase separation is essentially a one-dimensional event (each cloud point being typically conjugated with just one another phase). Similarly, when computing the CPC for a given polymer mixture, only one σ parameter can be chosen independently, while the other one (just as ϕ and T) is fixed by equilibrium conditions. Solution to this problem is found by introducing transformed variables

$$\eta^2 = \sigma_1\sigma_2 \; , \; \xi^2 = \sigma_2/\sigma_1 \; , \text{ i.e.,} \qquad (3)$$

$$\sigma_1 = \eta/\xi \quad , \; \sigma_2 = \eta\xi$$

Note that η and ξ become imaginary if one of the σ_i's is negative and the other is positive, i.e., if each polymer concentrates in a different phase. η is an analogue of σ of quasibinary systems: $\eta = 0$ at the critical point (CP), and generally it grows in absolute value as one moves away from the CP. Parameter ξ is a relative measure of tendency of polymer 2 over that of 1 to concentrate in the incipient phase, and depends only on the reduced interaction term $\tilde{W} \equiv W/\eta$:

$$\xi^2 - 2\tilde{W}\xi - 1 = 0 \quad (4)$$

where

$$2W = (g_2-g_1)\ \Delta\phi + g_x(\Delta\phi_2-\Delta\phi_1)$$

and $\phi = \phi_1 + \phi_2$. Mapping of ξ onto the axes of the complex plane of $\tilde{W}$ thus provides a one-dimensional space of all ternary systems, with three prominent locations: (i) if $\tilde{W} = 0$, then $\xi^2 = 1$, and $\sigma_1 = \sigma_2$ (no interaction-induced enrichment of one polymer over the other; this is also the location of a quasibinary system); (ii) if $\tilde{W} \to -\infty$ (or $\tilde{W} \to +\infty$), then $\xi^2 \to 0$ (or $\xi^2 \to \infty$) and $\sigma_2 \to 0$ (or $\sigma_1 \to 0$) (one of the polymers shows the same concentration in both phases); (iii) if $\tilde{W} = \pm i$, then $\xi^2 = -1$, and $\sigma_1 = -\sigma_2$ (polymers show the same tendency to partition, but concentrate in opposite phases).

Interestingly, in most cases the value of ξ changes very little over the entire span of the CPC. Hence, the numerical iteration of cloud points proceeds better and faster in η,ξ space than in the traditional σ_1,σ_2 space. The most important advantage of the new parameters, however, is in an enormous simplification of series expansions leading to the criteria for the existence of multiple critical points.

3 CRITICAL POINTS

The classical Gibbs method of critical determinants[7,8] yields a criterion for the CP, but the result is very unwieldy and physically unintelligible.

One can generalize instead the σ-expansion successfully used in the past for quasibinary systems,[2,4,5] and write a two-dimensional series in σ_1 and σ_2 for the cloud point function F

$$F(\sigma_1,\sigma_2,\phi) \equiv \sigma_1(\phi_1+\phi_1^*) + \sigma_2(\phi_2+\phi_2^*) - 2\Delta A + \frac{\phi_0+\phi_0^*}{r_0} \ln \frac{\phi_0^*}{\phi_0} = 0 \quad (5)$$

As expected, the expanded function is cubic in σ_i, and its first coefficient yields the relation for the critical concentration that can be written compactly as

$$r_0\langle r^2\xi^3\rangle - \langle r\xi\rangle^3 [\phi/(1-\phi)]^2 = 0 \quad (6)$$

where the average $\langle \ \rangle$ is defined by

$$\langle r^i \xi^j \rangle = w_1 r_1^i \xi^{-j} + w_2 r_2^i \xi^j$$

and the weight fractions are $w_i = \phi_i/\phi$. The critical point equation (6) is (i) extraordinarily simple; (ii) it transparently reduces to the classical quasibinary formula[5] for ξ=1; and (iii) it succintly isolates the only parameter - ξ - affecting the critical concentration in ternary systems, in addition to the polydispersity effect known from earlier studies. The critical value of ξ needed in eq. (6) has to be computed from two other relations for critical state. It specifies the limiting direction of tie-lines at the CP, given as $\Delta\phi_2/\Delta\phi_1 = \xi(w_2 r_2/w_1 r_1)$.

While the σ_1, σ_2 formalism was adequate for the single critical point criterion above, it becomes akward to handle for critical points of higher multiplicities. Also, it is unnatural (and wasteful) to carry on a formally two-dimensional expansion for an essentially one-dimensional problem. Fortunately it can be proven that for our purpose, the "complete" expansion in terms of σ_1 and σ_2 can be replaced by a simple one-dimensional expansion in terms of only η (while ξ is kept constant), and the criteria for higher critical points can be recovered from partial derivatives $(\partial^n F/\partial \eta^n)_{\xi,T,\phi}$.

Specifically, a double critical point has to satisfy also the relation

$$3\,\frac{\langle r^2\xi^2\rangle}{\langle r\xi\rangle} + 2r_0^{\frac{1}{2}}\,\frac{\langle r^2\xi^3\rangle^{\frac{1}{2}}}{\langle r\xi\rangle^{\frac{1}{2}}} - \frac{\langle r^3\xi^4\rangle}{\langle r^2\xi^3\rangle} = 0 \qquad (7)$$

while a triple critical point is subjected to an additional condition

$$15\,\frac{\langle r^2\xi^2\rangle^2}{\langle r\xi\rangle^2} + 10\frac{\langle r^3\xi^3\rangle}{\langle r\xi\rangle} + 40\frac{\langle r^2\xi^2\rangle\langle r^2\xi^3\rangle^{\frac{1}{2}}}{\langle r\xi\rangle^{3/2}}r_0^{\frac{1}{2}} + 18r_0\frac{\langle r^2\xi^3\rangle}{\langle r\xi\rangle} - 3\frac{\langle r^4\xi^5\rangle}{\langle r^2\xi^3\rangle} = 0 \qquad (8)$$

For a quasibinary system (ξ=1), eqs. (7) and (8) again reduce to the known criteria.[2,4]

It is interesting to explore the range of existence of various critical points. Equation (6) indicates that, for a given set of chain lengths r_1, r_2, one can choose <u>both</u> the polymer mixture composition $\underset{\sim}{w}$ and the parameter ξ^2 to compute the critical concentration ϕ. The triangular composition space is thus filled with an infinite set of iso-ξ^2 lines, each being a locus of critical points with a given ξ^2 (some of them are displayed in fig. 1). Double critical points form a line, crossing the countour lines of ξ^2 and marking the boundary between (meta)stable and unstable critical points (not shown in fig.1).

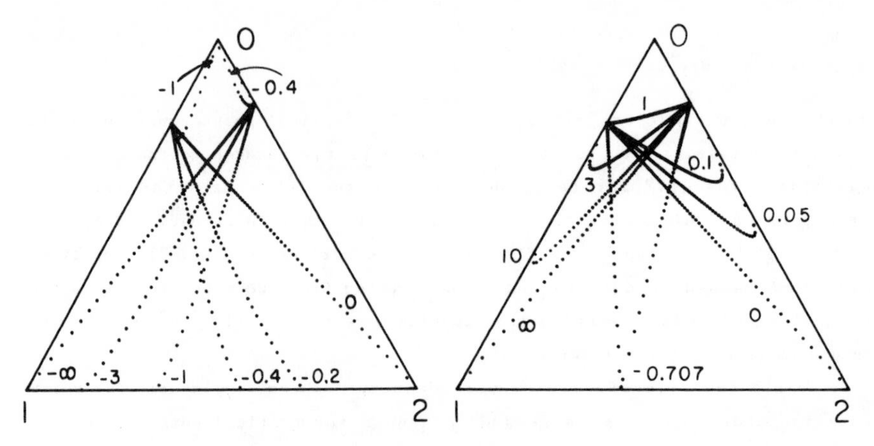

Fig. 1. Example of critical ξ^2 contour lines (with ξ^2 values indicated) for a system $r_0/r_1/r_2$ = 1/10/20. Note that lines with negative ξ^2 are discontinuous.

Finally, a triple critical point would exist in this diagram just as a single point (a "lucky accident"). Note, however, that this is a global projection of all ternary systems. For each of the points in the diagram, two out of three interaction parameters are fixed by other equilibrium relations (e.g., a chosen g_x determines g_1 and g_2). Thus some of the locations of critical points may be improbable in real systems, if they result in uncommon values of interaction parameters.

Equation (6) also yields Scott's approximate relations for the plait points of ternary systems,[9] with tie-lines "nearly parallel to the 1 - 2 axis", which lead to the often quoted conclusion that "the main contribution of the solvent is purely that of lowering the critical solution temperature by dilution". They arise as a special class of solutions when both averages of eq. (6) assume zero values. Contrary to previous belief, this class of systems has an exact analytical solution

$$w_i = \frac{\rho_j}{\rho_1 + \rho_2} \quad , \qquad \xi^2 = -\frac{\rho_1}{\rho_2} \tag{9}$$

$$g_x = \frac{(\rho_1+\rho_2)^2}{2\phi r_1 r_2} \quad , \qquad \frac{g_2-g_1}{g_x} = \frac{\rho_1-\rho_2}{\rho_1+\rho_2}$$

where $\rho_i = r_i^{1/2}$. Thus Scott's approximation for concentrations ϕ_1 and ϕ_2 is in this particular case exact, and tie-lines at the CP become truly (not only "nearly") parallel to the 1 - 2 axis irrespective of r_1 and r_2. Novel in eq. (9) is the relation for g_2-g_1 revealing that, in general, this difference does not have to be small in magnitude as assumed by Scott; on the contrary, it has to be proportional to g_x. On the other hand, eqs. (9) are independent of the average $(g_1+g_2)/2$ which may be viewed as support of Scott's claim of "the exact nature of the solvent being of only secondary importance".

ACKNOWLEDGEMENT

It is a pleasure to thank Ron Koningsveld for the original impulse to this investigation as well as for his continuing interest expressed through many discussions.

REFERENCES

1. Flory, P. J. 'Principles of Polymer Chemistry', Chapter XIII, Cornell University Press: Ithaca, NY 1953.
2. Šolc, K. Macromolecules 3, 665 (1970).
3. Šolc, K. Macromolecules 8, 819 (1975).
4. Šolc, K., Kleintjens, L.A. and Koningsveld, R. Macromolecules 17, 573 (1984).
5. Stockmayer, W. H. J. Chem. Phys. 17, 588 (1949).
6. Šolc, K. Macromolecules 16, 236 (1983).
7. 'The Scientific Papers of J. Williard Gibbs,' Vol. I, p. 132, Dover Publications: New York, NY 1961.
8. Koningsveld, R. and Staverman, A. J. J. Polymer Sci., Part A2 6, 325 (1968).
9. Scott, R. L. J. Chem. Phys. 17, 279 (1949).

POLYMER - POLYMER INTERACTIONS AND PHASE DIAGRAMS OF COMPATIBLE POLYBLENDS BY GAS - CHROMATOGRAPHY

S. KLOTZ, H. GRÄTER and H.-J. CANTOW

Institut für Makromolekulare Chemie der Albert-Ludwigs Universität Freiburg, Hermann-Staudinger-Haus, D-7800 Freiburg i. Br.

SYNOPSIS

The paper presents the temperature and composition dependence of interaction in polymer blends in terms of Flory - Huggins[1,2] theory as well as according Flory's equation of state theory[3]. From these data the phase diagram and the interaction surface of the system poly(styrene) and poly(vinylmethylether) were obtained. First measurements of poly(styrene) and poly(phenyleneoxide) blends reveal lower critical solution behaviour at blends compositions with high ratio of the components.

1 INTRODUCTION

Patterson et al.[4] and Olabisi[5] have shown that the gas - chromatographic method of determining specific retention times of a vapor phase probe in a mixed stationary phase is capable to measure thermodynamic interaction between two polymers in the stationary phase. We have applied this method to investigate interaction in blends of PS - PVME[6] and PS - PPO.

2 THERMODYNAMICS

The retention behaviour of volatile substances on gas - chromatographic columns is usually described by means of the reduced specific retention volume V_g^o . In the case of infinite dilution of the solvent the retention volume is related to the infinite weight fraction activity coefficient and to other thermodynamic parameters. Expanding the classical Flory - Huggins theory with three component approach by Scott[7] and combining with the theory of gas - chromatography, consequently we conclude for the polymer - polymer interaction parameter, χ'_{23} ($= \chi_{23} V_1 / V_2$) ,

$$\chi'_{23} = \chi_{12}\Phi_3^{-1} + \chi_{13}\Phi_2^{-1} - \chi_{1(23)}\Phi_2^{-1}\Phi_3^{-1} \tag{1}$$

The more advanced Flory equation of state theory (EST) differs from the Flory - Huggins theory in certain aspects. Straight forward development of the EST leads to the expression

$$X_{23} = (\frac{S_2}{S_1} X_{12})\Theta_3^{-1} + (\frac{S_2}{S_1} X_{13})\Theta_2^{-1} - (\frac{S_2}{S_1} \chi^*_{1(23)}) \cdot \frac{R\,T\,\tilde{v}}{V^*_1} \cdot (\Theta_2\Theta_3)^{-1}$$

$$+ p^*_1\tilde{v}\,\frac{S_2}{S_1}\left[3\,\tilde{T}_1 \ln\left(\frac{\tilde{v}_1^{1/3} - 1}{\tilde{v}^{1/3} - 1}\right) + \frac{1}{\tilde{v}_1} - \frac{1}{\tilde{v}}\right](\Theta_2\Theta_3)^{-1} \tag{2}$$

$\chi^*_{1(23)}$ is a newly defined interaction parameter in the Flory approximation which is based on conditions of a hypothetical liquid at 0^oK. $\chi^*_{1(23)}$ is related to the retention volume V_g^o of the volatile solvent probe (1) in a column with a mixed stationary phase (23). The exchange energy parameters X_{12} and X_{13} describe the binary solvent - polymer interaction. They have to be calculated separately. The surface fraction Θ and the molecular surface (volume ratio) S were estimated according Bondi's method[8]. The equation of state parameters, v^*_1, $\tilde{v}_1$, $\tilde{T}_1$, p^*_1 and $\tilde{v}$, were obtained from the expansion coefficients and the isothermal compressibilities. Finally the polymer - polymer exchange interaction parameter, X_{23}, may be determined, which is a more consistent quantity than the Flory - Huggins χ'_{23}.

3 RESULTS AND DISCUSSION

The results for PS - PVME blends show that polymer - polymer mutual solubility decreases with increasing temperature and that the polymer interaction parameter, χ'_{23}, tends towards positive values as required for a lower critical solution behaviour (table 1). Trend to positive values of χ'_{23} indicates phase separation.

Table 1:
Temperature dependence of χ'_{23} and X_{23}

T (^oC)	135	150	165	180
χ'_{23}	-0.45	-0.33	-0.27	-0.14
X_{23}	-18.4	-15.2	-13.0	-9.40

Because lower critical solution behaviour is not involved in the classical theory these results have to be verified applying the more advanced Flory equation of state theory. Negative exchange energy parameters X_{23} (table 1) reveal strong specific interaction of PS and PVME as expected for a miscible polymer pair[9]. X_{23} increases with increasing temperature, and the change in the sign of the two parameters is associated with the phase separation temperature. To compare these results, χ'_{23} (Flory - Huggins) and X_{23} Flory EST) are plotted versus temperature for the 50%PS / 50%PVME mixture in fig. 1.

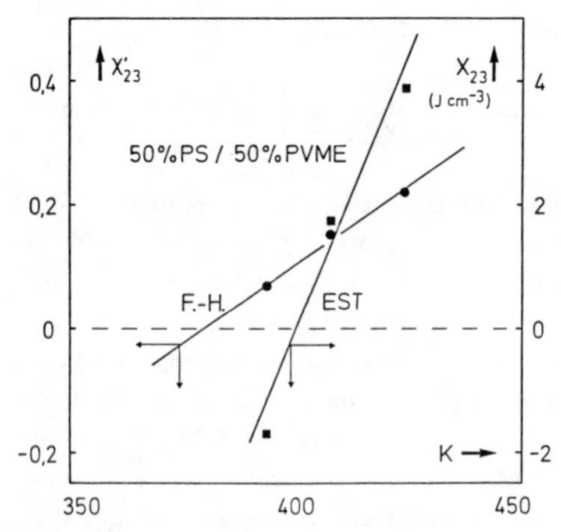

Figure 1. χ'_{23} and X_{23} (50%PS / 50%PVME) versus temperature

It is evident that the phase separation temperature derived from the EST is shifted around 15° to 20° towards higher temperature when compared with data on the base of the Flory - Huggins approach. This difference may be explained by the influence of the free volume term in eq. 2, which is not involved in the Flory - Huggins theory. Although data treatment according to EST is more consistent, a lot of physical data have to be known a priori. The much simpler Flory - Huggins theory offers an acceptable approximation for straight forward investigations for polymer - polymer interaction by gas - chromatography, taking into account that the consequences of the systematical deviation are not too serious. If the Flory - Huggins polymer - polymer interaction parameters are plotted versus temperature and blend composition one obtains the polymer - polymer interaction surface of the system PS - PVME. It includes the phase diagram (fig. 2), which is in good agreement with literature data.

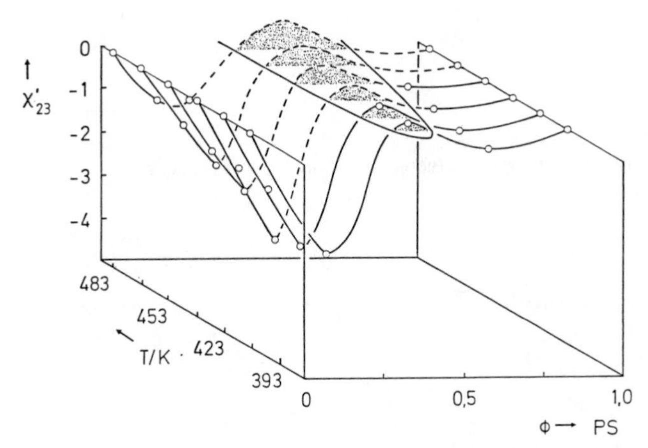

Figure 2. Polymer - polymer interaction parameter versus temperature and blend composition

ACKNOWLEDGEMENTS

The authors acknowledge support of this project by Deutsche Forschungsgemeinschaft within SFB 60. They are most grateful to Dr. B. Schmitt, BASF, for supplying PVME samples.

REFERENCES

1. Stavermann, A.J. and Van Santen, J.H.; Rec. Trav. Chim., 60, 76, 1941

2. a) Flory, P.J.; J. Chem. Phys., 10, 51, 1942
   b) Huggins, M.L.; Ann. N.Y. Acad. Sci., 43, 1, 1942

3. Flory, P.J., Orwoll, R.A. and Vrij, A.; J. Amer. Chem. Soc., 86, 3507, 1964

4. Patterson, D., Schreiber, H.P., Su, C.S. and Desphande, D.D.; Macromolecules, 7, 530, 1974

5. Olabisi, O.; Macromolecules, 8, 317, 1975

6. Klotz, S. and Cantow, H.-J.; Makrom. Chem., (submitted)

7. Scott, R.L.; J. Chem. Phys., 17, 268, 1949

8. Bondi, A.; J. Phys. Chem., 68, 441, 1964

9. McMaster, L.P.; Macromolecules, 6, 760, 1973

APPLICATION OF THE MEAN-FIELD LATTICE-GAS MODEL TO PARTIALLY-MISCIBLE POLYMER SYSTEMS

L.A. Kleintjens

DSM, Research and Patents, Geleen, NL

SYNOPSIS

The present MFLG-model assumes the number of interacting contacts of each kind of polymer segment or molecule to be related to its specific surface area. The model further introduces free volume in the polymer and thus can deal with upper- and lower critical demixing and with volume and pressure variations of the system.

Expressions for the Helmholtz free energy of mixing, equation of state, spinodal and critical condition were derived and do not contain mixing rules; It is shown that such expressions can describe the phase behaviour of molten (co-)polymers quite satisfactory, as they do for polymer solutions under pressure and polymer blends. The model can be extended to block-copolymer systems and deals with solid-liquid equilibria.

INTRODUCTION

The behaviour of mixtures under high pressure has attracted interest since the end of the past century. Van der Waals in particular has already shown that the behaviour of fluid mixtures at elevated pressure may be very perculiar if the components differ much in critical properties. One of his most remarkable predictions was the possibility of gas-gas demixing[1)], a phenomenon experimentally confirmed much later in several gas mixtures[2,3)].

Elevated pressures and temperatures are nowadays often applied in the production of polymers or during the processing of polymer blends. A study of the thermodynamic behaviour of polymer systems under such conditions is needed to describe and predict the state of a blend, or of the systems used in solution- and/or bulkpolymerization. In the latter case the monomer may act as solvent. Finally, it is nowadays well established that blockcopolymers also show a change in volume upon the order-disorder phase transition.

Profound theoretical models for systems showing volume changes have in general terms been developed by Prigogine et al[4)], by Flory et al[5)] and by Patterson et al[6)]. These treatments make use of the constituents' PVT data and try to describe the mixture with smallest number of adaptable parameters. Bonner et al[7,8)] and Bogdanović et al[9)] applied such a procedure to the system polyethylene-ethylene and calculated fluid phase relations at moderate pressure (approximately up to 20 MPa).

In the mean-field lattice-gas model (MFLG) we follow an approach outlined in principle by Guggenheim[10)] and Trappeniers et al[11,12)]. Guggenheim developed a lattice model for a two component system. With this model, the critical miscibility behaviour of low molar mass mixtures can be qualitatively described. The assumptions underlying his lattice model involve the partial specific densities of each constituent to be same in the two phases. Such a scheme is not suitable for systems varying in volume.

Trappeniers et al[12)] circumvented this problem by combining Guggenheims two-component lattice model with a lattice-gas model developed for one component e.g. by Mermin[13)] and by Mulholland et al[14)]. A similar approach was followed by Sanchez and Lacombe[15)] and by Arai and Saito[16)].

In the MFLG model, Trappeniers' model is extended so as to cover polymers in supercritical solvents and polymer blends. This relatively simple treatment was chosen here in preference to the more complex alternative theories[5,15)] because the relations derived are straightforward and easily allow effects of molar mass and chain branching in the polymer to be incorportated. The present study aims primarily at a qualitative understanding of the influence exerted by pressure on thermodynamic properties of linear and branched polymer solutions and blends, and the model used proved to be adequate for that purpose. The parameters obtained by optimization procedures are used to show the general pattern rather than fine details.

THEORY

Pure substances

In hole theories the liquid state of a pure substance is often represented by a lattice with occupied and vacant sites (holes). Hence, a pure substance can be viewed as a binary mixture in which one of the constituents (holes) contributes to the entropy of mixing but not to the internal energy. The only interactions to accounted for exist between the molecules. Temperature and pressure cause the hole concentration to change but the volume per lattice site (v_o) is kept constant for convenience, thus departing from the scheme of Gibbs and Di Marzio's theory of the glass transition. In order to deal with solutions and mixtures later on in the simplest way, i.e. without the need to design mixing rules for v_o, we postulate in this model that all constituents of a mixture shall have the same volume per lattice site. We further assume that only nearest-neighbour molecules contribute significantly to the internal energy.

Since we also want to describe the thermodynamics of supercritical systems, the model should be valid for liquids and gases alike. A gas-liquid equilibrium of a pure substance is then a fluid 1 - fluid 2 equilibrium where the two fluid phases only differ in the concentration of holes. However, by changing the hole concentration (volume) one also alters the number of contacts between the sites of the substance considered and this will cause a change in the total interaction energy.

If we have n_1 molecules of substance 1 occupying $n_1 m_1$ sites and n_o empty sites, with contact surface area of σ_1 and σ_o resp., we have for the total change in energy of mixing holes and molecules 1, ΔE,

$$\frac{\Delta E}{N_\phi RT} = g_{11} (1 - \gamma) \phi_o \phi_1 (1 - \gamma_1 \phi_1)^{-1}$$

where $g_{11} = -\frac{1}{2} w_{11} \sigma_o/RT$, $\gamma_1 = 1 - \sigma_1/\sigma_o$, w_{11} is the interaction energy involved in a 1-1 contact and

$\phi_o = \frac{n_o}{N_\phi} = \frac{n_o}{n_o + n_1 m_1} = (1 - \phi_1)$, m_1 is a measure of the 'chainlength' of the substance (number of sites per molecule).

Since the fluid mixtures considered here undergo volume changes upon changes in pressure, it is convenient to derive the equation of state from the Helmholtz free energy.

To obtain a complete expression for the Helmholtz free energy of mixing we add to ΔE the usual athermal (combinatorial) entropy of mixing terms. However, it is well known[87] that the athermal entropy of mixing generally is not sufficient to fit experiments. To this end we add an entropy-correction term of the form $\phi_o \phi_1 \alpha$, origination from the fact that the entropy calculation is based upon volume statistics instead upon segment surface statistics. We thus arrive at the following expression for the change in Helmholtz free energy (ΔF) upon mixing n_o moles of vacant lattice sites with n_1 moles of substance 1 of m_1 segments (sites)

$$\frac{\Delta F}{N_\phi RT} = \phi_o \ln \phi_o + \frac{\phi_1}{m_1} \ln \phi_1 + \phi_o \phi_1 \left\{ \alpha_1 + \frac{g_{11} (1 - \gamma_1)}{(1 - \gamma_1 \phi_1)} \right\}$$

A similar version of this model for the polymer liquid state (large m_1) has been explored by Simha and coworkers[17,18]. Our interest here being concentrated on polymer solutions and blends, we abandon such detailed and thorough considerations and restrict ourselves to the present, much simpler version of the hole theory.

Following Trapeniers one can derive equations for the gas-liquid equilibria ($\mu' = \mu'' = \left(\frac{\partial \Delta F'}{\partial n_1}\right)_{T,N}$ and critical point $((\partial^2 \Delta F/\partial \phi_1{}^2)_{V,T} = 0$ and $(\partial^3 \Delta F/\partial \phi_1{}^3)_{V,T} = 0)$. Within the framework of the model, the 'concentration' variable ϕ_1 is directly related to the density d of the system. For a substance with molecular mass M_1 this relation is given by:

$$d = \frac{\text{total mass}}{\text{total volume}} = \frac{n_1 M_1}{(n_o + n_1 m_1)\, v_o} = \frac{\phi_1}{v_o\, c_i}$$

where $c_1 = m_1/M_1$

The full equation of state for the fluid follows conventionally from:

$p = -(\partial \Delta F/\partial V)_{T,n_1}$

which results in:

$$\frac{-pv_o}{RT} = \ln \phi_o + (1 - \frac{1}{m_1})\, \phi_1 + \phi_1{}^2 \left\{ \alpha + g_{11}\, (1 - \gamma_1)^2\, (1 - \gamma_1 \phi_1)^{-2} \right\}$$

For polydisperse polymers one can replace $\phi_1/m_1 \ln \phi_1$, in ΔF, by the summation over all polymer species which will lead to the usual moments of the distribution m_n, m_w and m_z in the corresponding derivatives of ΔF.

Binary mixtures

In the lattice-gas description, a binary mixture is represented by a ternary system composed of constituent 1, constituent 2 and holes. In such a system we thus have three concentration variables (ϕ_o, ϕ_1, ϕ_2) and three interaction terms (g_{01}, g_{02} and g_{12}).

We assume v_o to be identical throughout the system and thus avoid problems arising in the design of mixing rules for v_o[5,6]. With the method given in the previous section we can calculate $g_{01} = g_{01}$ (α_1, γ_1, g_{11}, c_1) and $g_{02} = g_{02}$ (α_2,

γ_2, g_{22}, c_2) from experimental data on 'pure' 1 and 'pure' 2. To obtain the appropriate values for g_{12} one needs extra data (critical loci or vapour-liquid or liquid-liquid equilibria on binary mixtures). As in the previous section liquid-vapour separation is, within the model, considered as fluid-fluid demixing.

The MFLG-expression for the change in Helmholtz free energy upon mixing n_1 molecules 1 and n_2 molecules 2 has been derived by Trappeniers et al[12]. In the present terminology it reads:

$$\frac{\Delta F_{mix}}{RT\ N_\phi} = \phi_o \ln \phi_o + \frac{\phi_1}{m_1} \ln \phi_1 + \frac{\phi_2}{m_2} \ln \phi_2 + \phi_o \phi_1 \left\{ \alpha_1 + g_{11} (1 - \gamma_1) Q^{-1} \right\}$$

$$+ \phi_o \phi_2 \left\{ \alpha_2 + g_{22} (1 - \gamma_2) Q^{-1} \right\} + \phi_1 \phi_2 \left\{ \alpha_{12} + g_m (1 - \gamma_2) Q^{-1} \right\}$$

where:

$\gamma_1 \equiv 1 - \sigma_1/\sigma_o$

$\gamma_2 \equiv 1 - \sigma_2/\sigma_o$

$Q \equiv 1 - \gamma_1 \phi_1 - \gamma_2 \phi_2; \ \phi_1 = n_1 m_1/(n_o + n_1 m_1 + n_2 m_2)$ etc.

The expression or the pressure of the system is derived as in section 1 of this chapter and reads:

$$\frac{-pv_o}{RT} = \ln \phi_o + (1 - \frac{1}{m_1}) \phi_1 + (1 - \frac{1}{m_2}) \phi_2 + (\alpha_1 \phi_1 + \alpha_2 \phi_2)(\phi_1 + \phi_2)$$

$$- \left\{ \alpha_{12} + g_m (1 - \gamma_2) Q^{-2} \right\} \phi_1 \phi_2 +$$

$$\left\{ g_{11} (1 - \gamma_1) \phi_1 + g_{22} (1 - \gamma_2) \phi_2 \right\} (Q - \phi_o) Q^{-2}$$

For critical points we can derive two equations viz. the spinodal expression and the critical condition. The spinodal is defined by:

$$J_{sp} = A_{\phi_1 \phi_1} \cdot A_{\phi_2 \phi_2} - A_{\phi_1 \phi_2}{}^2 = 0$$

where:

$A_{\phi_1 \phi_1} \equiv \partial^2 (\Delta F/N_\phi)/\partial \phi_1{}^2$

$A_{\phi_1 \phi_2} \equiv \partial^2 (\Delta F/N_\phi)/\partial \phi_1 \partial \phi_2$ etc.

For the critical condition we can write:

$$J_{cr} = \partial J_{sp}/\partial \phi_1 \cdot A_{\phi_2\phi_2} - \partial J_{sp}/\partial \phi_2 \cdot A_{\phi_1\phi_2} = 0$$

The volume fraction ϕ_2 is related to the mole fraction x_2 by:

$$x_2 = \left\{ 1 + ({\phi_2^o}^{-1} - 1)\, m_2 m_1^{-1} \right\}^{-1}; \text{ where } \phi_2^o = \phi_2/(1 - \phi_o)$$

The compositions of coexisting gas and/or liquid phases are completely determined by the three equalities of chemical potentials ($\mu_1' = \mu_1''$; $\mu_2' = \mu_2''$ and $\mu_3' = \mu_3''$). The relevant expressions are given before[19].

APPLICATIONS OF THE MFLG-MODEL

Fluid phase behaviour of pure substances

As was shown before[20-23] the MFLG-model is well-able to describe quantitatively the fluid phase behaviour of non-polar as well as polar substances over quite a pressure and temperature range. Densities of polymers as a function of p and T proved to be well represented[24,25,35]. Another example is given in Fig. 1.

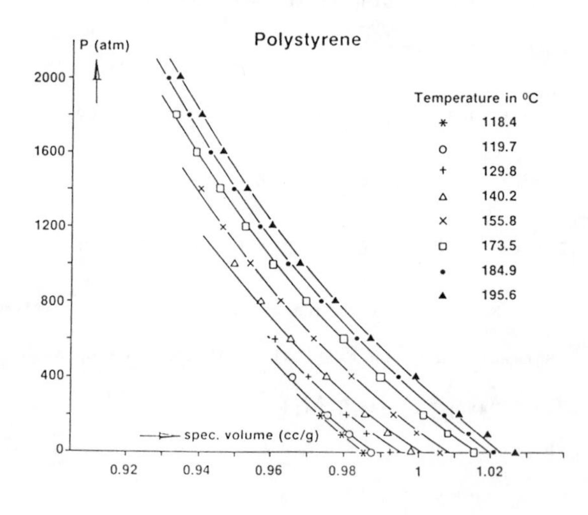

Fig. 1. MFLG-description of density data of Polystyrene by Simha et al.

Polymer solutions under pressure

a) Polyethylene in n-alkanes: For this system one may assume the segment of the polymer chain to be identical with those in the (not too short) n-alkane chain. One thus may set α_{12} and g_m equal to zero. It was shown[26)] that a quantitative prediction of the LCST phase behaviour of such systems (as a function of p,T and chain lengths of solvent and polymer) can be obtained if one adjusts $\alpha_{11}(m)$ $(= \alpha_{22}(m))$ to a series of n-alkanes.

b) Solutions of other polymers can be described if one adjusts α_{12} and $g_m(T)$ to experimental data on the quasi binary system. As was shown for the system ethylene-polyethylene[19)] one obtains satisfying descriptions of the UCST-phase behaviour up to a pressure of 2000 atm.

The effect of chain-branching of the polymer can be included by assuming the contact surface ratios σ_i to be different for end- and middle segments in the polymer chain[19)]. Further examples of descriptions of polymer solutions are given[19-26)].

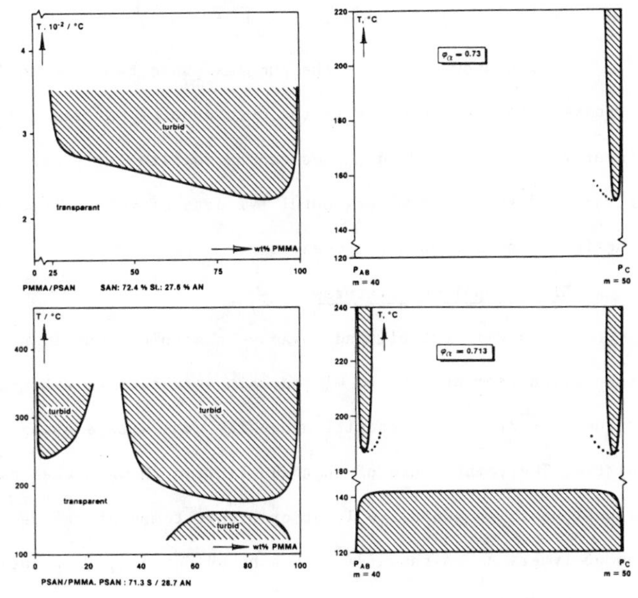

Fig. 2. left: Experimental miscibility behaviour of PMMA/SAN by Schmitt[28] right: spinodals calculated with the copolymer version of MFLG-model. ..... improvement by adjustion of contact surface ratios

Polymer blends

Phase diagrams of polymer blends often show a complex behaviour, shoulders and two extremes in the cloud point curve are quite usual phenomena[27]. Some systems show both upper and lower critical demixing[28]. Rigid-lattice descriptions of such a behaviour are only possible with very complex T and concentration dependences of the interaction parameters. As was shown before a qualitative description of such a behaviour can be obtained easily with the copolymer version of the mean-field lattice gas model[29].

As is shown in fig. 2 one can improve the description by adjustion of the interacting contact surface ratios to the experimental data. A quantitative description calls for a extensive parameter estimation program which is currently being developed[30].

For the description of copolymer mixtures the MFLG expression for the interaction in the copolymer ($\alpha\beta$) is extended to:

$$g_{\alpha\beta} = g_{o\alpha}\phi_\alpha + g_{o\beta}\phi_\beta - g_{\alpha\beta}\phi_\alpha\phi_\beta \left(\frac{\sigma_\alpha\sigma_\beta}{\sigma_\alpha\phi_\alpha + \sigma_\beta\phi_\beta} \right)$$

As was shown before[3] the complex phase behaviour of binary and ternary copolymer ($\alpha\beta$) mixtures, with the occurance of several critical point and separation into 3 liquid phases can be qualitatively described. For a quantitative description experimental pVT data of molten copolymers as a function of chemical composition are needed.

Block-copolymer mixtures

It is well established nowadays that block-copolymers may have a phase transition from an ordered microdomain structure to a homogeneous disordered structure[32]. Such a transition usually goes with a change in density of the system. The temperature of that equilibrium changes when one of the homopolymer constituents is added, to lower or higher temperatures depending on the homopolymer. As is shown[33] such a behaviour can, in principle, be described with the MFLG-model.

Glass transition and S-L equilibria

As was shown quite some years ago[34] lattice-gas models are well able to correlate the glass-transition temperature in binary systems. Quite recently we showed[35] that also solid-liquid equilibria in binary systems can be included in the MFLG-model. The essential information is a good description of both pure substances and of the fluid phase behaviour of the mixture. An experimental value for the heat of melting of the pure substance then results in a quantitative description of the liquidus of the mixture.

ACKNOWLEDGEMENT

The author thanks Dr. Koningsveld for many stimulating discussions.

REFERENCES

1. J.D. van der Waals, Zittingsverslag Kon. Akad. v. Wetensch. Amsterdam 1894, p. 133 (see: J. de Boer, Physica, 73, 1 (1974))

2. I.R. Krichevskii, Acta, Phys. Chim. USSR, 12, 480 (1940)

3. J.A. Schouten, Thesis, Amsterdam 1969; and N.J. Trappeniers and J.A. Schouten, Phys. Letters, 27 A, 340 (1968)

4. I. Prigogine, 'Molecular theory of solutions', North Holland Press, Amsterdam, 1957

5. P.J. Flory, J. Am. Chem. Soc., 86, 3507, 3515 (1964)

6. D. Patterson, J. Pol. Sci., C 16, 3379 (1968)

7. D.C. Bonner, D.P. Maloney and J.M. Prausnitz, Ind. Eng. Chem. Process Des. Develop., 13 (1), 91 (1974)

8. S.C. Harmony, D.C. Bonner and H.R. Heichelheim, AIChE J., 23 (5), 758 (1977)

9. H. Bogdanović, B. Djordević, A. Tasić and Z. Krnić, Adv. Sep. Sci., p. 185 (1978)

10. E.A. Guggenheim, 'Mixtures', Clarendonpress, Oxford 1952

11. J.A. Schouten, C.A. Ten Seldam and N.J. Trappeniers, Physics, 73, 556 (1974)

12. N.J. Trappeniers, J.A. Schouten and C.A. Ten Seldam, Chem. Phys. Letters, 5, 541 (1970)

13. N.D. Mermin, Phys. Rev. Letters, 26, 957 (1971)

14. G.W. Mulholland and J.J. Rehr, J. Chem. Phys., 60, 1297 (1974)

15. I.C. Sanchez and R.H. Lacombe, J. Phys. Chem., 80 (21), 2352 (1976); 80 (23), 2568 (1976); Macromolecules, 11 (6), 1145 (1978)

16. Y. Arai and S. Saito, J. Chem. Eng. Jap., 5 (2), 9 (1972)

17. V.S. Nanda and R. Simha, J. Chem. Phys., 41 (12), 3870 (1964); J. Phys. Chem., 68, 3158 (1964)

18. V.S. Nanda, R. Simha and T. Somcynsky, J. Pol. Sci., C 12, 277 (1966)

19. L.A. Kleintjens, Ph.D Thesis, Essix Univ., Colchester UK, 1979

20. L.A. Kleintjens and R. Koningsveld, J. Electrochem. Soc. Electrochem. Sci & Techn. 127 (11), 2352 (1980)

21. L.A. Kleintjens and R. Koningsveld, Sep. Sci. & Techn. 17 (1), 215 (1982)

22. L.A. Kleintjens and R. Koningsveld, in: Chemical Engineering at supercritical fluid conditions, Ann Arbon Sci, p. 245 (1983)

23. E. Nies, L.A. Kleintjens, R. Koningsveld, R. Simha and R.K. Jain, Fluid Phase Equil. 12, 11 (1983)

24. L.A. Kleintjens, Fluid Phase Equil. 10, 183 (1983)

25. E. Nies, R. Koningsveld, L.A. Kleintjens, 5th Conf. on mixtures of non-electrolytes and intermolecular interactions, Halle (1983), in press

26. L.A. Kleintjens and R. Koningsveld, Colloid & Pol. Sci. 258, 711 (1980)

27. R. Koningsveld, L.A. Kleintjens, J. Pol. Sci., Pol. Symp., 61, 221 (1977)

28. B.J. Schmitt, Angewandte Chem., Int. Ed. Eng., 18, 273 (1979)

29. H.M. Onclin, L.A. Kleintjens, R. Koningsveld, Makromol. Chem. Suppl. 3, 197 (1979)

30. A. Swenker, T. Hillegers, to be published

31. R. Koningsveld and L.A. Kleintjens, Macromolecules, 18, 243 (1985)

32. R.J. Roe and W.C. Zin, Macromolecules, 17, 189 (1984)

33. R. Koningsveld, this book

34. G. Kanig, Kolloid Z. & Z. Polymere 190, 1 (1963)

35. R. Koningsveld, L.A. Kleintjens and G.A.M. Diepen, Ber. Bunsenges. Phys. Chem. 88, 848 (1984)

LIQUID-LIQUID PHASE SEPARATION IN MIXTURES OF STATISTICAL COPOLYMERS

R. Van der Haegen
University of Antwerp, Dept. of Chemistry, B-2160 Wilrijk, Belgium
and DSM, Research and Patents, 6160 MD Geleen, The Netherlands

SYNOPSIS

A thermodynamic framework for statistical copolymers has recently been developed by Koningsveld and Kleintjens on the basis of the rigid lattice model, in order to investigate liquid-liquid phase separation in mixtures containing statistical copolymers. Limits of thermodynamic stability (spinodals) and critical points have been calculated and the results prove to agree in a qualitative manner with the scarce experimental data available. The Pulse Induced Critical Scattering method (PICS) proves to be an elegant method and a useful tool to collect accurate data on the present subject. We report here some experimental results concerning liquid-liquid phase separation in solutions containing one or several etheenvinylacetate random copolymers.

1 INTRODUCTION

Mean-field lattice model for statistical copolymers

Several years ago, Scott (1) theoretically treated mixtures of two random copolymers $P_{\alpha\beta 1}$ and $P_{\alpha\beta 2}$ composed of the same monomers α and β but differing in chemical composition (ϕ_α). It has been verified (2) that the miscibility range is affected by the α-β interaction, the average chain length of $P_{\alpha\beta 1}$ and $P_{\alpha\beta 2}$ and the difference in chemical composition $d_{12} = \phi_{\alpha 1} - \phi_{\alpha 2}$, all of which again depend on the system on hand.

Stockmayer et al. (3) suggested for the effective interaction parameter in this case:

$$g = g_{o\alpha}\phi_\alpha + g_{o\beta}\phi_\beta - g_{\alpha\beta}\phi_\alpha\phi_\beta$$

where $g_{o\alpha}$ and $g_{o\beta}$ are the solvent-homopolymer interaction parameters and ϕ_α (= $1-\phi_\beta$) is the α content of the copolymer. In the case of a system containing a solvent (index o) and an n-component statistical copolymer $P_{\alpha\beta i}$, application of regular solution theory (4) yields for the Gibbs free energy of mixing:

$$\frac{\Delta G_m}{N_\phi k_B T} = \frac{\phi_o}{m_o} \ln \phi_o + \sum_{i=1}^{n} \left(\frac{\phi_i}{m_i}\right) \ln \phi_i + Q^{-1}\phi_o \left\{ g_{o\alpha} \Sigma \phi_{\alpha i}\phi_i + g_{o\beta}\Sigma \phi_{\beta i}\phi_i - s_\alpha s_\beta g_{\alpha\beta} \Sigma \frac{\phi_{\alpha i}\phi_{\beta i}\phi_i}{\delta_i} \right\} + s_\alpha^2 s_\beta^2 Q^{-1} g_{\alpha\beta} \sum_{i=1}^{n-1} \sum_{j=i+1}^{n} d_{ij}^2 \phi_i\phi_j (\delta_i\delta_j)^{-1} \quad (1)$$

where ϕ_o and ϕ_i are the volume fractions of solvent molecules and copolymer molecules occupying m_o and m_i lattice sites respectively; $s_\alpha = \frac{\sigma_\alpha}{\sigma_o}$ and $s_\beta = \frac{\sigma_\beta}{\sigma_o}$ are the ratio's of interacting surface areas of repeat units and solvent molecules. Further:

$$\delta_i = s_\alpha\phi_{\alpha i} + s_\beta\phi_{\beta i}$$

$$Q = \phi_o + \Sigma\delta_i\phi_i$$

$$g_{o\alpha} = \sigma_\alpha \frac{\Delta w_{o\alpha}}{k_B T} \qquad g_{o\beta} = \sigma_b \frac{\Delta w_{o\beta}}{k_B T} \qquad g_{\alpha\beta} = \sigma_o \frac{\Delta w_{\alpha\beta}}{k_B T}$$

where Δw_{hk} is the interchange (free) energy per h-k contact, $k_B T$ has its usual meaning and N_ϕ is the total number of sites in the lattice under consideration. Investigation of miscibility behaviour is, rather than analysing equilibrium conditions, provided by the spinodal, which separates a stable from an unstable region in the phase diagram, and which is according to Gibbs (5) defined by:

$$J_s = // \partial^2 \frac{\Delta_m G}{N_\phi k_B T}/\partial\phi_i \cdot \partial\phi_j //_{P,T} = 0 \quad (2)$$

where ϕ_i and ϕ_j are the independent concentration variables (all ϕ_i in eq. (1)). Further exploration requires determination of critical points through the conditions (2) in conjunction with:

$$J_c = 0 \tag{3}$$

where J_c is the determinant obtained from J_s by replacing any row or column by:

$\frac{\partial J_s}{\partial \phi_1}$, $\frac{\partial J_s}{\partial \phi_2}$, etc... Investigation of equations (1), (2) and (3) for systems of more than two components requires of course the aid of a computer.

2 EXPERIMENTAL APPARATUS AND INVESTIGATED SYSTEMS

The PICS instrument we use to determine spinodal- and cloud point curves (CPC's) was developed by Gordon et al. in close cooperation with Koningsveld and co-workers (6), (7). The Mark II instrument is fully automated and computerised. Essentially, the technique consist of two operations, explained in the following for a simple hypothetic homodisperse polymer solution showing upper critical solution behaviour (see fig. 1a) for a plot of the typical phase diagram).

i) The system is exposed to a thermal pulse going from a temperature T_1 (above the cloud point T_{c1}) to T_2 situated in the metastable region of the phase diagram (between T_{c1} and the spinodal temperature T_s) (see Fig. 1a) and 1b)).

ii) During the thermal pulse, the capillary cell containing the polymer solution is axially enlightened by a low-power He-Ne laser beam. Due to concentration fluctuations, and consequently fluctuations in refractive index, light is scattered from the blend and is observed at selected angles (30° and 90°) (see fig. 1b)).

Intensities are measured immediately after immersion and again when equilibrium is reached (see fig. 1c)). Before nucleation and macroscopic

Figure 1a
Schematic phase diagram for a quasi-binary polymer solution showing Upper Critical Solution Temperature behaviour.

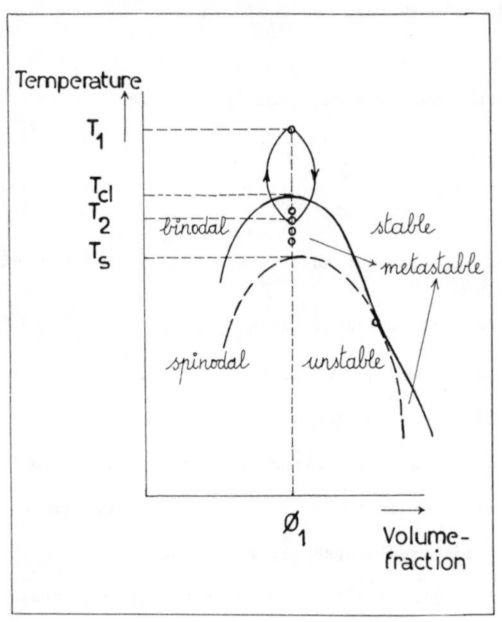

Figure 1b
Principle of Pulse Induced Critical Scattering.
The cell, mounted on a carrousel is transferred from the stirred air box (at temperature T_1 above the cloud point of the solution) into the stirred oil bath (temperature T_2) in about 1 msec with the aid of a stepping motor.

Figure 1c
Typical recorder measurement of light scattering in the metastable region of the phase diagram at scattering angles of 30° and 90° during thermal pulse in the oil bath.

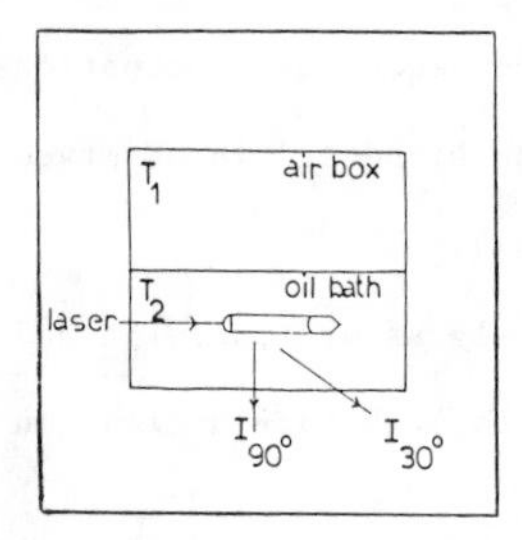

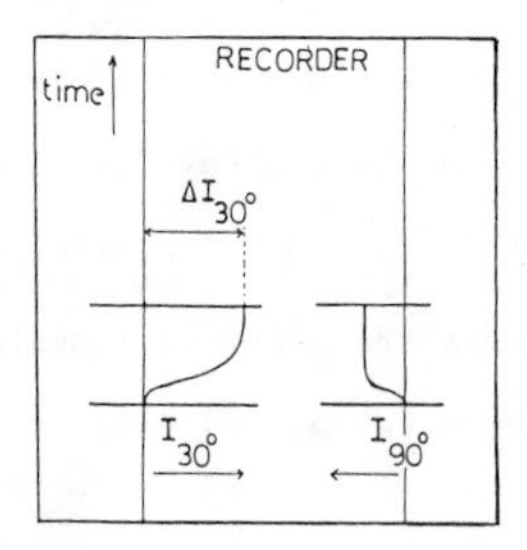

growth of emulsion particles lead to phase separation, the sample is returned, after a selected time, into the high temperature compartiment Following Debye (8) and Scholte (9), the reciprocal difference in scattering intensity at low angle is then plotted for various successive thermal pulses, to accurately locate T_s as intercept on the temperature axis. An example of this extrapolation procedure is shown in Fig. 2. Automisation makes it possible to measure four cells at the same time.

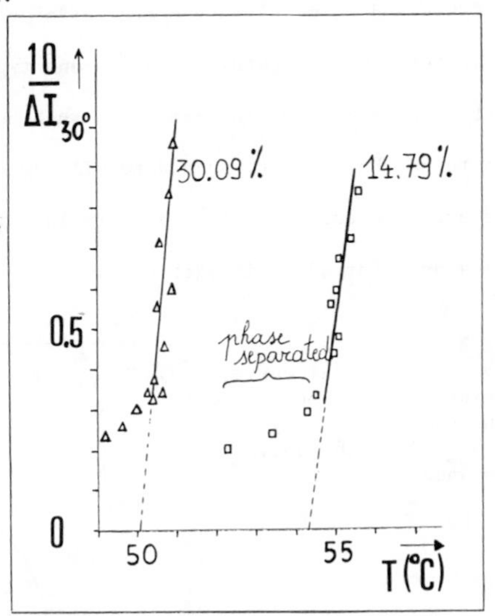

Figure 2
Typical experimental Debye-Scholte plot for the system EVAc (26)/2-heptanone at weight percentage concentration shown. Reciprocal light scattering intensity is plotted against temperature to determine the spinodal point by short extrapolation to the T-axis.

The cloud point temperature T_{c1} can simply be obtained by holding the sample in the laser beam and letting it slowly cool down, starting from a homogeneous situation at temperature T_1. The slightest increase in intensity of the scattered light indicates one has reached T_{c1}.

The sample cells are capillary tubes containing the viscous polymers under high vacuum. In order to make it possible to handle even high molar mass polymers, Gordon et al. developed the 'centrifugal homogeniser' (10). Its principle is explained before[11]. In this machine, twenty samples can simultaneously be homogenised under controlled conditions and it also permits fast measurements of phase volume ratio's (11).

The systems we experimentally deal with are high molar mass random copolymers of the monomers etheen and vinylacetate and are from now on denoted as EVAc(5), EVAc(29) and EVAc(26) where the numbers between brackets denote the weight percentage of vinylacetate.

3 RESULTS

a. Quasi-binary copolymer-solvent mixtures

CPC's and spinodal curves are, after centrifugal homogenisation, determined in the PICS apparatus for different types of EVAc copolymers mixed in a solvent. Figure 3 shows an example of such measurements where 2-heptanone is chosen to be the solvent. Such data will be used to determine interaction parameters and interacting surface area ratio's from modern statistical mechanical theories as referred to in section 1.

Figure 3
CPC (-) and spinodal curve (---) determined by PICS measurements for the quasi-binary system copolymer EVAc (26)/solvent (2-heptanone).

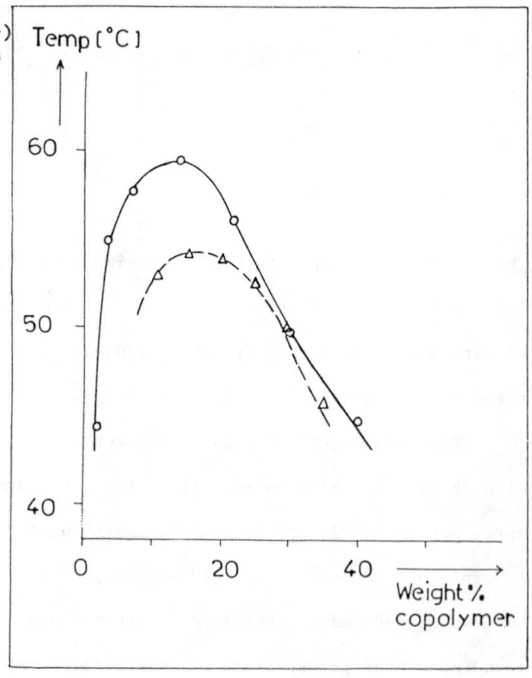

b. Copolymer $P_{\alpha\beta 1}$ - copolymer $P_{\alpha\beta 2}$ - solvent

Setting $m_o = 1$ in equation (1) we obtain the free enthalpy of mixing for solutions of double polydisperse copolymers in a small single solvent. The polymer chains may differ on two points; length and chemical composition. Molau (12) investigated phase separation in mixtures of two styrene/acrylonitrile copolymers, differing in percentage acrylonitrile, as a function of the solvent (methylethyl-ketone) concentration, with a fixed weight ratio of the copolymers (1:1). He found that the, towards demixing, tolerable difference in chemical composition $\Delta = \phi_{1\alpha} - \phi_{2\alpha}$ decreases with increasing total concentration of the copolymers.
This means that the solvent acts merely as a diluent to substantially increase the entropy of mixing per unit volume. Agreement of both theory and experiment is only possible in this case with negative (quite unrealistic) values of $g_{o\alpha}$ and $g_{o\beta}$ (4). Although spinodal points should be compared with theoretical predictions, the latter suggests the important role surface areas play in copolymer mixtures (4).

Similar to Molau's experiment, one expects to observe the same behaviour towards demixing in a binary 1:1 copolymer solution, i.e. the more solvent added to the solution, the lower its cloud point (for a fixed value of Δ). This expectation is experimentally confirmed here for EVAc mixtures as can be seen from figure 4. Figure 5 shows a CPC for EVAc mixtures ($\Delta = 0.21$) of 40 weight percentage overall copolymer concentration in 2-heptanone.

c. Copolymer $P_{\alpha\beta 1}$ - copolymer $P_{\alpha\beta 2}$ - copolymer $P_{\alpha\beta 3}$ - solvent

Molau (12) also investigated in a significant manner solutions containing three copolymers, roughly equal in molar mass, but differing in chemical composition. He observed a 25 % solution of a 1 : 1 : 1 mixture and found that phase separation in one, two or even three phases occurred depending only on the composition of the copolymers. Theoretical considerations (14) indicate that extreme sensitivity of phase separation towards composition changes is to be expected and the same does hold for appearance or disappearance of critical points. In the near future, PICS measurements will be performed for solutions containing three copolymers, different in chemical composition.

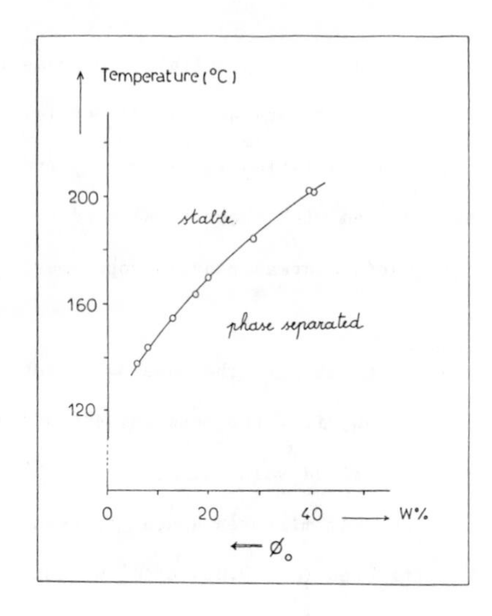

Figure 4
CPC measured by PICS method for 1 : 1 mixtures of EVAc (5) - EVAc (26) in 2-heptanone (Δ = 0.21).

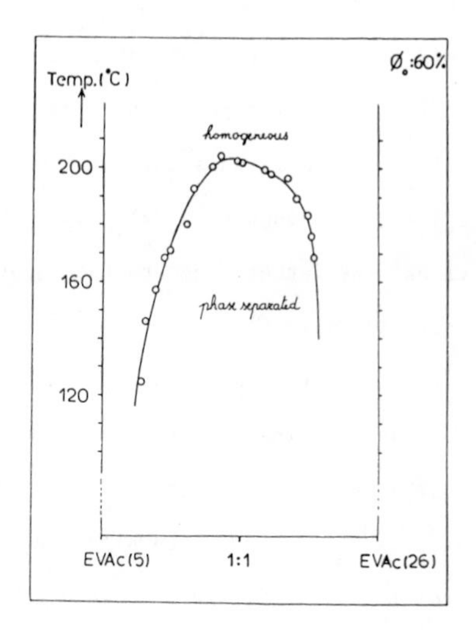

Figure 5
CPC for random copolymer mixtures, containing different ratios of EVAc (5) - EVAc (26), and having 40 weight percentage overall copolymer concentration.

d. Critical behaviour in mixtures containing random copolymers

Another striking feature, shown from theoretically calculations (4) concerns the behaviour of the critical temperature of quasi-binary copolymer solutions as a function of the chemical composition of the dissolved copolymer ϕ_α. It appears that if the two limiting homopolymer solutions (solvent/$P_{\alpha\alpha}$ and solvent/$P_{\beta\beta}$) are of the upper consolute type, miscibility in the intermediate region is enlarged or diminished depending whether the monomers α and β repel or attract each other. An interesting case is presented by a system in which the two limiting homopolymers show opposite consolute behaviour. It follows from theoretical considerations (4) that in the investigated case extensive areas of miscibility are expected to be formed if the monomers repel each other and reversily.

Figure 6 shows measurements of the CPC's for various solutions of copolymer EVAc/solvent, differing in ϕ_α (vinylacetate content). For the sake of comparison, CPC's of the limiting homopolymer solutions have been included.

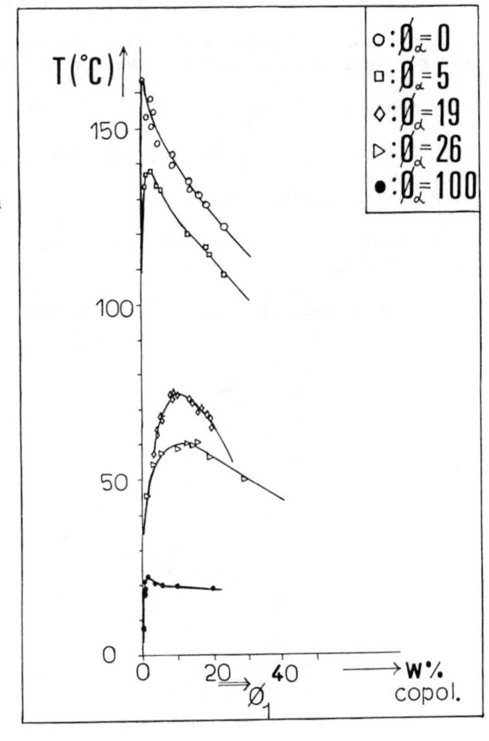

Figure 6
CPC's for various quasi-binary mixtures of EVAc random copolymers in 2-heptanone (ϕ_α is the weight percentage of vinylacetate) ϕ_α = o denotes a polyetheen solution and ϕ_α = 100 relies to a mixture of high molar mass polyvinylacetate in 2-heptanone.

4 ACKNOWLEDGEMENT

The author wants to thank Prof. Koningsveld and Dr. Kleintjens for stimulating and helpful discussions.

Technical assistance of H. Grooten of DSM-HPL was also much appreciated.

5 REFERENCES

1. R.L. Scott, J. Polym. Sci. 9 (1952) 423.

2. F. Kollinsky and G. Markert, Makromol. Chem. 121 (1969) 117.

3. W.H. Stockmayer, L.J. Moore, M. Fixman and B.N. Epstein, J. Polym. Sci 16 (1955) 517.

4. R. Koningsveld and L.A. Kleintjens, Macromolecules 18 (1985) 243.

5. J.W. Gibbs, Collected Works, Vol. I, Yale University Press Reprint, 1948.

6. K.W. Derham, J. Goldsbrough and M. Gordon, Pure and Appl. Chem. 38 (1947) 97.

7. H. Galina, M. Gordon, B.W. Ready and L.A. Kleintjens, 'PICS, Its method and its role in characterisation'. Chapter in book by Forsman, in press.

8. P. Debeye, J. Chem. Phys. 31 (1959) 680.

9. Th.G. Scholte, J. Polym. Sci. A28 (1970) 841; A2 9 (1971) 1553.

10. M. Gordon and B.W. Ready, US Patents 4131369 (December 26, 1978).

11. M. Gordon, L.A. Kleintjens, B.W. Ready and J.A. Torkington, Br. Polym. J., 10 (1978) 170.

12. G. Molau, Polymer Letters 3 (1965) 1007.

CHARACTERIZATION OF INDUSTRIAL POLYMERS AND POLYMER MIXTURES BY TURBIDIMETRIC MEASUREMENTS AT THE LOWER CRITICAL SOLUTION TEMPERATURE

Phi-Thi Tu-Anh, H. Phuong-Nguyen and G. Delmas
Chemistry Department, UQAM, C.P. 8888, Succ. A, Montreal H3C 3P8 Canada

SYNOPSIS

The paper presents a new method for characterizing polymers and polymer mixtures based on the determination a critical temperature and on the measurement of turbidity during phase separation. Due to its low cost and rapidity, it could be used either in small industrial laboratories or for mixtures difficult to characterize otherwise.

1 INTRODUCTION

The actual method for characterizing industrial polymers such as the polyolefins, and various polymers and copolymers is to separate the different molecular weights on a high temperature gel permeation chromatograph and to characterize them in line by viscosimetry or light scattering. Absolute molecular weights, molecular weight distribution and degree of branching have been obtained successfully in this way. The high cost of the equipment (investment and upkeep) restricts the use of this technique to the fairly large industrial laboratories. Therefore, the availability of a new technique of characterization of modest cost would be beneficial for small companies.

The aim of this paper is to present a method of characterizing polymers based on the measurement of a critical temperature, the Lower Critical Solution Temperature (LCST) and a critical turbidity which could be used routinely. This technique is particularly suited for industrial polymers such as the block copolymers which contain sizable amounts of polymeric impureties. On the other hand, the industrial recuperation and separation of polymers and polymer mixtures at the LCST seems very promising particularly for the more expensive polymers.

2 THEORY

2.1 The Lower Critical Solution Temperature (LCST)

Polyethylene (PE) dissolved, for instance, in a branched heptane (namely 2,4 dimethyl pentane(2,4 DMP)) forms two separate phases, when heated above the boiling point of the solvent at around 140°C. This diminished solubility between the two components does not come from a chemical difference between polymer and solvent but is due to their difference of volatility. The loss of entropy or increase of order of the volatile solvent when it is mixed with the less expanded polymer is the physical reason for the phase separation.

2.2 Special Case of PE

The trends of the LCST for different polymer-solvent couples has been well predicted by equation of state theories expressed in some simple cases by a function of the solvent and polymer densities (ref. 1a). However, the LCST of PE of different densities have been found much lower than expected by comparison of their physico-chemical parameters with those of other polymers. The explanation for these results (ref. 1a) has been found in the presence of correlations of molecular orientations among the regularly shaped segments of the PE chains. Order is created in the concentrated phase formed at the LCST not only for the volatile solvent but inbetween the polymeric chains. Such effect does not occur with polypropylene (PP) whose LCST in 2,4 DMP is 90°C higher than that of PE.

2.3 Turbidity at a critical temperature

The use of critical turbidity data for the characterization of polymer samples has been extensively made at the θ-temperature but not at the LCST. Theoretical and experimental work (refs. 2) have shown the interest of this technique, notwithstanding its inherent difficulties.

3. EXPERIMENTAL

A typical thermogram (figs. 3-6) is obtained by recording the light transmitted across the solution of the sample to be analysed (volume fraction $\emptyset_2$ such as $0.005 < \emptyset_2 < 0.04$) versus the temperature. The temperature at which a sudden change of light occurs is the critical temperature, Tc, of the sample. The minimum in the Tc ($\emptyset_2$) curve will be identified with the LCST . The thermogram can give two other parameters: one is the temperature interval, ΔT, for recovering transparency, and the other is the value, ξ, of the turbidity, normalized by the total incident light and integrated over ΔT.

The value of the LCST has the advantages to be absolute, reproducible to 0.2-0.5 degrees in most systems, not sensitive to the rate of temperature increase and hardly dependant on the thermal history of the solution. Consequently, a LCST determination is not a difficult and time-consuming operation once the polymer is dissolved. On the other hand, the value of ξ depends on the solution characteristics ($\emptyset_2$, sample polydispersity) but also on experimental conditions.

4. INFORMATIONS GIVEN BY A THERMOGRAM

4.1 Binary systems

4.1.1 Dependance on molecular weight: The correlation between $M_n^{-1/2}$ and the LCST in 2,4DMP has been plotted on fig. 1 for monodisperse PE samples of low molecular weights. A similar curve has been obtained for the LCST of ten non-fractionnated samples of HDPE using Mv (ref. 1b). The LCST of LDPE and LLDPE are higher than those HDPE, understandly so, because, in the branched samples, and the copolymers, correlations of molecular orientations are less extensive than in the linear PE.

4.1.2 Dependance of the LCST on polymer composiiton: The LCST in n-pentane, (n-C_5), of a series of random copolymers of PE and PP (EP Cop) having a molecular weight around 150,000 are plotted in fig. 2 versus the sample composition. Due to the large difference in LCST between the two homopolymers, this method can be an accurate way of obtained the polymer composition.

4.1.3 Dependance of the thermogram on polydispersity: The thermograms taken at a slow rate of temperature increase in n-C_5 of the solutions at the same $\emptyset_2$ of two polyisobutylene (PIB) samples are shown in fig. 3a, A is a monodisperse fraction (given by Dr. H. Fujita) and B a non-fractionnated sample, their Mw/Mn being respectively 1.01 and 4.5. Due to the presence of low molecular weights in B, one would expect that the complete phase separation requires a larger temperature interval for B than for A. Indeed ΔT are respectively 2.5 and 6.5° for A and B. A similar increase in ΔT is found for two PE samples (fig. 3b). The shape of the cloud-point curve and the correlation ξ ($\emptyset_2$) are also useful for comparing the polydispersity of two samples. A quantitative evaluation of the polydispersity requires a calibration.

4.1.4 Thermogram of polymer fractions: The conditions of a satisfactory fractionnation can be readily checked by taking the thermogram of a fraction obtained in a small scale experiment. A commercial sample of PE (Du Pont 2916, d=0.959, Mw/Mn=3.0) was separated into the two phases obtained at 153°C, namely 11° above the LCST. As seen in Fig. 4, the polymers recovered from the two phases have different LCST respectively 132°C (conc. phase) and 145°C (dilute phase).

4.1.5 Thermogram and thermal history of the solution: The shape of the thermogram and ξ have been found to be a sensitive parameter for measuring the good dissolution of the polymer. The presence of aggregates gives a large ξ value which diminishes for a longer time of dissolution. Also in the case of ternary systems (fig. 5) the separability of the peaks is much improved by a longer thermal history or a preliminary temperature scan.

4.2 Ternary Systems

Many ternary systems (two polymers, P_1, P_2 + solvent) show two peaks of turbidity corresponding to the succesive formation of a concentrated phase of P_1 and P_2. The thermogram shown in fig. 5a is that of a mixture of PIB and polydimethylsiloxane (PDMS) in n-C_5 whose LCST are 100° apart. The industrial block copolymers of PE and PP are produced with different overall ethylene content (between 1% and 60%). The thermogram of fig. 5b is that of a Hercules block copolymer ($\emptyset_2$=0.05 in n-C_5) with a nominal ethylene content of 56%. The first turbidity at low temperature starts at the LCST of a ethylene-rich copolymer in this solvent, which, from fractionnation and IR analysis contains 85% of ethylene. The second small peak must correspond to a small amount of a block copolymer less rich in ethylene and the third peak which occurs at the LCST of PP corresponds to this homopolymer. Fig. 5c is the thermogram in 2,4 DMP of the part of another block copolymer extracted from the concentrated phase formed below the LCST of PP. The PP has remained, for most part, in the dilute phase. These block copolymers, as revealed by thermograms (fig. 5b, c), are indeed mixtures of copolymers of different composition and homopolymers.

4.3 Pseudo-ternary systems

This label is used for the mixtures of different samples of the same polymer. Mixtures are purposefully made in companies, from different polymerisation batches to reach the desirable melt index for processing. The number of phases of two samples A and B in solution or the general features of the thermogram may give an indication of the mixing characteristics of A and B in the melt phase. Theoretical calculations of the phase behavior of polymer mixtures have shown that it is very complex and depends critically on the M.W. distribution (ref. 3). Three examples of such mixtures are given in fig. 6. The value of the LCST of (A+B) compared with those of A and of B indicates that a weight average and not a number average of A and B must be used to predict the LCST of the mixture from those of A and B. On the other hand, the thermograms of A+B show distinctive features of the mixture having two LCST since two turbidity peaks can distinctly be seen. Fig. 6 a, b, and c are the thermograms of the pure samples and of the mixtures of PE and PIB. The M.W., densities and polydispersity indexes of the samples are respectively 6a: PE, A(40,000, 0.959, 3.1) B(7.2 10^4, 0.937, 3.1). 6b: PE, A(3.11 10^4, 0.96, 1.1). B(4.7 10^3, 0.96, 1.1) 6c: PIB, A(1.4 10^6, -, 1.05), B(3.9 10^5 -, 1.05).

5. RECUPERATION OF POLYMERS ABOVE THE LCST

5.1 Binary Systems

The concentration of the phase above the LCST depends only of the temperature. Solutions of different $\emptyset_2$ will have different values of the ratio Vc/Vd, the volumes of the concentrated and dilute phases. As: $\emptyset_2 = \emptyset_{2_d} + \emptyset_{2_c}$ (Vc/Vd) the values of $\emptyset_{2_c}$ and $\emptyset_{2_d}$, the polymer volume fractions in the two phases can be calculated from the measurement of Vc/Vd for different $\emptyset_2$. Such measurements have been done for PE in 2,4 DMC_5 for $0.009<\emptyset_2<0.048$. The % of the initial polymer recuperated in the concentrated phase near the LCST at 140°C and at 160°C are given below:

$10^3\emptyset_2$		20	40	60	$\emptyset_{2_d}$	$\emptyset_{2_c}$
%Recovery in	140°C	35	67	86	610^{-3}	8810^{-3}
Concentrated phase	160°C	75	90	95	810^{-3}	8910^{-3}

5.2 Ternary Systems

The number of phases observed in 4% solutions in 2,4DMP of two polymers, each at a 2% concentration are listed below:

T°C	120	140	160	180	220
PE + PP	1	2	2	2	3
PE + EP Cop.	1	2	2	3	

When three phases coexist, the volume of the PP or EP Cop. phase is twice as large as that of the PE phase, due to the fact that the minimum of the cloud-point curve is at a lower temperature for PE. In the two systems, the volume of the two concentrated phases is about 20% of the total volume.

The examples of the thermograms above show that this simple technique gives important information concerning industrial polymers. Polymeric impureties difficult to see by DSC if they are not crystalline or by GPC if they have the same molecular weight than the main polymer can be seen easily on the thermogram.

REFERENCES

1. a) Charlet, G. and Delmas, G. Polymer 22, 1181, 1981.
   b) Charlet, G., Varennes, S. and Delmas, G. Polym. Eng. Sci., 24, 98, 1984.
2. a) Beattie, W.H. and Jung, H.C. J. of Colloid and Interface Science 27, 581, 1968.
   b) Kajiwara, K., Burchard, W., Kleintjens, L.A., Koningsveld, R. Polym. Bull., 7, 191, 1982.
3. Šolc, K., Kleintjens, L.A. and Koningsveld, R. Macromolecules 17, 573, 1984.

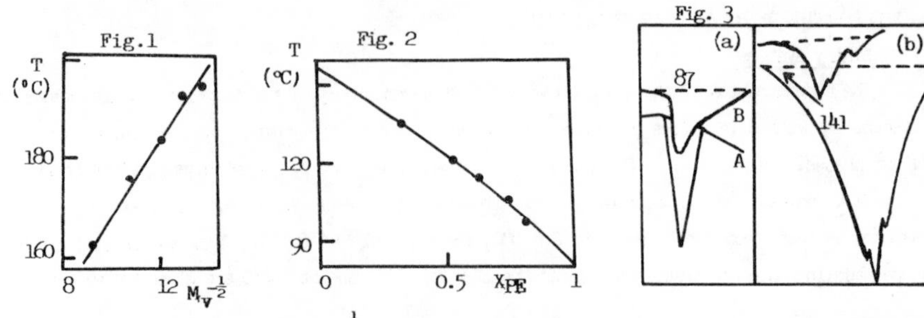

Fig. 1. Dependance of the LCST on $M_v^{-\frac{1}{2}}$ for PE in 2,4DMP. Fig. 2. Dependance of the LCST on polymer composition. EP random copolymers in n-C_5. Fig. 3. Effect of polydispersity on the width of thermograms; polydispersity indexes as follows: a. PIB (1.01 and 4.3) in n-C_5 ; b. LLDPE (2.3 and 4.8) in 2,4DMP. Scanning rate 0.1 (a) and 0.2 (b) °/min.

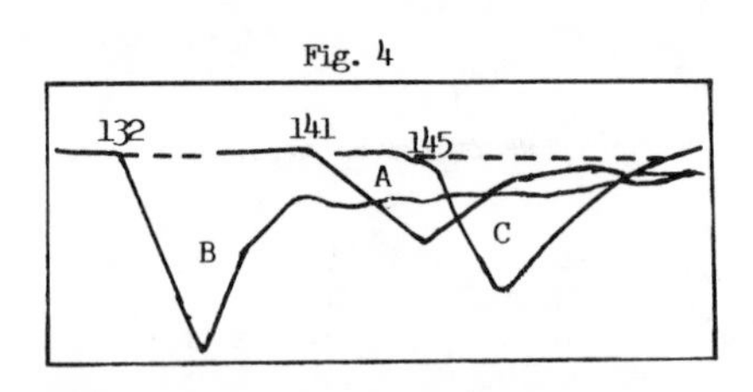

Fig. 4. Thermograms of polymer fractions: A initial sample, B and C concentrated and dilute phase

Fig. 5. Thermograms of ternary systems. a. PIB + PDMS in n-C_5; b. EP block blocks copolymer; c. concentrated phase (see text).

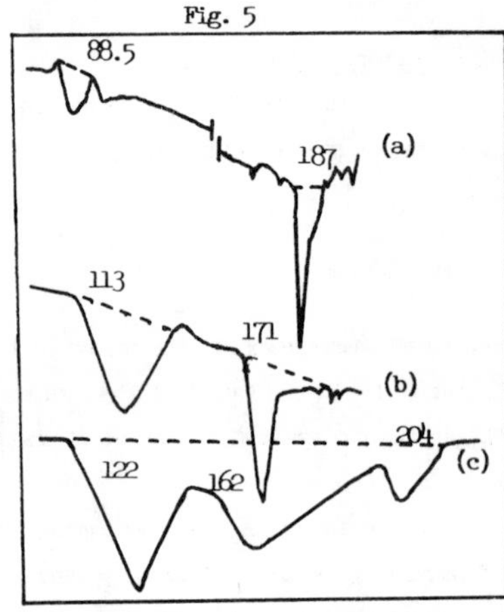

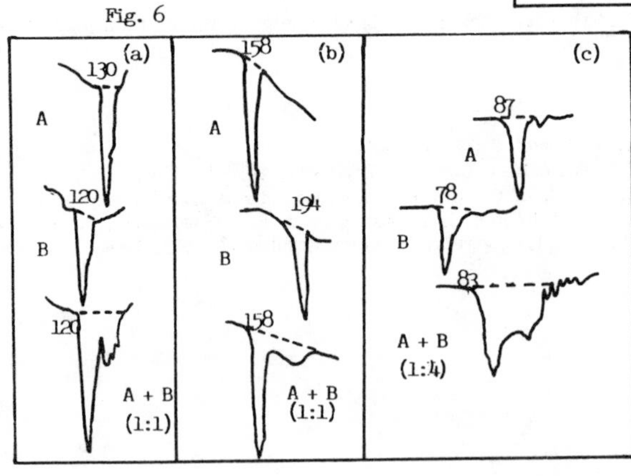

Fig. 6. Thermograms of pseudo-ternary systems. a., b. PE in 2,4 DMP; c. PIB in n-C_5.

Note: Numbers on the thermograms are the LCSST in °C

Part II

CHARACTERIZATION/SOLUTION BEHAVIOUR

CHARACTERIZATION OF COPOLYMERS
CHROMATOGRAPHIC CROSS-FRACTIONATION ANALYSIS OF STYRENE-ACRYLONITRILE COPOLYMERS

G. Glöckner[1)], J.H.M. van den Berg[2)], N.L.J. Meijerink[2)] and Th.G. Scholte[2)]

1. Dresden University of Technology, Department of Chemistry, Dresden, DDR
2. DSM Research and Patents, Geleen, The Netherlands

SYNOPSIS

The combination of size exclusion chromatography (SEC) and high performance precipitation liquid chromatography (HPPLC) of the SEC fractions is performed on a mixture of five styrene-acrylonitrile copolymers (SAN) of strongly different composition and on a commercial SAN sample. HPPLC has proved to give a good separation by chemical composition. The cross-fractionation method has proved to be able to detect the combined molar mass- and chemical composition distribution very accurately. Conditions for an accurate and reliable application of this method are described. Results for the average composition of a commercial SAN as a function of molar mass are compared with those from SEC with dual detection and SEC with pyrolysis gas chromatography.

1 INTRODUCTION

One of the multifarious topics of integration between fundamental polymer science and technology is the evaluation (and eventually prediction) of the effect of technological parameters on the molecular structure of polymers. Molecular structure often determines the level of supermolecular structure that will be obtained in the final product. Thus, molecular structure is an important link between polymerization technology and product quality and its evaluation contributes to the 'know why' of performing a technological process.

One of the fundamental problems with copolymers is the evaluation of both CCD and MMD and of the possible connections between CCD and MMD. Although it was stated decades ago that such an evaluation requires CF (ref. 1) the list of papers dealing with that kind of work on amorphous copolymers is rather short (ref. 2-15). The reason might be that with classical techniques CF requires several weeks work for one analysis. Many investigations of distributions in copolymers have been performed by SEC using multiple detection. The basic idea is that one detector should indicate the total amount of polymer while another selectively monitors the content of one kind of chemical units. However, multi-detection SEC can never reveal the two-dimensional distribution of a copolymer completely, since at each molar mass it gives only the average composition.

A few modern approaches of CF are reported in literature, e.g., the SEC investigation of copolymer fractions obtained by TLC (ref. 16) or by a similar technique employing a column packed with dry silica (ref. 17). Turbidimetric titration of SEC fractions has also been performed (ref. 18-20). The combination of copolymer SEC in pure THF with the chromatography of the fractions in another column packed with μBondagel and using a mixture of THF and n-heptane as an eluent is referred to as orthogonal chromatography (ref. 21,22). The two-dimensional separation of poly(styrene-co-methyl methacrylate) by SEC of fractions from column adsorption chromatography has been reported recently (ref. 23).

In CF starting with an SEC fractionation the second analysis has to reveal the CCD in the eluate slices taken from the SEC. We used a chromatographic technique based on the solubility differences between species of different composition. In our method (HPPLC) in the beginning of a run the eluent in the column is so rich in non-solvent that the polymer sample is precipitated at the top of the column when the solution injected gradually replaces with the non-solvent in the pores of the packing material. The elution is performed with the help of an elution gradient of increasing solvent concentration, i.e., increasing in thermodynamic quality. To some degree this technique resembles the fractionation method suggested by Baker and Williams in 1956 (ref. 24). In BWF, repeated steps of dissolution and precipitation were obtained by means of the temperature gradient. The separations reported here have been obtained with a column held at uniform temperature. Repeated precipitation and dissolution is possible by the use of packing material with pores so small that the polymer is excluded and only the solvent components can fill the pore volume. This leads to an overall velocity of the gradient smaller than that of the dissolved polymer.

In previous investigations (ref. 25-28) we mainly used RP columns, i.e., columns packed with silica having a bonded layer of hydrocarbon chains of e.g. 8 or 18 carbon atoms. These columns are widely used in reversed phase HPLC, which mostly employs an elution gradient decreasing in solvent polarity. In contrast to the RP technique, we used a solubility gradient with increasing concentration of THF in 2,2,4 trimethylpentane, i.e., running from a low level of polarity to a high one. Thus it can be concluded that in our case the column packing acts mainly as a passive support, as a rest for the precipitated polymer.

In this work we shall report on the separation of SAN copolymers both on RP packing and on a column with a suitable activity (CN-packing). 'Suitable' means that the activity must not be so strong that severe changes in conformation occur; collapse of coil conformation might result in irreversible adsorption. The underlying idea was that the surface of the packing material might improve the separation. But even with a stationary phase of suitable activity the additional interaction does not necessarily improve the separation of SAN by AN content. The improvement of separation by copolymer composition will not be obtained if either the specimen under investigation or the calibration standards contain extra groups of very high adsorption power. Another point in favour of RP columns is the fact that an inactive chromatographic support will hardly be affected by concomitants which might disturb the stationary phase by adsorption.

2 EXPERIMENTAL

2.1 Size exclusion chromatography

The SEC equipment and technique resembled those already described (ref. 26-28). Calibration took place with polystyrene samples with narrow MMDs. A volume of 500 µl of a solution of 0.235 mass % of the copolymer mixture (see table 1) in THF was injected on the SEC apparatus. For sample VI a concentration of 0.125 mass % in THF was used.

2.2 High performance precipitation liquid chromatography

2.2.1 Columns and gradient programmes: A RP column (dimensions 150 x 4.6 mm ID) packed with Polygosil 60-5 C18 (d_p = 5 µm) was used with a solvent gradient as shown in fig. 1.a. Solvent A is 2,2,4 trimethylpentane with about 10 ppm toluene, while solvent B is THF with 10 vol.% methanol. With the CN column (dimensions 150 x 4.6 mm ID) packed with Nucleosil 5 CN (d_p = 5 µm), a solvent gradient was used as shown in fig. 1.b. In this case solvent A is 2,2,4 trimethylpentane (with about 10 ppm toluene), solvent B is THF and solvent C is methanol. The time-lag (t_{lag}) values are 2.3 and 2.1 min. for the RP column and CN column respectively. The time-lag values indicated include both the contribution of the column and the apparatus.

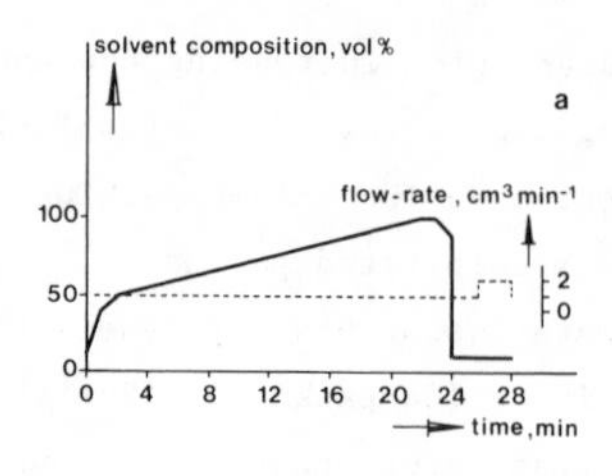

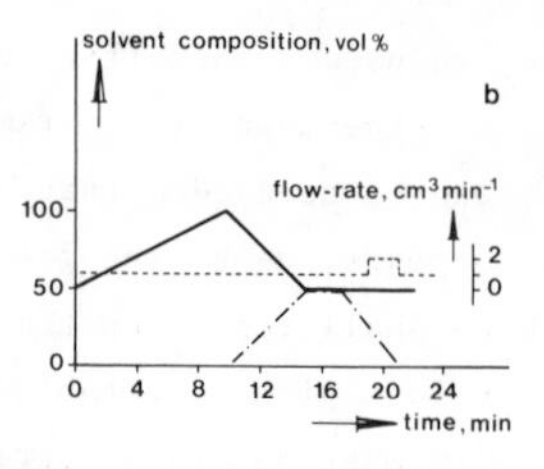

Fig. 1 Gradient elution and flow programme patterns
The dashed line shows the flow rate
(a) RP column
solvent A: 2,2,4 trimethylpentane with a trace toluene
solvent B: THF with 10 vol.% methanol (————)
(b) CN column
solvent A: 2,2,4 trimethylpentane with a trace toluene
solvent B: THF (————)
solvent C: methanol (—·—·—·)

2.2.2 <u>Apparatus:</u> The HPPLC experiments were performed on a Hewlett Packard liquid chromatograph, type HP 1090A, equipped with a ternary solvent delivery system type DR5, a thermostatted column compartment, an autosampler, an autoinjector, a diode array detector type HP 1040A, an integrator type HP 3392A, a personal computer type HP 85B with double disc drive type HP 9121D and a plotter type HP 7470A. The injection volume was 200 µl and the UV detector was set at 259 nm. All experiments were performed at 50 °C and at a volumetric flow rate of 1 $cm^3 min^{-1}$.

2.3 Pyrolysis gas chromatography

The PGC determinations were carried out on a Perkin Elmer gas chromatograph type F17, with a 85 m x 0.25 mm SS capillary column coated with Carbowax 1540 at a linear velocity of 30 cm s^{-1} of the carrier gas He. About 10 µg SAN were injected with a split ratio of 1 : 40. The temperature programme was 12 min. at 40 °C, from 40 to 90 °C in 10 min. and 20 min. at 90 °C. The Fisher high frequency Curie-point pyrolysator type 310 was used at 600 °C for 10 s. The SEC fractions (THF as solvent) were dried with nitrogen and dissolved in 40 µl of acetone and subsequently analysed. The area ratio of the styrene and acrylonitrile peaks was determined and compared with that of a reference sample of known composition.

2.4 Samples and solvents

2.4.1 SAN copolymers: The specimens I to V were prepared by bulk polymerisation to a low degree of conversion and reprecipitated twice with methanol. Sample VI is a commercial product. Further details of the samples are given in Table 1.

Table 1. Samples and mixture investigated

Sample	I	II	III	IV	V	VI
AN, mass-%	16.1	23.0	29.1	36.4	42.7	27
M_n, kg mol^{-1}	325	480	510	380	340	130+)
M_w, kg mol^{-1}		825				220+)
Conversion, %	3.6	6.7		7.3	10.3	
Presence in mixture, mass %	11.5	22.5	17.7	26.1	22.2	

Molar mass values indicated by +) were calculated from SEC using polystyrene calibration. The others were measured by osmometry or light scattering.

2.4.2 Solvents: All organic solvents used were chromatographic grade; they were obtained from several suppliers. The solvents were degassed by nitrogen purge immediately before use. The nitrogen was oxygen free (ref. 28).

3 RESULTS

3.1 High performance precipitation liquid chromatography

Experiments were performed on a mixture of five SAN samples with strongly different AN contents and on a commercial SAN (for mixture and sample VI, see 2.4.1). For the mixture, two different columns and gradient programmes (see 2.2.1) were used for the HPPLC separation.

The precision with which the gradient is established determines the reproducibility of the chromatographic results. Figure 2 shows the good reproducibility that could be obtained with the apparatus used. The two curves were measured in different overnight runs using the autosampler. In one case the gradient programme was exactly the same as the programme given in Fig. 1.a for the RP column, in the other it was slightly different in the introductory part between 0 and 2 minutes. In the crucial stretch between 2 and 22 minutes the programmes were identical.

Figure 3 enables comparison of the columns used. The time scales of the two chromatograms differ by a factor of two in order to compensate for the dif-

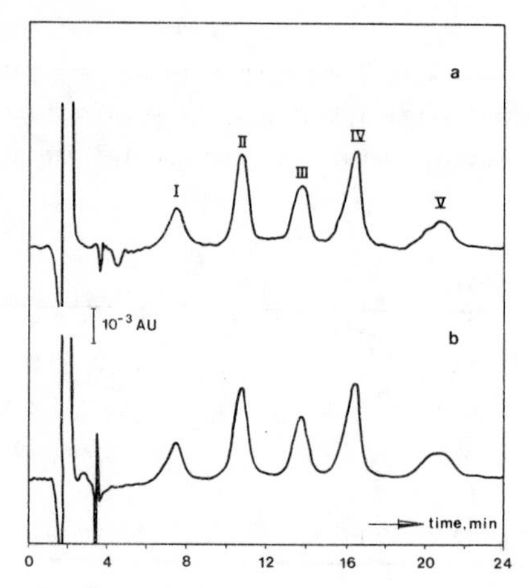

Fig. 2 Repeated HPPLC analysis of a SEC fraction of the SAN copolymer mixture on a RP column
(a) september 17, 1984, 23.25 p.m.
(b) september 18, 1984, 22.44 p.m.

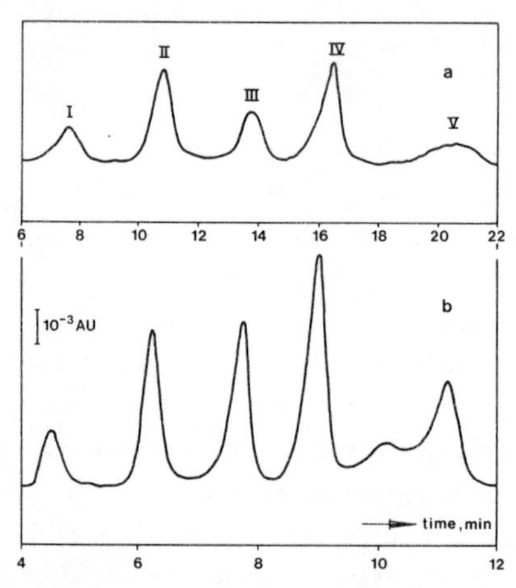

Fig. 3 HPPLC chromatograms of a SEC fraction of the SAN copolymer mixture on a RP column (a) and on a CN column (b).

ferent gradient steepnesses (RP column: 2.5 % B/min; CN column: 5 % B/min.). The resolution of the CN column seems to be slightly better than that of the RP column for sample V. The peak areas of corresponding samples are different due to evaporation of solvent in the fractions before injection into the HPPLC apparatus. The difference in the peak area ratios on the two columns for the various samples cannot be explained yet. From this experiment it can be concluded that both column types are suitable for HPPLC of SAN and that there is only a small difference in retention behaviour (ref. 28).

The principle of chromatographic CF by combination of SEC and HPPLC is shown in Fig. 4 (for the CN column). The example gives an impression of the separation power of HPPLC in a direction which is roughly perpendicular to the direction of SEC.

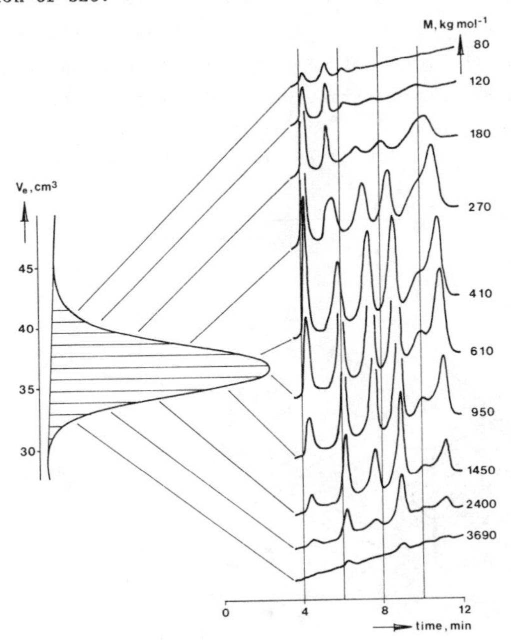

Fig. 4 SEC chromatogram of the mixture and HPPLC traces (on the CN column) of the fractions indicated.
Amount of the material used: 1.04 mg. Time required for performing the experiments: 7 hours.

The results obtained by these measurements have been used for calibrating the CN column. For each SEC fraction (with known molar mass) the elution volume and hence the percentage $\varphi(M,w)$ of THF in an eluent which elutes a SAN with w mass % AN could be determined for a number of w values. This led to a linear $\varphi(M,w)$ vs $\sqrt{M}$ relationship for each w value (ref. 27) from which $\varphi(140,w)$ could be calculated (for molar mass 140 kg mol^{-1}, the maximum of the MMD of the commercial sample VI). For w = 26.4 for example, the molar mass dependence reads

$$\varphi(M,w) = \varphi(140,w) + 2255\ (140000^{-0.5} - M^{-0.5}) \qquad (1)$$

The w vs $\varphi(M,w)$ relation for M = 140 kg mol^{-1} reads

$$w = 33.124 + 0.849\ \varphi(140,w) \qquad (r = 0.994) \qquad (2)$$

On the basis of this calibration we elaborated the CF of the commercial sample VI. The areas of the HPPLC peaks were corrected for loss of solvent by evaporation in such a way that the total area of the HPPLC peaks for each SEC fraction is proportional to the amount of copolymer in that fraction calculated from the slice area in the SEC chromatogram. The result of the analysis is shown in fig. 5.

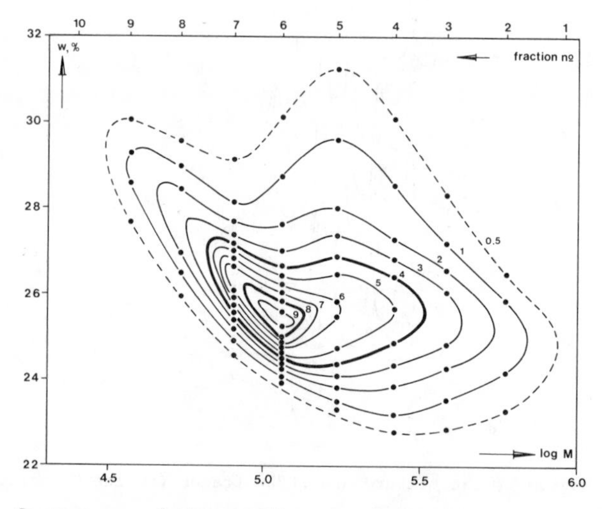

Fig. 5 Contour map of the two-dimensional distribution in sample VI as obtained by chromatographic CF with SEC and HPPLC (CN column). The contour lines are indicated by relative units.

3.2 Size exclusion chromatography with dual detection

Since the SEC apparatus used in this work is equipped with both RI and UV detection, and only the styrene component shows UV absorption, it was possible to derive the styrene content of each fraction by comparing the RI and UV signals. Data from literature (ref. 29) show that for SAN copolymers the RI increment is the sum of the products of mass fraction and dn/dc of the components. To the UV absorption only the styrene component contributes, but not in proportion to its mass fraction in the copolymer. According to Hamielec and Garcia-Rubio (ref. 29,30) the extinction coefficient of the styrene part (ε_S) is a function of the number average sequence length of the styrene blocks N_S:

$$\varepsilon_S = \varepsilon_{PS} (1 - a/N_S) \qquad (3)$$

In strongly alternating copolymers like most commercial SANs this yields for x (mol fraction styrene) $> \frac{1}{2}$

$$N_S = x/(1 - x) = \alpha M_{AN}/(1 - \alpha) M_S \qquad (4)$$

where α is the mass fraction styrene and M_S and M_{AN} are the molar masses of the S and AN monomers. From data given by Brüssau and Stein (ref. 31) and by Garcia-Rubio (ref. 29) it follows that ε_S can be represented as a linear function of $(1 - \alpha)/\alpha$:

$$\varepsilon_S = \varepsilon_{PS} (1 - 0.28 (1 - \alpha)/\alpha) \qquad (5)$$

The best way to determine α as a function of M is to start from the normalized refractive index chromatogram ($p = f(V_e)$), the normalized UV chromatogram ($q = f(V_e)$) and the styrene content of the whole copolymer α_o, and to use the equations:

$$dn/dc = 0.195 \alpha + 0.070 (1 - \alpha) \quad \text{and} \qquad (6)$$

$$\varepsilon/\varepsilon_{PS} = \alpha - 0.28 (1 - \alpha) \qquad (7)$$

Then for each elution volume i, the styrene content α_i can be calculated from the chromatogram values p_i and q_i by

$$\alpha_i = \frac{A + B\, p_i/q_i}{-C + D\, p_i/q_i} \qquad (8)$$

with

$A = 4.57\ \alpha_o - 1$

$B = 1.79\ \alpha_o + 1$

$C = 8.18\ \alpha_o - 1.79$

$D = 8.18\ \alpha_o + 4.57$

For the determination of p_i and q_i at exactly the same elution volume it is necessary to correct accurately for the effective volume between UV and RI detector cell. This was determined from the difference in elution volume between RI and UV chromatograms of a narrow polystyrene sample (Pressure Chemical la, $M = 160$ kg mol^{-1}) and was found to be 0.305 cm^3. The values of the acrylonitrile content $(1 - \alpha_i)$ are inserted in fig. 6.

3.3 Pyrolysis Gas Chromatography

The results of the PGC analysis of the SEC fractions of sample VI are also given in fig. 6. The repeatability was about 1 %.

4 DISCUSSION

The average AN content of the fractions of sample VI has been evaluated by the methods mentioned in the previous section. From HPPLC results the average values were calculated through:

$$\bar{w} = \sum (w_i \cdot h_i)/\sum h_i \tag{9}$$

All results are plotted in fig. 6. Dual detection SEC implies the greatest drift in composition, but this method requires precise knowledge of the time shift between UV and RI monitoring and is sensitive to any uncertainty concerning the base line. The latter limitation influences especially the edges of the distribution curve. In an investigation by Mori and Suzuki (ref. 32) a 'chemical heterogeneity' was found by dual detection SEC even for polystyrene homopolymer.

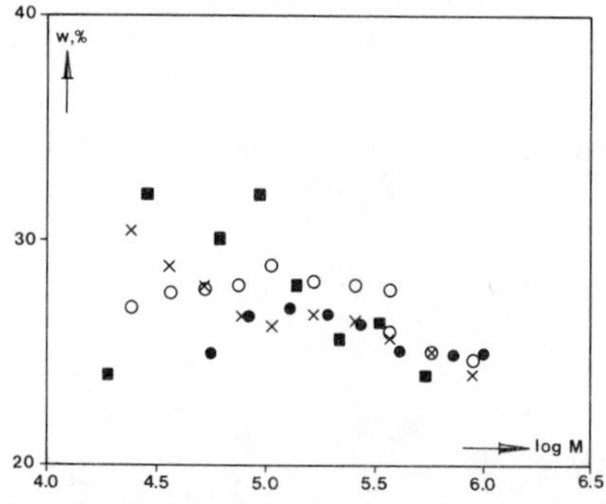

Fig. 6 Acrylonitrile content vs molar mass for the methods used

- ● SEC-HPPLC (RP column), $\bar{w}$ = 26.3 %
- × SEC-HPPLC (CN column), $\bar{w}$ = 26.4 %
- ■ SEC-DD
- O SEC-PGC , $\bar{w}$ = 28.0 %

Such a strong effect (originating from band broadening between the RI and the UV detector) was not found in our experiments.

Increase in AN content of the fractions of sample VI with decreasing M is shown also by the HPPLC investigation using the CN column. The data obtained with the RP column correspond very well in the high molecular region. Three fractions of the series obtained with the CN column with molar mass values lower than 60 $kg.mol^{-1}$ are found to differ by more than 2 mass % AN from the data expected by the tendency of CF on the RP column. Whether this discrepancy (in less than 10 % sample material) is significant cannot be decided at the moment because we have no traces from the two lowest fractions with the RP column. It is possible that the high retention of the low molecular fractions with the CN column is partly due to extra adsorption effects. This suspicion is supported by the results from PGC, which do not indicate the continuous increase of AN content with decreasing M. Further, it should be noted that statistical analysis of the accuracy of the CF methods has not been performed. In summary we conclude that HPPLC is not less reliable than PGC and is more reliable than DD SEC. In comparison with these methods it has the unique advantage that it reveals the distribution in the fractions.

SEC with DD does not yield the CCD over the fractions, but has the advantage that it can be performed with just the normal SEC apparatus. The combination of SEC and HPPLC enables chromatographic CFs of SAN to be performed without any pretreatment of the SEC eluate slices. All experimental work for a complete CF can be performed within a few hours. With a HPLC apparatus equipped with an autosampler and running continuously, several copolymers can be cross-fractionated within 24 hours. The total sample amount needed for a complete analysis is 1 mg. In comparison with classical CF, which requires several grams of material and several weeks work the chromatographic procedure considerably reduces analysis time and material consumption.

The principle is not restricted to SAN. It has already been used for analyses of poly(α-methyl styrene-_co_-acrylonitrile). With some modification it can be adapted to any soluble copolymer that can be monitored by a sensitive detector in a solvent/nonsolvent combination which does not disturb the signal from the polymer.

ACKNOWLEDGMENTS

The authors are indebted to Mrs. E. Claus and Mrs. Ch. Meissner (TUD), and Mrs. B.G.P. Limpens and Mr. H.L. Nelissen (DSM) for technical assistance.

They also thank the Hewlett Packard company for placing the liquid chromatography equipment at their disposal. The whole work was stimulated and supported by Dr. R. Koningsveld.

NOTATION

AN	acrylonitrile
BWF	Baker-Williams fractionation
CCD	chemical composition distribution
CF	cross-fractionation
CN	nitrile bonded phase
DD	dual detection
HPLC	high performance liquid chromatography
HPPLC	high performance precipitation liquid chromatography
MMD	molar mass distribution
PGC	pyrolysis gas chromatography
RI	refractive index
RP	reversed phase
S	styrene
SAN	poly(styrene-_co_-acrylonitrile)
SEC	size exclusion chromatography
THF	tetrahydrofuran
TLC	thin layer chromatography
UV	ultraviolet
a	constant
d_p	mean particle size, μm
h_i	height of HPPLC trace over the base line at position i
N_S	number average sequence length of styrene blocks
p_i	height at position i in normalized RI chromatogram (SEC)
q_i	height at position i in normalized UV chromatogram (SEC)
r	correlation coefficient
t_{lag}	time difference between command and corresponding change in eluent composition at the exit of the column, min.
V_e	elution volume, cm^3
w	mass % AN in a copolymer ($\overline{w}$ average over whole polymer)
w_i	mass % AN in a subfraction i
x	mol fraction styrene in copolymer
α	mass fraction styrene in copolymer

α_i	mass fraction styrene at position i in SEC chromatogram
α_o	mass fraction styrene of original copolymer
ε	extinction coefficient of copolymer, $cm^2\ g^{-1}$
ε_{PS}	extinction coefficient of polystyrene, $cm^2\ g^{-1}$
ε_S	extinction coefficient of styrene in copolymer, $cm^2\ g^{-1}$
$\varphi(M,w)$	volume % of eluent component B in the mixture that just eluates a SAN with molar mass M and w % AN
$\varphi(140,w)$	the same for M = 140 kg mol^{-1}
dn/dc	refractive index increment, $cm^3\ g^{-1}$

REFERENCES

1. A.V. Topčiev, A.D. Litmanovič, V.Ya. Štern
Dokl. Akad. Nauk. SSSR, 147 (1962) 1389-1391

2. A.J. Rosenthal, B.B. White
Ind. Eng. Chem., 44 (952) 2693-2696

3. V.A. Agansandyan, L.G. Kudryavtseva, A.D. Litmanovič, V.Ya. Štern
Vysokomol. Soed., Ser. A 9 (1967) 2634

4. A.D. Litmanovič, V. Ya. Štern
J. Polymer Sci., Part C 16 (1967) 1375

5. S. Teramachi, M. Nagasawa
J. Macromol. Sci., Chem., A 2 (1968) 1169-1179

6. S. Teramachi, Y. Kato
J. Macromol. Sci., Chem., A 4 (1970) 1785-1796

7. S. Teramachi, Y. Kato
Macromolecules 4 (1971) 54-56

8. S. Teramachi, H. Tomioka, M. Sotokawa
J. Macromol. Sci., A 6 (1972) 97-107

9. S. Teramchi, T. Fukao
Polymer J., 6 (1974) 532-536

10. S. Teramachi, R. Obata, K. Yamashita, N. Takemoto
J. Macromol. Sci., Chem., A 11 (1977) 535-545

11. J.J. Bourguignon, H. Bellissent, J.C. Galin
Polymer 18 (1977) 937-944

12. J. Stejskal, P. Kratochvil
Macromolecules 11 (1978) 1097-1103

13. J. Stejskal, P. Kratochvil, D. Straková
Macromolecules 14 (1981) 150-153

14. S. Teramachi, A. Hasegawa, S. Hasegawa, T. Ishibe
Polymer J., 13 (1981) 319-323

15. S. Teramachi, A. Hasegawa, S. Yoshida
Polymer J., 14 (1982) 161-164

16. T. Taga, H. Inagaki
Angew. Makromol. Chem., 33 (1973) 129-142

17. H. Inagaki, T. Tanaka
Pure & Appl. Chem., 54 (1982) 309-322

18. M. Hoffmann, H. Urban
Makromol. Chem., 178 (1977) 2683-2696

19. G. Glöckner, V. Albrecht, F. Francuskiewicz
Angew. Makromol. Chem., 127 (1984) 153-169

20. G. Glöckner, V. Albrecht, F. Francuskiewicz, D. Ilchmann
Angew. Makromol. Chem., in press

21. S.T. Balke, R.D. Patel
J. Polymer Sci., Polymer Letters Ed., 18 (1980), 453-456

22. S.T. Balke, R.D. Patel
Abstracts of papers presented at the 181th National Meeting of the American Chemical Society, Atlanta, GA, March 29-April 3, 1981, MACR 33

23. S. Mori
Paper presented at POLYMER 85, Melbourne, Australia, 11th-14th February, 1985

24. C.A. Baker, R.J.P. Williams
J. Chem. Soc. (London), (1956) 2352-2362

25. G. Glöckner
Pure & Appl. Chem. 55 (1983) 1553-1562

26. G. Glöckner, J.H.M. van den Berg, N.L.J. Meijerink, Th.G. Scholte, R. Koningsveld
Macromolecules 17 (1984) 962-967

27. G. Glöckner, J.H.M. van den Berg, N.L.J. Meijerink, Th.G. Scholte, R. Koningsveld
J. Chromatogr., 317 (1984) 615-624

28. G. Glöckner, J.H.M. van den Berg
Chromatographia, in press

29. L.H. Garcia-Rubio
Ph.D. thesis, Hamilton (Canada), 1980

30. L.H. Garcia-Rubio, A.E. Hamielec and J.F. MacGregor
ACS, Symposium Series 197 (1982) 151

31. R.J. Brüssau, D.J. Stein
Angew. Makromol. Chem., 12 (1970) 59-72

32. S. Mori, T. Suzuki
J. Liq. Chromatogr., 4 (1981) 1685-1696

CPF: A NEW METHOD FOR LARGE SCALE FRACTIONATION

H. GEERISSEN, J. ROOS, B.A. WOLF

Institut für Phyikalische Chemie der Universität
D-65 Mainz, Germany

SYNOPSIS

The paper demonstrates how the Continous Polymer Fractionation (CPF) functions and how it has been applied to polyvinylchlorid.

1 INTRODUCTION

In case a polymeric substance cannot be synthesized with narrow molecular weight distribution, fractionation is the only way to achieve such a material. There exist numerous ways to fractionate[1], but all of them on comparatively small scales only. Many attempts have been made to develop large-scale methods, preferably on the basis of a continous counter-current extraction, but the partition of the polymer according to the chain length between two immissible low molecular liquids turned out to be impracticable[2]. For this reason the CPF makes use of the miscibility gap between the polymer and a suitable mixed or single solvent.

2 PRINCIPLE OF THE CPF

The solvent component(s) and the composition of the feed (FD) and the extracting agent (EA) are chosen such that (i) the entire system formed by the starting polymer and the solvent exhibits a miscibility gap at the temperature of operation, (ii) that, in the Gibbs phase triangle, the compositions of FD and EA correspond to points outside the solubility gap, and (iii) that the straight line between FD and EA intersects the solubility gap.

If these requirements are fulfilled and the corresponding flows $\dot{V}$ selected properly, the FD (depleted of the low mol.wt. polymeric material) leaves the counter-current column as the gel phase (GL) and the EA (now containing the short polymer chains) as the sol phase (SL); the situation is shown in fig. 1.

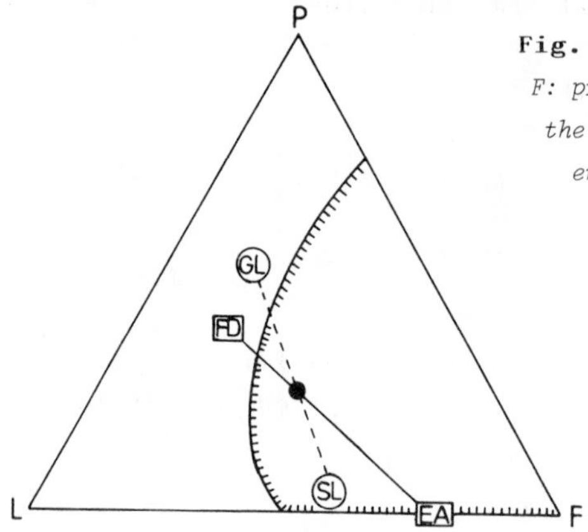

Fig. 1: *Phase diagram (P: polymer, F: precipitant, L: solvent) showing the typical composition of the phases entering and leaving the counter-current column.*
• : composition of the entire content of the column.
For abbreviations cf. text.

Apart from the compostion of EA and FD, the main independent variable of the CPF is $\dot{q} = \dot{V}^{EA}/\dot{V}^{FD}$, the ratio of the streams entering the column, followed by the amplitude and frequency in the case of pulsation.
Among the important dependent variables are the corresponding ratio $\dot{r} = \dot{V}^{SL}/\dot{V}^{GL}$, $\dot{G} = \dot{m}_P^{SL}/\dot{m}_P^{GL}$, the ratio of the polymer flows $\dot{m}_P$, and $\widetilde{M}_w$ the wt.-av. mol.wt. of the fractions, reduced to that of the starting material.

3 APPLICATION TO PVC

With the majority of polymers it turned out that the components of the mixed solvents used for the CPF should thermodynamically be as similar as possible. PVC, however, constitutes an exception; in this case we had to employ tetrahydrofurane (THF) as solvent and water as precipitant in order to prevent gelling during the CPF. The special aptitude of this mixed solvent can be explained in terms of a pronounced preferential solvation of PVC by THF, which hinderes the formation of a physical network via direct intersegmental contacts.

The PVC sample used for fractionation was Vestolit M 5867 (Chemische Werke Hüls) polymerized in bulk; its M_w-value is 67 000 and its polymolecularity index M_w/M_n 2. Prior to the application of the CPF to the production of larger amounts of material with narrow mol.wt. distribution, systematic explorative fractionations were carried out.

For given compositions of the EA (determined by a certain minimum solubilty of the polymer in it) and of the FD (limited by a maximum handable viscosity of the solution), the decisive variable is $\dot{q}$, the ratio of the flows entering the column. Fig. 2 shows how $\dot{q}$ governs $\dot{G}$, the ratio of the polymer discharged in the SL and in the GL, and to what extent the sample has been fractionated, i.e. how much the $\tilde{M}_w$-values deviate from unity.

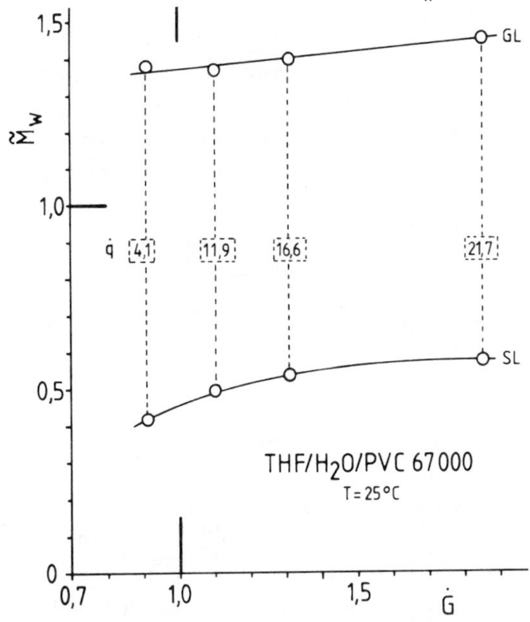

Fig. 2: *Partition ($\dot{G}$) and fractionation ($\tilde{M}_w$) of the original polymer sample by means of the CPF at different $\dot{q}$-values (cf. text).*
The composition of the FD (wt.%) is: THF 69,5 and PVC 15,0.
$\dot{V}^{FD}$ = 2,8 $cm^3.min^{-1}$.
For the apparatus used, and the particular experimental procedure (like the pulsation conditions) see ref.[3,4].

As can be seen from fig. 2, an increase in $\dot{q}$ (i.e. predominance of the stream of EA) increases $\dot{G}$ and therefore the fraction of the original polymer sample discharged in the SL. With rising $\dot{q}$ the reduced mol.wts. $\tilde{M}_w$ become larger for both fractions; the M_w/M_n values range from 1,3 to 1,5 in the present case.

Instead of the variation of $\dot{q}$ for a given composition of EA, the thermodynamic quality of EA can be varied for fixed $\dot{q}$. Fig. 3 demonstrates how the CPF works under such conditions. In this graph w_2^{EA}, the weight fraction of water in EA replaces $\dot{q}$, but else the situation is very similar: An increase of solvent power of EA grossly corresponds to an increase of $\dot{q}$.

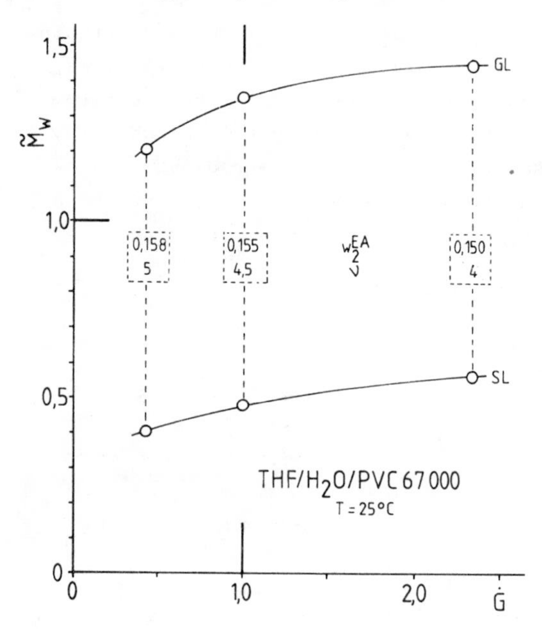

Fig. 3: *Effect of variation of the thermodynamic quality of EA on the results of the CPF. Except* w_2^{EA}, *the wt.fraction of water in EA, the variables are the same as in fig. 2.* ν *denotes the pulsation frequency, which had to be optimized in order to prevent damming up of the FD.*

Preparative CPFs were carried out, choosing conditions identical with those described in the context of fig. 1 and operating at $\dot{q}$ = 12 for ca. 20 hours. By this, two fractions (ca. 200 g each) were obtained with M_w^{SL} = 32 000 and M_w^{GL} = 90 000. A further CPF of these products yielded four PVC samples with the following M_w and M_w/M_n values: 20 000 and 1,20; 43 000 and 1,16; 76 000 and 1,16; 102 000 and 1,25.

1. *M.J.R. Cantow: Polymer Fractionation; Acad. Press N.Y. 1976*
2. *A. Dobry, Makromol.Chem.* ***120****,58(1968)*
3. *B.A. Wolf, H. Geerißen, J. Roos, P. Amareshwar; Dt.Pat.Anm. 32 42 130.3*
4. *H. Geerißen, J. Roos, B.A. Wolf; Makromol.Chem., April 1985*

FLOW BIREFRINGENCE OF ASSOCIATIONS OF POLYMERS IN SOLUTION

D. DUPUIS and C. WOLFF

Laboratoire de Physique et Méchanique Textiles, Equipe "Méchanique et Rhéologie"
E.N.S.I.T.M.
11 rue Alfred Werner
F68093 MULHOUSE-CEDEX

SYNOPSIS

In diluted solutions, macromolecules may form reversible associations which change drastically the rheological behaviour. Assuming the formation of one to one associations of 2 macromolecules according to the mass action law, the extinction angle, the birefringence and the average extension ratio of flexible macromolecules in solution have been calculated as a function of the concentration and the shear rate. It is shown that the associations display some unusual effects and that the birefringence may exceed widely that of the individual macromolecules at same concentration. Some comparisons with experimental results are presented.

1 INTRODUCTION

In poor solvents, i.e. at θ temperature or slightly above, diluted solutions of flexible macromolecules of high molecular weight M display a shear-thickening behaviour (1). Several attempts of explanation have been proposed for this phenomena. One of them is the formation of reversible associations of macromolecules in the flow (2,3). This theoretical result is in good agreement with the experimental ones (4,5). Using the same assumption, we have extended this calculation to the quantities which are obtained from flow birefringence: extinction angle χ magnitude of the birefringence Δn, and average coil expansion E; $E = h^2/h_0^2 - 1$, h_0^2 and h^2 are the mean square end to end distance of the macromolecule respectively at zero shear rate and at shear rate $\dot{\gamma}$.

2 THEORY

2.1. Fundamental relations and assumptions:

According to Zimm (6) and Peterlin (7)

$$\Delta n = \frac{2}{3} B\beta \sqrt{1 + (\lambda\beta)^2} \quad ; \quad \cot 2\chi = \lambda\beta \tag{1}$$

where $\beta = M[\eta]\, \eta_s\, \dot{\gamma} / R_g T$ is a Deborah number, $[\eta]$ is the intrinsic viscosity, η_s the viscosity of the solvent, R_g the gas constant and T the absolute temperature. λ is a numerical parameter depending on the model choosen to describe the coil ($\lambda = 0.204$ for a non draining Rouse's coil)

$$B = \frac{4\pi}{45} \frac{(n^2+2)^2}{n} (\alpha_1 - \alpha_2) \frac{N_a c}{M}$$

N_a is the Avogadro number, c the concentration, n the refractive index of the solution (nearly identical to that of the solvent), $(\alpha_1 - \alpha_2)$ the optical anisotropiy of the polymer segment.

Then $E = 2\Delta n \cos 2\chi / 3(\Delta n/\beta)_{\beta=0}$ (2)

Initially, the polymer is monodisperse of molecular weight M_1, and its total concentration is c. In flow, we assume that associations of 2 macromolecules may be formed, according to the equilibrium reaction $M_1 + M_2 \Leftrightarrow M_2$ (3) (The subscripts 1 and 2 refer respectively to the single macromolecule and to the associations).

The mass action law's constant is K (β). Experimental results (5,8) have shown that: $K = K_0\, \beta_{10}^x$ (4) where K_0 and x are constants depending on the polymer-solvent pair. Using the dimensionless parameter $\varepsilon = Kc/M_1$, the respective concentrations c_1 and c_2 are then

$$c_1 = c\left(\sqrt{1 + 8\varepsilon} - 1\right)/4\varepsilon \quad ; \quad c_2 = c - c_1 = 2\, kc_1^2/M_1 \tag{5}$$

It has also been shown experimentally (5,8) that

$$[\eta]_2/[\eta]_1 = R = R_0 + H\beta_{10} = \beta_2/2\beta_{10} \quad ; \quad \beta_{10} = M_1[\eta]_{10}\eta_s\dot{\gamma}/RT \tag{6}$$

The shear rate dependance of $[\eta]_1$ is given (9) bij:

$$[\eta]_1/[\eta]_{10} = 1 - \dot{\gamma}/(a_1\dot{\gamma} + b_1) = 1 - \beta_{10}/(A_1\beta_{10} + B_1) \tag{7}$$

with $a_1 = A_1$; $B_1 = b_1 M_1 [\eta]_{10} \eta_s / R_g T$

$[\eta]_{10}$ is the intrinsic viscosity of the single macromolecule at $\dot{\gamma} = 0$
A_1 and B_1 are constants depending on the polymer-solvent pair.

2.2. Calculation:

For a 2 components solution the Sadron's relations (10) become:

$$\tan 2\chi = (\Delta n_1 \sin 2\chi_1 + \Delta n_2 \sin 2\chi_2)/(\Delta n_1 \cos 2\chi_1 + \Delta n_2 \cos 2\chi_2)$$
$$(\Delta n)^2 = (\Delta n_1 \sin 2\chi_1 + \Delta n_2 \sin 2\chi_2) + (\Delta n_1 \cos 2\chi_1 + \Delta n_2 \cos 2\chi_2)^2 \qquad (8)$$

By introducing in these relations eq. 1,4 and 5, it becomes, after some simple calculations:

$$\tan 2\chi = N/D \quad ; \quad \Delta n = \frac{8\pi}{135} \frac{(n^2+2)^2}{n} N_a (\alpha_1 - \alpha_2) \frac{c}{M_1} \frac{(\sqrt{1+8\varepsilon} - 1)}{4\varepsilon} \sqrt{N^2+D^2}\, \beta_1 \qquad (9)$$

$$E = \frac{2}{3} \frac{(\sqrt{1+8\varepsilon} - 1)}{4\varepsilon} \sqrt{N^2+D^2}\, \beta_1 \cos 2\chi \qquad (10)$$

$$N = 1+(\lambda\beta_1)^2 \sin 2\chi_1 + (c_2/2c_1)(2R) \sqrt{1+(\lambda\beta_2)^2} \sin 2\chi_2$$

$$D = 1+(\lambda\beta_1)^2 \cos 2\chi_1 + (c^2/2c_1)(2R) \sqrt{1+(\lambda\beta_2)^2} \cos 2\chi_2$$

3 RESULTS AND DISCUSSION

With eq. 1, 9 and 10 and 10 and suitable values of the parameters, numerical values of χ, $\frac{(\sqrt{1+8\varepsilon}-1)}{4\varepsilon} \beta_1 \quad N^2+D^2$ which is proportionnal to $\Delta n/c_1$ ($\propto \Delta n/c$) and E can be easily calculated for solutions containing associations and compared with the values obtained with solutions of single macromolecules of same (initial) concentration and molecular weight. As an example fig. 1 displays the variation of $\chi \propto \Delta n/c$ and E versus β_{10} for $\lambda = 0.4$, $A_1 = 2$, $B_1 = 5$, $R_0 = 1.4$, $x = 3$, $H = 0.02$, $\varepsilon_0 = \frac{K_0 c}{M_1} = 5\ 10^{-5}$ which are typical values for flexible macromolecules (8,11). It can be seen that for $\beta_{10} > 5$, $\propto \frac{\Delta n}{c}$, E have much larger values that in absence of associations (dotted lines), whereas χ is slightly smaller. Fig. 2 represents the influence of ε_0, i.e. of the concentration c; the values of the parameters are the same as for fig. 1; $\beta_{10} = 6$. The influence of the presence of associations is very important. This calculation has been restricted at the present time to simple shear flows and to initially monodisperse polymers. Experimental results already published (12) and obtained in simple shear with a polydisperse polyethylene oxide are in quite good agreement with the present theory. On the other hand, it is not excluded that the large birefringence observed with dilute polymer solutions in laminar flow with an extensional flow component (13) may also been explained partly by the formation of associations.

REFERENCES

1. Layec-Raphalen, M.N. and Wolff C., J.non Newt.Fl.Mech., vol.1, 159-173, 1976
2. Wolff C., Silberberg A., Priel Z. and Layec-Raphalen M.N., Polymer, vol. 20, 281-287, 1979
3. Wolff C., Adv. Colloïd and Interface Sci., vol. 17, 263-274, 1982
4. Layec-Raphalen M.N., Proc. IUTAM Symp. "The influence of Polymer additives on velocity and temperature fields", Essen 1984, to appear
5. Layec-Raphalen M.N., Dr.Sci. Thesis, Brest 1985
6. Zimm B.H., J. Chem. Phys., vol. 24, 2-7, 1956
7. Peterlin A., J. Chem. Phys., vol. 39, 224-228, 1963
8. Layec-Raphalen M.N. and Wolff C., in "Rheology", Ed. G. Astarita et al., Plenum Press, 303-308, 1980
9. Wolff C., J. Chim. Phys., vol. 63, 1174-1178, 1962
10. Sadron C., J. Physique-Radium, vol. 9, 381-385, 1938
11. Wolff C., J. Physique, vol. 32, n°10-C5a, 263-268, 1971
12. Dandridge A., Meeten G., Layec-Raphalen M.N. and Wolff C., Rheol.Acta, vol. 18, 275-278, 1979
13. Cressely R., Dr.Sci. Thesis, Metz, 1982

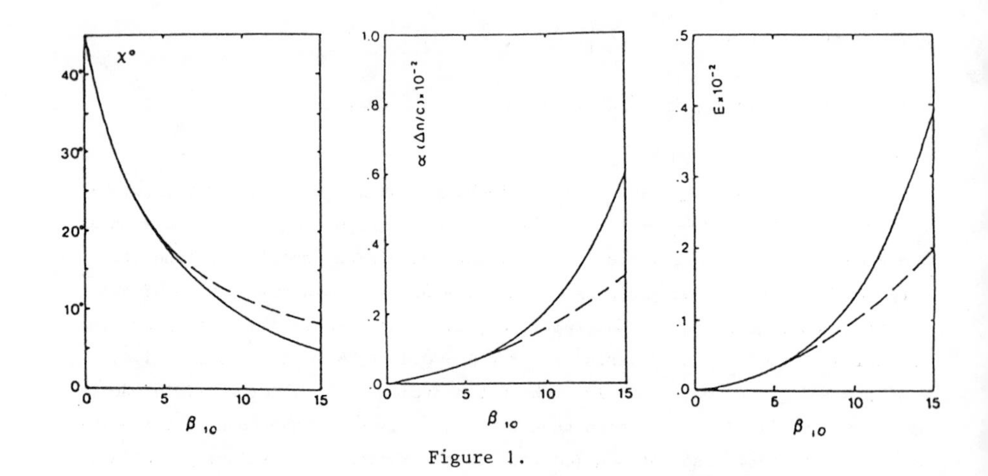

Figure 1.

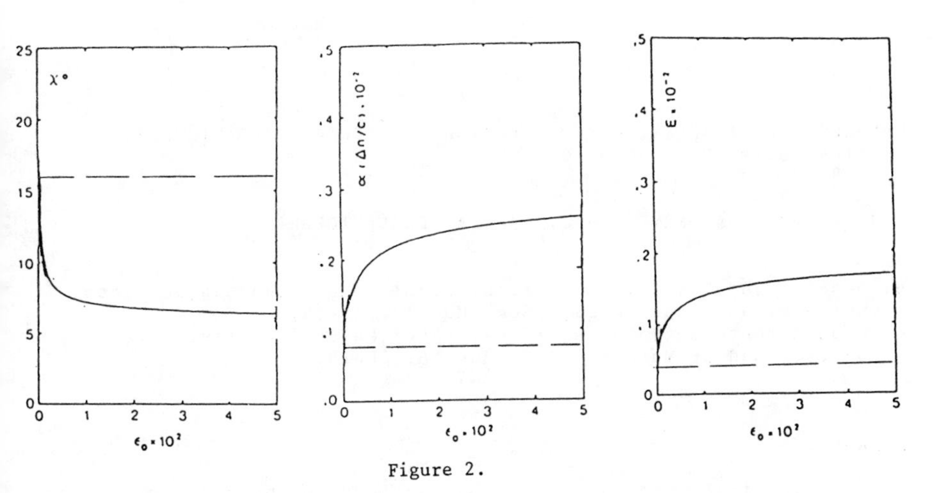

Figure 2.

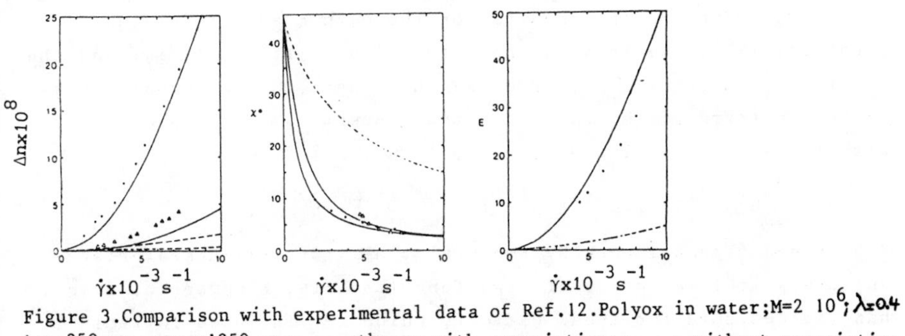

Figure 3.Comparison with experimental data of Ref.12.Polyox in water;M=2 10^6; λ=0.4
Δ:c=250 ppm ;●:c=1350 ppm;——:theory with associations;- - -: without association

THEORETICAL CALCULATION OF DIFFUSION COEFFICIENT AND VISCOSITY OF STAR POLYMERS IN SOLUTION

J.J. FREIRE[a], A. REY[a] and J. GARCIA DE LA TORRE[b]

a) Departamento de Química Física, Facultad de Ciencias Químicas, Universidad Complutense, 28040 Madrid, SPAIN.
b) Departamento de Química Física, Facultad de Ciencias, Universidad de Murcia, 30001 Murcia, SPAIN.

SYPNOSIS

Numerical calculations for dimensions and hydrodynamic properties of star-like chains are performed, considering a model with intramolecular potential and avoiding preaveraging in the hydrodynamic treatment. A good agreement is found in the comparison of theoretical results with existing experimental data.

1 INTRODUCTION AND THEORETICAL BACKGROUND

The experimental results for the mean quadratic radius of gyration, $\langle S^2 \rangle$, the translational friction coefficient, f_t, and the intrinsic viscosity, $[\eta]$, of a branched chain, b, are usually described in terms of the ratios to the properties of a linear chain with the same molecular weight, l,

$$g=\langle S^2\rangle_b/\langle S^2\rangle_l; \quad h=(f_t)_b/(f_t)_l; \quad g'=[\eta]_b/[\eta]_l$$

Recent experimental work by Roovers et al. for well charaterized uniform star-like chains of high functionality, F(number of arms), has yielded data for these ratios[1], which are included in Table 1. Then, these data constitute a good reference to test molecular theories on branched polymers.

The ratio g can be theoretically estimated from analytical formulas derived with a Gaussian model for these structures in the high molecular weight limit[2]. The estimations are included in Table 1

Table 1: Summary of theoretical and experimental[1] (EX) results. (GC)Gaussian model with the conventional treatments for hydrodynamic properties, hp, mentioned in text; (GP)Gaussian model with the KR theory for hp; (GZ)Gaussian model with the Zimm method for hp; (LJ)Chain with LJ intramolecular potential and the Zimm method for hp.

	F	GC	GP	GZ	LJ	EX
g						
	6	0.444	0.444	0.444	0.49±0.03	0.46
	12	0.236	0.236	0.236	0.33±0.02	0.35
	18	0.160	0.160	0.160	0.21±0.01	0.23
h						
	6	0.798	0.860	0.86±0.01	0.89±0.02	0.89
	12	0.626	0.727	0.75±0.01	0.75±0.01	0.75;0.81
	18	0.529	0.647	0.66±0.01	0.73±0.02	0.71;0.76
g'						
	6	0.667	0.694	0.58±0.02	0.59±0.02	0.63
	12	0.486	0.510	0.38±0.03	0.40±0.05	0.41
	18	0.400	0.421	0.24±0.03	0.28±0.03	0.31-0.35

where it can be observed that they are considerably lower than the data. The cause of these discrepancies can be attributed to repulsive forces in the central part of the chain which tend to expand the arms as functionality increases, even though the star is in its unperturbed state. Simulation calculations performed in lattices seem to confirm this effect[3].

Nonetheless, simple formulas can only be obtained for hydrodynamic properties through drastic approximations[2]. Conventionally, the Kirkwood double-sum formula[2] is used to calculate the friction coefficient while g' is approximated as $g'=g^{\gamma}$, γ being a parameter which is frequently set as $\gamma=0.5$. Discrepancies between theoretical and experimental values for these properties are also very significant, as shown in Table 1.

In order to establish how much of the difference can be attributed to approximations in the theory and how much corresponds to deficiencies in the model, we have recently performed different types of calculations with the Gaussian chain[4,5]. First, we used the Kirkwood-Riseman theory (KR)[2], introducing non-draining effects through a conveniently chosen parameter[4]. The KR theory avoids some approximations implicit in the Kirkwood and other double-sum formulas which concern to the way of solving the hydrodynamic interaction equations. However, the KR theory performs a previous orientational and conformational average of the relative locations of the theoretical internal units which describe such interactions. This approximation allows us to obtain numerical results for the properties in terms of simple averages of quadratic and reciprocal distances between the chain units. The results for different values of the number of units, N, corresponding to a given type of star were included in a linear regression analysis versus N^{-1} in order to estimate the extrapolated value in the high molecular weight limit, $N^{-1}=0$. The final values obtained this way are presented in Table 1. Though the results for ratio h are closer to the experimental values, the results for g' are higher than the data. Since a realistic modification of the model in which the expansion of arms is considered would give still higher values of the viscosity for stars, it is clear that only preaveraging can explain the overestimation of g' in the KR theory.

Consequently, our next calculations have been carried out with a procedure that avoids preaveraging[5]. Thus, a simple Monte Carlo generation of Gaussianly distributed conformations was performed and the hydrodynamic properties of these conformations were obtained as though they were instantaneously rigid chains (Zimm method[6]), according to the algorithm developed for rigid structures by García de la Torre and Bloomfield[7]. As it is shown in Table 1, these calculations yield results closer to the experimental results and, moreover, they are consistently smaller than the data. Then, the remaining differences should be mainly attributed to chain expansion effects.

2. MODEL, NUMERICAL PROCEDURES AND DISCUSSION

According to the ideas sketched above, a model that accounts for expansion of stars in their unperturbed state should be able to yield good results for dimensions and hydrodynamic properties if the latter are calculated according to the Zimm simulation method that avoids preaveraging. We have investigated a model in which intramolecular interactions are introduced by means of a Lennard-Jones (LJ) potential. Details of the model and calculation methods and a broader analysis of results are given elsewhere[8]. For this type of chain the generation of Monte Carlo conformations should be carried out by using the Metropolis criterion. Due to computational time limitations the hydrodynamic properties have been obtained only for a given fraction of the conformations, the rest being only used to drive the stochastic process. The LJ parameter σ has been set as $\sigma=0.8$ in statistical segment length units. This value predicts a reasonable expansion of linear chains at high temperatures. As to parameter ε, we have obtained results for $\varepsilon/k_BT=0.3$. For this value we have verified that the theta state law $\langle S^2\rangle \propto N^{1/2}$ is obeyed in the case of linear chains. The final extrapolated values are contained in Table 1. Though small discrepancies can be observed in certain cases, the results provide the best simultaneous estimation of the three ratios for the three high functionalities here studied. On the other hand, it should be noted that the Gaussian model with an adequate hydrodynamic treatment is able to give reasonable results for the hydrodynamic properties, mainly depending on the external chain units. However the performance of the model is poor for the radius of gyration,for which internal units have an important role.

ACKNOWLEDGEMENT: Grant 1409/82 of the CAICYT.

REFERENCES

1. a) Roovers, J.; Hadjichristidis, N. and Fetters, L.J.,Macromolecules, 16, 214 (1983) and earlier papers of this group; b) Huber, K.; Burchard, W. and Fetters, L.J., Macromolecules, 17, 541 (1984) (lower data for h); the lowest datum of g' for F=18 corresponds to unpublished work of N. Hadjichristidis and L.J. Fetters.

2. Yamakawa H. 'Modern Theory of Polymer Solutions', Harper and Row, 1971.

3. McCrackin, F.L. and Mazur, J., Macromolecules, 14, 1214 (1981).

4. Prats, R.; Pla, J. and Freire, J.J., Macromolecules, 16, 1701 (1983).

5. Freire, J.J.; Prats, R.; Pla, J. and García de la Torre, J., Macromolecules, 17, 1815 (1984).

6. Zimm, B.H., Macromolecules, 13, 592 (1980).

7. García de la Torre, J. and Bloomfield V.A., Q. Rev. Biophys., 14 81 (1981).

8. a) Freire, J.J.; Pla, J.; Rey, A. and Prats, R. ; b) Freire, J.J.; Rey, A. and García de la Torre, J. Companion papers submitted for publication.

A PHOTON CORRELATION SPECTROSCOPY INVESTIGATION OF PRECIPITATION POLYMERIZATION IN LIQUID VINYL CHLORIDE

F M Willmouth, D G Rance and K M Henman

Imperial Chemical Industries PLC, Petrochemicals and Plastics Division, Wilton, Cleveland, England.

SYNOPSIS

Photon correlation and light transmittance measurements are used to follow changes in the size and number density of poly(vinyl chloride) particles precipitating from liquid vinyl chloride during the early stages of bulk polymerization. The magnitude of the interparticle forces are estimated, and their nature and role in colloidal stability are discussed.

INTRODUCTION

The phase separation of poly(vinyl chloride) (PVC) in liquid monomer (VCM) is common both to the suspension and mass (bulk) polymerization processes used for PVC manufacture. In the early stages of polymerization, polymer appears as finely divided colloidal particles which by growth and coagulation eventually fuse to create a continuous polymer particle network. In the present study, the formation of polymer particles during the earliest stages of VCM polymerization has been investigated using light scattering techniques. The data is analysed to provide information about the forces acting between monomer-swollen PVC particles in liquid VCM.

EXPERIMENTAL PRINCIPLES

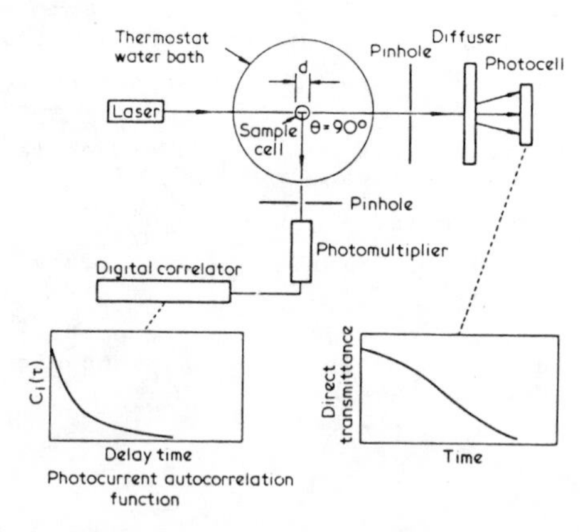

Fig. 1 Schematic Diagram of Light Scattering Experiments

Figure 1 shows the experimental layout. The polymerizing VCM is contained in a sealed cylindrical cell immersed in a thermostatically controlled water bath. The intensity of the beam directly transmitted through the cell is monitored continuously. Light scattered through an angle Θ (=90°) is detected by a photomultiplier which feeds a digital correlator. Details of the theoretical analysis are given in Ref. 1. Briefly, the photocurrent autocorrelation function is analysed to give the translational diffusion coefficient of the PVC particles, from which their size may be derived using the Stokes-Einstein relation. This enables the particle scattering cross section to be calculated using standard Mie theory, which in turn permits the particle number density to be deduced from the direct transmittance. No attempt has been made to allow for particle size polydispersity. The effects of multiple scattering were investigated in separate experiments using aqueous poly(butyl acrylate) latices[1]. These showed that serious errors in particle size arise for transmissions less than about 0.15, which limit the study to PVC particle volume fractions less than a few tenths of a percent.

RESULTS

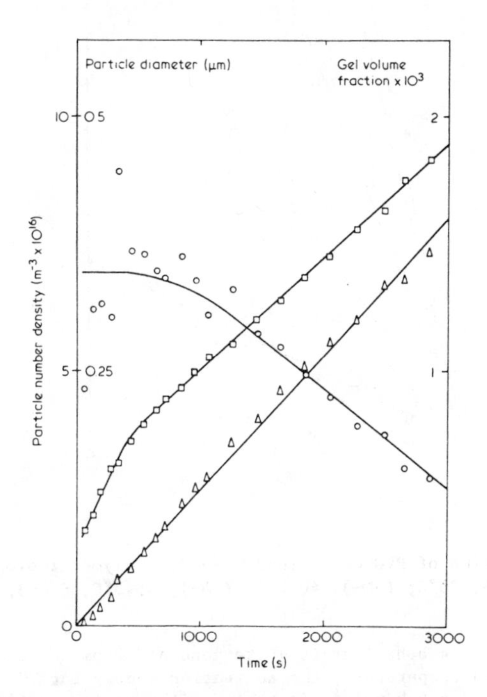

Fig. 2 Data on Monomer-Swollen PVC Particles as a Function of Polymerization Time at 35°C : (-O-), Number Density; (-□-), Diameter; (-Δ-), Gel Volume Fraction.

Figure 2 shows typical results obtained at a polymerization temperature of 35°C using 0.1% lauroyl peroxide as initiator. Plots of particle size consist of two regions, both showing a linear increase in diameter with time and separated by a clear-cut transition. At long times, there is a well-defined decrease in the particle number density, again approximately linear with time. At shorter times, the data points are scattered, but a rough estimate of the initial number density may be made. The change in the rate of increase of particle diameter coincides with the onset of the reduction in number density, to within the limits of experimental error, and is associated with the beginning of an agglomeration process. The overall gel volume fraction is a simple linear function of time.

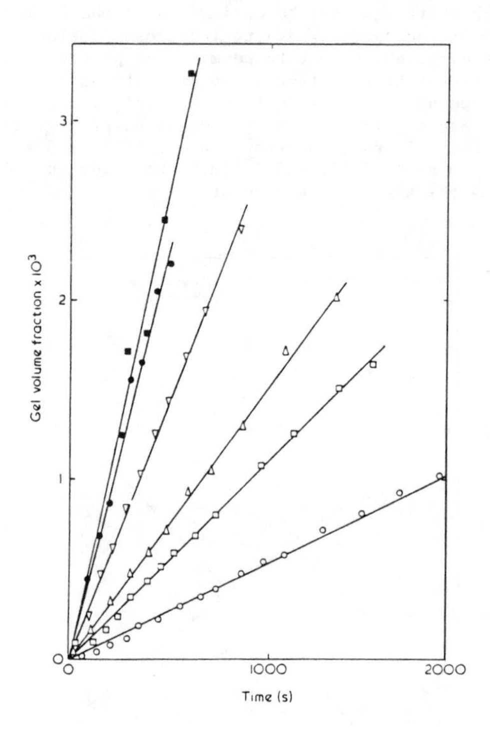

Fig. 3 Volume Fraction of PVC Gel Versus Time for Polymerization at Different Temperatures : (-O-), 35°C; (-□-), 40.2°C; (-Δ-), 45.3°C, (-∇-), 50°C; (-●-), 55.1°C; (-■-), 60°C.

Figure 3 shows that this behaviour is quite general. From the change in the rate of gel formation with temperature, the activation energy for VCM polymerization is determined to be 87 ± 6 $kJmol^{-1}$, in accord with previous estimates[2]. Table 1 summarises results from a number of polymerizations.

TABLE 1 Results for VCM Polymerization Using 0.1% Lauroyl Peroxide

Polymerization Temperature (°C)	Rate of Increase of Gel Volume Fraction ($s^{-1}x10^{-6}$)	Initial Particle Number Density ($m^{-3}x10^{17}$)	Agglomeration Onset Time (s)	Gel Volume Fraction At Agglomeration $x10^4$	Particle Diameter At Agglomeration (um)
35.0	0.53	0.7	450	2.4	0.19
40.0	0.76	3	670	5.1	0.16
40.2	1.11	1.2	380	4.2	0.19
45.3	1.53	1.0	270	4.1	0.19
45.9	1.93	3.0	300	5.8	0.16
50.0	1.89	1	Poorly defined transition		
50.0	2.90	2.9	290	8.4	0.18
50.6	2.74	2	160	4.4	0.18
55.1	4.64	2.2	190	8.8	0.22
55.1	5.28	2.8	190	10.0	0.19
60.0	5.32	-	Poorly defined transition		

DISCUSSION

A mechanism for PVC particle formation based on this and other work has been given[3]. The polymer is believed to precipitate initially as 10nm diameter 'microdomains', which agglomerate 10 - 20s after initiation to give 0.1um diameter 'domains'. In the present experiments, we have not detected the ephemeral microdomains, and are following the growth and subsequent coagulation of domains. This coagulation process is analysed in detail in Ref. 1. Briefly, the second order rate constant for coagulation is derived from plots of number density versus time, from which it is concluded that repulsive forces act between the particles, the height of the potential barrier being estimated to be in the range 10kT - 15kT (k is Boltzmann's constant, T is absolute temperature). This in turn is consistant with an electrostatic stabilization mechanism, with each 0.1um diameter particle carrying about 15 elementary charges.

REFERENCES

1. Willmouth, F. M., Rance, D. G and Henman, K. M., Polymer, Vol. 25, 1185, 1984.

2. Bengough, W. I and Norrish, R. G. W., Proc. Roy. Soc. London, Vol. A200, 301, 1950.

3. Rance, D. G. and Zichy, E. L., Pure Appl. Chem., Vol 53, 377, 1981.

Part III

BLENDS

THE ROLE OF SPECIFIC INTERACTIONS IN POLYMER MISCIBILITY

D.J. WALSH

Department of Chemical Engineering and Chemical Technology, Imperial College, London SW7 2BY
Address from 1st May 1985: Experimental Station, E.I. Du Pont de Nemours & Co., Wilmington, Delaware 19898.

SYNOPSIS

When two polymers are miscible it is normally because they are of low molecular weight, because they are similar chemically, or because they exhibit specific interactions of an electron transfer nature. Various theories exist which describe the phenomena of polymer miscibility but because they do not directly address themselves to the problem of specific interactions they do not necessarily represent the properties of such systems adequately.

Examples are given of how theory can predict the phase diagrams of mixtures and the effect of variables such as molecular weight, chemical composition and external pressure on the phase diagrams. It is shown that in some cases theory works quite well but in others it fails badly to conform to the observed behaviour. This can be understood when one considers how the specific interactions contribute to the properties of the various mixtures.

1 INTRODUCTION

In its simplest form the Flory-Huggins theory is unsatisfactory in describing the behaviour of high molecular weight polymer mixtures since it cannot predict the appearance of phase separation on heating (lower critical solution temperature, LCST, behaviour). The free energy of mixing can only become more favourable at higher temperatures. There is another problem in that the interaction parameter, χ, is experimentally found to be composition dependent.

χ has often been replaced by a parameter g such that[1]

$$\frac{\Delta G_m}{RT} = \frac{\phi_1}{m_1} \ln\phi_1 + \frac{\phi_2}{m_2} \ln\phi_2 + g\phi_1\phi_2 \qquad \ldots (1)$$

and g can be expanded in terms of its composition and temperature dependence

$$g = g_o(T) + g_1\phi_2 + g_2\phi_2^2 \qquad \ldots (2)$$

The form of this dependence could lead one to suggest the origin of the functions.

Others have developed models which do not make all the assumptions inherent in the Flory-Huggins theory. Most of these have concentrated on the effect of allowing volume changes to accompany mixing. The equation-of-state theory as developed by Flory and co-workers[2-4] is possibly that which has received most attention. It is capable of predicting LCST behaviour and explaining the compositional dependence of χ. One should however not lose sight of the fact that there are many other assumptions still remaining in the theory which may not be valid for all systems.

2 SIMULATION OF PHASE DIAGRAMS

Using a modified version of the equation-of-state theory developed by Flory and his collaborators, McMaster[5] examined the contribution of the pure- and binary-state parameters to the miscibility of hypothetical polymer-polymer mixtures. He observed that the theory is capable of predicting both lower and upper critical solution temperature behaviour individually or simultaneously. Olabisi[6] has applied McMaster's treatment to a real system of polycaprolactone and poly(vinyl chloride). Using a form of the above theory we have simulated the spinodal and binodal curves of the phase diagrams for mixtures of chlorinated polyethylene (CPE) with poly(methyl methacrylate)[7], CPE with poly(butyl acrylate)[8], and CPE with ethylene-vinyl acetate copolymers (EVA)[9]. The equation for the simulation of the spinodal curve was

$$\begin{aligned}&-1/\phi_1 + (1 - r_1/r_2) + (P_1^*V_1^*/(RT_1^*))(-D/(\tilde{v} - \tilde{v}^{2/3}))\\&+ P_1^*V_1^*D/(RT\tilde{v}^2) + PV_1^*D/(RT)\\&+V_1^*X_{12}2\theta_2^2\theta_1/(RTv\phi_1\phi_2) - V_1^*X_{12}D\theta_2^2/(RT\tilde{v}^2)\\&- V_1^*Q_{12}2\theta_2^2\theta_1/(R\phi_1\phi_2) = 0\end{aligned} \qquad \ldots (3)$$

where

$$\partial\tilde{v}/\partial\phi_2 = D \qquad \ldots (4)$$

In each case a value of the interaction parameter X_{12} was found from heat of mixing measurements on analogue materials. Q_{12} (the interaction entropy para-

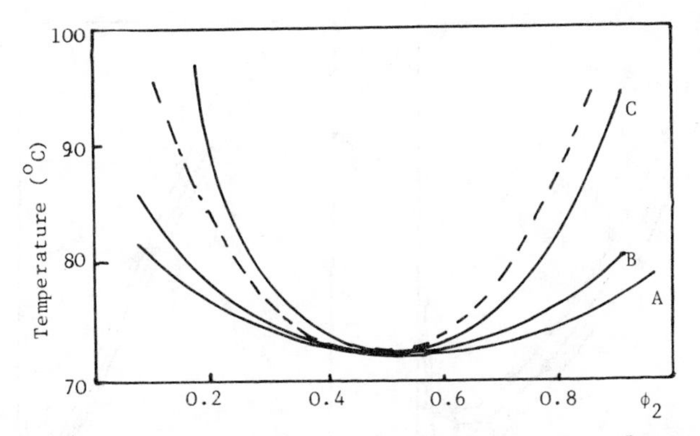

Figure 1. Simulated spinodal curves for EVA-CPE mixtures using Flory's equation of state theory at the following conditions. (A) X_{12} = 14.2 J.cm^{-3}, Q_{12} = -0.0108 J.cm^{-3} deg^{-1}. (B) X_{12} = -2.63 J.cm^{-3}, Q_{12} = -0.00678 J.cm^{-3} deg^{-1}. (C) X_{12} = -0.5 J.cm^{-3}, Q_{12} = -0.00138 J.cm^{-3} deg^{-1}. The dotted line is the experimental cloud point curve.

meter to make the simulated spinodal coincide with the cloud point curve at its minimum. A negative Q_{12}, unfavourable for mixing was always required. The full spinodal was then calculated using this value. An example of the results are shown in Fig. 1 for mixtures of CPE having 52 wt.% chlorine with EVA copolymer having 45 wt.% vinyl acetate. From heats of mixing of analogues X_{12} was found to have a value of 14.2 Jcm^{-3}. In this and the other examples the results are very similar. With the calculated X_{12} the curves are far too flat-bottomed. In order to fit the results it is necessary to use a much smaller X_{12} value with a much smaller Q_{12} value being used to balance it out.

In a further example for mixtures of a polyether sulphone with poly (ethylene oxide) the results are more satisfactory as shown in Fig. 2[10,11]. This system is unusual in the extremely large values of X_{12}, suggesting a very strong interaction between the polymers. It is suggested below that the theory may be better able to accommodate a strong specific interaction. In all cases the use of the surface area per unit volume ratio of the polymers appears to be necessary to simulate the phase diagrams but how this is to be interpreted in terms of the specific interactions is not clear.

Other workers have studied mixtures of non-polar polymers, both high molecular weight[12] and low molecular weight[13]. In such mixtures, as expected, positive (unfavourable) values of X_{12} are calculated. Another common feature

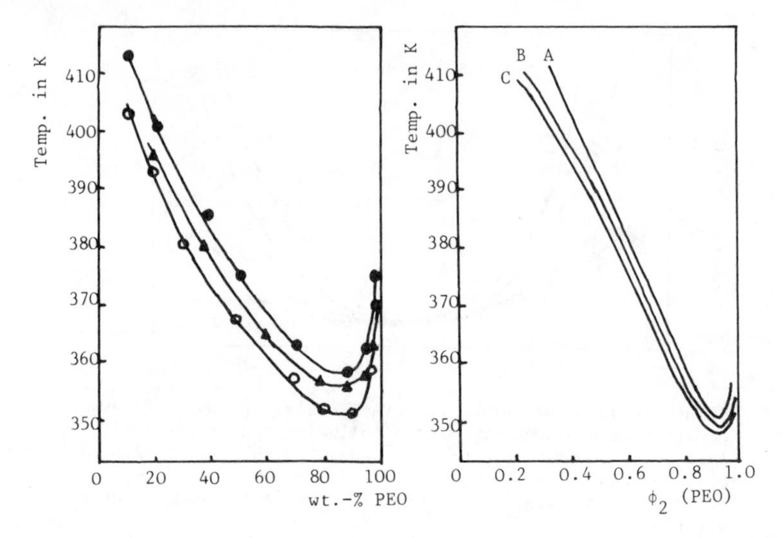

Fig. 2a. Plots of the cloud point curves for PEO-PES mixtures. The curves are for PEO of molar mass 4000g.mol^{-1} (●), 20000g.mol^{-1}(▲), and 200000g.mol^{-1}(O)$^{-1}$

Fig. 2b. Plots of the simulated spinodal curves for PEO-PES mixtures with PEO of molar mass 4000 (A), 20000 (B), and 200000 g.mol^{-1} (C). The spinodal curves were calculated using a value of X_{12} = -40 J.cm^{-3} and a value of Q_{12} = -0,048 J.cm^{-3}.K^{-1} adjusted to fit the cloud point curve at its minimum.

of such systems is the need to use positive (favourable) values of Q_{12} to fit the theory to observed cloud points. What can be the significance of this? For systems having specific interactions and negative (favourable) X_{12} values, an unfavourable Q_{12} can be interpreted as a loss of entropy associated with the interaction. How can you interpret a favourable Q_{12}? It could be interpreted as a loss of order in one of the pure components on mixing but, in the absence of evidence for such order, it may more likely be due to errors in the calculation of the combinatorial entropy due to, for example, the assumption of equal molecular sizes and surface areas as has been suggested by Koningsveld[14].

3 THE EFFECT OF PRESSURE ON PHASE DIAGRAMS

Since some advanced theories try to take account of the volume change on mixing, we may ask how well they do this. There is evidence concerning how well the theories predict the PVT properties of pure components but this does not tell one how satisfactory they may be in blends for the purpose of simulating phase diagrams. This is made difficult by uncertainties in the interactional and combinatorial contributions to the free energy.

Measurement of the effect of pressure on the phase diagram is useful as it gives another piece of information more heavily slanted towards the free volume effects. Figures 3 and 4 show examples of experimental results and theoretical calculations of the effect of pressure on a high molecular weight pair {polyether sulphone and poly(ethylene oxide)} and a low molecular weight pair {polystyrene and polybutadiene}[15]. In the first case the heat of mixing and the free volume effects are both large. The effect of pressure is to increase the range of miscibility since the heat of mixing is negative. In the second case both the heat of mixing and the free volume effects are smaller (the contribution of the latter at atmospheric pressure is almost negligible). The effect of pressure is to reduce the range of miscibility since the heat of mixing is positive. In both cases the theory satisfactorily predicts the effect of pressure on the phase diagram.

Our conclusions are that Flory's equation-of-state theory satisfactorily describes the effect of volume changes on the free energy of mixing especially

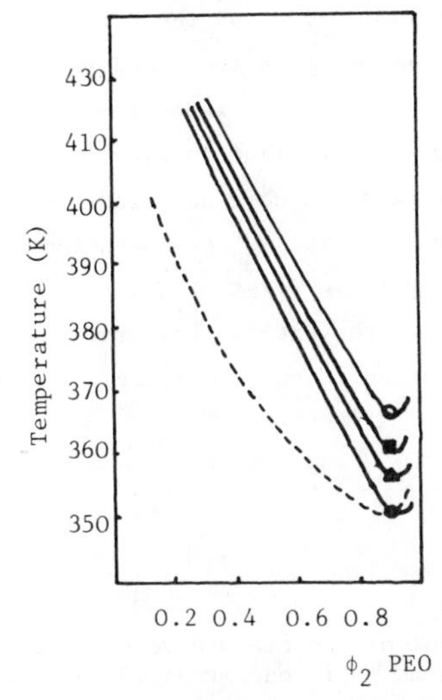

Fig. 3 Simulated spinodals (solid lines) and measured cloud points for PES/PEO mixtures at:
1 atm., (●) ; 108 atm., (▲) ;
221 atm., (■) ; 348 atm., (O).
The dotted line shows the full experimental cloud point curve at 1 atm.

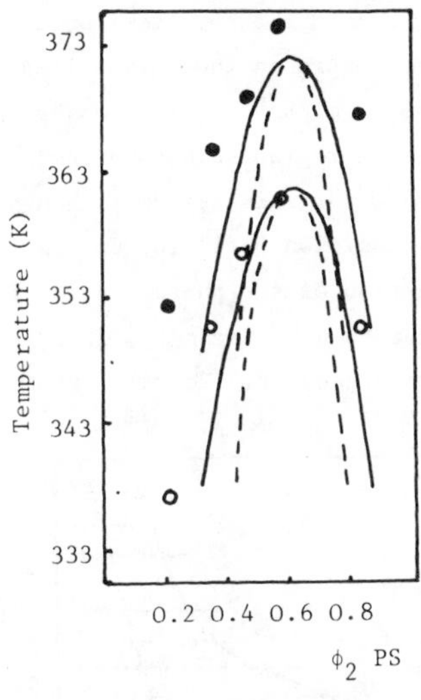

Fig. 4 The simulated binodals (solid lines) and spinodals (dashed lines) for mixtures of Polybutadiene (2350 mol.wt.) and Polystyrene (1520 mol.wt.) and experimental cloud point data at 1 atm. (O) and 1000 atm. (●).

when we consider the very much greater uncertainties which exist in the contributions of the interactional and combinatorial terms.

4 THE SPECIFIC INTERACTION

The theories which we have been discussing were never designed to apply to systems dominated by specific interactions. Since the contribution of the specific interaction to the heat of mixing can be shown to roughly depend on ϕ_1 times ϕ_2 for a weak interaction, this should not cause much trouble as far as its contribution to the interaction parameter is concerned. A very strong interaction with the groups concerned 'paired' to a large extent will not give this composition dependence however. A much greater problem is the temperature dependence. In the above simulations we have assumed X_{12} to be temperature independent but experimental evidence suggests that this is not correct.

In blends of EVA with CPE the heat of mixing was found to be temperature dependent[9]. A more direct observation of the behaviour of the specific interactions comes from infra-red studies. In these blends a weak hydrogen bond is postulated between the carbonyl of EVA and the α hydrogen of CPE. A shift in the infra-red carbonyl adsorption confirms this. Figure 5 shows the effect of increased temperature on the carbonyl adsorption frequency in the blend relative to that in the pure EVA[16]. It is observed that the difference reduces and almost disappears over a region below and above the LCST. The minimum in the cloud point is at 85°C for this system although phase separation does not occur for this blend composition until 100°C. This observation raises several problems for the simulation of the phase diagram for this system. First, X_{12} should be temperature dependent. Secondly, Q_{12}, if related to the interaction, must also be temperature dependent. Thirdly, if the interaction is very sensitive to tem-

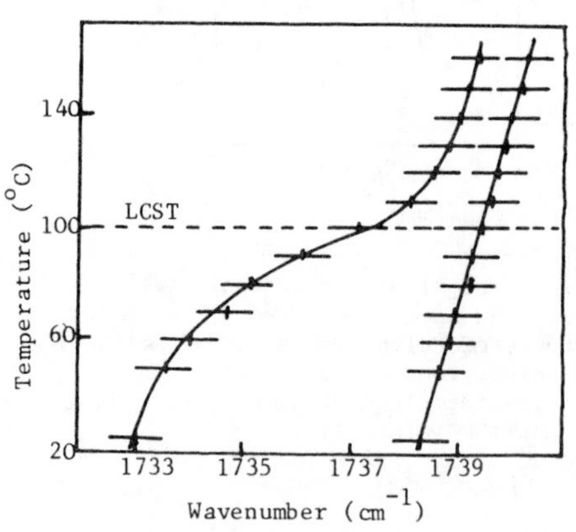

Fig. 5 Plot of temperature versus the carbonyl peak position for an 80:20 wt.% CPE-EVA blend.

perature it may also be sensitive to molecular weight (due to its effect on density) and the heat of mixing measurements for the low molecular weight analogues may not be appropriate to the polymer at any given temperature. This probably explains the discrepancies in the simulated spinodals reported earlier.

In the case of the blends of polyether sulphone with poly(ethylene oxide) the interactions are much stronger. If they are large relative to kT we might expect that their contribution to the free energy is not temperature dependent. This particular blend phase separates in part due to the large unfavourable free volume contributions which arise due to the large differences in the expansion coefficients of the base polymers. The theory much more successfully describes the observed phase boundary.

5 THE MIXING OF COPOLYMERS

In the last few years there has been much discussion of the contribution of 'cross terms', unfavourable interactions between different segments in the same polymer, to the miscibility of copolymers[18-20]. Such terms have often been considered in the literature and previously have been used in, for example, describing the solution properties of copolymers[17]. It is suggested that a copolymer is more likely to be miscible with other polymers since mixing can reduce the number of unfavourable contacts between the two types of segments. This has been used to explain the presence of the "miscibility windows" often observed in mixing involving copolymers whereby, for example, EVA is miscible with PVC over a certain copolymer composition range whereas neither polyethylene nor poly (vinyl acetate) are miscible[21]. Similarly neither poly(o-chlorostyrene) nor poly(p-chlorostyrene) are miscible with poly(phenylene oxide) whereas some copolymers are[22].

This treatment has been extended to the miscibility of homopolymers[19], for example to the miscibility of various linear polyesters with the polyhydroxy ether of bisphenol A and also with PVC. The polyesters are treated as alternating copolymers of aliphatic units and -O-CO- units. Miscibility is observed to occur within a certain range of the ratio of the units. This is explained in terms of unfavourable cross terms without recourse to any specific interaction.

One might ask therefore whether one needs to consider specific interactions in all these systems. First, it has been pointed out that as long as each segment/component can have attributed to it a single solubility parameter, then the use of the cross term cannot generate a negative heat of mixing[19]. Systems must be very non-ideal for the cross term to generate large negative heats of mixing. Secondly, we have direct spectroscopic evidence for the specific interactions[16] and the suggested energies of the interaction are considerable.

An alternative possible explanation for the window of miscibility has been suggested. The specific interactions do contribute substantially to the free energy of mixing but in some cases the unfavourable remaining dispersive forces are sufficient to balance them out and produce an unfavourable overall interaction[23]. Crudely speaking, as one varies the ratio of the segments in a copolymer there is a point where the solubility parameter is equal to that of the other polymer. Around that value the specific interactions are able to produce an overall favourable interaction and the two polymers will be miscible.

The 'cross terms' in the interaction parameter certainly exist but in systems where there are specific interactions they are probably of minor significance. In mixtures of non-polar polymers and of polymers without specific interactions all the contributions to the free energy are smaller and the cross terms could be of great importance.

6 CONCLUSION

The various contributions to the free energy of mixing, and how well these can be described by theory, have been considered.

The free volume contribution can be treated by the equation-of-state theory of Flory and co-workers. The effect of this contribution can be highlighted by studies of the effect of pressure on phase diagrams. Although the theory may not exactly describe the properties of pure polymers it appears to be adequate in describing the free volume contributions to the free energy of mixing. This is especially true considering the great uncertainties in the much larger contributions from the interaction energy.

Uncertainties in the interaction parameters may arise in both their concentration and temperature dependence. This may be particularly true in systems dominated by specific interactions. Other uncertainties exist in the extra entropic contributions to the free energy which appear to be present in all mixtures whether these exhibit specific interactions or not. In the absence of a satisfactory model we have no way of predicting their composition or temperature dependence.

REFERENCES

1. Koningsveld, R., Onclin, M.H. and Kleintjens, L.A. in 'Polymer compatibility and incompatibility', Harwood Academic Publishers, 1982.
2. Flory, P.J., Orwoll, R.A. and Vriji, A., J.Am.Chem.Soc., Vol. 86, 3507,1964.
3. Flory, P.J., J.Am.Chem.Soc., Vol. 87, 1833, 1965.
4. Eichinger, B.F. and Flory, P.J., Trans.Farad.Soc., Vol.64, 2035, 1968.
5. McMaster, L.P., Macromolecules, Vol.6, 760, 1973.
6. Olabisi, O., Macromolecules, Vol.8, 316, 1975.

7. Zhikuan, Chai, Sun, Ruona, Walsh, D.J. and Higgins, J.S., Polymer, Vol.24, 263, 1983.
8. Zhikuan, Chai and Walsh, D.J., Makromol.Chem., Vol.184, 1459, 1983.
9. Rostami, S. and Walsh, D.J., Macromolecules, Vol.17, 315, 1984.
10. Walsh, D.J. and Singh, V.B., Makromol.Chem., Vol.185, 179, 1984.
11. Walsh, D.J., Rostami, S. and Singh, V.B., Makromol.Chem., Vol.186,145,1985.
12. Zacharius, S.L., ten Brinke, G., MacKnight, W.J. and Karasz, F.E., Macromolecules, Vol.16, 381, 1983.
13. Allen, G., Chai, Z., Chong, C.L., Higgins, J.S. and Tripathi, J., Polymer, Vol. 25, 239, 1984.
14. Koningsveld, R., private communication, also see the Chapter by Koningsveld in this volume.
15. Walsh, D.J. and Rostami, S., Macromolecules, in press.
16. Coleman, M.M., Maskala, E.J., Painter, P.C., Walsh, D.J. and Rostami, S., Polymer, Vol.24, 1410, 1983.
17. Stockmayer, W.H., Moore, L.D. Jr., Fixman, M. and Epstein, B.N., J.Pol.Sci., Vol.16, 517, 1955.
18. Kambour, R.P., Bendler, J.T. and Bopp, R.C., PRI Conference on Polymer Blends, Univ. of Warwick, 1981.
19. Paul. D.R. and Barlow, J.W., Polymer, Vol.25, 487, 1984.
20. ten Brinke, G., Karasz, F.E. and MacKnight, W.J., Macromolecules, Vol.16, 1827, 1983.
21. Hammer, C.F., Macromolecules, Vol.4, 69, 1971.
22. Alexandrovich, P., Karasz, F.E. and MacKnight, W.J., Polymer, Vol.18, 1022, 1977.
23. Walsh, D.J. and Cheng, G.L., Polymer, Vol. 25, 499, 1984.

NOTATION

G_M free energy of mixing

P pressure

P^*_i hard core pressure of species i

P^* hard core pressure of mixture

Q_{12} interaction entropy parameter

R gas constant

r_i chain length of molecule i

T Temperature

T^*_i hard core temperature of species i

V^*_i molar hard core colume of component i

$\tilde{v}$ reduced volume of mixture

X_{12} interactional parameter

ϕ_i segment fraction of species i

θ_i site fraction of species i

RELATION OF INTERDIFFUSION AND SELF-DIFFUSION IN POLYMER MIXTURES

H. SILLESCU

Institut für Physikalische Chemie der Universität Mainz

SYNOPSIS

From different assumptions one obtains linear relationships either between the interdiffusion coefficient and the self-diffusion coefficients or between the inverse interdiffusion coefficient and the inverse self-diffusion coefficients. The nature of these assumptions and possible experiments for discriminating between them are discussed.

INTRODUCTION

The theory of interdiffusion (mutual diffusion) in compatible polymer mixtures has been developed along the lines of a proposal by de Gennes[1] with assumptions similar to those made in the theory of interdiffusion in solids resulting in a linear relationship between the inverse interdiffusion coefficient D^{-1} and the inverse self-diffusion coefficients ${D_A}^{-1}$ and ${D_B}^{-1}$ where A and B are the components of a binary mixture.[2,3] Kramer et al.[4] have introduced vacancies into the binary model system, and from particular assumptions about the chemical potential and the flux density of these vacancies they arrive at a linear relationship between D, and Onsager mobilities M_A and M_B which are proportional to D_A and D_B, respectively. The same relationship has been derived[5] without the assumption of vacancies by simply placing the Flory-Huggins chemical potential of a compatible polymer mixture into the standard expression of the interdiffusion coefficient in binary liquid mixtures, and by an assumption (see below) usually made in derivations of the Hartley-Crank equation:[6,7]

$$D = (x_B D_A + x_A D_B)\, Q \tag{1}$$

$$Q = \partial \ln a_i / \partial \ln x_i \tag{2}$$

(a_i, x_i: activity and mole fraction of component i = A, B)
Though, only future experiments will decide on the validity of the assumptions made in the different theoretical approaches, we try to contribute to understanding their nature in the following discussion.

THEORY AND DISCUSSION

In a binary liquid mixture at constant pressure and temperature there is only one independent flux $\vec{J} = \vec{J}_A$ related to the gradient of the chemical potential difference by the phenomenological equation defining the Onsager coefficient $\Omega = \Omega_{AA} = \Omega_{BB}$ (see refs. 5 and 8). By introducing the Fickian center of volume reference frame one arrives at Fick's law, $\vec{J}_i = - D \nabla c_i (i = A,B)$, with the interdiffusion coefficient[5-8]

$$D = \Omega Q R T \bar{V} (c_A M_A + c_B M_B)^2 / c_A c_B M_A^2 M_B^2 \quad (3)$$

$\bar{V}$ is the mean molecular volume; c_i and M_i are the molar concentrations and masses, respectively. Within the framework of linear response theory, Ω is related with the integral of the flux correlation function which can be decomposed into velocity correlation coefficients

$$D_i^{(\alpha)} = \frac{1}{3} \int_0^\infty \langle \vec{v}_\alpha^{(i)}(0) \cdot \vec{v}_\alpha^{(i)}(t) \rangle dt \quad (4)$$

$$f_i^{(\alpha)} = \frac{1}{3} \int_0^\infty \sum_{\beta \neq \alpha} \langle \vec{v}_\alpha^{(i)}(0) \cdot \vec{v}_\beta^{(i)}(t) \rangle dt \quad (5)$$

$\vec{v}_\alpha^{(i)}$ is the velocity of molecule α of component i relative to the center of mass velocity of the volume element (local equilibrium). In order to derive eq.(1) from eq.(3) one assumes[7] that $D_i^{(\alpha)}$ and $f_i^{(\alpha)}$ are independent of α, and that the relation

$$f_i = - w_i D_i \quad (6)$$

is valid. w_i is the mass fraction, and $D_i = D_i^{(\alpha)}$ the self diffusion coefficient of component i. One can readily show that these assumptions imply

$$D_A M_A = D_B M_B \quad (7)$$

A similar relation was obtained long ago by Bearman.[9] If the usual Florv-Huggins expression for the chemical potential is placed into eqns.(1-2) one obtains[5]

$$D = (\phi_B N_A D_A + \phi_A N_B D_B)(\phi_B / N_A + \phi_A / N_B - 2 \chi \phi_A \phi_B) \quad (8)$$

where ϕ_A, ϕ_B and N_A, N_B are volume fractions and chain lengths of A and B, respectively, and $\chi < 0$ is the interaction parameter. Eq.(8) is identical with eq.(13) of ref. 4 if the segment mobilities B_i defined in their eq. (19) are identified with $N_i D_i/kT$. The agreement is rather surprising since the assumptions of refs. 4 and 5 differ in many respects. Our eq. (6) is justified in liquids on the basis of momentum conservation arguments[7] which cannot be applied to flexible macromolecules. Thus, eq. (7) is inconsistent with self-diffusion by reptation, $D \sim M^{-2}$. In refs. 1-4, the phenomenological equations relate segment fluxes with the gradients of reduced chemical potentials obtained by dividing the Flory-Huggins potential by the segment number N_i. Kramer et al.[4] assume in their treatment that the vacancies have a constant chemical potential whereas the vacancy flux balances the segment fluxes ($-J_V = J_A + J_B$). This appears rather unphysical in view of typical vacancy mechanisms applied in the theory of interdiffusion in solids.[10] The treatments following de Gennes,[2,3] which are closer in spirit to diffusion in solids, arrive at

$$D^{-1} = [\phi_B (N_A D_A)^{-1} + \phi_A (N_B D_B)^{-1}](\phi_B/N_A + \phi_A/N_B - 2\chi\phi_A\phi_B)^{-1} \qquad (9)$$

The difference between eqns. (8) and (9) will be most pronounced in systems with very different mobilities, say, $D_A \ll D_B$. Thus, eq. (9) cannot describe the dissolution of a polymer A in a low molecular weight solvent B since $D = 0$ in the limit $D_A = 0$.

Let us discuss a possible experiment which exhibits many difficulties of the present status of polymer diffusion theory. In the tube shown in the figure, let A be non-linear polystyrene (PS), e.g., rings, stars, or intra-crosslinked PS molecules,[11] whereas B is linear PS of the same molecular weight. Thus, $N_A = N_B$, $M_A = M_B$, and $\chi = 0$. Furthermore, the center of volume, center of mass, and laboratory coordinate systems coincide. Since the non-linear A molecules cannot move by reptation we have $D_A \ll D_B$ in contrast to eq.(7).

A ← → B

Since the segment mobilities of A and B are identical it is difficult to rationalize different Onsager segment mobilities which is a necessary assumption in refs. 1-4 in order to obtain $D_A \neq D_B$. A further difficulty is related with the "bulk flow"[12] caused by the rapid penetration of B into A at constant pressure. This flow of A and B (to the right in the figure) affects the "self-diffusion" coefficients D_A and D_B if they are defined as usual by eq. (4) or equivalently by mean square displacements of A and B. "Intrinsic diffusion coefficients" have been introduced[6,12] in order to separate "true" diffusion from molecular displacements by the bulk flow. The holographic grating techniques presently applied in our laboratory[11] should allow for separating the diffusive and flow motions of photo-labeled molecules. This technique has been applied recently for monitoring electrophoretic flow.[13]

REFERENCES

1. De Gennes, P.G., J. Chem. Phys. 72, 4756 (1980)
2. Brochard, F., Jouffroy, J., and Levinson, P., Macromolecules 16,1638 (1983)
3. Binder, K., J. Chem. Phys. 79, 6387 (1983)
4. Kramer, E.J., Green, P., and Palmstrøm, Ch.J., Polymer 25, 473 (1984)
5. Sillescu, H., Makromol.Chem., Rapid Commun. 5, 519 (1984)
6. Hartley, G.S. and Crank, J., Trans. Faraday Soc. 45, 801 (1949)
7. Harris, K.R. and Tyrrell, H.J.V., J. Chem. Soc., Faraday Trans. 1, 78, 957 (1982), and references therein.
8. Fitts, D.D., 'Nonequilibrium Thermodynamics', McGraw-Hill, New York 1962
9. Bearman, R.J., J. Phys. Chem. 65, 1961 (1961)
10. Binder, K., private communication.
11. Antonietti, M. and Sillescu, H., Macromolecules, in press.
12. Crank, J., 'The Mathematics of Diffusion', 2nd Ed., Clarendon, Oxford 1975
13. Kim, H., Chang, T., Yu, H., J. Phys. Chem. 88, 3946 (1984)

CRYSTALLIZATION AND MELTING STUDIES ON POLY(ETHYLENE OXIDE)/POLY(METHYL METHACRYLATE) MIXTURES

G.C. ALFONSO* and T.P. RUSSELL

IBM Research Laboratory, San Jose, California 96195 (USA)
*Present address : Istituto di Chimica Industriale, Corso Europa 30,
I-16132 Genova (Italy)

SYNOPSIS

The melting and crystallization behaviours in mixtures of monodisperse fractions of PEO and PMMA have been studied as a function of composition and molecular weight of each component. Crystal annealing during crystallization and heating prohibited a reliable evaluation of the equilibrium melting point and cast serious doubts on the determination of the interaction parameter via a melting point depression analysis.

To describe the measured radial growth rates in the range 10-60°C, it is proposed a kinetic equation that takes into account the thickness of the crystalline lamellae, a cooperative diffusion coefficient and the free energy for the formation of secondary crystal nuclei in a mixture.

1 INTRODUCTION

Mixtures of a semicrystalline polymer with an amorphous one have been of considerable interest for some time. Twenty years ago, Keith and Padden[1] proposed the use of a parameter $\delta = D/G$, to account for the effect of the non-crystallizable component on the kinetics of crystallization and the spherulitic texture of these systems. Basically, this parameter is related to the rate at which the lamellae growth and the rate at which the amorphous polymer can diffuse away from the growth front.

With the advent of newer theories of polymer diffusion[2] and from some recent results of Keith and Padden[3], this semplified approach using the δ parameter has been questioned. In order to address this issue, we have studied mixtures of very sharp molecular weight fractions of poly(ethylene oxide), PEO and poly(methyl methacrylate), PMMA that are believed to be miscible in the melt at any composition[4-10].

2 EXPERIMENTAL

A series of mixtures composed of narrow molecular weight distribution fraction of PEO (M_w/M_n = 1.05) from Toyo Soda Manufacturing Co. with PMMA (M_w/M_n = 1.17) from Polymer Laboratories Ltd. and Rohm and Haas Co. have been prepared by freeze drying from 2.5% solution in benzene. The molecular weights have been varied from 10^4 to 10^6 for PEO and from 10^3 to $5x10^5$ for PMMA and we have concentrated our attention to compositions with more then 50% of PEO.

Thin films of the blends have been prepared by compression molding at 100-140°C and have been studied mainly by a polarizing microscope equipped with a Mettler FP52 hot stage.

The melting points, corresponding to the disappearance of birefringence, have been measured at an heating rate of 3°C/min. Isothermal conditions before crystallization have been obtained at low temperatures by means of a stage with circulation of a cooling fluid. Fast crystallizations have been recorded and played back in a frame by frame mode in order to get precise measurements of the radial growth rate.

3 RESULTS AND DISCUSSION

In order to gain a good understanding of the kinetic aspects of the crystallization in polymer blends, the evaluation of the thermodynamic interaction parameter χ ia mandatory. Among the available techniques for the determination of χ in a bulk system, the analysis of melting point depression[11] has been applied by several different groups on PEO/PMMA mixtures. Unfortunately, while all the values cited are negative, there is no consistency in the results[4,5,8].

We have carried out the Hoffman and Weeks analysis[12] on our mixtures and have found that crystal annealing during the crystallization, and possibly also during melting, dramatically affect the value of the melting point, T_m, thus preventing us from obtaining reliable values of the equilibrium melting temperature, T°_m, and consequently of χ.

Our data show that T_m increases linearly with the logarithm of the age of the crystals at a rate that depends both on temperature and composition. As an example, in fig.1 the results obtained for a mixture containing 30 wt.% of high molecular weight PMMA are reported. However, despite the inaccuracy of the method, our results suggest that χ is very small.

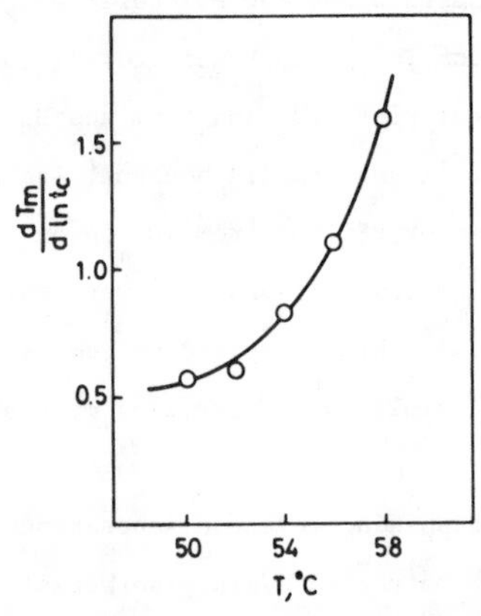

Figure 1 : Logarithmic rate of melting point elevation as function of crystallization temperature. $\phi_2 = 0.7$; $M_{PEO} = 145,000$, $M_{PMMA} = 525,000$.

Typical radial growth rates, G_{mix}, as a function of temperature are reported in fig.2 for the blends of PEO (145,000) with two different molecular weights PMMA. The most peculiar feature is that, at high T_c, the crystallization rate is depressed to a greater extent by the high molecular weight PMMA than by the lower. A purely thermodynamic reasoning would suggest that the contrary should be true since shorter chains must be more effective in lowering the melting point of the crystallizable component. On the other hand, the application of the criterion of Keith and Padden would lead to a non-linear growth rate and to a marked effect of the molecular weight of the diluent on the crystallization kinetics at low temperatures. Neither effect has been observed in our systems.

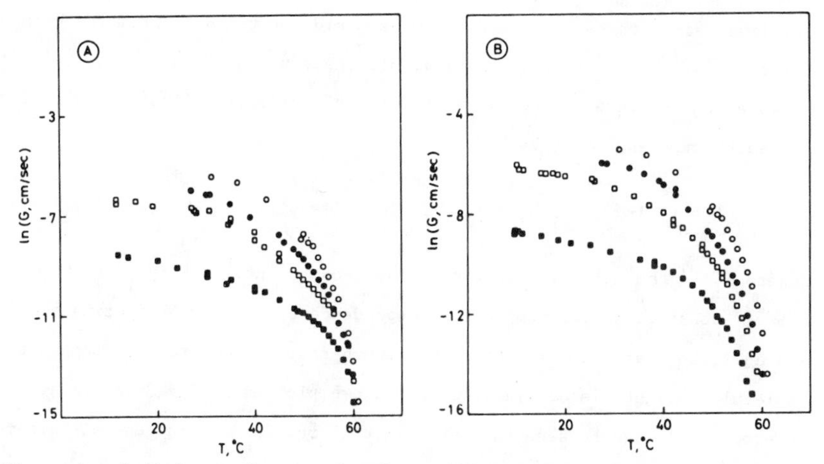

Figure 2 : Radial growth rate of PEO 145,000/PMMA blends at different temperatures. O : ϕ_2 = 1; O = ϕ_2 = 0.925; : ϕ_2 = 0.85; = ϕ_2 = 0.70. *Molecular weight of PMMA : A = 29,700; B = 525,000.*

In general, the results of our experiments can be described by[13]

$$G_{mix} = \phi_2 \frac{G_\circ (D/L)}{G_\circ + (D/L)} \exp(-\Delta F^*/kT) \quad \ldots \quad (1)$$

where D is a cooperative diffusion coefficient, L is the average lamellar thickness of the crystals and $G_\circ$ is the growth rate of the pure PEO at the same undercooling. ΔF^* is the critical nucleation free energy evaluated from the extension to polymer-polymer mixtures of the treatment of Flory[14] and Mandelkern[15] for polymer-solvent systems :

$$\Delta F^* = 2b\sigma\sigma_e / (\frac{\Delta h \Delta T}{T^\circ_m} + RT\chi (\bar{V}_2/\bar{V}_1)(1 - \phi_2)^2) \quad \ldots \quad (2)$$

where b is the width of the secondary nucleus, σ and σ_e are the lateral and fold surface free energies, respectively; Δh is the enthalpy of fusion per monomer unit, $\bar{V}_i$ is the molar volume of component i and ϕ_2 is the volume fraction of the crystallizable component.

The application of eq.1 requires the evaluation of L and D for the polymer mixtures as a function of temperature, composition and molecular weights. Crystallization theories foresee that the crystal thickness is related to undercooling; since we have assumed $\chi \cong 0$ we could use the data from the work of Arlie et al.[16] on pure PEO. Kramer et al.[17] have recently shown that D in an homogeneous mixture can be given in terms of the self diffusion coefficients of each component, D_{si}, by :

$$D = ((1-\phi_2) D_{s1} N_1 + \phi_2 D_{s2} N_2) ((1-\phi_2)/N_1+\phi_2/N_2+ 2\phi_2(1-\phi_2)|\chi|) \quad ..($$

where N_i is the molecular weight of the i-component.

Considering the temperature dependence of L and substituting eqs.2 and in eq.1, an expression relating the growth rate to temperature, composition a molecular weights is obtained. Our data are plotted in fig.3 according to thi treatment. It can be seen that experimental results are in general agreement with the proposed eqaution in all range of conditions studied.

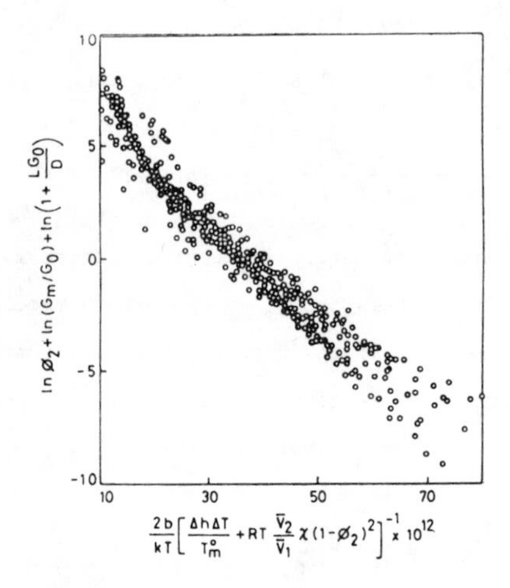

Figure 3 : Plot of the radial growth rate of various mixtures according to eq

REFERENCES

1. Keith, H.D. and Padden, F.P. J.Appl.Phys., 35, 1270,1286, 1964
2. de Gennes, P.G. "Scaling Concepts in Polymer Physics", Cornell Universi Press Ithaca, New York, 1979
3. Keith, H.D. and Padden, F.P. Bull.Am.Phys.Soc. 29(3), 452, 1984

4. Martuscelli, E. and Gemma, G. in "Polymer Blends: Processing, Morphology and Properties", Martuscelli, E. Palumbo, R. Kryszewski, M. Eds., Plenum Press, New York, 1980

5. Calahorra, E. Cortazar, M. and Guzman, G.M. Polymer, 23, 1322, 1982

6. Cortazar, M.M. Calahorra, M.E. and Guzman, G.M. Eur.Polym.J., 18, 165, 1982

7. Martuscelli, E. Demma, G. Rossi, E. and Segre, A.L. Polymer, 24, 266, 1983

8. Martuscelli, E. Pracella, M. and Yue, W.P. Polymer, 25, 1097, 1984

9. Liberman, S.A. De S.Gomes, A. and Macchi, E.M. J.Polym.Sci., Polym. Chem.Ed.. 22, 2809, 1984

10. Li, X. and Hsu, J.L. J.Polym.Sci., Polym.Phys.Ed., 22, 1331, 1984

11. Nishi, T. and Wang, T.T. Macromolecules, 8, 909, 1975

12. Hoffman, J.D. and Weeks, J.J. J.Res.Natl.Bur.Stand. U.S., 66, 13, 1962

13. Alfonso, G.C. and Russell, T.P. to be published, 1984

14. Flory, P.J. J.Chem.Phys., 17, 223, 1949

15. Mandelkern J.Appl.Phys., 26, 443, 1955

16. Arlie, P. Spegt, P. and Skulios, A. Makromol.Chem., 104, 212, 1967

17. Kramer, E.J. Green, P. and Palmostrøm Polymer, 25, 473, 1984

SPECIFIC INTERMOLECULAR INTERACTIONS IN POLYMER BLENDS

J M G COWIE

Department of Chemistry, University of Stirling, Stirling FK9 4LA, Scotland.

SYNOPSIS

The formation of miscible quasi-binary polymer blends is not a commonly observed phenomenon. With a few exceptions, one phase blends will only form if there are some specific intermolecular interactions (SII) between the two components such as, hydrogen bonding, ion-dipole, dipole-dipole, π-bonding, or charge transfer interactions. In certain copolymer blends, however, miscibility can be achieved by making use of repulsion effects rather than SII.

1 INTRODUCTION

It is rather unfortunate from a technological point of view that attempts to form miscible blends from the commonly available polymers usually results in two phase systems which rarely display any advantageous properties. As the formation of miscible polymer blends is an attractive goal, methods of overcoming the difficulties in producing one phase polymer mixtures have been explored by applying fundamental principles. A suitable point of departure is to consider the simple Flory-Huggins equation for the free energy of mixing ΔG^M,

$$\Delta G^M/NRT = (\emptyset/m_1)\ln\emptyset_1 + (\emptyset_2/m_2)\ln\emptyset_2 + \chi\emptyset_1\emptyset_2 \tag{1}$$

which is the sum of a combinatorial entropy of mixing contribution and an interaction parameter χ, related to the enthalpy of mixing ΔH^M. Here $\emptyset_i$ is the volume fraction and m_i the number of chain segments of component i. For polymers, $m_i >> 1$, and so the favourable entropy term is very small. This means that the sign of ΔG^M will depend largely on χ (or ΔH^M).

Formation of one phase polymer mixtures will then be restricted to systems for which ΔH^M is a very small positive, zero, or negative value. For many polymer pairs ΔH^M is positive, and stable one phase mixtures are either impossible to obtain or can only be formed from low molar mass components where the residual entropy of mixing is sufficient to overcome the unfavourable ΔH^M. The majority of blends which are miscible over a wide concentration and molecular weight range, have an exothermic heat of mixing resulting from some specific intermolecular interactions (SII) operating between the components. A number of these have now been identified and used to enhance polymer miscibility.

2 SPECIFIC INTERMOLECULAR INTERACTIONS

2.1 Hydrogen bonding

Pearce et alia[1] have demonstrated that if polystyrene (PS) is modified by introducing a number of hexafluoro dimethyl carbinol groups into the phenyl rings then miscible blends can be formed with polymers containing proton acceptor groups, which are normally immiscible with the un-modified polystyrene.

The modification means that the polystyrene is actually a copolymer (MPS) with the structure (I), but when there are less than 10% of the

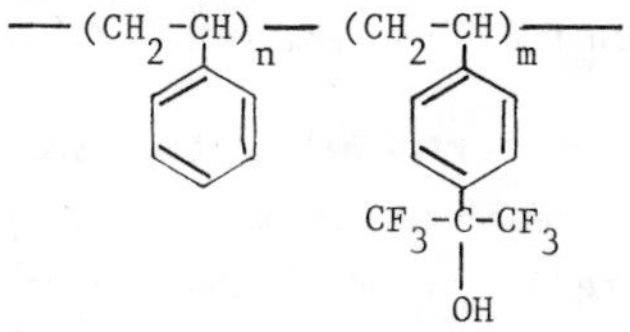

$$—(CH_2\text{-}CH)_n—(CH_2\text{-}CH)_m— \qquad \text{I}$$

$CF_3\text{-}C\text{-}CF_3$

OH

hydroxyl groups present the glass transition temperatures, Tg, are similar to polystyrene. Miscible blends of MPS with poly(vinyl acetate), poly(vinyl methyl ether), (PVME), poly(alkyl methacrylate)s and poly(2,6 dimethyl-1,4-phenyleneoxide) (PPO) have been reported.

The main evidence for hydrogen bonding in the blends has come from FTIR spectroscopy. The MPS has two absorption peaks at 3600 cm^{-1} and 3520 cm^{-1}, characteristic of the hydroxyl group, which disappear on blending with, for example, poly(alkyl methacrylate)s and are replaced by a new absorption at 3400 cm^{-1}, see Figures 1 and 2. Similar shifts are recorded in other blends and indicate that H-bonding takes place between the -OH of the MPS and either the carbonyl or the ether groups in the acceptor polymers.

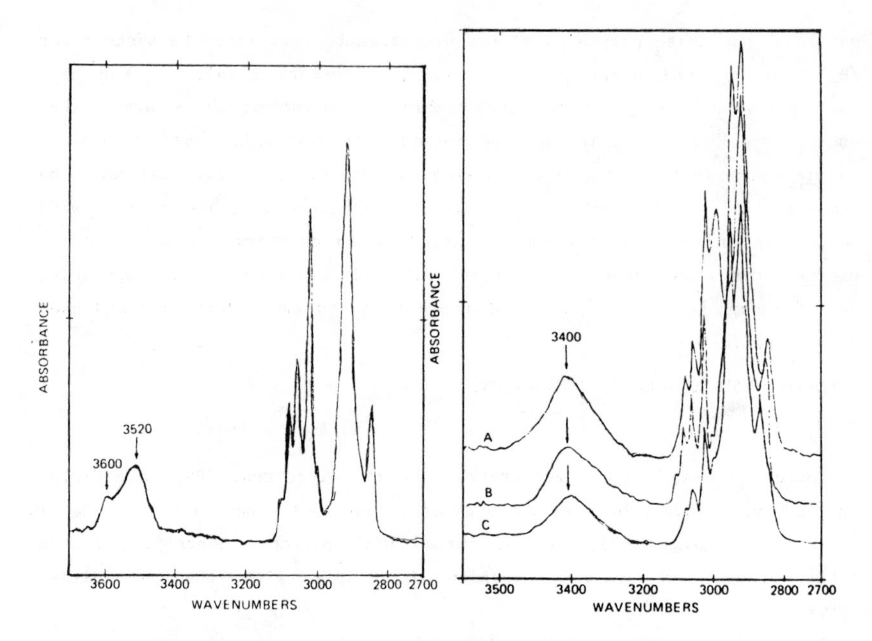

Figure 1. The FTIR spectrum of copolymer MPS with 9.7 mole % of hydroxyl groups.

Figure 2. FTIR spectra of blends of MPS (9.7) with A - poly(methyl methacrylate); B - poly(ethyl methacrylate); C - poly(butyl methacrylate).

(Reproduced from ref 1 with the kind permission of Marcel Dekker Inc.)

Examination of the lower critical solution cloud point curves (LCST) for the blends, shows that the temperatures of phase separation are raised as the hydrogen bonding sites are increased. Thus introduction of only 0.4 mol% hydroxyl into the MPS raises the LCST by about $50^{o}C$ in the MPS/PVME blends as seen in Fig 3, while 4 mol% is sufficient to change the immiscible PS/poly(methyl methacrylate) system into one with an LCST $>200^{o}C$. Thus remarkably low levels of H-bonding sites in the MPS are sufficient to produce miscibility in these systems.

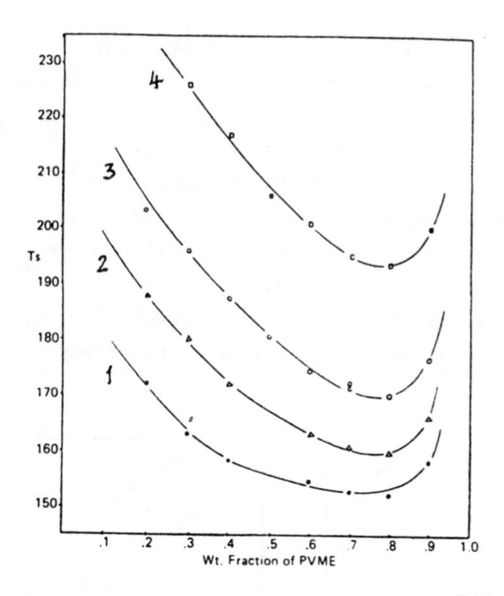

Figure 3. LCST curves for blends of PVME with pure polystyrene (curve 1) and MPS samples: curve 2 - MPS (0.1); curve 3 - MPS (0.2); and curve 4 - MPS (0.4).
(Reproduced from ref 1 with the kind permission of Marcel Dekker Inc.)

Coleman and his associates[2,3] studied blends of poly(vinyl chloride), (PVC), with poly(ε-caprolactone)[2], (PCL), and ethylene/vinyl acetate copolymer, (EVA),[3] using FTIR. Spectral shifts detected for the carbonyl groups in PCL and EVA indicated that this unit was involved in the SII, but it could not be determined unequivocally that the interaction was hydrogen bonding with the methine proton of PVC (C=O....H-C-Cl) as the C-H stretching vibration is weak and often submerged in the methylene absorbances, although this was thought to be the most likely explanation. This was resolved by deuteration of the α-hydrogen in PVC to allow the C-D stretching vibrations to be observed; these were seen to shift in the blends, substantiating the hydrogen bond hypothesis.[4] This does not rule out the possibility of dipole-dipole interactions also operating in such systems involving the (C-Cl) and (C=O) groups or perhaps even H-bonding involving the β-hydrogens in PVC.

2.2 Ionic and dipolar interactions

Eisenberg and coworkers[5-8] have taken a lead from observations by

Michaels and Miekka[9] that polymer complexes can form through ionic interactions. They have examined the effect of introducing groups into polymer chains capable of ion-ion, ion-dipole, and dipole-dipole interactions, on the miscibility of normally immiscible polymer pairs.

Poly(ethyl acrylate), PEA, and PS are normally immiscible, but if PS is modified by inclusion of styrene sulphonic acid units in the chain and ethylacrylate is copolymerized with small quantities of 4-vinyl pyridine, then the two resulting copolymers form miscible blends.[5] This can be achieved, as can be seen from the dynamic mechanical spectra in Figures 4 and 5 of unmodified and copolymer blends, by incorporating at least 5 mol% of the

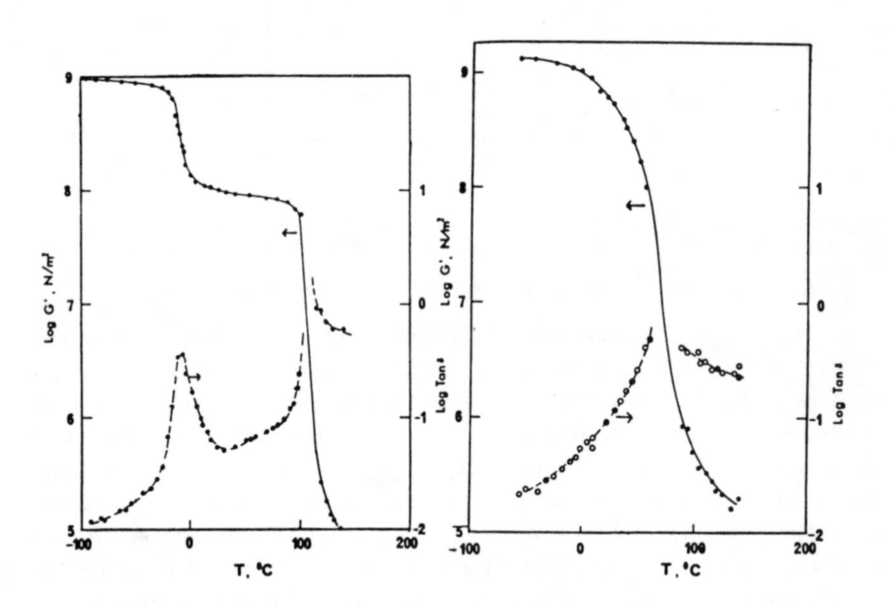

Figure 4. Temperature dependence of the shear storage modulus G' and the loss tangent tan δ for a 50/50 blend of polystyrene and poly(ethyl acrylate).

Figure 5. Temperature dependence of the storage modulus, G', and the loss tangent, tan δ, for a blend of the modified polystyrene and poly(ethyl acrylate) (see text).

(Reproduced from ref 6 with the kind permission of John Wiley and Sons Inc.)

($-SO_3H$) and 4-vinyl pyridine units respectively in each copolymer. The SII in this system is a coulombic attraction, but the mere introduction of any coulombic force into the polymer can be insufficient to promote miscibility. This was demonstrated by mixing poly(styrene-_co_- methacrylic acid) with poly(ethyl acrylate-co-acrylic acid) which proved to be an immiscible system, so introduction of carboxyl groups only did nothing to enhance the miscibility.

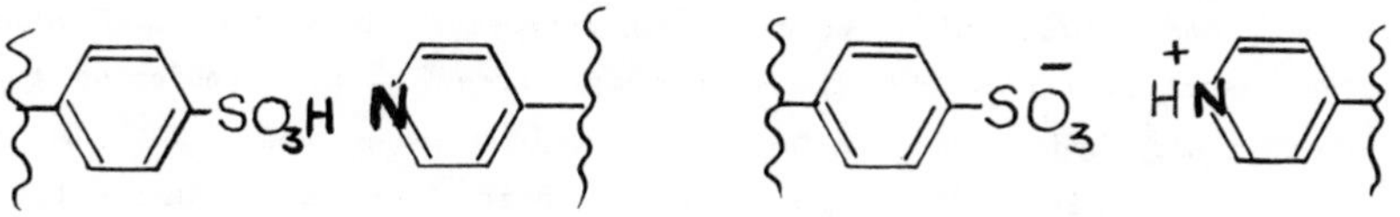

Strong coulombic interactions between an acidic group and a non ionic basic group provide an effective means of increasing the miscibility in polymer blends, but weaker ion-dipole interactions can also prove useful.[7] Poly(ethylene oxide), (PEO), and PS form two phase mixtures but this can be improved by incorporating up to 10 mol% of the lithium salt of methacrylic acid in the polystyrene chain. A greater level of miscibility is obtained when this ionomer is blended with PEO which is encouraged by the ion-dipole interactions in the system.

Dipole-dipole interactions[10-12] have also been identified as SII in certain blends. Challa and Roerdink[13] found evidence for the interaction between the carbonyl groups in poly(methyl methacrylate) and the dipoles of the poly(vinylidene fluoride) segments, in blends of these two components. A similar interaction has been suggested as the operative SII in blends of poly(vinyl chloride) and several other polymers containing carbonyl units.[8,10,12]

2.3 Interactions with π electron systems

FTIR has been found to be a useful tool in identifying SII in polymer blends. The digital subtraction method leaves a residual spectrum from which band broadening or shifting can provide information on the groups involved. Studies on PS/PPO blends[14] have revealed the presence of strong favourable dispersion forces between the phenyl rings of PS and PPO. Several workers[15-18] have used the technique to examine PS/PVME blends and have found evidence for an interaction between the π electrons of the phenyl ring in PS and the lone pairs of the ether oxygen in PVME. Garcia[17] demonstrated the involvement of the latter by observing spectral changes when the temperature was raised above the LCST. The ether absorption bands at

1183.6 cm^{-1} and 1107.6 cm^{-1} increased in intensity above the LCST after remaining essentially constant at lower temperatures. This indicated that the ether group had gained rotational freedom above the LCST due to elimination of the SII. The strength of the interaction was estimated to be 6300 $Jmol^{-1}$.

Similar conclusions concerning the origins of the SII in PS/PVME blends have been arrived at by Porter et alia[19] from ^{1}H nmr and ^{13}C nmr studies. The miscibility in PS/PVME blends is known to be solvent dependent and casting from $CHCl_3$ results in a two phase structure with no evidence of SII.[15] Porter's work has shown that there could be competitive interaction between the π-electron system and the electron deficient hydrogen in $CHCl_3$ which would prevent efficient mixing of the two polymers. This is certainly a possibility and highlights the importance of solvent in such cases.

2.4 Charge-transfer complexes

The use of electron donor-acceptor groups to form intermolecular charge-transfer complexes in polymer blends was first studied by Sulzberg and Cotter[20] who incorporated donor and acceptor residues into the main polymer chain. More recently polymers containing pendant donor and acceptor groups have been prepared[21,22] using the type of group shown in Table 1.

Table 1

Type	Donor	Acceptor	Polymer
1	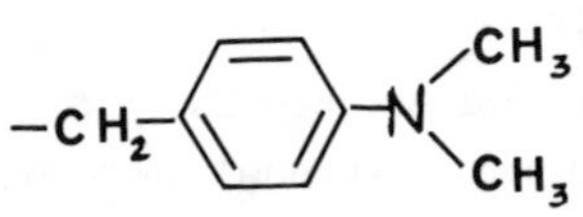 $-CH_2-C_6H_4-N(CH_3)_2$	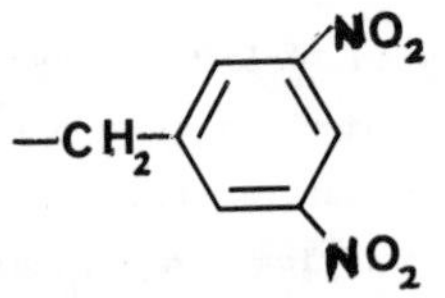 $-CH_2-C_6H_3(NO_2)_2$	Polyesters
2	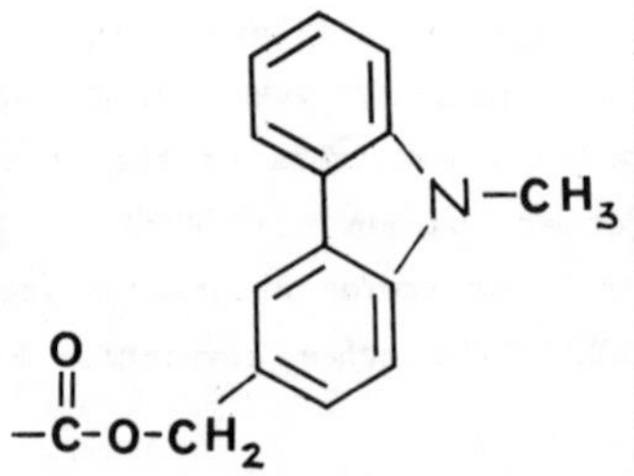 $-C(=O)-O-CH_2-$(N-methylcarbazolyl) ($N-CH_3$)	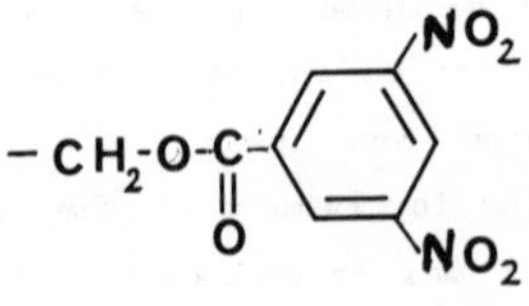 $-CH_2-O-C(=O)-C_6H_3(NO_2)_2$	Acrylates

Blends of type 1 polyesters, each containing high levels of donor and acceptor groups were not completely miscible but did show some improvement due to the formation of charge-transfer complexes. Melt mixing of type 2 polyacrylates was not successful, but (1:1) mixtures cast from tetrahydrofuran solutions gave homogeneous one phase blends. Dynamic mechanical studies indicated that crosslinking, through the charge-transfer complex, had occurred in the blend.

3. INTRAMOLECULAR REPULSIVE INTERACTIONS

The systems mentioned so far have relied on the presence of intermolecular interactions to promote miscibility. In blends where the components are statistical copolymers one should also consider the effect of intermolecular interactions between the different monomeric units comprising the copolymer. Interesting examples have been reported where a homopolymer such as PMMA, which is neither miscible with PS nor polyacrylonitrile, forms miscible blends with a styrene/acrylonitrile copolymer[23] within a restricted range of copolymer compositions. This "miscibility window" has been reported in several systems; PPO blends with copolymers of o- and p-chlorostyrene and other halogenated styrenes,[24,25] PVC with EVA[26] and with butadiene/acrylonitrile copolymers,[27] are a few. Karasz and MacKnight,[28] Kambour et alia,[29] and Paul and Barlow[30] all agree that the restricted miscibility in these systems is a result not of SII but of the repulsive forces in the copolymer, which cause it to "dissolve" in a polymeric solvent when the net forces of interaction are most favourable. This is in many ways similar to the dissolution of a polymer in a liquid cosolvent pair where the cosolvent liquids often have an unfavourable free energy of mixing.[31] For a binary mixture of two statistical copolymers, one composed of monomers (1) and (2), and with a composition $[(1)_\alpha\ (2)_{(1-\alpha)}]$, and a second with monomers (3) and (4) of composition $[(3)_\beta\ (4)_{1-\beta}]$, where α and β are volume fractions, the interaction parameter for the blend is given by[28],

$$\chi_{blend} = \alpha\beta\chi_{13} + \beta(1-\alpha)\chi_{23} + \alpha(1-\beta)\chi_{14} + (1-\alpha)(1-\beta)\chi_{24} - \alpha(1-\alpha)\chi_{12} - \beta(1-\beta)\chi_{34} \qquad (2)$$

Here χ_{ij} are the component binary interaction parameters.

This can be simplified for the special case of a homopolymer (1) and a copolymer (3) and (4) to

$$\chi_{blend} = \beta\chi_{13} + (1-\beta)\chi_{14} - \beta(1-\beta)\chi_{34} \qquad (3)$$

These equations have the same general form as those proposed by Kambour,[29] and Paul and Barlow[30] and they show that it is possible to obtain a favourable value of χ_{blend}, from a combination of χ_{ij} parameters each of which can be unfavourable to mixing. Obviously a large (repulsion) value for χ_{34} in equation (3) will make the existence of a "miscibility window" more likely. A more sophisticated version of equation (2) has been developed by Koningsveld and Kleintjens[32] which introduces parameters to account for the interacting surface areas of the repeat units. This appears to be a very promising development and a fruitful area for further work.

REFERENCES

1. Pearce, E.M., Kwei, T.K. and Min, B.Y. J.Macromol.Sci. - Chem. A21, 1181 (1984).
2. Coleman, M.M. and Zarian, J. J.Polym.Sci.Phys.Ed., 17, 837 (1979).
3. Coleman, M.M., Moskala, E.J., Painter, P.C., Walsh, D.J. and Rostami, S. Polymer 24, 1410 (1983).
4. Varnell, D.F., Moskala, E.J., Painter, P.C. and Coleman, M.M. Polym. Eng.Sci., 23, 658 (1983).
5. Eisenberg, A., Smith, P. and Zhou, Z.L. Polym.Eng.Sci., 22, 1117 (1982).
6. Smith, P. and Eisenberg, A. J.Polym.Sci.Polym.Lett.Ed., 21, 223 (1983).
7. Hara, M. and Eisenberg, A. Macromolecules, 17, 1335 (1984).
8. Clas, S.D. and Eisenberg, A. J.Polym.Sci.Polym.Phys.Ed., 22, 1529 (1984).
9. Michaels, A.S. and Miekka, R.G. J.Phys.Chem., 65, 1765 (1961).
10. Prud'homme, R.E. Polym.Eng.Sci., 22, 90 (1982).
11. Tremblay, C. and Prud'homme, R.E. J.Polym.Sci.Polym.Phys.Ed., 22, 1857 (1984).
12. Woo, E.M., Barlow, J.W. and Paul, D.R. J.Polym.Sci.Polym.Symp., 71, 137 (1984).
13. Roerdink, E. PhD Thesis, Groningen 1980.
14. Wellinghoff, S.T., Koenig, J.L. and Baer, E. J.Polym.Sci.Polym.Phys.Ed. 15, 1913 (1977).
15. Lu, F.J., Benedetti, E. and Hsu, S.L. Macromolecules, 16, 1525 (1983).
16. Garcia, D. J.Polym.Sci.Polym.Phys.Ed., 22, 107 (1984).
17. Garcia, D. J.Polym.Sci.Polym.Phys.Ed., 22. 1773 (1984).
18. Ventkatesh, G.M., Gilbert, R.D. and Fornes, R.E. Polymer, 26, 45 (1985).
19. Djordjevic, M.B. and Porter, R.S. Polym.Eng.Sci., 22, 1109 (1982).
20. Sulzberg, T. and Cotter, R.J. J.Polym.Sci. A1, 8 , 2747 (1970).
21. Ohno, N. and Kumanotani, J. Polym.J., 11, 947 (1979).

22. Schneider, H.A., Cantow, H.J. and Percec, V. Polym.Bull., 6, 617 (1982).

23. Stein, D.J., Jung, R.H., Illers, K.H. and Hendus, H. Angew.Makromol. Chem., 36, 89 (1974).

24. Alexandrovich, P.R., Karasz, F.E. and MacKnight, W.J. Polymer, 18, 1022 (1977).

25. Vukovic, R., Karasz, F.E. and MacKnight, W.J. Polymer, 24, 529 (1983).

26. Hammer, C.F. Macromolecules, 4, 69 (1971).

27. Zakrzewski, G.A. Polymer, 14, 347 (1973).

28. ten Brinke, G., Karasz, F.E. and MacKnight, W.J. Macromolecules, 16, 1827 (1983).

29. Kambour, R.P., Bendler, J.T. and Bopp, R.C. Macromolecules, 16, 753 (1983).

30. Paul, D.R. and Barlow, J.W. Polymer, 25, 487 (1984).

31. Cowie, J.M.G. and McEwen, I.J. J.Chem.Soc.Farad.Trans.I, 70, 171 (1974).

32. Koningsveld, R. and Kleintjens, L.A. Macromolecules, in press.

THERMAL AND MORPHOLOGICAL ANALYSIS OF POLY(ε-CAPROLACTAM)-POLY(ETHERESTER) MIXTURES

E.GATTIGLIA, E.PEDEMONTE and A.TURTURRO

Centro Studi Chimico-Fisici di Macromolecole Sintetiche e Naturali, CNR, Genova, Italy

SYNOPSIS

Results concerning mixtures of poly(ε-caprolactam) with a copoly(ether-ester) are reported. Thermal analysis data lead to the conclusion that the two components are not mixable but a transcrystallization process is induced by the previously crystallized polyamide spherulites on the phase transition of the copolymer.

1 INTRODUCTION

Poly(ε-caprolactam) is a relevant plastic material from the technological point of view; therefore it seems to be of general interest to study mixtures where polyamide 6 (PA6) is the main component. The improvement of the impact resistance of PA6, particularly at low temperatures, is at present a relevant goal to achieve;the addition of a thermoplastic copolymer seems to be a good approach to the solution of the problem.

Moreover the improvement of the knowledge on polymer-polymer mixibility in binary systems containing at least one crystalline component need more foundamental investigations.

In this communication some results concerning mixtures of PA6 with a copoly(etherester) (PEE) are reported.

2 EXPERIMENTAL

2.1 Materials

PA6 = Polyamide 6 Nivionplast 2.7 (Enichimica)

$\bar{M}_w = 68.9 \times 10^3$ $\bar{M}_n = 20.4 \times 10^3$

PEE = Poly(etherester) Hytrel 5556 (Dupont)

General formula $(A_n B_m)_z$

hard segments A = $-CO-C_6H_4-CO-O-(CH_2)_4-O-$

soft segments B = $-CO-C_6H_4-CO-O-|(CH_2)_4-O-|_x$

2.2 Specimen preparation and techniques

Blends have been prepared by dissolution of the two polymers in a common solvent (hexafluorobuthanol) and controlled evaporation at room temperature. Specimens have been dried under vacuum to constant weight and stored in desiccator over P_2O_5 until used.

The thermal behaviour has been investigated with a Perkin Elmer DSC 2 calorimeter equipped with the mod.3500 Data Station. The isothermal crystallizations have been carried out by melting the copolymer at 267°C for 5 mins and rapidly cooling down the specimen at the selected temperature.

The morphological observations have been carried out on thin sections by transmission electron microscopy.

3 RESULTS AND DISCUSSION

The thermograms obtained by isothermal crystallizations have been elaborated according to the well known Avrami equation (Table I).

Since the exothermic effect decreases by increasing the crystallization temperatures T_c, the Avrami constant K is more and more over-evaluated and the Avrami index n is more and more under-evaluated[1]. This "area effect" implies that K will decrease more rapidly and is able to explain the variation of n with T_c.

The rather low value of n measured for PEE can originate both from the broad crystallization curve and from the actual morphology of the copolymer[2].

PA6 follows the same kinetics both in the pure polymer and in the blends. Values of K and n have the same order of magnitude; small differences can be accounted for by the area effect. These results lead to the conclusion that PEE

Table 1 : Kinetic behaviour of PA6 and PEE

T_c, °C	ΔH_c cal/g	K	n	ΔH_c cal/g	K	n
	PURE PA6			PA6 IN PA6/PEE 75/25		
190	14.6	0.75	3.4	12.4	0.87	2.6
194	11.1	0.20	3.1	8.9	0.24	2.4
198	8.4	0.07	2.3	-	-	-
	PURE PEE			PEE IN PA6/PEE 25/75		
180	6.2	0.20	2.4	-	-	-
182	4.1	0.12	2.2	4.8	7.12	3.3
186	3.5	0.02	2.1	4.0	1.6	3.1

is not mixible with PA6 in the melt.

The crystallization of the PEE in PA6/PEE blends seems to follow a kinetics rather different from that one of the pure polymer. Since the heat of crystallization is higher, K is under evaluated more than in the pure polymer and n is more over evaluated. In conclusion n has probably the same value in all the systems investigated but K in the mixture is two order of magnitude higher. It means that the spherulites of the previously crystallized PA6 remarkably influence the crystallization of the copolymer; this lead to a faster crystallization and to a larger number of smaller and less regular aggregates.

These conclusions are supported both by the morphological observations and by the values of the melting temperatures.

In Table 2 the experimental values of T_m for the pure polymers and the single components of the mixtures are compared for the same crystallization temperatures T_c. The agreement is excellent and it can be inferred that the equilibrium melting temperatures both of PA6 and PEE are unaffected by the second component.

According to the Hoffman-Weeks treatment, the following results are obtained :

PA6 $T^\circ_m = 261°C$

PEE $T^\circ_m = 242°C$

Table 2 : Thermodynamic behaviour of PA6 and PEE

Polymer	T_c, °C	T_m, °C	
		Pure Polymer	Polymer in Blends
PA6	178	186.5	-
	182	190.0	-
	186	193.6	-
	188	-	195.4
	190	197.5	-
	193	-	197.6
	194	201.0	200.0
PEE	171	178.3	-
	174	181.1	180.9
	177	183.7	183.7
	180	186.5	186.2
	182	188.2	188.2
	184	190.0	190.0
	186	-	192.0

REFERENCES

1. Alfonso, G.C., Pedemonte, E., Re, M. and Turturro, A. Gaz.Chim.It., 112, 99, 1982

2. Pedemonte, E., Leva, M., Gattiglia, E. and Turturro, A. Polymer, in press

ISOCHRONE VISCOELASTIC FUNCTIONS VIA ACTIVATION ENERGY OF FLOW - CHARGE TRANSFER COMPATIBILIZED POLYBLENDS

M.-J. BREKNER, H.A. SCHNEIDER and H.-J. CANTOW

Institut für Makromolekulare Chemie der Universität Freiburg - Hermann-Staudinger Haus - D-7800 Freiburg, F.R.Germany

SYNOPSIS

It is shown that charge transfer interaction in polymer systems enlarges the rubber plateau and contributes to stabilization of incompatible polymer blends.

1 BACKGROUND TO THE ACTIVATION ENERGY OF FLOW

Based on Boltzmann's superposition principle viscoelastic functions are expressed in terms of the relaxation time distribution function - relaxation time spectrum - ,$H(\tau)$, which is normalized by the zero shear viscosity

$$\eta_o = \int_0^\infty H(\tau)d\tau = \int_0^\infty \tau H(\tau)d\ln\tau \quad . \tag{1}$$

Because of the time scale involved the logarithmic formalism is prefered.

The temperature invariance of the relaxation time spectrum is the main condition of the time-temperature superposition principle which governs the shift procedures for geting composite curves of viscoelastic functions. Taking into account the temperature function of the zero shear viscosity predicted by Eyring's transition state theory

$$\eta_o = A \exp(E/RT) \tag{2}$$

the temperature invariance of $H(\tau)$ implies identical temperature dependence for all relaxation times.

The shift factor of isotherms, $a_T=\tau(T)/\tau(T_o)$, can then be expressed:

$$\log a_T = (\log\tau - \log\tau_0) = (\log\omega_0 - \log\omega) = \frac{E}{2.3R}\left(\frac{1}{T} - \frac{1}{T_0}\right). \quad (3)$$

It is evident that a logarithmic time difference - $\log a_T$ - for isotherms, IT, is equivalent to a reciprocal temperature difference - $a_F = \Delta(T)^{-1}$ - for isochrones, IC[1,2]. This is demonstrated in Fig.1 for the storage modulus, G', in the terminal zone of an acceptor group containing poly(butylmethacrylate). Both the shifts are performed assuming constant activation energy as predicted by the Eyring model of viscous flow.

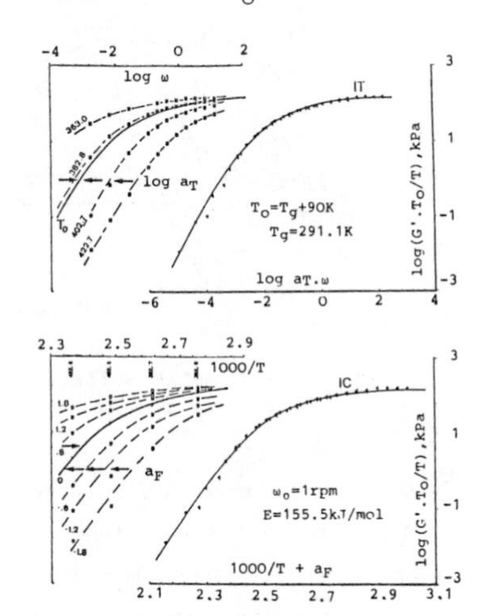

Fig.1 Composite curves of the storage modulus G' of PBMA-$A_{10\%}$

Starting with the observation that in accordance with (3) the activation energy of flow is given by the ratio of the two shift factors, $\log a_T/ a_F$, it can be easily shown (Fig.2) that this corresponds also to the slopes in a given point of the crossing isochrone and isotherm viscoelastic curves, respectively[3]

$$\frac{(\delta \log G/\delta(1/T))_\omega}{(\delta \log G/\delta \log\omega)_T} = \frac{E(T)}{2.3\ R} . \quad (4)$$

Temperature invariant activation energy is not imposed by relation (4), and taking into account that approaching Tg temperature dependent activation energy is predicted by the WLF relation

$$E(T) = 2.3RT_0 c_1^0 c_2^0 T^2/(c_2^0+T-T_0)^2 \quad (5)$$

equation (4) is formulated correspondingly.

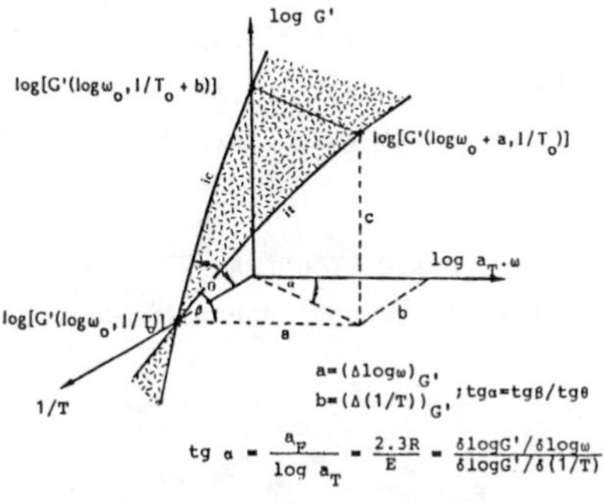

Fig.2 Interrelation between IT and IC curves of viscoelastic functions

In the terminal zone of constant activation energy a smooth surface of the viscoelastic functions in the log frequency - reciprocal temperature space is obtained[3]. Approaching Tg the surfaces become wavy as evidenced in Fig.3 for a head-to-head PVC of M_W = 30,000.

Although the WLF relation predicts, in accordance with (5), a steady in-

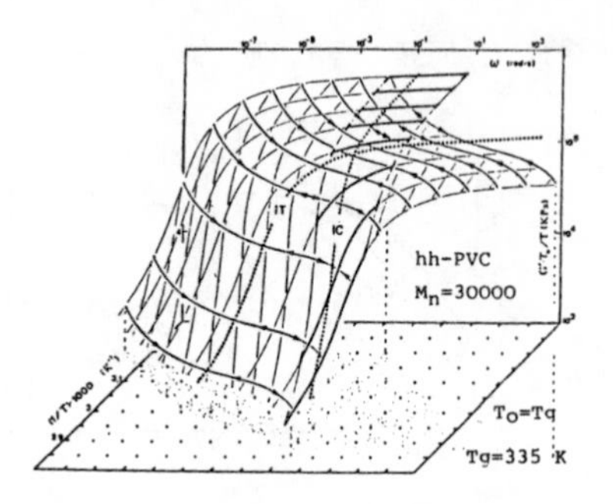

Fig.3 Viscoelastic surface of hh-PVC in the Tg range

crease only of the activation energy, a decrease of E(T) is observed near Tg[2]. This is in accordance with suggestions of failure of the WLF relation very near Tg[4].

As it has been shown, E(T) relates also the relaxation time spectrum with a corresponding reciprocal relaxation - temperature spectrum[5].

2 RESULTS AND DISCUSSION

2.1 Compatible blends

Data concerning the rheological behaviour of a compatible polyblend, illustrated for the storage modulus, are shown in Fig.4 for a 1:3 w/w blend of Polystyrene (PS, M_w=75,000, Tg=373K) and Poly(vinylmethylether) (PVME, M_w=73,000, Tg=237K). It is demonstrating that the isochrone composite curve of the blend is situated in the temperature range between the two components. Analogous the isotherm composite curve is situated in a frequency range between the components, with a tendency towards PVME.

If the IT composite curves are shifted to contact at the same G_N-value, (Fig.5), the flater slope of the blend suggests an increase of the polydispersity of the relaxation process in the compatible blend. The respective characteristic $\log(J_e^o G_N)$ values are 0.02 for the PS, 1.27 for the PVME and 2.86 for the blend.

Charge transfer (CT) - interaction of electron donor (D) and electron acceptor (A) groups generally extends the rubber plateau in compatible blends[6].

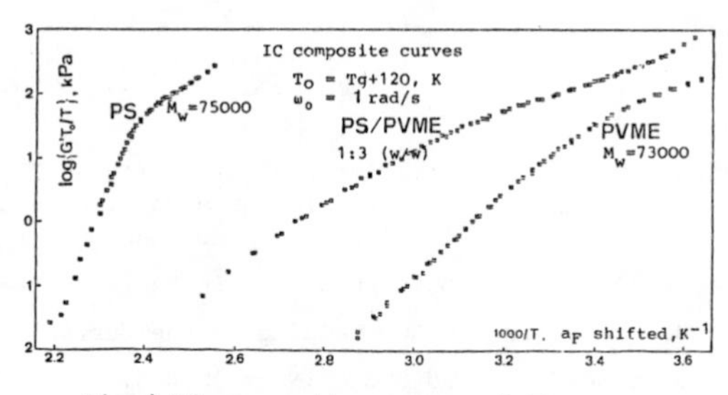

Fig.4 IC composite curves of the compatible PS/PVME blend

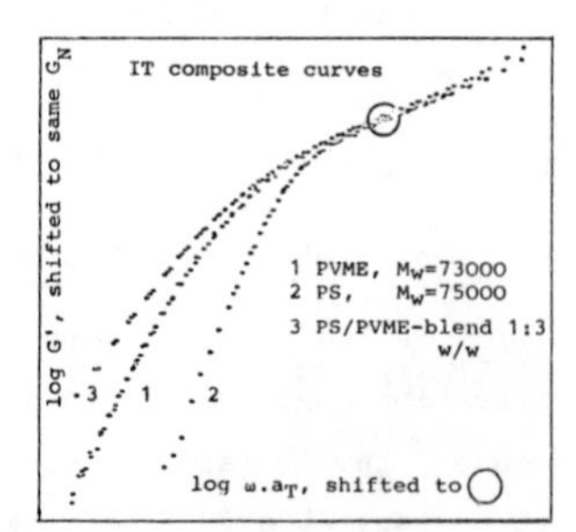

Fig.5 IT composite curves of the PS/PVME system shifted to contact at same G_N-value.

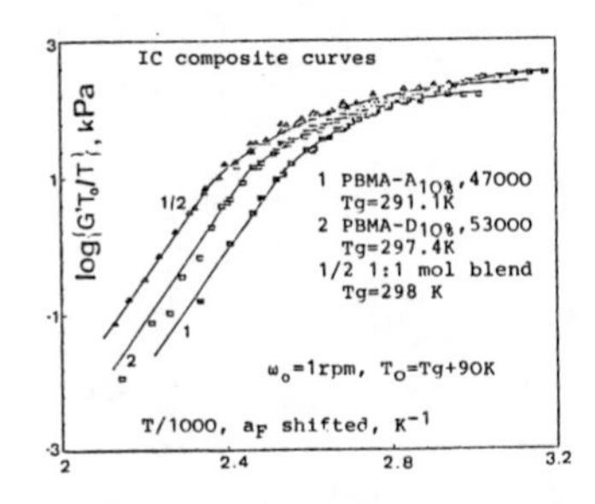

Fig.6 IC composite curves of the PBMA-system in the terminal zone

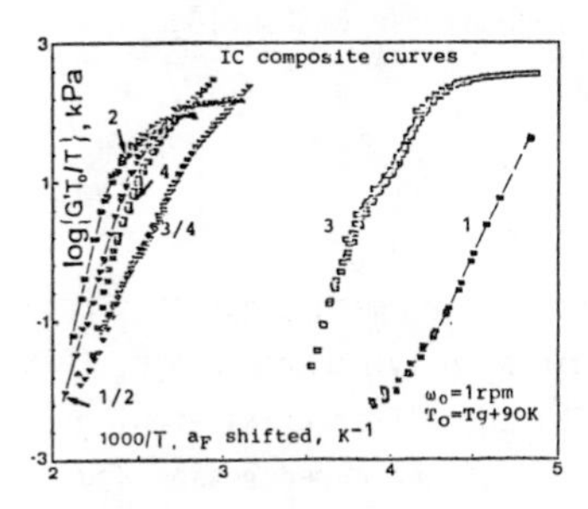

Fig.7 IC composite curves of the incompatible PBMA/PDMS system

For illustration data of the 1:1 mole blend of random copolymers of Poly(butylmethacrylate) - PBMA - with 10% mole of D 2-(3,5-dinitrobenzoyl) ethyl methacrylate and A 2-(9-carbazolyl)ethyl-methacrylate, respectively, are presented in Fig.6.

2.2 CT-compatibilization

CT-interaction may be applied also for stabilization of incompatible polyblends. These is demonstrated for both the blends of PBMA with Poly(methylmethacrylate) -PMMA- and Poly-(dimethylsiloxane) -PDMS[7]. Comparison of the rheological behaviour of incompatible poly-blends of random and blockcopolymers containing D and A groups respectively, suggests a stronger CT-effect in the blend of random copolymers as seen in Fig.7. The notations stand for:
1- random PDMS-$A_{10\%}$, M_n=7,500, Tg=157.5K,
2- PBMA-$D_{10\%}$, M_w=53,000, Tg=291.1K, 3- $\frac{A}{A}$PDMS$\frac{A}{A}$ a PDMS, M_w=30,000 with two A-endgroups on both the chainends and 4- a triblock $(D)_{15}$-PBMA-$(D)_{15}$ M_w of the PBMA of about 50,000. Both the blends 1/2 and 3/4 are of 1:1 A/D-ratio.

ACKNOWLEDGEMENTS

Financial support by AIF is gratefully acknowledged.

REFERENCES

1. Schneider, H.A. and Cantow, H.-J., Polymer Bull., 9 (1983) 361-368.
2. Schneider, H.A., Cantow, H.-J. and Brekner, M.-J. Polymer Bull., 11 (1984) 383-390.
3. Brekner, M.-J., Cantow, H.-J. and Schneider H.A., Polymer Bull., 10 (1983) 328-335.
4. Schwarzl, F.R. and Zahradnik, F., Rheol. Acta 19 (1980) 137-152.
5. Brekner,M.-J., Cantow, H.-J. and Schneider H.A., Polymer Bull., 13(1985) 51-56
6. Schneider,H.A., Cantow, H.-J. and Percec V., Polymer Prepr., 23 (1982) 203-204
7. Schneider, H.A., Cantow, H.-J., Lutz, Pierre and Northfleet Neto, H., Makromol. Chem., Suppl.8 (1984) 89-100

MODIFICATION OF THERMOSETTING RESINS BY THERMOPLASTICS

C. B. BUCKNALL[*], P. DAVIES[+] and I. K. PARTRIDGE[*]

[*]School of Industrial Science, Cranfield Institute of Technology,
Cranfield, Bedford, MK43 OAL, U.K.
+Division Polymères et Composites, Université de Technologie de Compiègne,
60206 Compiègne, France.

SYNOPSIS

Two systems are considered: a styrenated polyester resin containing added poly(vinyl acetate) (PVA), and epoxy resins containing poly(ethersulphone) (PES). A combination of scanning electron microscopy (SEM) and differential etching procedures was employed to examine fracture surfaces of the cured resins and to determine the composition of phase-separated morphological features. The effects of morphology upon physical properties are discussed.

1 INTRODUCTION

Addition of thermoplastics such as PVA, poly(caprolactones), polystyrene and acrylics to unsaturated polyesters has produced extremely beneficial results; the high shrinkage of unmodified polyesters during cure, typically 8% by volume, may be reduced or eliminated when thermoplastic in the form of a 'low profile additive' is included. The commercial importance of unsaturated polyesters in sheet and dough moulding compounds and their potential for automobile applications has stimulated much research into the mechanisms by which shrinkage is controlled.

In the late 1960's, Bartkus and Kroekel[1] were among the first to follow the cure of acrylic-modified polyesters quantitatively. They suggested that boiling of liquid monomer, differential thermal expansion between phases and migration of monomer to leave voids could explain shrinkage reduction. The acrylic-modified system is initially two-phase whereas PVA is initially compatible with many polyesters. The mechanisms acting are therefore not necessarily the same in the two systems. Atkins et al.[2] have suggested that the separation of a thermoplastic phase, which expands more than the matrix, is sufficient to impede matrix shrinkage in the PVA modified system. Other combinations of these mechanisms have also been proposed[3]. It is clear however that the morphology produced during cure will be critical in determining the shrinkage observed.

Thermoplastics are added to epoxy resins for other reasons, such as ease of processing and improved impact resistance. For example, PES is included in certain epoxy resin formulations to improve flow characteristics during preparation of carbon fibre-epoxy laminates. The different morphologies observed in epoxy resins modified with reactive rubbers are known to affect the mechanical properties of these blends and particularly, their fracture behaviour; the present work describes the structures observed in cured epoxy-PES blends.

2 EXPERIMENTAL

The polyester system was based on 1 mol. maleic anhydride, 1 mol. isophthalic acid, 1.5 mol. propylene glycol and 0.6 mol. diethylene glycol. This resin was dissolved in styrene monomer, 60 parts resin to 40 parts styrene. The low profile additive was a solution of 40% by wt. of acrylic modified PVA in styrene. Resin, low profile additive and benzoyl peroxide initiator were thoroughly mixed and cured in 30g batches at 130°C for 4h. Further details are available elsewhere[4].

The epoxy system was a mixture of trifunctional and tetrafunctional glycidyl amine resins. The resin blend was modified by the addition of low molecular weight PES and cured using dicyanodiamide (Dicy) and a stepwise cure regime involving temperatures rising from 93°C to 200°C[5].

Fracture surfaces were obtained by impact and coated with a thin layer of Au/Pd alloy before examination by SEM.

3. RESULTS AND DISCUSSION

Although the thermoplastic additive was initially compatible with the polyester-styrene solution, the mixture became cloudy on heating. The opacity of cured modified polyesters has been noted by several authors[3,6] and may be explained in terms of a separation of phases. A typical fracture surface of a resin containing 8% by wt. of PVA is shown in Fig. 1 (a). Clusters of spheres are visible and appear to be covered by a coating. Etching with methyl ethyl ketone (MEK) removed this coating to leave the structure shown in Fig. 1 (b). The amount and solubility of the coating indicate that it is largely composed of PVA.

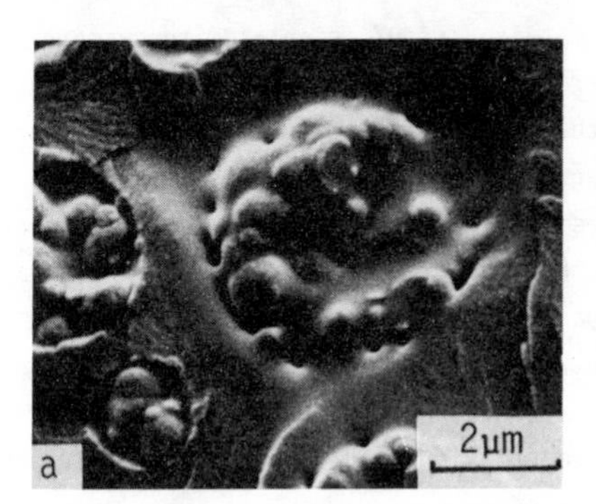

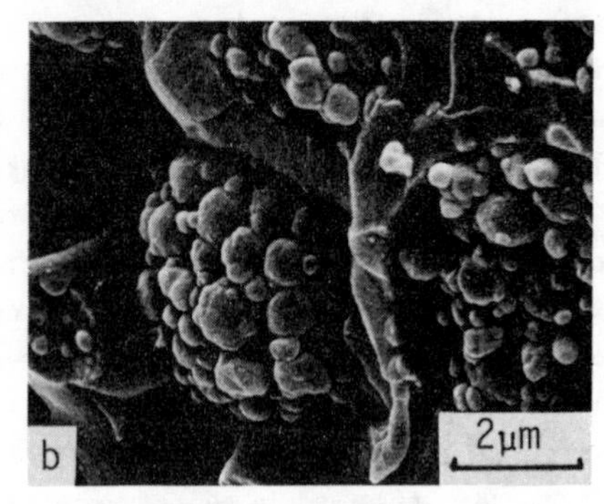

Fig. 1 Polyester containing 8% by wt. PVA (a) unetched, (b) etched.

Increasing the PVA concentration to 16% by wt. results in a phase inversion and a structure after etching with MEK as shown in Fig. 2. A similar structure, shown in Fig. 3, was obtained by curing a blend of 80 pbw of trifunctional epoxy, 20 pbw of tetrafunctional epoxy and 31 pbw of PES with 5 pbw Dicy.

Fig. 2 Polyester containing 16% by wt. PVA, etched.

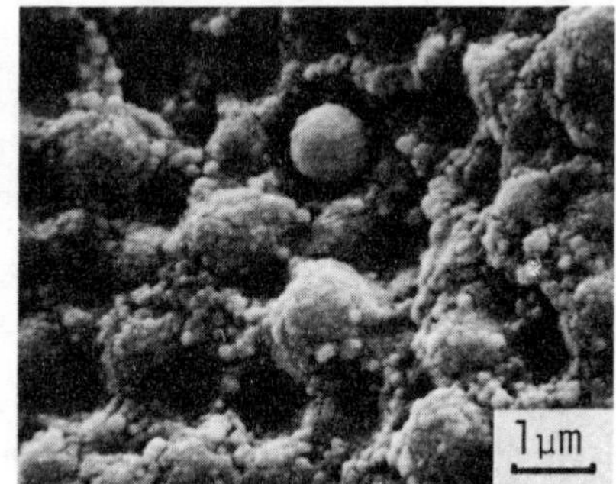

Fig. 3 80 Epoxy III/20 Epoxy IV/ /5 Dicy/31 PES

However, a combination of X-ray microanalysis and differential etching techniques showed that the location of the thermoplastic component is entirely different for these two systems. For the epoxy-PES system, spot X-ray microanalysis for sulphur has revealed the 1 μm nodules to be areas of high PES concentration[5]. Volume fraction considerations coupled with the insolubility of the nodules in methylene chloride indicate that epoxy must also be present in the nodules. For the 16% PVA modified polyester system on the other hand, the solubility in MEK of some of the interconnecting polymer matrix reveals the location of the thermoplastic between the spherical particles.

The type of structure shown in Fig. 2 was also observed by Pattison et al.[6], who detected a coating on small (0.2 to 1μm) beads fused together in a different polyester system modified by 12% by wt. PVA. This led these authors to introduce a fourth possible shrinkage compensation mechanism in addition to those listed earlier: namely, the relief of shrinkage stresses built up during cure by the propagation of cracks through the weak thermoplastic matrix.

Theories incorporating the various mechanisms have been proposed to explain the shrinkage behaviour of a wide range of thermoplastic modified polyester systems. Interpretation of the available experimental data has been hindered by variations in polyester composition and variations in additive type and concentration. Phase separation has been frequently observed. The structure seen here at low (8% by wt.) PVA concentration is similar to that in polyester systems modified with polystyrene or with acrylic polymer and these additives have been shown to be less effective in controlling shrinkage than PVA[2]. Most commercial resin formulations include PVA at higher concentrations and at these levels the structure observed is that shown in Fig. 2[2,6]. A modifier level sufficient for the formation of a continuous thermoplastic matrix would therefore seem desirable. Possible reasons for a phase inversion occurring in these systems at additive concentrations between 8 and 16% by wt. are proposed in ref. 4. While these considerations do not clarify the question of which shrinkage control mechanism is acting, they do indicate how critical the resin formulation may be.

References

1. Bartkus,E.J. and Kroekel,C.H. "Low shrink reinforced polyester systems",Appl.Polym.Symp. 15 ,113,1970

2. Atkins,K.E.,Koleske,J.V.,Smith,P.L.,Walter,E.R. and Matthews,V.E. "Mechanism of low profile behaviour",31st Annual Techn.Conf.Reinf. Plastics/Composites,2-E,1976

3. Siegmann,A.Narkis,M.,Kost,J.DiBenedetto,A.T."Mechanism of low profile behaviour in unsaturated polyester systems".Int.J.Polymeric Mater.,6,217-31 , 1978

4. Bucknall,C.B.,Davies,P.Partridge,I.K. "Phase separation in styrenated polyester resin containing a poly(vinyl acetate) low-profile additive",Polymer, 26 ,109-12,1985

5. Bucknall,C.B. and Partridge,I.K. "Phase separation in epoxy containing polyethersulphone",Polymer, 24 , 639-44,1983

6. Pattison,V.A.,Hindersinn,R.R.,Schwartz,W.T. "Mechanism of low-profile behaviour in single-phase unsaturated polyester systems",J.Appl. Polymer Sci. , 19 , 3045-50 , 1975

THE TOUGHNESS BEHAVIOR OF EMULSION ABS. EFFECT OF RUBBER CONCENTRATION AND ACRYLONITRILE CONTENT ON THE DEFORMATION MODES

D. MAES[a], G. GROENINCKX[a], N. ALLE[b] and J. RAVENSTIJN[b]

(a) University of Leuven, Laboratory of Macromolecular Structure Chemistry, Belgium
(b) DOW Chemical B.V., 4530 AA Terneuzen, Nederland

SYNOPSIS

The tensile stress-strain properties of emulsion ABS with different rubber concentrations (4 to 20 wt.%) and different acrylonitrile contents (24 to 33 wt. % AN) have been investigated. It has been observed that the toughness exhibits a specific pattern as function of the strain rate as well as a function of the temperature. A peak is observed for the toughness which shifts to higher strain rates if the rubber concentration, the acrylonitrile content or the temperature is increased.

1 INTRODUCTION

ABS is a toughened polymer consisting of a rubbery dispersed phase in a glassy matrix. The dispersed particles in emulsion ABS are relatively small (0.1 μm). The matrix is a statistical copolymer of styrene and acrylonitrile. Under loading the material deforms by voiding of the rubber particles and by crazing and/or shear yielding of the matrix.

2 MATERIAL AND EXPERIMENTAL PROCEDURES

The test specimens were obtained by melt blending and solvent blending, respectively, of concentrated emulsion particles (rubber particles size = 0.1 μm, 25 wt.% AN, 40 wt.% rubber) and SAN materials. Thin test specimens (films) were made by compression moulding to a thickness of 0.2 ± .06 mm at 200°C. Tensile tests were carried out with an Instron 1121 apparatus.

3 EXPERIMENTAL RESULTS AND DISCUSSION

In fig.1 the influence of the rubber concentration is shown. A toughness peak is observed which shifts to higher strain rates; at constant strain rate, the toughness increases with increasing rubber concentration (0-10 wt.%). For higher rubber concentrations (16-20 wt.%) one can expect the toughness peak under impact rate conditions (fig.2).

The influence of the strain rate and the temperature (tensile test conditions) on the toughness is presented in a three-dimensional plot in fig.3. The toughness decreases with increasing strain rate and decreasing temperature and a peak can be observed which shifts to higher strain rates with increasing temperature. From fig.4 it can be seen that the toughness peak increases with increasing acrylonitrile content; at the same time, the peak shifts to higher strain rates.

3.1 Influence of rubber concentration on the toughness

By increasing the rubber concentration at constant particle size, more particles will be available to initiate crazes and/or shear bands, and as a consequence the local strain rate for each deformation mode (crazes or shear bands) will be smaller. If many rubber particles induce crazes, a high amount of crazes will contribute to the total specimen deformation so that the local strain rate for one particle or for one craze will be small. This means that the local strain rate will determine the craze propagation stress and the stability of the craze. In the case of such behaviour, the toughness-strain rate curve has to shift to higher strain rates when more rubber particles are present. The propagation stress of crazes is less strain rate dependent than that of shear bands; this can be explained by the higher activation volume for crazing than for shear deformation ($V = 5.0\ nm^3$ for crazing in PS (1) compared with $4.6\ nm^3$ for shear during compression of PS (2). From the Eyring equation, it can be deduced that the propagation stress will be more influenced by the strain rate if the activation volume is low.

Within the strain rate range studied, the two deformation modes play a role and under certain conditions we can expect crazing and shear deformation appearing together. As is known from the literature, the interaction between crazes and shear bands in ABS materials appears to be relatively weak, probably on account of the fact that shear deformation is of the diffuse type (3). Due to the absence of interaction, both mechanisms contribute to the deformation so that the local strain rate will be lower. With regard to crazing, Bucknall has shown that there is a lower strain rate at decreasing craze propagation stress (4). If the local strain rate decreases, the local craze propagation stress will decrease accordingly and will become much lower than the critical stress of craze breakdown.

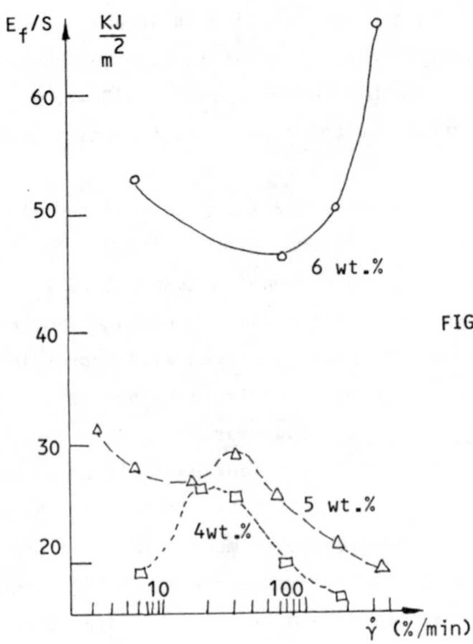

FIG.1. INFLUENCE OF RUBBER CONCENTRATION ON THE TOUGHNESS.
AN-CONTENT = 28 wt.%
$\bar{M}_W$ = 140 000

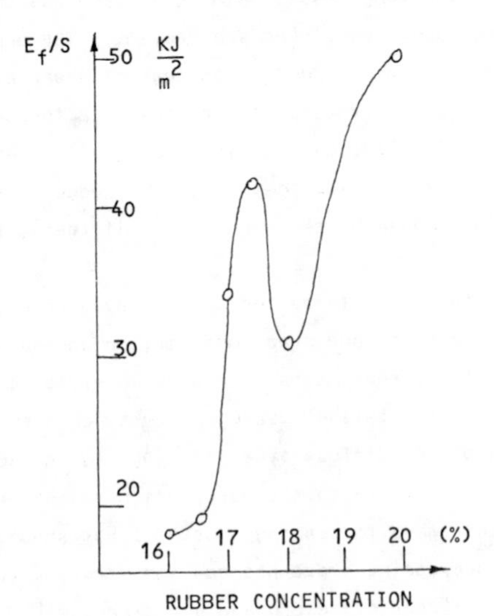

FIG.2. THE IMPACT TOUGHNESS MEASURED IN TRACTION.
AN-CONTENT : 28 wt.%
$\bar{M}_W$ = 140 000

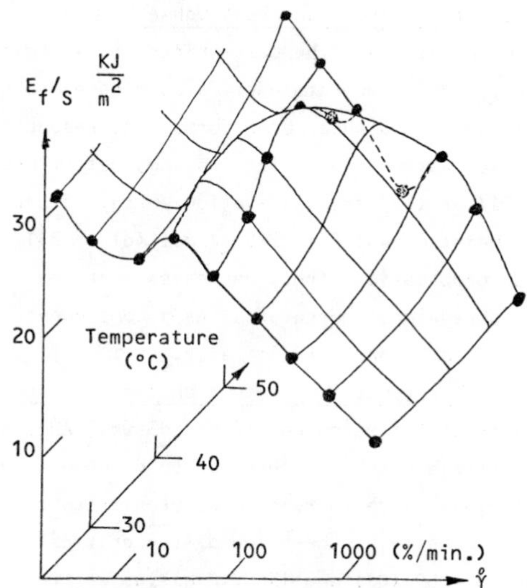

FIG.3. INFLUENCE OF TEST CONDITIONS ON THE TOUGHNESS. RUBBER CONC.: 5 wt.%; AN-CONTENT = 28 wt.%; $\bar{M}_W$ = 140 000.

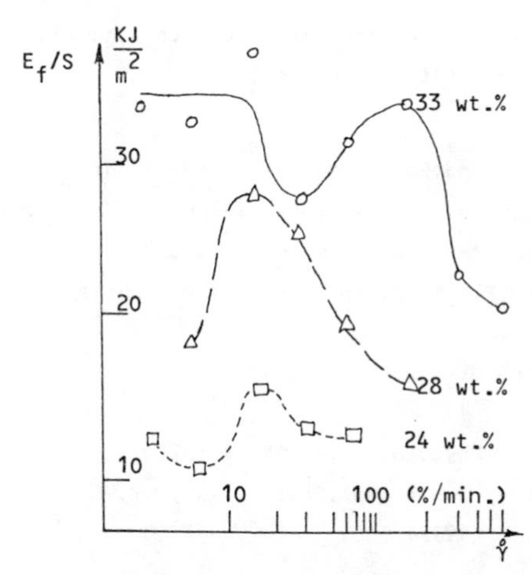

FIG.4. INFLUENCE OF ACRYLONITRILE CONTENT ON THE TOUGHNESS. RUBBER CONC.: 4 wt.% $\bar{M}_W$ = 140 000

3.2 The influence of the temperature on the toughness

From fig.3, it can be seen that the peak shifts to higher strain rates as the temperature increases. This temperature effect can be explained using the activation energies for crazing and shear band formation, respectively. In the case of polystyrene, the activation energy for crazing, determined using the Eyring equation, is 175 kJ/mol (5); this activation energy is lower than that obtained for shear band formation which is 270 kJ/mol (6) or 251 kJ/mol (7). For that reason the shear propagation stress decreases more than the craze propagation stress with increasing temperature. As a consequence, the two mechanisms will act together at higher strain rates.

3.3. The influence of the acrylonitrile content on the toughness

With increasing acrylonitrile content (24 to 33 wt.% AN), a shift of the toughness peak to higher strain rates was observed together with an increase of the absolute value (fig.4). With increasing acrylonitrile content, an increase of the entanglement density has been found (the critical molecular weight M_{cr} for SAN being between 35 000 (PS) and 1300 (PAN)),as well as a decrease of the amount of chains crossing a unit-surface. As a consequence, it will be more difficult to create voids in the SAN matrix in the case of a high acrylonitrile content because a higher number of chain scissions has to occur. Hence, the craze propagation stress will increase more than the shear propagation stress and this will lead to a shift of the toughness peak to higher strain rates as a function of increasing acrylonitrile content.

REFERENCES

1. B. Maxwell and L.F. Rahn, Ind.Eng.Chem., 41 (1949) 1988.
2. E.M.R. Andrews and P.E. Reed, In : Advances in polymer science, vol.27 (1978) Ed. by J.D. Ferry, Springer-Verlag, New York, p.1.
3. C.B. Bucknall, Toughened Plastics, Applied Science Publishers Ltd., London, (1977) p. 214.
4. see ref.3, p. 214, figure 8.2.
5. see ref.3, p. 164.
6. E.J. Kramer, J.Poly.Sci. (Phys.Ed.), 19 (1975) 509.
7. T.E. Brady and G.S.Y. Yeh, J.Macromol.Sci., Phys., B9, 659.
8. P.I. Vincent, Polymer, 13 (1972) 558.

(Approval number : B 85 132)

Part IV

NETWORKS

THERMODYNAMICS OF CASEIN GELS AND THE UNIVERSALITY OF NETWORK THEORIES

Manfred Gordon

Statistical Laboratory, University of Cambridge

SYNOPSIS

Quantitative fittings of literature data on the equilibrium moduli of weak aqueous casein gels to four variants of a classical polycondensation model are compared with the fitting of the same data by their originators to the modern scaling model. By no means unexpectedly, all five fittings are statistically almost equivalently good, though the differences between the classical and modern approaches are profound. Scaling assumes the dominance of a 3-dimensional excluded-volume effect, while the dimension-invariant classical theory relies on this effect cancelling out. The methodology for making progress in this situation depends on the refinement of the initial models by the interplay of theory and experiment in the light of statistical analysis.

1. INTRODUCTION

The statistical theory of rubber elasticity, is "one of the simplest and most successful theories in polymer science"[1]. On the basis of a few rather straightforward measurements, it can predict elastic and other physical properties of vulcanisates and other polymer networks, sometimes without adjustment of parameters. Given a suitable structural model, it can deal with systems of great chemical and even biochemical complexity, so that recently the theory of rubber elasticity has found fruitful applications in Biorheology[2]. Using elastic measurements on weak casein gels by Niki et al.[3] as an example, the status of rival approaches to the theory of rubber elasticity is reviewed, with due emphasis on the refinement of structural models by the interplay of theory and experiments.

The two competing theoretical approaches are:

1] CLASSICAL MEAN-FIELD NETWORK THEORY OF PHANTOM CHAINS (e.g. Kuhn, Flory, Guth and James, for references see Treloar[4]).

2] MODERN SCALING THEORY, BASED ON LATTICE OR FRACTAL PERCOLATION AND RENORMALISATION GROUP METHODS (De Gennes[5], Stanley et al.[6], Stauffer et al[7]).

As in a previous review[8], we show again that measurements of modulus against degree of cross-linking can be fitted to both types of theory. This does not mean that both types of theory are usefully confirmed by experiment. Scaling theory makes the excluded volume effect dominate the shape of such curves, while the classical phantom chain treatment implies that the excluded volume effect cancels out. Both claims cannot be fruitful: if the theory is to be refined further, dominant effects must be attended to first. Scaling models can be made to fit essentially all experiments. A proper statistical basis for the validation of abstract models by physical experiments is indispensable.

2. MODELS FOR CASEIN GELS

Niki et al.[3] developed advanced techniques and equipment for measuring the shear moduli of weak gels G as a function of concentration Ø and time. In 1982 they fitted preliminary measurements[9] to the classical phantom-chain formalism; in 1984 they had become persuaded to fit instead a scaling law:

$$G = 17.3 * [(Ø/Ø_c)-1]^{1.89} \text{ dyn/cm}^{-2} \qquad (1)$$

as the continuous curve in fig. 1 to their their new equilibrium measurements[3] (the crosses). The left-hand side shows the early points close to the critical gel concentration (arrow) on a scale expanded 3.5 times (hence the large experimental scatter). While the fit itself is quite satisfactory, four variants of the classical phantom-chain model are shown below to fit at least equally well! Two of them are shown, as the circles and rhombs. The variants arise from a generalisable classical working model (see, e.g. Gordon & Parker[10]), introduced by Parker into the field of the gelation of aqueous casein, including the coagulation of milk, and worked out by him and Dalgleish[11]. Their postulates are summarised as follows:

(i) Ca^{++} binds to casein and creates reactive units acting as 'monomers' for random f-functional polycondensation (RFFP) or VULCANISATION.

(ii) The functionalities thus created on these 'monomers' interact in pairs randomly to form links,

(iii) until an equilibrium is established between reacted and unreacted functionalities.

Six possible structural variants of this general model are generated as follows: first, the functionalities created by Ca^{++} may be located on a more or less branched star-like 'monomer' structure, or along linear chain-'monomers'. The random pairwise linking under (ii) of functionalities borne by these 'monomers' then constitutes respectively

(A) Random f-functional polycondensation (RFFP), or

(B) a classical vulcanisation reaction

Secondly, each of these two classes may be subdivided into three cases, by tentatively identifying the casein 'momomer' units with one of the three kinds of particles characterised by previous workers in aqueous casein solutions:

(a) The original huge and notoriously heterodisperse casein micelles of estimated weight-average molar mass[3] $M_w = 0.2$ to $1.8*10^6$ kg/mol.

(b) The casein submicelles[12] of $M_w = 300 \pm 60$, reasonably taken as homodisperse, $M_n=M_w$, which are easily set free from the micelles by dialysis.

(c) The mixture of primary polypeptide casein chains of approx. $M_n=M_w=20$ of which the submicelles are made up.

The case A-c is eliminated because the primary polypeptide chains are known to be linear in structure, so that RFFP is not applicable to them. The case B-a is eliminated since it is hardly conceivable that the huge micelles carry their

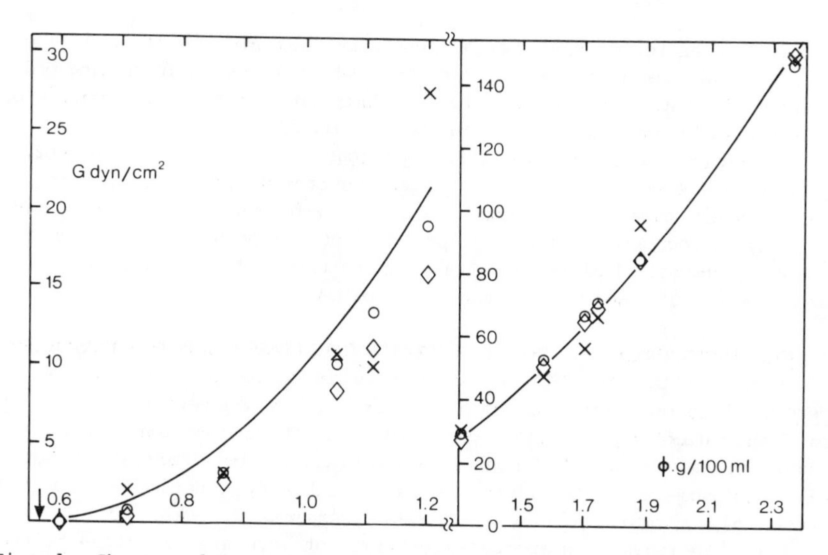

Fig. 1 Shear modulus of casein gels vs. concentration. Curve drawn: percolation law eq. 1 with arbitrary parameter value 17.3 dyn/cm^2. Crosses: experiments[3], first cross after gel point (arrow) omitted for clarity, it coincides with circle and rhomb. Rhombs: classical RFFP-model A-b for submicelles; circles: classical vulcanisation model B-b for submicelles (see below).

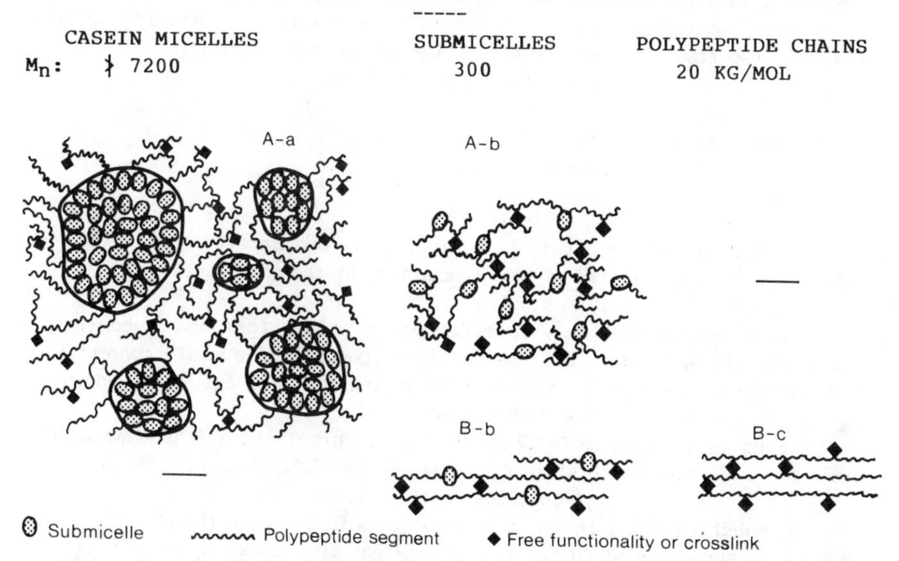

Fig 2. Four classical model variants for aqueous casein gels.

functionalities for cross-linking along flexible chain segments in a linear array. Thus only the four variants A-a, A-b, B-b, and B-c, in increasing order of extent of denaturation, need to be considered. These structural variants of the Dalgleish-Parker model are schematized in fig. 2.

The cross-linking reaction (ii) is triggered by addition of the enzyme chymosin[3]. The extent of the denaturation occuring during cross-linking by chymosin is not known; and the models A-a, A-b, B-b, and B-c implicitly arise from assuming increasing extents of unfolding of the primary micellar structure. The mathematical theory for the shear modulus of these model systems is over 20 years old[13], and briefly summarised below.

3. STATISTICAL THEORY FOR THE SHEAR MODULUS G IN CLASSICAL RUBBER ELASTICITY

To adapt either the RFFP or the chain vulcanisation model to the case of reversible bond formation (postulate (iii) above), requires merely the calculation of the parameter p, the fractional conversion of the functionalities, from an equilibrium constant K of bond scission (= 1/K for bond formation). Moreover, K can subsequently be eliminated, as first done by J. Hermans[14], provided the critical conversion p_c (gel point) can be measured.

His solution was in approximate form, not accurate too close to the critical point, and restricted to the mean functionality $f_m \gg 1$, i.e. $p_c \ll 1$ (eq. 4 below). This Hermans model was therefore a typical weak-binding model for a gel ($K \gg 1$). It has been fitted successfully to gels of agar and gelatin[2]. Strong-binding models, with p_c near unity, f_m just above 2, and $K \ll 1$, have frequently been favoured for gels from denatured globular proteins (see later).

The dissociation equilibrium constant for a link (cf. Ostwald's dilution law) into two free sites is given by and equation solvable for p:

$$K = [(1-p)^2/p]\emptyset = y\ \emptyset, \text{ say} \tag{2}$$

$$p = [2 + y - (y(y+4))^{1/2}]/2 \tag{3}$$

The critical value being given by Flory's theory[15]:

$$p_c = 1/(f-1), \tag{4}$$

one finds:

$$y_c = (f-2)^2/((f-1). \tag{5}$$

To fit the data by Niki et al. in fig. 1, one determines K from their measurement of the critical volume fraction $\emptyset_c$ at the gel point:

$$\emptyset_c = 0.0056_9 \tag{6}$$

(In model variant B-c, specific polypeptides form the 'monomers'. We shall simplify by treating all the casein polypeptides of overall concentration $\emptyset$ in the mixture as equally active in crosslinking). From eqs. 2, 5, and 6,

$$K = 0.0056_9(f-2)^2/(f-1) \tag{7}$$

Using as an example f=2.2 as in fig. 1, this gives a dissociation constant K = 0.17, i.e. quite strong binding (1/K = 5.9). From (2), (6) and (7):

$$y = 0.0056_9(f-2)^2/(f-1)\emptyset. \tag{8}$$

By substituting y in (3), p is known as a function of $\emptyset$ and f. Then p can in turn be substituted in the formulae[13] for the number N_e of elastically active network chains (EANCs), either per repeat ('momomer') unit in f-functional random polycondensation (RFFP) or per primary chain in vulcanisation:

(A) RFFP $N_e = fp(1-v)^2(1-B)/2$ (9)

with $B = (f-1)pv/(1-p+pv)$ (10)

and with the extinction probability:

$$v = (1-p+pv)^{f-1} \quad (11)$$

Equation 11, solved iteratively for given p and f in seconds on a Sirius microcomputer, applies equally to case B, where simply eq. (12) replaces (9):

(B) VULCANISATION OF UNIFORM CHAINS

$$N_e = [(f-2)pv+fp-2][fp-(2f-1)pv+1]/fp \quad (12)$$

The shear modulus is inversely proportional to M_n of the 'monomer' repeat (A) or primary chain (B), and given, for comparison with experiment by

$$G = FRTN_e\emptyset/M_n \quad (13)$$

The front factor F is now accepted[16] to be unity, independently of f, for low-strain measurements like those in fig. 1. This agrees satisfactorily with measurements on well characterised covalently bonded gels[17].

4. TESTING 5 MODELS AGAINST DATA[3] FOR CASEIN GELS

The fittings of the four classical models (A-a, A-b, B-b and B-c) defined above, and of the modern scaling equation (1) used by Niki et al.[3], are now tested against their data (cf. fig 1, and table 1). The goodness of fit is expressed by the standard deviation s of the data points in dyn/cm^2 from the calculated value of the shear modulus G. Weighting individual deviations, e.g. through division by the measured G, would not affect the conclusions.

All five models fit the modulus/concentration data (fig. 1) almost equally well, with a standard deviation s just above 5 dyn/cm^2 (last column, table 1). That represents an experimental scatter of about 3% of the maximum modulus found at the highest concentration measured, which prevents discrimination between the four classical models and the modern percolation model. However, the quantitative fits of the four classical variants to the modulus impose different testable constraints. Thus for variant A-a (random f-functional condensation of the micelles), the average number of functionalities f_m per micelle must be large ($f_m \geq 1000$; column 5), and the binding between these would necessarily be very weak (column 2), since a minute conversion p would produce a gel. The other three variants require strong binding to fit the data, with a mean number of functionalities per 'monomer' unit f_m only just over 2, so as to produce a gel. This finding is supported by the biochemical background information on casein precipitation by Ca^{++} and gelling of serum albumin, for which RFFP models have successfully been fitted by previous workers, as outlined in section 5. The model A-b, submicellar polycondensation, is therefore the most likely, and gives marginally the best fit to the data (col. 6).

Turning to the modern percolation model (the curve in fig. 1), the fit to the data is not really significantly worse (table 1, col. 6). In fact, such

MODEL	RANGE OF p implied by data (eq.3,8)	N_e(max)	M_n kg/mol	f_m	Standard Dev. s
A-a RFFP (micelles) (eqs. 9-11,13, F=1)	0.001 - 0.005	1.9	$\geq$7200	$\geq$1000	5.08
A-b RFFP (submicelles) (eqs. 9-11,13, F=1)	0.83 - 0.91	0.0788	300 ±60 ref[12]	2.209 ±0.045	5.005 ±0.004
B-b VULC. (submicelles) (eqs. 11-13, F=1)	0.83 - 0.91	0.0722	300 ±60 ref[12]	2.247 ±0.055	5.147 ±0.002
B-c VULC. (polypeptide) (eqs. 11-13, F=1)	0.985 - 0.993	0.00488	20	2.015	5.14
PERCOLATION eq. 1	___	___	___	___	5.18

Table 1. Fitting various model variants to data[3] on casein gels.

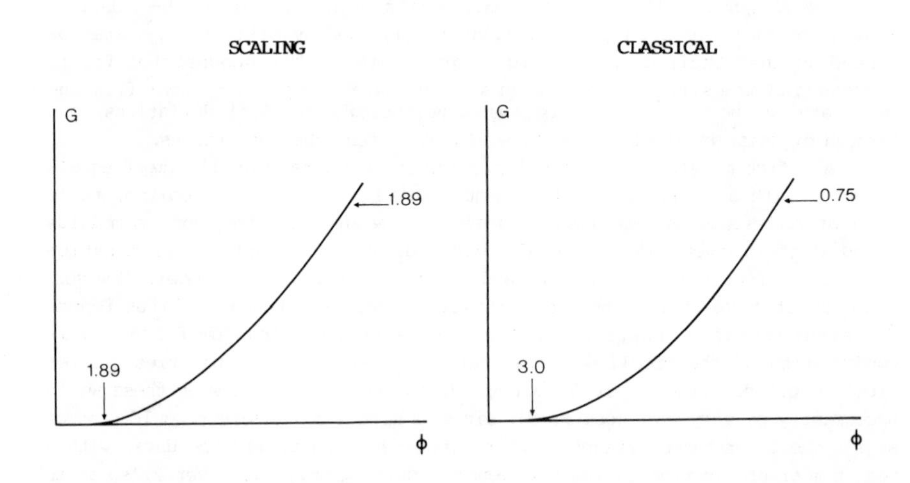

Fig. 3. Schematic modulus/concentration plots: scaling (eq. 1). on left, classical RFFP (eq. 9-11, 13) on right. Numbers shown: the apparent local exponents (dlog G/dlog Ø) at the beginning and end of the experimental range[3]. The exact curves agree within the experimental scatter (line and rhombs, fig. 1)!

modern universal theories are self-validating, and can be fitted to all data[8]. Unlike the classical model, they adjust a parameter freely (17.3 in eq. 1), and have other adjustable coefficients of their series expansions in reserve. Fig. 3 reveals just how the fits of quite different theories, classical and modern, happen to agree so closely. The modern theory adjusts the modulus scale freely,

after which the optimised and approximately theoretical value 1.89 of the exponent determines the whole curve. By contrast, the classical model A-b, which predicts its curve essentially without adjusting parameters (see near end of section 5), produces an apparent exponent (d log G/d log Ø) which varies widely, from 3.0 at the gel point, to 0.75 at the upper end of the range of moduli chosen by the experimenters. So the percolation exponent is just the mean of the range of the rapidly decreasing classical exponent. Yet the use of the mean exponent hardly affects the fit, even with the rather good data of Niki et al.

Besides, the percolation model leaves the gaps in the bottom row of table 1, because it provides no further testable predictions. It is unfortunate, therefore, that the interest in scaling theories has temporarily eclipsed the development of the sound classical approach (see section 6).

5. BIOCHEMICAL BACKGROUND INFORMATION AS ADDITIONAL TESTIMONY

Direct enzymatic aggregation of casein micelles has been much discussed. Payens[18] favoured a diffusion-controlled model, but Niki et al.[9] found this in conflict with their kinetic and viscoelastic measurements, which they interpreted by a classical network model. Our analysis of their equilibrium modulus measurements suggests the need to measure M_n of the micelles, since the classical polycondensation of whole micelles (table 1, row 1) cannot be fitted if M_n is much larger than 7200 kg/mol, however big we assume f_m to be. (M_w is estimated[3] to be ∢200,000; but the micelles are very heterodisperse).

The polycondensation model variant A-b for submicelles (table 1, row 2) gives marginally the best fit and has some colloid- and bio-chemical plausibility. The micelles are known to disintegrate readily into submicelles. The mere dialysis of aqueous solutions against Ca-free buffers produces clear solutions of submicelles which are rather uniform in composition, but with substantial separation between the submicelles. They are found to be swollen with water to the extent[12] of only 4-5.5 ml/g. To be linked into a continuous gel network in solutions containing >150 ml of water per gram of casein, they must undergo a substantial conformational expansion, so that at least some strands of each unit overlap with those of other units. This is not unlikely, since the liganding of a few Ca^{++}, or even of a single one, to many polypeptides causes considerable conformational changes. Indeed such changes are often of much biological significance[18]. The strength of a ligand bond of Ca^{++} to protein can lower the free energy by as much as 8.1 kcal/mol (in thermolysine[19]). These facts confer much plausibility on the proposal[20] by Dalgleish and Parker as to the nature of the functionalities for crosslinking created when [alpha]$_s$-casein is precipitated by Ca^{++}: "The functionalities on the protein molecules are produced as a consequence of the binding of Ca^{++}". The exact nature of the cross-linking functionalities remains to be elucidated both for the precipitation process and for the enzymatically induced gelation studied in table 1, where calcium is, of course, also present. The initial step of the gelation process is known[21]: the addition of chymosin destabilises the existing micelles of casein by removing the hydrophylic part of the [kappa]-casein component on the submicelles lying at the micellar surface, because the bond between phenyl alanine at position 105 and methionine at 106 is hydrolysed.

Dalgleish and Parker[20] fitted their measurements of gel points and sol fractions as functions of concentration in the Calcium-induced precipitation reaction, to an exact RFFP model with a narrow distribution of the functionality around its mean of f_m sites per 'monomer' unit. They found the equilibrium value of f_m to be "very close to 2" (but necessarily >2). This finding was compared by Richardson & Ross-Murphy[22] to similar values of f_m reported by several teams who had fitted measurements on gelation of serum albumin to the RFFP model. From their measurements of gel times, in excellent agreement with those of Bibkov et al.[23], these workers deduced f_m for albumen to lie between 2.05 and 2.5. They quote the pioneering paper of Kratochvil et al.[24], who deduced f_m=2.05, and Müller and Burchard[25], who recently obtained f_m~2.1.

It appears, then, that Nature uses structurally rather similar models to produce gels from compact proteins - here casein and serum albumin - whereby they undergo unfolding (denaturation) and liberate about 2.2 strongly binding functionalities per particle. Armed with this information, we may claim that the RFFP model A-b could tentatively enter the value 2.2 in row 2 of table 1 from previous knowledge of similar reactions. This amounts to the "most stringent test" of Hermans[14], viz. the absolute prediction of the modulus curve (fig. 1) without adjustment of parameters.

The model B-c (row 4) seems much less attractive. Here the micelles would have to disintegrate completely into their component casein polypeptide chains before or during network formation. But even if all four casein species were to bear sites available for random crosslinking, these short chains would have to swell very strongly in the aqueous medium to achieve a coherent gel at Ø=0.6 gm/100 ml. At the prevailing ionic strength, this seems unlikely to occur, but the possibility cannot be excluded a priori.

6. CLASSICAL AND MODERN STRATEGIES FOR MODELLING NETWORK POLYMERS

The random f-functional polycondensation (RFFP) model of Flory[15] has been nominated as the basic paradigm for the whole of chemistry[26]. It is fitting that it is found to describe well the gelation of casein, on which depends the clotting of milk, which is not only the basic reaction in the technology of cheese manufacture, but also the first reaction which initiates the absorption of food by the newborn human baby! The universal significance of the Stockmayer distribution of molecular species, which underlies the RFFP model, arises because it results precisely from a partition function which incorporates no more that Flory's equireactivity postulate for the functionalities plus the effects of molecular rotations (including symmetry)[27]. Further refinements (cyclisation and substitution effects) have successfully been grafted on to these two postulates; but many attempts have also been made to start models from radically different premises. Gordon and Judd[28] argued against recurring attacks on the Flory-Stockmayer theory. More recently, Klonowski[29] claimed that Flory and his followers had 'missed some easily attainable topological information' in setting up their model theory. These proposed constraints on equireactivity are graph-theoretical (not 3-dimensional) and consists simply in replacing the requisite statistical law (his eq. 17) by a different one (his 16). Such constraint is not found in the theory of statistical mechanics, and what experimental support he has adduced is, in the previous literature, traced by quantitative analysis to substitution and cyclisation effects.

As in the study of casein gels, the claims of scaling theorists have been widely influential in rejecting the classical models such as RFFP in the range of conversion close to the gel point, which is sensitive to structure (table 1). Scaling theory bases itself primarily on percolation models. It thus implicitly jettisons the dependence, in the vicinity of the gel point, of the physical properties of gels on rotational molecular motions, which clearly dominate polymer science elsewhere. Molecular rotations and their symmetry are solely responsible for the existence of the critical point in the RFFP model and a proven invariance renders the partition function, and hence the whole thermodynamics, universal for any dimensionality of space[27]. Conversely, scaling theory, on the basis of a statistically all but self-fulfilling hypothesis of universality, abandons dimensional invariance for an invariance to molecular symmetry. Thus it breaks new ground by claiming that the 3-dimensionality of space is fundamental for this part of physics[5] (cf. [27]). It is accepted that, except at most very close to the gel point, the excluded volume effect is absorbed[27] in the conversion parameter p. Any specifically 3-dimensional effect remaining near the critical point would involve only a minute fraction of the total Hamiltonian[30]. Scaling theorists have often claimed that this minute part is "qualitatively" out of reach for matching by refinements of the classical model approach, such as RFFP. Statistical arguments for[5,7] and against[8,27,30] such a radical claim have been pressed. At the end of their review, Stauffer et al.[7] have generously agreed to a dual program: "Improve percolation theory step by step in extending it to the region farther away from p_c and improve classical theory step by step in extending it close to the gel point." Some implications, based on a vital and technologically important example, for both subprograms are evident in fig. 3 and table 1. Other aspects concerning the excluded-volume effect have been, and more will be, published[31].

DEDICATION: On his sixtieth birthday, this paper is offered to Professor Ron Koningsveld, of the Netherlands, who has kept bright for so many in his generation the torch of classical thermodynamics, and added to its lustre.

REFERENCES

1. E.M.Valles and C.W.Macosko, in Chemistry and Properties of Crosslinked Polymers, (Ed. S.S.Labana), Academic Press, New York, 1977.
2. A.H.Clark, R.K.Richardson, S.B.Ross-Murphy and J.M.Stubbs, Macromol., **16**, 1367, 1983.
3. M.Tokita, R.Niki,and K.Hikichi, J. Phys. Soc. Japan, **53**, 480, 1984.
4. L.R.G.Treloar, The Physics of Rubber Elasticity, 3rd. ed., Clarendon Press, Oxford 1977.
5. P.G.de Gennes, Scaling Concepts in Polymer Physics, Cornell University Press, Ithaca, N.Y., 1979.
6. D.C.Hong and H.E.Stanley, J.Phys.A:Math.Gen.,**16**, L525, 1983, and references listed there.
7. D.Stauffer, A.Coniglio and M.Adam, Adv. Polymer Sci., **44**, 103, 1982.
8. M. Gordon and J.A.Torkington, Pure Appl. Chem.,**18**, 1461, 1981.
9. M.Tokita, K.Hikichi, R.Niki and S.Arima, Biorheol., **19**, 695, 1982.
10. M.Gordon and T.G.Parker, Proc. Roy. Soc. (Edin.), **A69**, 181, 1970/1.
11. D.G.Dalgleish & T.G.Parker, J. Dairy Res., **46**, 259, 1979.
12. P.H.Stothart and D.J.Cebula, J. Mol. Biol.,**160**, 391, 1982.
13. G.R.Dobson and M.Gordon, J.Chem.Phys., **43**, 705, 1965.
14. J.Hermans, Jr., J.Polymer Sci..**A3**, 1859 1965.
15. P.J.Flory, Principles of Polymer Chemistry, Cornell University Press, Ithaca, N.Y., 1953.
16. J.E.Mark and M.A.Llorente, J.Amer.Chem.Soc.,**102:2**, 632, 1980.
17. M.Gordon and K.R.Roberts, Polymer, **20**, 1349, 1979.
18. R.H.Kretsinger, in Calcium-binding Proteins and Calcium Function, (Ed. R.H.Wasserman et al., North Holland, New York, 1977, p. 63.
19. B.A.Levine, R.J.P. Williams, C.S. Fulmer and R.H. Wasserman, ibid. p. 29
20. T.G.Parker and D.G.Dalgleish, J.Dairy Research, **44**, 79, 85, 1977.
21. D.G.Dalgleish, in Developments in Dairy Chemistry, (Ed. P.F.Fox), Appl. Science Publishers, London and New York, 1982.
22. R.K.Richardson and S.B.Ross-Murphy, Int.J.Biol.Macromol., **3**, 315, 1981.
23. T.M.Bikbov, V.Ya.Grinberg, Yu.A.Antonov, V.B.Tolstoynzov, and H.Schmandke, Polym. Bull., **1**, 865, 1979.
24. P.Kratochvil, P.Munk and B.Sedláček, Coll.Czech.Chem.Commun.,**27**, 288, 1961
25. M.Müller and W.Burchard, Int.J.Biol.Macromol., **2**, 225, 1980.
26. M.Gordon and W.B.Temple, in Chemical Applications of Graph Theory, (Ed. A.T.Balaban), Academic Press, New York, 1976, p. 304
27. M. Gordon, Macromol., **17**, 514, 1984.
28. W.Klonowski, Bull. Acad. Roy. Belg., Classe des Sciences, 5th series., **64**, 568, 1978-9.
29. M.Gordon & M.Judd, Nature (London), **234**, 96, 1971.
30. M.Gordon and J.A.Torkington, Ferroelectrics, **30**, 237, 1980
31. R.G.Cowell, M.Gordon and and P.Kapadia, Polymer Reprints Japan, **31**, 29, 1982; full version to be published.

CROSSLINKING THEORY APPLIED TO INDUSTRIALLY IMPORTANT POLYMERS

K.DUŠEK

Institute of Macromolecular Chemistry, Czechoslovak Academy of Sciences,
162 06 Praha 6, Czechoslovakia

SYNOPSIS

Application of the crosslinking theory to macromolecular systems important in practice has required extension of the theory to multicomponent systems and complex reaction mechanism. The applicability and limitations of the statistical and kinetic (coagulation) theories are discussed. Success in the application of the theory is demonstrated by results obtained for amine and acid curing of epoxy resins and formation of polyurethane networks. The experimental verification is based on the determination of the gel point conversions, changes in the molecular weight averages before the gel point, and on the sol fraction and equilibrium modulus data in the rubbery state beyond the gel point. It is shown how to deal with important "side" reactions such as polyetherification in the curing of epoxy resins or allophanate group formation in the case of polyurethanes.

1 INTRODUCTION

The elucidation and understanding of the formation-structure-properties relationships represents a modern basis for the research and development of polymeric materials in industry. The development of predicitive relationships is possible only by an integration of theory and experiment. The role of the theory is crucial, particularly for systems undergoing branching and crosslinking.

Which information offered by the crosslinking theory is closely related to technology and materials science? In the pregel stage, it is particularly that on molecular weight distribution and averages and radii of gyration which determine the rheology and processibility of the reacting system up to its limit - the gel point. Beyond the gel point, the leakage of extractable substances (sol fraction) is a limiting factor in many applications; the concentration of elastically active network chains (effective crosslinking density) determines the modulus of elasticity in the rubbery state, and is thus one of the most important application characteristics of vulcanized elastomers. The time-temperature dependence of mechanical properties is determined by the network architecture characterized by such parameters available in the theory as molecular weight distribution of network chains and dangling chains. Also, in glassy thermosets, the temperature, ageing and environmental resistance are dependent on the crosslinking density.

Because of the central role of the crosslinking theory in the understanding of the network structure, the applicability of the current theories will be briefly analyzed in the next section.

2 THE THEORETICAL TOOLS

The theoretical models that describe formation of polymer networks can be divided into two groups:

(a) Graph-like (especially tree-like) models not associated with the dimensionality of space,

(b). lattice or off-lattice simulation in the n-dimensional space.

The present state of applicability of these theories to the elucidation of formation-structure-properties relationships has been briefly analyzed recently[1]. The most suitable model is the tree-like model which is conceptually and mathematically simple and can absorb the majority of information important for the chemist or chemical engineer, such as functionality and reactivity of monomers, reaction paths, etc. It can treat spatial correlations, such as cyclization, as a perturbation of the ring-free state. It is therefore suitable for systems with a weak tendency for cyclization, but is not suitable, if the cyclization is strong. If we speak about cyclization, we have in mind the pregel cyclization and formation of elastically inactive loops in the gel. The extensive circuit closing by bond formation between two functionalities in the gel governed by mass action law is automatically included in a way used by Flory for the first time[2]. The percolation technique seems to be relatively well applicable for systems with strong cyclization in which the polymerization rate is so fast that conformational rearrangements cannot occur between the formation of successive bonds[3]. Since (perturbed) tree-like models have been applied to almost all systems of technological importance, their variants will be briefly discussed. The Flory-Stockmayer model was generalized by Gordon and collaborators[4-7] and can be extended to multicomponent systems with complicated reaction paths. The branched and crosslinked structures are generated from building blocks (monomer or initial units) which corresponds to equilibrium controlled reactions but may be an approximation for reactions controlled by chemical kinetics. This type of multifunctional reactions can be described properly by coagulation (kinetic) equations, the mathematical handling of which is more difficult. Moreover, this method cannot be used for a description of the gel structure which appears in the theory as one particle without internal structure. It has been shown (cf.[1]) that, by using the statistical method of generation from building blocks, one can lose information on long-range statistical correlations inherent in the particular network build-up process. The magnitude of the deviation of the (approximate) statistical solution from the correct kinetic solution determines the decision whether the use of the simpler and more versatile statistical method is tolerable. The stochastic long-range correlation can be brought in by substitution effects and, what is more serious, by an initiated type of reactions as has been shown recently[1]. Fortunately, in many cases a combination of kinetic and statistical methods is rigorous, if at least some of the polyfunctional monomers bear groups of independent reactivity. An example of such approach will be demonstrated below for the case of polyetherification in polyepoxide-polyamine systems.

3 APPLICATION

3.1 Polyurethane Networks

Crosslinked polyurethanes can be obtained from polyols and diisocyanates, or diols and polyisocyanates, or from diols and diisocyanates employing additional crosslinking by allophanate formation. The features of the endlinking process to which formation of polyurethanes belongs has been discussed in details elsewhere[7]. The case macrotriol-diisocyanate is simple and can be used for a demonstration of the application of the theory.

The building blocks in this case are represented by macrotriol (e.g., polyoxypropylene triol) units differing in the number of reacted OH groups. Let the molar fraction of these units be denoted by a_0, a_1, a_2, a_3, where the subscript denotes the number of the reacted groups. For diisocyanate units we have three options the probability of which is given by b_0, b_1, b_2. Since the reactivity of the OH groups in macrotriol is independent and the state of each group is determined by the overall conversion of OH groups α_H, the fractions a_i are given by the binominal expansion of $(1-\alpha_H+\alpha_H)^3$. The same approach can be used for some diisocyanates, e.g., for 4,4´-diphenylmethane diisocyanate (MDI). The only information needed are the conversions of OH and NCO groups which are interrelated. For a detailed explanation of the statistical treatment using generating functions it is referred to the general (e.g.[4-6]) and special [8-12] literature pertaining to the formation of polyurethane networks.

A comparison of the theory with experiments is demonstrated by the following examples obtained for poly(oxypropylene)triol (POPT) and 4,4´-diphenylmethane diisocyanate (MDI):

(a) critical conversion at the gel point is determined ont only by functionality (possible functionality distribution), but also by cyclization. The effect of cyclization can be eliminated by measuring the critical conversion as a function of dilution and by extrapolating the data to the imaginary state of "infinite" density. The extrapolated value yields information on the breadth of functionality distribution, and the effect of dilution on the gel point characterizes the extent of cyclization. The comparison of experiments with theory has confirmed the theoretical prediction that the extent of cyclization in endlinked networks is small[10].

(b) The sol fraction was examined[9] as a function of the molecular weight of POPT and molar ratio of OH and NCO groups r_H. The most densely crosslinked networks are expected to be formed at $r_H=1$, and the sol fraction is shown to increase with increasing excess of either OH or NCO groups. In the calculation, the effect of cyclization as well as of the possible incompleteness of the reaction has been taken into account. Fig. 1 shows the comparison with the theory. It is evident that the $r_H>1$ branch of the dependence fits very well the experimental points (good agreement with the theory), whereas considerable deviations in the $r_H<1$ branch point to additional crosslinking.

(c) The correlation of the calculated concentration of elastically active network chains, ν_e, with the equilibrium modulus depends not only on the network formation theory, but also on the rubber elasticity theory. A comparison shows again a good agreement between the shapes of the predicted and experimental dependence of the modulus on r_H at $r_H \geqq 1$ but systematic deviations to higher moduli at $r_H<1$. The superposition procedure described in[9] can answer the question whether interchain constraints are more effective in the increase in the modulus than only through the suppression of junction fluctuation. It has been found that the contribution to the modulus due to chain incrossability (trapped entanglements) is highly probable (Fig. 2). This conclusion follows from the fact that the reduced modulus for the Niax LG-56 networks considerably exceeds the upper limit of the junction fluctuation theory.

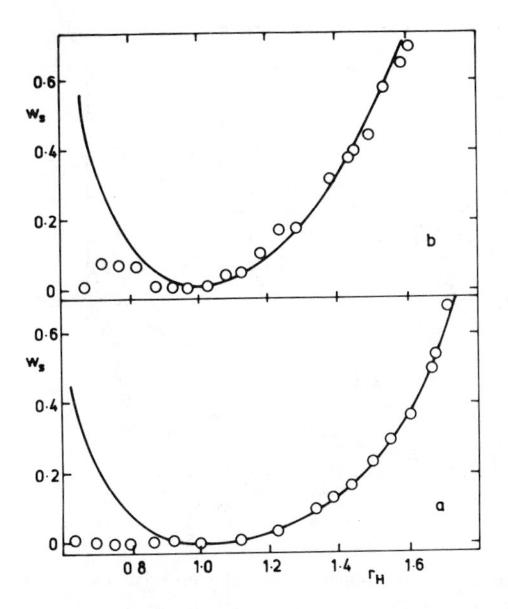

Fig.1. Dependence of the weight fraction of sol in poly(oxypropylene)-triol (POPT)-4,4´-diphenylmethane diisocyanate (MDI) networks on the molar ratio r_H=[OH]/[NCO][9].
a - networks from PPOT Niax-240 (M=708), b - networks from Niax LG-56 (M=2630). experimental data, — theory: numbers indicate final conversion of minority groups

The form of allophanate groups is a complicating factor which however could also be treated by the branching theory[13]. Allophanate groups are formed by a reaction of urethane with isocyanate groups following the isocyanate-hydroxyl addition

$$\text{-OH + OCN-} \xrightarrow{k_u} \text{-OCONH-}$$

$$\text{-OCONH- + OCN-} \underset{k_{-a}}{\overset{k_a}{\rightleftharpoons}} \begin{matrix}\text{-OCON-} \\ \quad\text{OCNH-}\end{matrix}$$

In the statistical generation[11] of structures from a POP triol and diisocyanate, one works again with the distribution of the triol units as in the simple isocyanate-hydroxyl addition, but one has to distinguish between the reacted isocyanate units in the urethane group and those in the allophanate group. In the allophanate group, one isocyanate unit becomes bifunctional and one is monofunctional, so that the whole allophanate group becomes a trifunctional branch point.

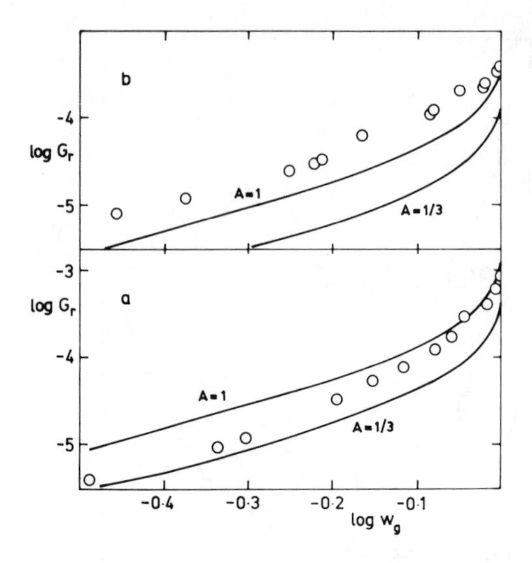

Fig.2. Dependence of the reduced equilibrium modulus of POPT-MDI networks on gel fraction[9].
a - networks from Niax LHT-240, b - networks from Niax LG-56;
o experimental data, — calculated dependence using the junction fluctuation theory of rubber elasticity with the value of front factor indicated

Using this approach, the deviations in the sol fraction and equilibrium modulus, if isocyanate is in excess, could be described quantitatively[12]. This approach is also useful for the prediction of stability (resistance against viscosity increase and gelation) of isocyanate endcapped macrodiols[14]. The stability depends on the initial OH/NCO ratio and on the equilibrium constant of allophanate formation k_a/k_{-a}. (Fig. 3).

3.2 Epoxy Networks

Epoxy resins are polymer systems of technological interest to which crosslinking statistics has been applied most extensively in many countries. A brief review of the published literature is given in Ref.13.

Curing of polyepoxides with polyamines is of the greatest technological importance and, fortunately, it is relatively simple from the theoretical point of view. The curing mechanism involves addition of the amino group to the epoxide ring in two steps

$$\underset{\diagdown O \diagup}{-CH-CH_2} + -NH_2 \xrightarrow{k_1} \underset{\underset{OH}{|}}{-CH}-CH_2NH-$$

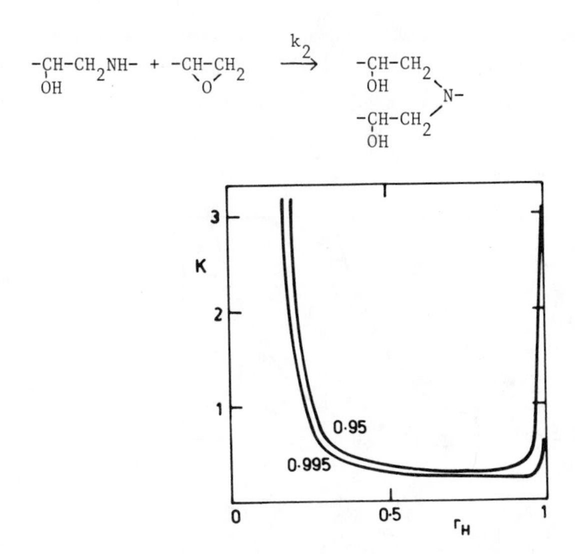

Fig.3. Stability of diols endcapped with diisocyanate in dependence on the initial ratio r_H = [OH]/[NCO]. $K = k_a/k_{-a}$ is equilibrium constant for allophanate formation; the numbers at curves indicate the conversion of OH groups. The region below the curves corresponds to a fully soluble system, the region above them to system with gel[11].

For curing of diepoxides with diamines, the branched and crosslinked structures are built-up from diepoxide units bearing 0,1 or 2 reacted epoxy groups (molar fraction e_0, e_1, e_2) and diamine units with 0-4 reacted hydrogens (molar fractions) a_0, a_1, a_2, a_3, a_4. These fractions are calculated using the chemical kinetics of the epoxy-amine reaction and taking into account the possible independence of the reactivities of epoxy groups in diepoxides and amine groups in diamine. The procedure has been outlined elsewhere[13-15] and it is applicable also for the autoaccelerated reactions with participation of various donor-acceptor complexes, provided these factors affect the apparent rate constants k_1 and k_2 to the same extent, so that they can be factored out.

The distributions of a_i and e_i as a function of conversion is the only information needed because in the absence of cyclization only one type of bonds can be formed and the reaction is strictly alternating. Due to the independence of the reactivity of groups in diepoxide and diamine, the application of the statistical generation from units is justified similarly to the case of polyurethane networks. Gelation and network properties are a function of conversion and are determined mainly by the epoxy/amine molar ratio and to some extent by the substitution effect in the amino group[13].

The results obtained for diepoxide-monoepoxide-diamine networks[14-15] show that the theory is well applicable. The sol fractions of the off-stoichiometric mixtures agree well with the theoretical prediction and also the dependence of the equilibrium modulus of dry and swollen networks in the rubbery state is close to that predicted (Fig. 4).

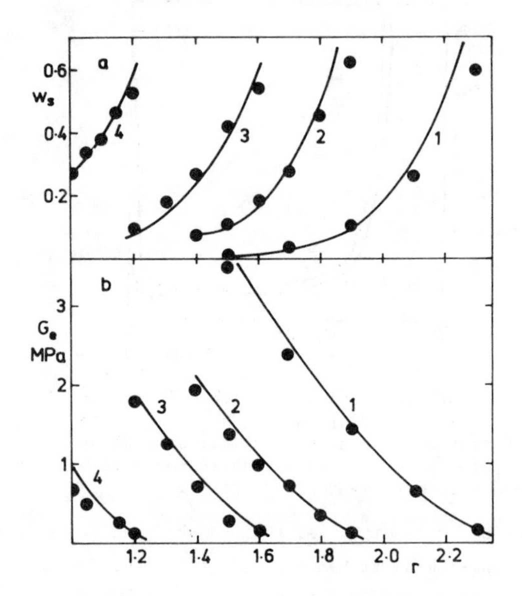

Fig.4. Dependence of the sol fraction (a) and equilibrium modulus for diglycidyl ether of Bispehnol A - phenylglycidyl ether - 4,4´-diaminodiphenylmethane networks on the initial molar ratio $r = 2[NH_2]/[EPOX]$. Initial molar fraction of epoxy groups in monoepoxide: 1 0, 2 0.2, 3 0.33, 4 0.5. ● experimental data, — calculated dependence[14,15].

If polyetherification interferes with the epoxy-amine addition, the situation becomes theoretically and experimentally more complicated. However, polyetherification is rather important technologically, since it is employed in the majority of formulations for high-performance composites. The theoretical complexity rests in the fact that polyetherification is induced by the OH group formed in the epoxy-amine addition and is thus a kind of initiated polyreaction. The statistical method cannot be used for initiated reactions[1] and one has to make use of the kinetic method. On the other hand, the independence of groups in diamine and diepoxide makes the situation simpler. The strategy is thus as follows:

The diamine and diepoxide are converted into monoamine and monoepoxide by cuts and the points of cut are specially labelled. Then kinetic rate equations are written down for the reaction

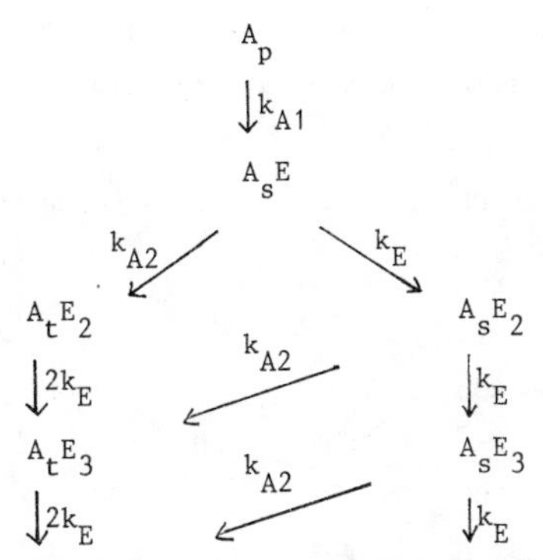

where A_p, A_s and A_t are primary, secondary and tertiary amine units. The distribution of A_p, A_sE_i and A_tE_i can be obtained by transformation of the set of the rate equations into a single differential equation for the probability generating function (pgf) of this distribution. In the final step, proper combination of the points of cut by convolution of the pgf´s gives the pgf for the diamine-diepoxide system, which can be processed by the conventional method.

Not yet published results in Fig. 5 shows the gel point conversion of epoxy group if k_{A1}, $k_{2A} >> k_E$ and if epoxy groups are in excess. It is a result of the interplay between the negative effect of off-stoichiometry (excess of epoxy groups) and formation of additional crosslinks by etherification. As can be seen,

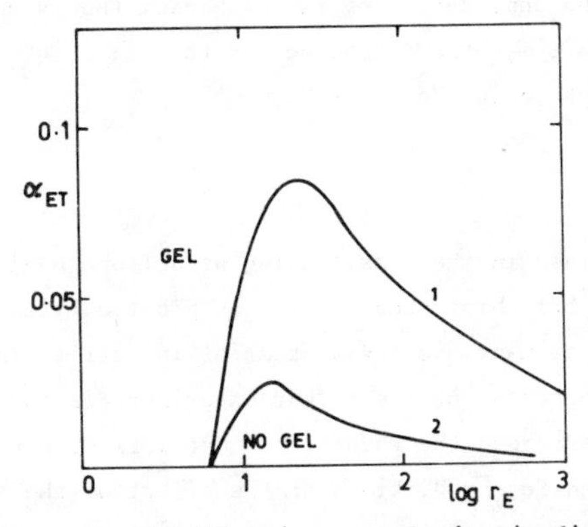

Fig.5. Effect of etherification on gelation in diepoxide-diamine networks. α_{ET} is conversion of excess epoxy groups in the etherification reaction, r_E is the initial molar ratio [EPOX]/[NH_2]. 1 - theory based on the kinetic method (rigorous), 2-theory based on the statistical method (approx.)

there exists only a certain range of amine/epoxy ratios in which the system can be fully soluble. Also, one can see that the neglect of stochastic correlations inherent in the statistical generation is serious.

The acid curing is even more complicated. If polyacids are used as curing agents, the addition esterification yielding polyhydroxyesters is accompanied by transesterification[16]

$$\mathrm{HOOC} \sim\sim \mathrm{COOH} + \underset{\backslash O /}{CH_2CH}\underset{\backslash O /}{RCHCH_2} \longrightarrow -\mathrm{OC} \sim\sim \mathrm{COOCH_2}\underset{OH}{\mathrm{CH}}\mathrm{R}\underset{OH}{\mathrm{CH}}\mathrm{CH_2O}-$$

$$\rightleftharpoons \quad \begin{matrix} -\mathrm{OC} \sim\sim \mathrm{COOCH_2} \searrow \\ \mathrm{CHR}- \\ -\mathrm{OC} \sim\sim \mathrm{COO} \nearrow \end{matrix} + -\mathrm{RCH}\begin{matrix} \nearrow \mathrm{CH_2OH} \\ \searrow \mathrm{OH} \end{matrix}$$

Transesterification does not affect the total number of bonds but causes their redistribution as a result of which the diepoxide units can issue 0-4 bonds in contrast to 1 or 2 bonds in the original polyhydroxyester. Due to transesterification, gelation is possible even for a system composed of diepoxide and diacid, which is important in the synthesis of bisunsaturated polyesters. The branching theory has been modified to take into account addition esterification and transesterification and the predictions agree well with experiments[16].

The study of the mechanism of curing of diepoxides with cyclic anhydrides catalyzed by tertiary amines has shown[17] that the tert. amine acts as an initiator and is built in covalently into the structure. The chain growth proceeds by anionic mechanism. The application of the branching theory has predicted and the experiments have confirmed the dependence of the gel point conversion on the concnetration of the tert. amine "catalyst".

4 CONCLUSIONS

The results obtained in the crosslinking of polyurethanes and epoxy resins have shown that (1) the branching theory based on the tree-like model is versatile and can be applied to many complex systems of industrial importance, and (2) that the agreement between the theory and experiment is good unless strong cyclization interferes. Because the theory does not work with any adjustable parameter, it can be used for predictions of the effect of the structure and reactivity of the components, initial composition of the system and crosslinking conditions on the network structure and its changes during the crosslinking reaction. In this connection, application of the branching theory to the crosslinking of silicone rubbers[18] or melamine/formaldehyde/acrylic coatings[19] should be mentioned.

There exists, however, an example where the application of the tree-like model fails due to strong cyclization. It is the (co)polymerization of polyvinyl monomers[20]. Extensive cyclization is responsible for the formation of internally crosslinked microgel-like particles in which only a fraction of reactive groups on the surface is available for network build-up.

REFERENCES

1. Dušek, K. Brit. Polym. J. in press
2. Flory, P.J. 'Principles of Polymer Chemistry', Cornell Univ. Press, Ithaca 1953.
3. Boots, H.M.J. and Pandey R.B. Polym. Bull. 11, 415, 1984.
4. Gordon, M. Proc. Roy. Soc. London A268, 240, 1962.
5. Gordon, M. and Malcolm, G.N. Proc. Roy. Soc. London A295, 29, 1966.
6. Gordon, M. and Ross-Murphy, S.B. Pure Appl. Chem. 43, 1, 1975.
7. Dušek, K. Rubber Chem. Technol. 55, 1, 1982.
8. Dušek, K., Hadhoud, M. and Ilavský, M. Brit. Polym. J. 9, 164, 1977.
9. Ilavský, M. and Dušek, K. Polymer 24, 981, 1984.
10. Matějka, L. and Dušek, K. Polym. Bull. 3, 489, 1980.
11. Dušek, K., Ilavský, M. and Matějka, L. Polym. Bull. 12, 33, 1984.
12. Ilavský, M., Bouchal, K. and Dušek, K. Proc. Internat. Conf. "Rubber 84" Moscow, Preprints vol.A3, Preprint A84, Moscow 1984.
13. Dušek, K. in 'Rubber-Modified Thermosets', Riew, C.K. and Gillham, J., Eds., Advan.Chem.Series, vol.208, p.1, Am. Chem. Soc. 1984.
14. Dušek, K. and Ilavský, M. J. Polym. Sci., Polym. Phys. Ed. 21, 1323, 1983.
15. Ilavský, M., Bogdanova, L. and Dušek, K. J. Polym. Sci., Polym. Phys. Ed. 22, 265, 1984.
16. Dušek, K. and Matějka, L. in 'Rubber-Modified Thermosets', Riew, C.K. and Gillham, J., Eds., Advan. Chem. Series, vol.208, p.15, Am. Chem. Soc. 1984.
17. Matějka, L., Lövy, J., Pokorný, S., Bouchal, K. and Dušek, K. J. Polym. Sci., Polym. Chem. Ed. 21, 2813, 1983.
18. Macosko, C.W. Pure Appl. Chem. 53, 1505, 1981.
19. Bauer, D.R. and Dickie, R.A. J. Polym. Sci., Polym. Phys. Ed. 18, 1997, 2015, 1980.
20. Dušek, K. in 'Developments in Polymerization 3', Haward, R.N., Ed., Appl. Sci. Publ., p. 143-206, Barking 1982.

REVERSIBLE AND IRREVERSIBLE DEFORMATION OF VAN DER WAALS NETWORKS

H.G.KILIAN

Universität Ulm, Abteilung Experimentelle Physik

1 INTRODUCTION

Since the appearance of Kuhn's original theory of rubber elasticity [1,2,3] a number of methods have been proposed for the solution of "the network problem". What is common to all of these theories is that the rubber is considered to have a "simple energetic structure" such as to show a definite number of energy-equivalent subunits of deformation [1-4]. For making the characterization complete there is need of defining the "perfect" strain-energy function.

To be at a loss with a molecular-statistical theory of such completeness, we have been motivated to formulate a phenomenological van der Waals theory[4]. Fortunately the attempt to modify the approximate "Gaussian" statistical theory in terms of a van der Waals approach involves no substantial sacrifice both of simplicity and of generality.

2 THE GAUSSIAN NETWORK

According to the Gaussian theory[1,2,3] kinetical energy can only be stored equiparted over the "phantom-chains" as the "energy-equivalent subsystems of deformation"[4]. This symmetry is the reason behind of having the strain-energy of the Gaussian network in proportion to the number of chains, N

$$W(\lambda) = N k_B T \Phi(\lambda) \quad (1)$$

where k_B is the Boltzman-constant and T the absolute temperature. This phantom network is purely entropy-elastic with the pecularity of being characterized by the "deformation function" of an incompressible elastic continuum [1,2,3,5]. In the mode of simple extension we then have

$$\Phi(\lambda) = (\lambda + 2\lambda^{-1} - 3)/2 \quad (2)$$

with the strain parameter $\lambda = L/L_o$ relating the actual to the initial length of the sample in direction of the force applied.

The strain-energy function of phantom network may be thought of as an abstract representation by which the "global" properties of a rubber are defined. In order to indicate its idealizing presumptions let us consider the Gaussian network as "the ideal gas-network".

3 VAN DER WAALS-NETWORKS

A correct description of real networks should imply "global effects" due to finite chain extensibility and to interactions between the chains[1,2,3,4]. Both of these aspects are accounted for by defining the strain-energy of a van der Waals network to be equal to[26]

$$W(\lambda) = N k_B T \left[2 D_m \left\{ -(\lambda-1) + \frac{\lambda_m^2 - 1}{\lambda_m^5 + 2\lambda_m^2} \left(-\lambda_m \ln \left| \frac{\lambda_m - \lambda}{\lambda_m - 1} \right| \right.\right.\right. \tag{3}$$

$$\left. - \ln \left| \frac{\lambda^2 + \lambda\lambda_m^{-2} + \lambda_m^{-1}}{1 + \lambda_m^{-2} + \lambda_m^{-1}} \right| \right) + \frac{2(1-\lambda_m^3)}{\lambda_m^2 W_u} \left(\operatorname{artan} \frac{2\lambda + \lambda_m^{-2}}{W_u} \right.$$

$$\left.\left.\left. - \operatorname{artan} \frac{2 + \lambda_m^{-2}}{W_u} \right) \right\} - \frac{a}{3} (\lambda^3 - \lambda_m^{-3} - 6 \ln \lambda) \right]$$

$$W_u = (4\lambda_m^{-1} - \lambda_m^{-4})^{1/2}; \quad D_m = \lambda_m - \lambda_m^{-2}$$

The van der Waals parameter λ_m is the maximum strain in the mode of simple extension. The parameter "a" is thought to cover global interactions[3,7,8].

Energy is assumed to be equiparted. Yet, the energy density of deformation depends in a "non-linear manner" on the macroscopical strain showing the typical pole

$$\lim_{\lambda \to \lambda_m} W(\lambda) = \infty \tag{4}$$

The universal features of these networks is fairly well intimated by calling them "weakly interacting van der Waals network gases".

The mechanical equation of state is then obtained to be given by

$$f = N k_B T \; D \left(\frac{1}{1 - D/D_m} - aD \right) \tag{5}$$

The existence of a reduced van der Waals equation of state[9] manifests a high degree of universality in accord with an appropriate scaling behavior [10].

In view of the rather perfect representation of the experimental data (see fig.1), the question arises how the density of the energy-equivalent subsystems of deformation is related to the total number of chains.

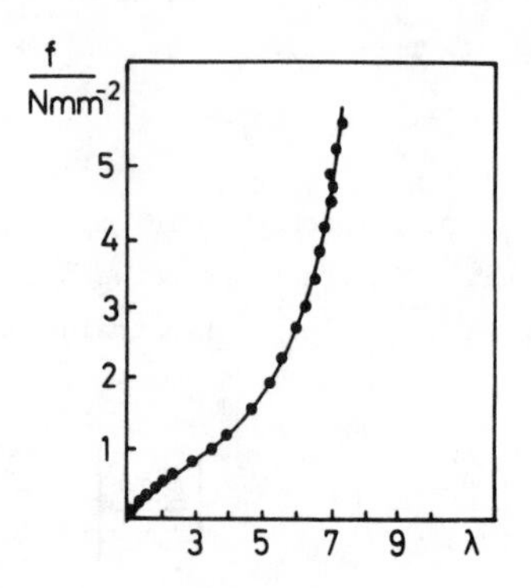

Fig.1a: Simple elongation for natural rubber at 295 K according to Treloar[2]. Solid line calculated with equation 5: $\rho RT = 22.28\ N\ mm^{-2}$; λ_m=9.7; a = .23; M=68 g/mole

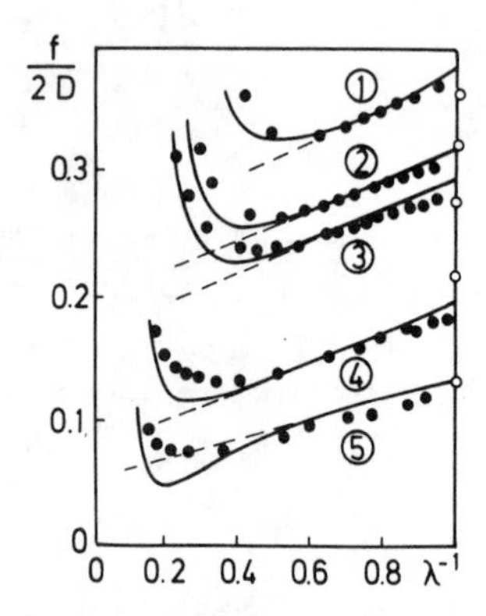

Fig. 1b: Mooney plot of experimental force-extension data for peroxide crosslinked rubbers containing (5) 1%, (4) 2%, (3) 3%, (2) 4%, (1) 5% peroxide according to Mullins[11]. Solid lines computed with eq.5: ρRT =22.28 N mm; a = .29; M = 68g/mole and λ_m as indicated with each curve. The solid circles are the moduli computed with the aid of eq.(13) assigning α and λ_{min} to the value of three[12].

4 CROSSLINKING AND MODULUS

For networks with not to short chains the maximum strain can be expressed by

$$\lambda_{mo} = y\, M_i / \sqrt{y}\, M_i = y^{1/2} \qquad (6)$$

where y is the number of "stretching invariant units per chain". With M_i as the molecular weight of the random link, M_c of the total chain can be written as

$$M_c = y\, M_i = \lambda_{mo}^2\, M_i \qquad (7)$$

such that the shear modulus G is in its simplest form given by [2,3]

$$G = \rho RT/M_c = \rho RT/\lambda_{mo}^2\, M_i \qquad (8)$$

The molar fraction of the crosslinks, x_{cr}, is related to the weight fraction of the crosslinking agent, φ_A, with molecular weight M_A

$$x_{cr} = \varphi_A/(\varphi_A + (1-\varphi_A)\, M_A/M_i/\alpha \qquad (9)$$

where α gives the number of cross-linkages produced by each of the agent molecules. With the aid of

$$x_{nc} = x_{nc}/(1 + x_c) \qquad (10)$$

the modulus G can be cast into the form

$$G = \rho RT/(x_c/x_{nc}\; M_i) \qquad (11)$$

Polydimethylenesiloxane networks are observed to obey this classical relationship under the assumption of $\alpha = 1$[13,14] (see fig. 2). Yet, for natural rubber, for polyethylene- or for styrene-butadien networks there is need of introducing the new parameter λ_{min}[12,14]

$$\lambda_m = \lambda_{mo} + \lambda_{min}: \; G = \rho RT \Big/ \lambda_m^2 M_i = \rho RT \Big/ \left\{ \left(\frac{x_c}{x_{nc}} \right)^{1/2} + \lambda_{min} \right\}^2 M_i \qquad (12)$$

Assigning α and λ_{min} to the value of three, we are indeed enabled to compute fairly well the experimental data as reported by Mullins[11,12] (full circles in fig.3). The typical bending of the modulus as a function of the density of the chemical cross-linkages [13,12] is nicely reproduced (see fig.3). For natural rubber at room-temperature we are led to $M_i = M_o$ whereby M_o is the molecular weight of the monomer unit while Mullins comes out with $M_i = 1.1\, M_o$[11].

It should not escape notice that λ_m determines together with "a" the relative course of the stress-strain curve. From the absolute value of the modulus G required to fit the experimental force-extension curve it was shown that in natural rubber M_i strongly depends on temperature [15].

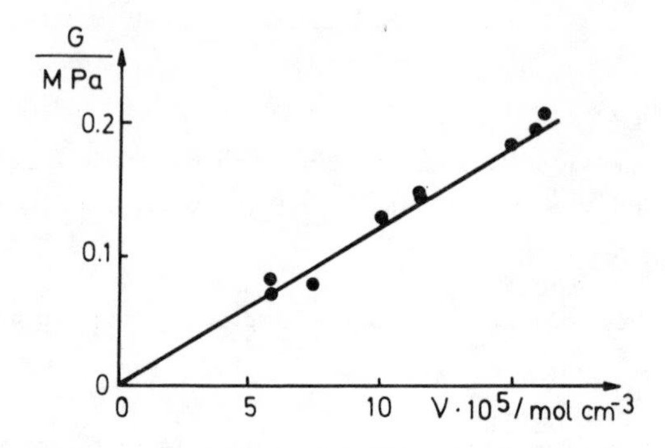

Fig.2: The modulus of polidimethylene-siloxane networks against the density of chemical crosslinkages according to Oppermann[13]. The solid is computed with the aid of eq.(11) using the parameters: $\rho RT \approx 22.28 Nmm^{-2}$ (T=295K), $M_i = 1.6M_o$ (M_o molecular weight of the monomer unit) $\alpha^1 = 1$

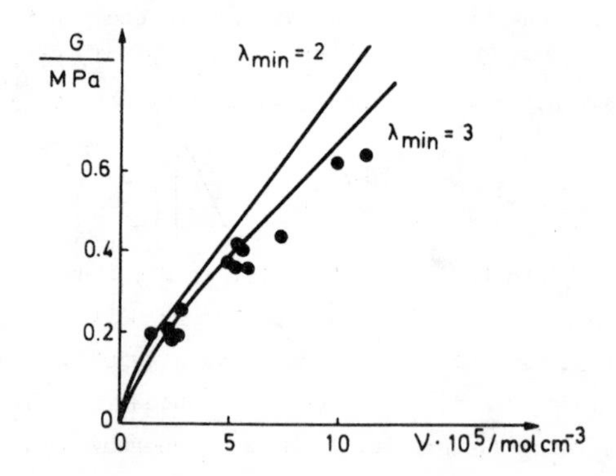

Fig.3: The modulus of NR against the density of the chemical cross-linkages according to Oppermann[13]. The solid computed with the aid of eq.(11) and the parameters: $\rho RT = 28.5 N mm^{-2}$ (T=423K); M=68 g/mole; $\lambda_{min=3}$; $\alpha = 3$ according to [12].

The global structure of all the systems reported on is in an universal manner characterized by permanent energy-equivalent units of deformation the number of which is in general smaller than the number of network chains. The reduction is likely to be related to the existence of a broad molecular weight distribution wherein the shortest chains do no more operate as autonomous energy-equivalent units [12,14].

Justification of not considering entanglements [2,3,12,14,23,24] is drawn from the fact that a satisfactory fit to the data over the whole range of extension is obtained with an invariant set of parameters. It appears to be very unlikely to have entanglements operating completely undependent upon the stress applied. In contradiction to what is discussed in literature [13,23-26], we are thus led to a production of three cross-linkages per each of the dicumyleperoxid molecules in natural rubber [1,14].

5 THERMOELASTIC PROPERTIES

Liquidlike properties of networks together with effects related to non-isoenergetical rotational-isomers [2,3] are found to be mirrored in the thermo-elastic phenomena. Internal equilibrium always present, thus having the Gibbs-function of the van der Waals network defined in the coordinates(T,p,L), it should be possible to compute the heat exchange during isothermal quasistatic simple extension [16,17,18]. This was shown to be true indeed (see fig.4).

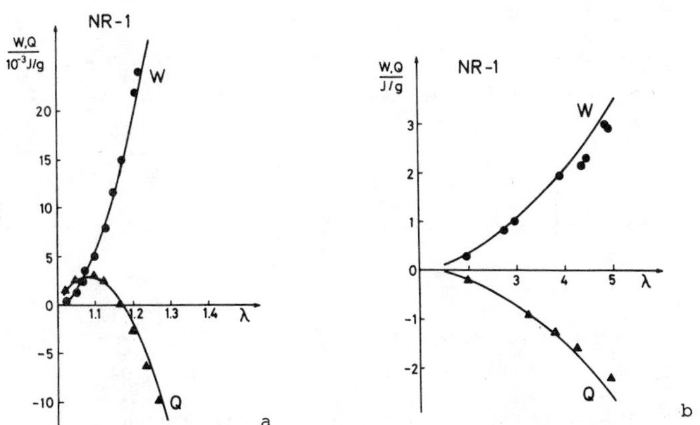

Fig. 4a: Plot of W(λ) and Q(λ) for NR in simple elongation at room temperature according to [17,18]. The solid lines have been computed with the thermoelastic equations of state of a van der Waals-network.

Fig. 4b: The same system as given in fig. 4a at larger extensions. The solid lines are computed [18].

Using the thermodynamical representation of the force [2,3,16] by adding up the "network term f^N and f^i related to the internal freedoms"

$$f = f^N + f^i = f^N + \frac{1}{L_o}\left[\left(\frac{\partial H^i}{\partial \lambda}\right)_{T,P} - T\left(\frac{\partial S^i}{\partial \lambda}\right)_{T,P}\right] = f^N \tag{13}$$

it is immediately seen that the stress-strain behavior should in fact only display the global phenomena in real networks.

6 IRREVERSIBLE DEFORMATION

So far we have been concerned with equilibrium properties of van der Waals networks. We would now like to illustrate that the "non-linear" viscoelastic behavior of networks far above the glass transition can fairly well be understood on using linear material equations (Onsager approach) [20-22]. The crucial point is that we have to define by introducing additional hidden variables ξ_k and extended Gibbs-function

$$g = \frac{f_{20}}{2} W + \sum_{k=1}^{N} f_{11}^{k} W^{1/2} \xi_k + \sum_{k=1}^{N} (f_{02}^{k}/2)\, \xi_k^2 \tag{14}$$

where the orthogonal relaxation mechanism are considered to be linearely coupled to the global state of deformation of the network. This way of coupling is characterized by the "moduli" f_{11}^{k}, the quasi-static network strain energy function by f_{20}^{k} and the time-dependent storage energy of the various mechanism by f_{02}. On defining the set of Onsager equations

$$\xi_k = \alpha_k A_k \tag{15}$$

A_k is the affinity which describes the distance to equilibrium while $_k$ determines the relaxation time of the mechanismus considered. The response given by [12,14]

$$\sigma(t) = \frac{G_o W'(t)}{2}\left[1 + \frac{\gamma}{G_o}\left(1 - W(t)^{-1/2}\right)\sum_{k=1}^{N} \frac{h_k}{\tau_\varepsilon^k} \int_o^t e^{-\frac{t-t'}{\tau_\varepsilon^k}} W^{1/2}(t')dt'\right] \tag{16}$$

$$W' \equiv \frac{\partial W}{\partial \lambda};\; G_o = f_{20} - \gamma;\; \gamma = \Sigma\left(f_{11}^{k}\right)^2 \Big/ f_{02}^{k};\; h_k = \gamma^{-1}\frac{\left(f_{11}^{k}\right)^2}{f_{02}^{k}};\; \Sigma h_k = 1$$

has been shown to display macroscopical non-linear relaxation. This is illustrated with the entropy production for isothermal extension at a constant strain rate (see fig.5). Hence, real networks as represented by van der Waals networks show strain- and rate-dependence relaxation always determined by that part of the "small-strain" relaxation spectrum that is of relevance. Calculations are depicted in fig. 6 using two dominant relaxation times only.

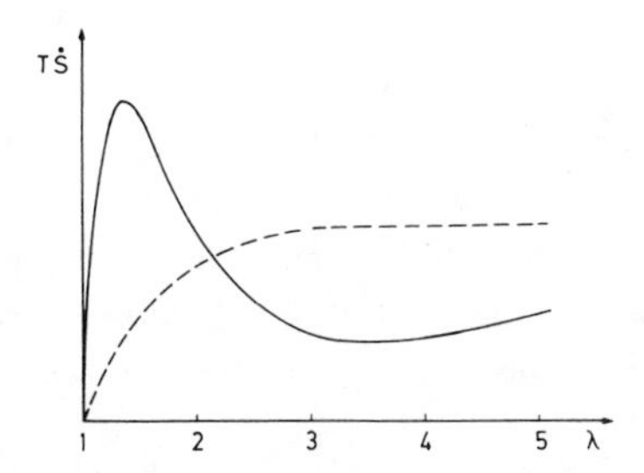

Fig.5: The rate of entropy production of a Hooke-Newton body (---) and a van der Waals network with Onsager material equations of state (——) [6].

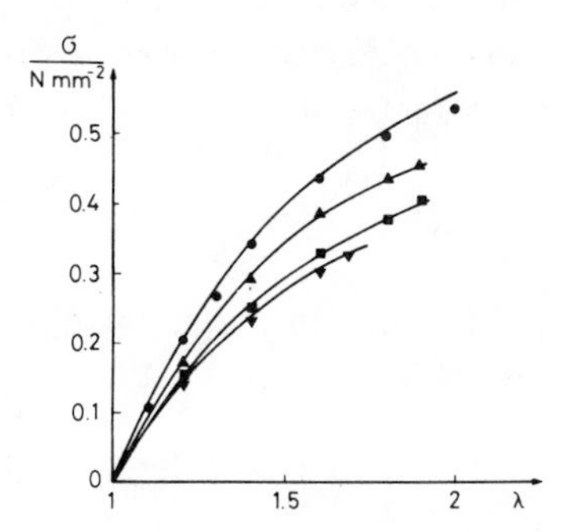

Fig. 6a: Strain-rate dependent measurements of force-elongation curves of SBR according to Tschoegl[27]: strain rates ● 4.54min^{-1}, ▲ .454 min^{-1}, ■ .0227 min^{-1}, ▼ .00123min^{-1}. The solid lines are calculated with the reaxation times as indicated in fig. 6b [6].

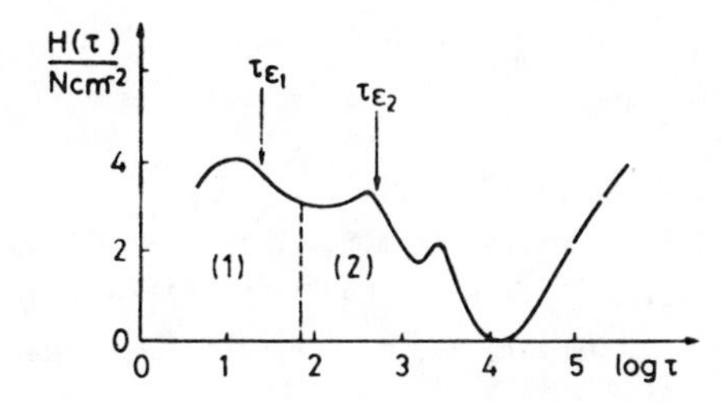

Fig. 6b: Relaxation time spectrum used in the calculations drawn out in fig. 6a according to Bartenev et al.[28].

7 FINAL REMARKS

In attempting to asses the significance of the above results, the idea of relating the deviations of the classical picture to effects due to finite chain length and "global" interactions proves to be fully justified. Satisfaction may be drawn from the fact that the modulus is found to be uniquely related to the density of the chemical cross-linkages. A single "systemtypical parameter" (λ_{min}) is in need only for defining a generalized relation between modulus and density of cross-linkages.

Taking a broad view, the lack of a molecular-theoretical interpretation of both of the van der Waals parameters is of less importance than the success of the van der Waals theory in describing the thermoelastic equilibrium properties of rubbers, in relating the modulus to the analytical concentration of the cross-linkages and in characterizing the "non-linear" viscoelastical properties of rubbers all of them in a quantitative and consistent manner.

REFERENCES

1. Kuhn, W. and Guen, F. Koll.Z.u.Z. 101, 248, 1942
2. Treloar, L.R.G. "The Physics of Rubber Elasticity" 3rd Ed., Clarendon Press, Oxford, 1975
3. Flory, P.J. "Principles of Polymer Chemistry", Cornell University Press, Ithaka-NY, 1953
4. Kilian, H.-G. Polymer, 22, 209, 1979
5. Green, A.G. and Atkins, J.E. "Large Elastic Deformations" 3rd Ed., Clarendon Press, Oxford
6. Enderle, H.F., Kilian, H.-G. and Vilgis, Th. Coll. & Polym.Sci. 262, 696, 1984
7. Kilian, H.-G. Polymer Bulletin 3, 151, 1980
8. Vilgis, Th. and Kilian, H.-G. to be published
9. Vilgis, Th. and Kilian, H.-G. Polymer 24, 949, 1982
10. de Gennes, P.G. "Scaling Concepts in Polymer Physics" Cornell University Press, Ithaka-London, 1979
11. Mullins, L. J.Appl.Polym.Sci. 2, 257, 1959
12. Kilian, H.-G., Unseld, K. Coll.& Polym.Sci., in press, 1985
13. Oppermann, R. Coll. & Polym.Sci. 259, 1177, 1981
14. Kilian, H.-G., Unseld, K., Jaeger, E., Müller, J., Jungnickel, B. Coll. & Polym. Sci. in press, 1985
15. Eisele, U., Heise, B., Kilian, H.-G., Pietralla, M., Angew.Makromol.Chemie 100, 67, 1981

16. Kilian, H.-G. Coll. & Polym.Sci. 260, 895, 1982
17. Godovsky, Yu.K. Polymer 22, 75, 1981
18. Kilian, H.-G. Coll. & Polym.Sci. 259, 1084, 1981
19. Kilian, H.-G., Vilgis, Th. Coll. & Polym.Sci. 262, 691, 1984
20. Valanis, K.C. J.Math.Phys. 46, 164, 1967
21. Biot, M.A. J.Appl.Phys. 25, 1385, 1954
22. Meixner, J. Ann. der Physik 5, 244, 1943
23. Graessley, W.W. Adv. Polym.Sci. 46, 47, 1982
24. Dusek, K. Internat. Rubber Conf., Moscow, 1984
25. Hummel, K. Koll.Z.u.Z.Polymere 182, 104, 1962
26. Vilgis, Th. Diplomarbeit, University of Ulm, 1982
27. Tschoegl, N.W. Polymer 20, 75, 1979
28. Bartenev, G.M., Stogova, Y.P. Polymer Sci.USSR 24, 322, 1982

PHOTOPOLYMERIZATION OF DIACRYLATES

J.G. KLOOSTERBOER, G.M.M. VAN DE HEI and G.F.C.M. LIJTEN

Philips Research Labs., P.O.Box 80 000, 5600 JA Eindhoven, The Netherlands.

SYNOPSIS

Experiments show that during the bulk polymerization of diacrylates at room temperature (i) the volume shrinkage cannot keep up with the chemical conversion; this explains the light intensity dependence of ultimate conversion; (ii) inhomogeneity persists far beyond the gel point; (iii) pendant double bonds initially show an enhanced reactivity due to local concentration effects; (iv) the size of domains of crystallizable monomer decreases strongly with increasing crosslink density.

1. INTRODUCTION

The photopolymerization of pure di(meth)acrylates is studied as a model system for various applications such as the coating of optical fibers (1), the replication of video discs (2) and of other optical components (3). All these applications require the control of optical and mechanical properties of the polymer. This demands a thorough understanding of bulk polymerization processes. Acrylates are preferred for their speed of polymerization and their versatility with respect to ultimate properties. Unless very long bridges are present between the acrylate moieties, polymerization at room temperature generally results in incomplete conversion of the double bonds due to vitrification of the sample. The distribution of unreacted groups between free monomer and pendant double bonds determines the crosslink density. Due to the high crosslink densities obtained, the distinction between cycles and active crosslinks breaks down, except for the smallest rings formed by cyclopolymerization.
We have investigated the bulk polymerization of 1,6-hexanediol diacrylate (HDDA) and bis(2-hydroxyethyl)bisphenol-A dimethacrylate (HEBDM) (Scheme 1) in some detail. Experimental details are given elsewhere (4,5).

SCHEME 1

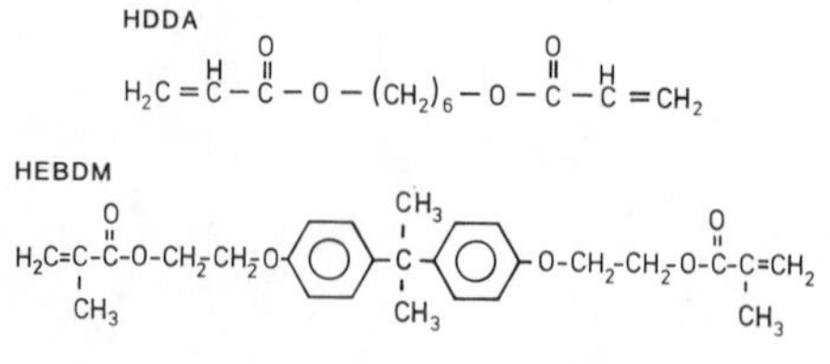

HDDA

$$H_2C=\overset{H}{C}-\overset{O}{\overset{\|}{C}}-O-(CH_2)_6-O-\overset{O}{\overset{\|}{C}}-\overset{H}{C}=CH_2$$

HEBDM

$$H_2C=\underset{CH_3}{C}-\overset{O}{\overset{\|}{C}}-O-CH_2-CH_2-O-C_6H_4-\underset{CH_3}{\overset{CH_3}{C}}-C_6H_4-O-CH_2-CH_2-O-\overset{O}{\overset{\|}{C}}-\underset{CH_3}{C}=CH_2$$

2. RESULTS AND DISCUSSION

2.1 Shrinkage and Conversion

During the polymerization of HDDA and HEBDM densely crosslinked networks are formed. Figure 1 shows the rate of polymerization for HDDA as measured with DSC, together with the rate of shrinkage, as measured with a thermomechanical system (TMS). The latter records the decrease of the thickness of a flat layer of monomer with an initial thickness of about 50 μm. Similar results were obtained with HEBDM.

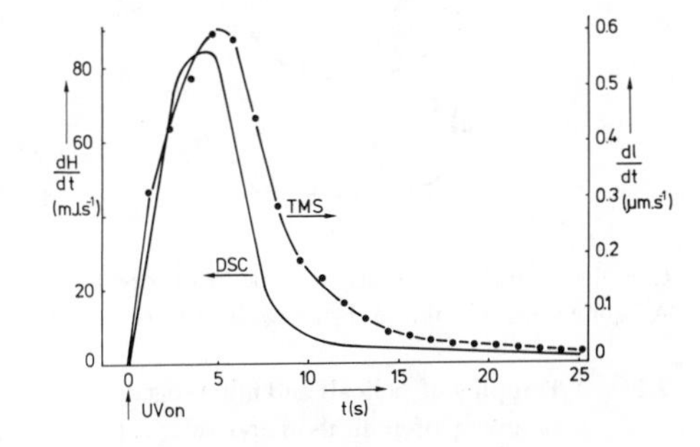

Figure 1.
Rate of polymerization (DSC) and rate of shrinkage (TMS) for the photopolymerization of HDDA at 20°C. Light intensity: 0.20 mW.cm^{-2}. Initiator: 4 w% α, α-dimethoxy-α-phenyl acetophenone. $[O_2] < 2$ ppm.

The distinct difference in decay of the polymerization and shrinkage rates beyond the maximum clearly shows that volume shrinkage cannot keep up with chemical conversion. The free volume created by the reaction is only slowly converted into over-all shrinkage. Upon increase of the rate of initiation the rate of polymerization also increases and so does the time lag until shrinkage occurs. In this way a temporary excess of free volume is generated which increases the mobility of unreacted C = C groups. The time lag found between conversion and shrinkage therefore explains our observation that ultimate conversion increases with light intensity, i.e. rate of initiation (Table 1).

TABLE 1. Influence of light intensity on final conversion at 20°C

	0.002 mW.cm^{-2}	0.020 mW.cm^{-2}	0.20 mW.cm^{-2}
HDDA	0.65	0.72	0.79
HEBDM	0.20[a]	0.47[a]	0.60

[a] inaccurate value (± 0.05 owing to extremely slow reaction; other figures ± 0.02)

The mutual dependence of shrinkage, conversion and their respective rates may be compared with physical aging of amorphous polymers (Fig. 2a). Physical aging is a self-decelerating process due to the closed-loop dependence of free volume V_f, segment mobility μ and rate of shrinkage dV_f/dt (6). The mobility of the polymer segments is the basic parameter. Physical aging is usually studied in polymers which have already been formed. However, the closed-loop dependence is expected to exist during isothermal bulk polymerization as well (Fig. 2b). During reaction there exists an additional and similar dependence of conversion x and rate of conversion dx/dt on the mobility μ of unreacted groups, either present as pendant double bonds or as free monomer (lower loop in Fig. 2b). Here too, the mobility is the basic parameter. Since chemical reaction immediately generates free volume there is also a direct relation between x and V_f.

a. Physical aging

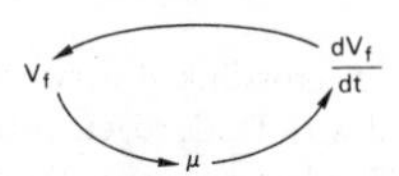

b. (Photo)polymerization

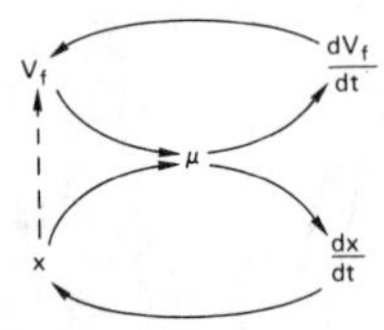

Figure 2.
Schematic relation between segmental mobility μ, free volume V_f and double bond conversion x.

Chemical reaction drives the shrinkage process but a high extent of reaction reduces the rate of shrinkage. A high extent of volume shrinkage in turn reduces the rate of chemical reaction.

2.2. Trapping of radicals and inhomogeneity

Trapping of radicals in polymeric glasses is a well-known phenomenon. We have established their presence by heating photopolymerized samples in the dark. This results in an additional conversion caused by remobilization of trapped radicals (Fig.3).
ESR spectra show a triplet and a nine line pattern which are identical with published spectra of acrylates and methacrylates, respectively (7,8).

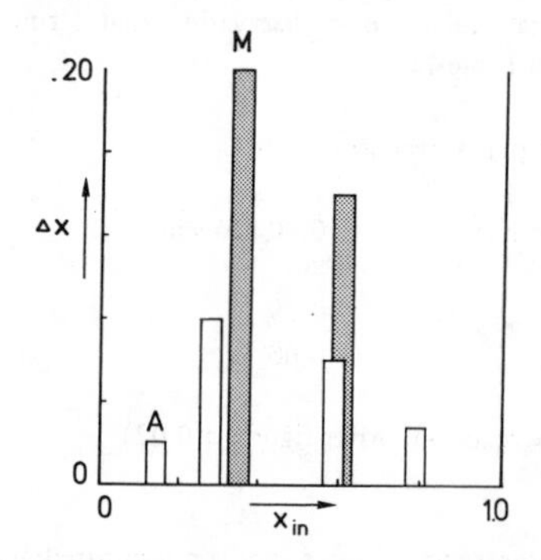

Figure 3.
Samples, polymerized in the DSC at 20°C to a double bond conversion x_{in} show an additional conversion Δx upon heating in the dark to 80°C. A (open bars): HDDA.
M (shaded bars): HEBDM.

The unexpected feature is the presence of trapped radicals in partly polymerized samples in which the polymerization is not yet hampered by over-all vitrification. In evacuated samples which still contained 30 per cent of unreacted HDDA the lifetime of the radicals exceeded 70 days at room temperature. The radicals thus have an extremely low mobility. At the same time continuation of irradiation leads to additional conversion, i.e. to the formation of mobile radical sites. This is interpreted as inhomogeneity, persisting far beyond the gel point (located near 1 per cent conversion). Pre-gel inhomogeneity is well-known in the bulk copolymerization of mono and divinyl compounds but it is generally assumed that rehomogenization occurs after gelation (9). This appears not to be the case with HDDA up to 80 and with HEBDM up to 60 per cent conversion.

The presence of regions of different mobility and, presumably, with different degrees of conversion, is not predicted by the classical assumption of equal reactivities of chemically identical groups. The Flory-Stockmayer-Gordon theory of gelation only describes step reactions and the crosslinking of polymer chains which already exist. However, the simultaneous chain polymerization and crosslinking can be simulated by using a percolation model (5,10). This model accounts for local structure variations, extensive ring formation and trapping of radicals.
Most of the HDDA radicals appear to be confined in deep traps.
During heating for 1 h at 80°C only 30 per cent of the radicals disappeared.
After an additional heating for 1 h at 120°C still 15 per cent of the radicals trapped at 20°C persisted. The introduction of oxygen (air) at 20°C caused a rapid decay of the ESR signal (several minutes or a few hours, depending on the initial amount of trapped radicals). Since no peroxyradicals could be detected it was concluded that rapid termination occurred. This requires an enhanced mobility of the radical sites.
Presumably, the mechanism established by Bresler and Kazbekov for the decay of radicals in PS, PMMA and PVAc (11) is also operative in our crosslinked systems. According to this mechanism rapid termination occurs through a chain process consisting of oxidation, hydrogen abstraction, oxidation of the newly formed radical and so on, until combination or disproportionation takes place.

2.3 Crosslink density

The way in which the crosslink density increases with conversion depends on the relative reactivities of pendant double bonds and those in free monomer, respectively. In Fig. 4 we have depicted the molality of the crosslinks as a function of double bond conversion for a few special cases. The molality was chosen in order to get a plot which is independent of density changes during polymerization.

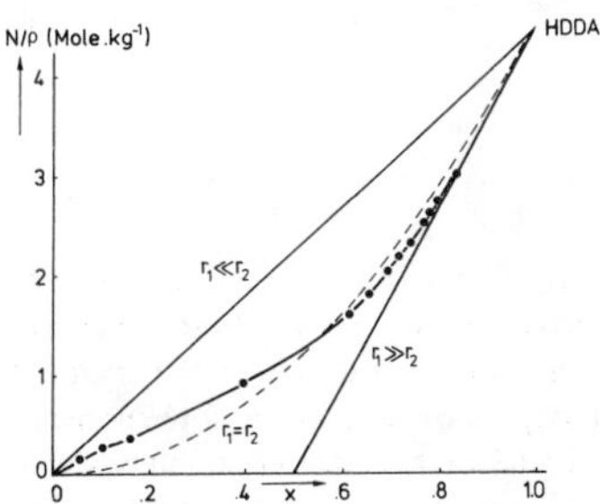

Figure 4.
Molality of crosslinks in HDDA versus double bond conversion x. r_1 is the reactivity of free monomer, r_2 that of pendant double bonds.

In the hypothetical case of $r_1 \gg r_2$ the lower path will be followed, i.e. crosslinking does not start until all molecules have reacted at one side. In the other hypothetical case of $r_1 \ll r_2$ the double bonds react in pairs and the crosslink density will be proportional to the conversion. All possible paths will be confined within the triangle, formed by the two limiting paths. In the classical case of $r_1 = r_2$ the dashed curve will be followed. The real path was determined by measurement of the concentration of pendant double bonds in the polymer. This was performed by polymerizing in the DSC to a known extent followed by extraction and determination of unreacted monomer with HPLC. The completeness of the extraction was monitored by adding an unreactive probe (5).
It can be seen that initially the pendant double bonds appear to exhibit an enhanced reactivity with respect to free monomer. This can be interpreted as being due to local concentration effects, arising from the inhomogeneous structure of the network (5,10).
In the final stage of the reaction a reversal of relative reactivities seems to occur. Here, over-all vitrification probably sets in, affecting the mobility of pendant double bonds more strongly than that of free monomer.

2.4 Melting of frozen, unreacted monomer

HDDA exhibits a sharp melting point at 284K which can easily be detected with DSC. In partially polymerized samples the transition broadens and shifts towards lower temperatures (Fig. 5).

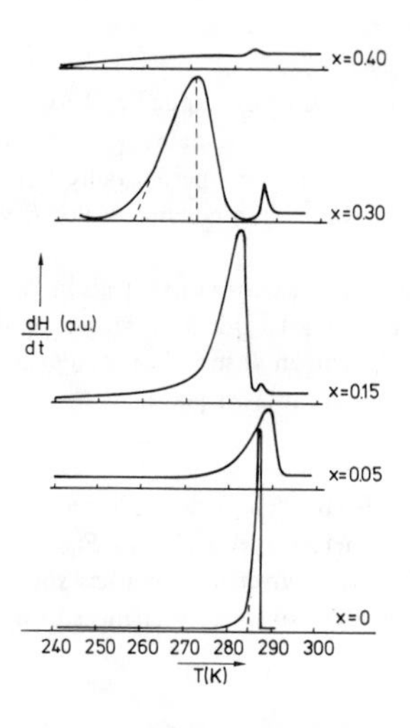

Figure 5.
Melting endotherms for HDDA at various double bond conversions x.

For x = 0.40, where at least 20 per cent of unreacted monomer is left, hardly anything can be observed. In the intermediate region a considerable melting point depression of up to 35K is observed. This is ascribed to a limitation of the size of the crystallites by the presence of the network. Size-dependent melting point depression is a classical phenomenon, similar to the increased vapour pressure of small droplets of a liquid or the enhanced solubility of a finely powdered solid in a solvent. The relation between size of a crystallite and its melting point depression is given by a special form of the Gibbs--Thomson equation

$$\Delta T = \frac{2\gamma M T_0}{\lambda_f \rho r}$$

in which γ is the interfacial tension between crystals and melt, M is the molecular weight, T_0 the normal melting point, λ_f the enthalpy of melting, ρ the density and r the radius of a spherical particle. With cubes a numerical factor of 4 instead of 2 is obtained and the edge size a appears instead of r. The interfacial tension has been estimated from (rather inaccurate) contact angle measurements. This enables an estimate of the crystal size from the melting point depression. Depending on whether the onset or the maximum of the melting endotherm is used, different sizes are obtained. In Table 2 these sizes are compared with the values obtained from X-ray diffraction line broadening. The order of magnitude seems to be correct and this is probably the best that can be hoped for using an approach as simple as followed here. It should further be stressed that probably a broad distribution of crystal sizes is formed and that the smallest ones are not even "seen" by X-ray diffraction.

TABLE 2. Crystallites in poly (HDDA)

C=C conversion	Melting point depression ΔT(K) onset	ΔT(K) peak	a (nm) onset	a (nm) peak	X-ray diffraction $\overline{a}$ (nm)
.05(?)					29
.08	9	3	8.1	24	
.15(?)					21
.16	21	10	3.3	7.5	
.30(?)					—*)
.33	34	20	2.2	3.7	
.40	—	—	—	—	
.70	—	—	—	—	

($\gamma = 10$ mN.m^{-1}) *) amorphous
(?) large sample; conversion estimated, not measured.

3. CONCLUSION

During the simultaneous chain polymerization and crosslinking of polyfunctional monomers densely crosslinked networks are formed. Their structure depends on the maximum extent of reaction but the latter depends on the rate of formation. This is caused by the mutual dependence of shrinkage, chemical conversion and their rates. Moreover, these networks tend to be inhomogeneous such that even at low conversion deeply trapped radicals will be produced. Oxygen restores the mobility of radical sites. Inhomogeneity enhances the reactivity of pendant double bonds with respect to free monomer. The size of regions with crystallizable monomer rapidly decreases with conversion.

ACKNOWLEDGMENT

The contributions of G.C.M. Dortant and R.G. Gossink who performed the experiments with HEBDM and of F.J.A. Greidanus who measured the ESR spectra is gratefully acknowledged.

REFERENCES

1-3. Broer, D.J., Lippits, G.J.M., and Zwiers, R.J.M., posters presented at this meeting.
4. Kloosterboer, J.G., Van de Hei, G.M.M., Gossink, R.G. and Dortant, G.C.M., Polym. Comm. **25**, 322 (1984).
5. Kloosterboer, J.G., Van de Hei, G.M.M. and Boots, H.M.J., Polym. Comm. **25**, 354 (1984).
6. Struik, L.C.E., "Physical Aging in Amorphous Polymers and Other Materials", Elsevier, N.Y. (1978).
7. Liang, R.H., Tsay, F.D., Kim, S.S. and Gupta, A., Polym. Mat. Sci. and Eng. **49**, 143 (1983).
8. Symons, M.C.R., J. Chem. Soc. 1186 (1963).
9. Dusek, K. in: Developments in polymerisation-3, Chapter 4 (R.N. Haward, Ed.) Appl. Science Publ., London (1982).
10. Boots, H.M.J., Paper presented at this meeting.
11. Bresler, S.E., and Kazbekov, E.N., Fortschr. Hochpolym. Forschung **6**, 688 (1964).

SIMULATION MODEL FOR DENSELY CROSS-LINKED NETWORKS FORMED BY CHAIN-REACTIONS

H.M.J. BOOTS

Philips Research Laboratories, P.O.Box 80 000, 5600 JA Eindhoven, The Netherlands.

SYNOPSIS

A simulation model is presented to study the structure of densely cross-linked networks as formed in the polymerization of, e.g., divinyl monomers. The model shows inhomogeneity in snapshots, which is consistent with indications of inhomogeneity in many experiments. The fraction of polymer units carrying a pendant double bond in the polymerization of a pure divinyl monomer is compared with experiment. At low conversion the model agrees qualitatively with experiment, in contrast to common analytic models. At high conversion the model underestimates the pendant-double-bond fraction.

1. INTRODUCTION

The standard model for network formation is the one proposed by Flory (1) and Stockmayer (2), which has been extended by Gordon and co-workers (3). This elegant model provides an adequate description of many network-forming systems. Two basic assumptions are:
a: the reaction probability of a reactive group depends on the average, rather than on the local concentrations of reaction partners (the mean-field assumption);
b: no (1,2) or few (3) rings are formed. For some classes of systems these assumptions are not fulfilled. Leung and Eichinger (4) concluded, that in tri- and tetrafunctional end-linking of Gaussian chains ring formation is important and they developed a simulation model for that case. In this paper we focus on the formation of densely cross-linked networks by a chain-reaction mechanism (e.g. free-radical polymerization of divinyl monomers or cationic polymerization of di-epoxides). Experiments on such systems (5-7) point to spatial inhomogeneity and strong cyclization. Therefore both assumptions are not fulfilled and it has often been pointed out (5) that the Flory-Stockmayer model does not apply to these experiments. In order to describe the network structure in this type of systems we use a simple lattice simulation model introduced for a different purpose by Manneville and De Sèze (8). It will be shown, that this model describes important features of the experimental systems in a qualitative way. At present there is no other model for the formation of densely cross-linked networks by chain-reaction polymerization.

In section 2 the model is presented. In section 3 snap-shots of typical network structures are shown and results on the fraction of polymer units carrying a pendant double bond are compared with experiment. The final section contains a discussion on the applicability of lattice simulation models to the formation of densely cross-linked networks.

2. THE MODEL

We start with a cubic lattice containing a point-like divinyl monomer molecule at each site. Then an arbitrary site is changed into a radical. Every unit of time a neighbour of the radical is chosen. If the neighbour has not fully reacted a chemical bond is formed and the radical function is transferred to this neighbouring site. Thus the radical performs a random walk on the lattice, thereby connecting a series of monomer units by chemical bonds. In the case of polymerization of divinyl monomer each site may be visited at most twice. The walk continues until the radical is trapped between fully reacted sites. Then a new radical is generated on an arbitrary free-monomer site, etc. This simulates the lowest possible initiation rate, which on the the largest lattices we used (90x90x90) was about equal to the usual experimental initiation rate. In other versions of the program higher initiation rates, termination by combination/disproportionation were used. Co-polymerization of monovinyl/divinyl mixtures was also studied.

The program is an adaptation of the programs developed by Manneville and De Sèze (8) and by Herrmann et al. (9). The model is usually called the Kinetic Gelation Model. Here 'kinetic' is meant as 'non-equilibrium' and should not be confused with chemical kinetics; in contrast to chemical kinetics the reaction probability depends on LOCAL concentrations of reaction partners. Most authors focus on the critical properties of the model very near to the gel point which in these systems occurs at a very low conversion. We concentrate on the (qualititative) description of the network structure before and after gelation (10).

3. RESULTS

In view of the experimental indications of inhomogeneity it is tempting to look first if spatial inhomogeneity is present in snap-shots of a two-dimensional version of the model. In Fig. 1a a snap-shot is presented of 25% bond conversion of divinyl monomer. The inhomogeneity of chain-reaction polymerizations, which is evident in Fig. 1a, becomes smaller and smaller, if one reduces the kinetic chain length. This may be done for example by increasing the initiation rate to unrealistically high values, or by artificially limiting the maximum kinetic chain-length as was done in Figs. 1b and 1c. The third snap-shot (Fig. 1c, for chain-lengths of at most two steps) is as homogeneous to the eye as the snap-shot of a step reaction of tetrafunctional monomer in Fig. 1d.

In Fig. 1a the occurrence of trapping of radicals inside regions of high polymer density is clearly observed. Also in three-dimensional simulations trapping is an important mechanism of radical termination, if the initiation rate is not very high. This is consistent with recent experimental observations on radical trapping (6).

To compare the model results with experimental data we calculated the fraction b of polymer units carrying a pendant double bond (an unreacted vinyl group) as a function of monomer conversion P (polymer mass fraction) in three dimensions; see Fig. 2. Experiments (5,6) show a sharp initial drop of b as a function of monomer conversion P (polymer mass fraction) and then a plateau up to very high conversion. The sharp drop is interpreted as an indication for the existence of highly internally cross-

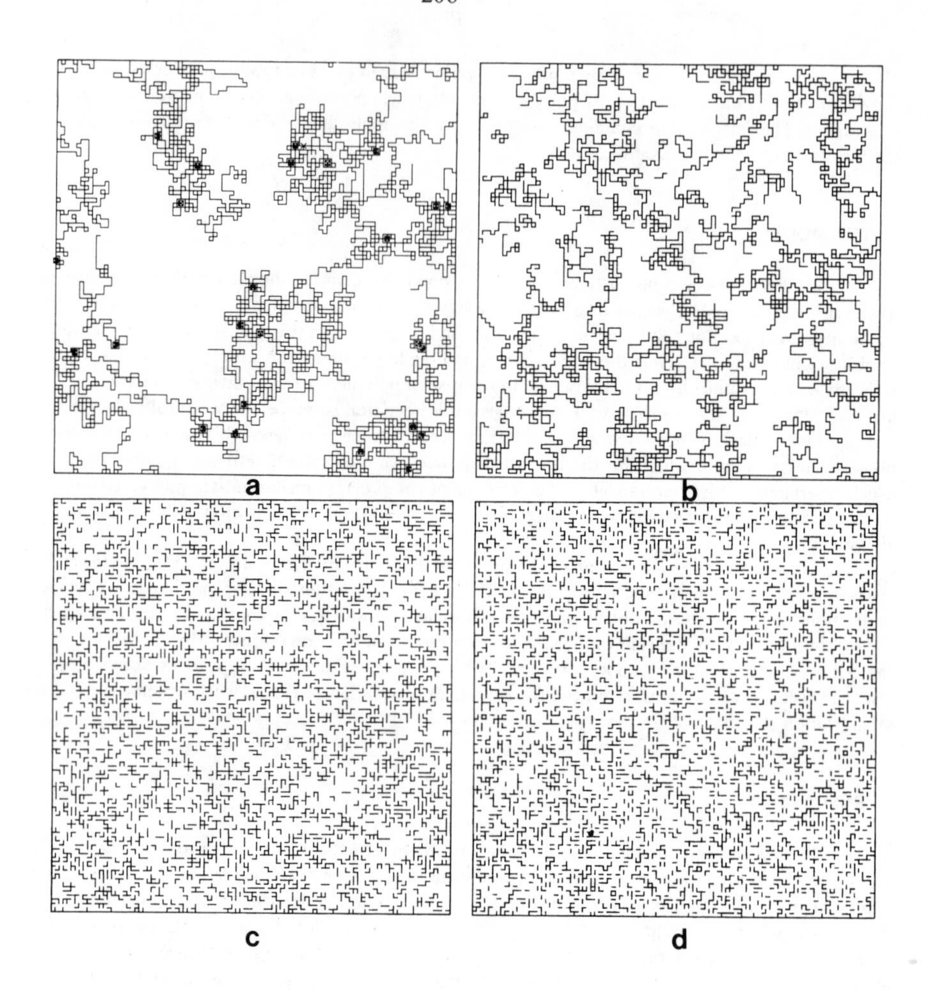

Figure 1.
Snap-shots of polymerization of tetrafunctional monomer in two dimensions on a 100x100 lattice at a bond conversion of 25%. Bonds between units are indicated, units themselves are not. Boundary conditions are such, that sites on the left (top) boundary are neighbours of sites on the right (bottom) boundary of the lattice. Fig. 1a is obtained by a chain-reaction of one 'living' radical at a time; when a radical is trapped (indicated by crossed squares) a new radical starts. The living radical is indicated by X. To show the decrease in inhomogeneity on decreasing the kinetic chain length, the maximum kinetic chain length in Figs. 1b and 1c is artificially limited to 20 steps and to 2 steps, respectively (here radicals are not indicated). Fig. 1c is as homogeneous as the snap-shot (Fig. 1d) for step reaction polymerization of tetrafunctional monomer after conversion of 25% of the reactive groups.

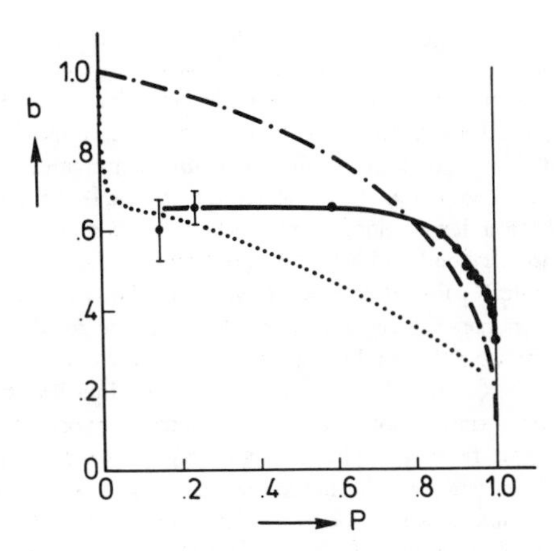

Figure 2.
The pendant-double-bond fraction b (the fraction of polymer units carrying an unreacted vinyl group) as a function of monomer conversion P. The drawn line guides the eye through experimental data by Kloosterboer et al. (6), the dashed-dotted curve results from the usual mean-field assumption and the dotted curve results from the present simulation model for the polymerization of a pure divinyl monomer in three dimensions.

-linked microgel-like macromolecules. The plateau is interpreted as the shielding (11) of pendant double bonds from further reaction inside densely cross-linked regions. Unlike the analytic theory, which uses the Flory-Stockmayer mean-field assumption, the simulation results provide a qualitatively correct description of the initial drop. At high conversion the model seems to underestimate the screening effect. Attempts to remove this discrepancy by changing the lattice size or the initiation rate, or by introducing some chain stiffness were in vain.

4. DISCUSSION

The main limitation of the model is the restriction of the units to a rigid lattice. This implies that effects which are due to the mobility of polymerized units are not incorporated in the model. Moreover polymerization shrinkage cannot be taken into account in a straightforward manner and radical functions move only by propagation reactions and not by diffusion. Therefore it is expected, that the model is adequate for systems which, at least locally, become densely cross-linked at early stages of the reaction. Such systems are soon (locally) very stiff and one expects that for most properties relaxation effects of polymer units may be neglected. Formally it is possible to simulate combined growth, cross-linking and relaxation off the lattice but even a simple non-trivial problem (one radical in divinyl monomer) demands a prohibitively large computer effort (12).

Thus densely cross-linked networks are the best candidates for the application of Kinetic Gelation Models, though relaxation effects, which are important before gelation, are disregarded. Then the question arises, why the model results for the pendant-double-bond fraction are qualitatively correct in the beginning of the reaction but not in the end. Relaxation effects in the beginning will tend to decrease the sizes of loosely cross-linked coils. This will enhance the cross-link probability on further reaction in the equilibrium ensemble, and the initial drop in b will only be sharper.

The discrepancy at high conversion is tentatively explained on the basis of the mobility of polymer units over distances of the order of the reaction distance. Before reaction between the reactive groups of neighbouring units can take place, reorientation and smal displacements of these units are in general required. Such motions are not accounted for by the model, which treats the units as points. This is a good approximation at low conversion, where these local motions are very rapid, but not at high conversion. Units which are attached to the network (pendant double bonds, radical functions) have a lower mobility than free-monomer units. This would explain that the reactivity of pendant double bonds is lower than predicted by the model. A similar effect is the immediate recombination of radical pairs after their creation in densely cross-linked regions. This favours polymer growth in monomer-rich regions. This effect is found to be small in the model, but may be of importance, if the shielding of radicals is enhanced by the bulk character of the units.

Though the present model certainly has its limitations, it is the only model available to study the formation of densely cross-linked networks by a chain-reaction mechanism. The model does not suffer from a wealth of adjustable parameters. It shows the appearance of inhomogeneity and its dependence on the initiation rate. It shows radical trapping during the polymerization. It compares favourably with experimental data on pendant double bonds at low conversion, but it overestimates the pendant-double-bond reactivity at high conversion.

ACKNOWLEDGMENT

The author thanks R.B. Pandey (Athens, GA, USA), D. Stauffer (Cologne), A. Baumgärtner (Julich), J.G. Kloosterboer and M.F.H. Schuurmans (Eindhoven) for stimulating discussions.

REFERENCES

1. Flory, P.J. 'Principles of Polymer Chemistry', p.124ff, Cornell University Press, Ithaca N.Y. 1953.
2. Stockmayer, W.H., J. Chem. Phys. **11,** 45 (1943).
3. See e.g., Gordon, M. and Scantlebury, G.R., J. Chem. Soc. B 1 (1967).
4. Leung, Y.-K. and Eichinger, B.E., J. Chem. Phys. **80,** 3877 and 3885 (1984).
5. Dušek, K. in 'Developments in Polymerization 3', Chapter 4 (R.N. Haward, Ed.) Applied Science Publ., London 1982 and refs. therein; Aso, C., J. Polymer Sci. **39,** 475 (1959).
6. Kloosterboer, J.G., van de Hei, G.M.M. and Lijten, G.F.C.M., paper in this conference; Kloosterboer, J.G., van de Hei, G.M.M. and Boots, H.M.J., Polym. Commun. **25,** 354 (1984).
7. Meijer, E.W., paper in this conference.
8. Manneville, P. and de Sèze, L. in 'Numerical Methods in the Study of Critical Phenomena' (I. della Dora, j. Demongeot and B. Lacolle, Eds.). Springer, Berlin 1981.
9. Herrmann, H.J., Stauffer, D. and Landau, D.P., J. Phys. **A16,** 1221 (1983).
10. Boots, H.M.J. and Pandey, R.B., Polymer Bulletin **11,** 415 (1984).
11. Minnema, L. and Staverman, A.J., J. Polym. Sci. **29,** 218 (1958).
12. Baumgärtner, A. and Boots, H.M.J., unpublished results.

NONLINEAR VISCOELASTICITY OF EPDM NETWORKS

Boudewijn J.R. Scholtens and Paul J.R. Leblans*

DSM, Research and Patents, PO Box 18, 6160 MD Geleen, Netherlands
* Department of Chemistry, University of Antwerp, Antwerp, Belgium

SYNOPSIS

The nonlinear viscoelastic behaviour of noncrystalline EPDM networks has been studied at different tensile velocities and measuring temperatures, for various crosslink densities and prepolymers.

1 INTRODUCTION

Recently two methods have been proposed for analyzing stress-strain measurements of elastomeric networks at constant tensile velocities.[1] Both methods are based on a well-known single integral constitutive equation in which the time and strain effects are separated. After some minor approximations the first, analytical, approach results in a simple equation similar to an empirical relation proposed by Smith.[2] The second, numerical, approach is exact for infinitely small summation steps, and yields results which are indistinguishable for practical summation step lengths.

It is the aim of the present study to apply these two methods to stress-strain measurements of various EPDM (ethylene propylene diene monomer) elastomeric networks which have been carefully characterized in the linear viscoelastic region.[3]

2 EXPERIMENTAL

Chemical and physical characteristics of the noncrosslinked elastomers are shown in Table 1, for more details see ref. 3. The permanent networks of these elastomers, obtained as described in ref. 3, are designated by the sample code followed by a number indicating the weight percentage of peroxide used

Table 1: Characteristics of the noncrosslinked EPDM elastomers.

sample code	composition (mol %) ethylene	diene monomer	GPC results (kg/mol) $\overline{M}_n$	$\overline{M}_m$	Mooney index[a)] $ML_w(1+4)$	T_i[b)] (K)
K	64.6	2.7	45	180	48	280
N	64.8	1.1	40	90	21	280
D	58.2	0.8	50	125	28	247

a) determined at T = 398 K.

b) initial crystallization temperature determined by DSC; for details see ref.3.

(0.2-1.6 wt.%), e.g. K 0.2. More details on these networks have been published elsewhere.[3,4]

The stress-strain measurements were performed with an Instron 1195 tensile tester equipped with a TNO nitrogen thermostat (± 0.1 K), and automated with an HP 85 calculator and an HP 6940 B multiprogrammer (both Hewlett Packard). Seven crosshead speeds were used between 5 and 500 mm/min, at temperatures T = 300, 345 and 390 K. Rectangular samples, about 4 mm wide and 2 mm thick, were fixed in the clamps, which were 50 mm apart at the start of an experiment. Consequently, the tensile velocity $\dot{\lambda} = d\lambda/dt$ ranged from 1.67 10^{-3} to 1.67 10^{-1} s^{-1}. The maximum extension ratio possible in the thermostat was $\lambda_{max} = 7.8$.

3 RESULTS

For the cured elastomeric networks, which exhibit a relatively low relaxation strength,[1-3] the nonlinear strain measure, $S_E(\lambda)$, can be calculated with high precision with the approximate relationship:[1]

$$S_E(\lambda) = 3\sigma_E[\lambda(t)]/F(t) \qquad [1]$$

where $\sigma_E[\lambda(t)]$ represents the true tensile stress and F(t) the constant-tensile-velocity-modulus.[1,2] As shown in fig. 1, $S_E(\lambda)$ is only slightly smaller if determined with eq. [1] than with the exact, numerical method, as is understandable.[1] In addition, its value equals the relative Finger strain, $C_E^{-1}[\lambda(t,t')]$, for small extension ratios. Apparently $S_E(\lambda)$ is independent of the tensile velocity, and hence also of time. This factorability of stress and strain effects is also obvious from the parallelism of the isostrain log $\sigma_E[\lambda(t)]$ vs. log t curves to log F(t) vs. log t, as shown in fig. 2. A comparison of these isostrain curves was already used by Smith to assess factorability.[2]

Whereas $S_E(\lambda)$ is independent of time, its value increases with rising measuring temperature, in particular for $\lambda \geqslant 2$, as shown in fig. 3 for a typical sample.

The variation of $S_E(\lambda)$ with the degree of crosslinking is less uniform. For both the N and D networks $S_E(\lambda)$ increases slightly with crosslink density, as exemplified in fig. 4, but no such dependence was observed in K 0.2 and 0.4.

Finally, the type of the prepolymer affects $S_E(\lambda)$ considerably, as demonstrated in fig. 5 for three networks prepared with different prepolymers but of comparable crosslink density.

4 DISCUSSION

The methods we used to determine $S_E(\lambda)$ do not impose any restrictions on its shape, except that $S_E(\lambda)$ must equal $C_E^{-1}(\lambda)$ at small λ values. As a consequence, these methods are more versatile than the approach of Tschoegl et al., which is based on a particular 1-parameter strain energy density function.[5-7] Indeed, the majority of the present $S_E(\lambda)$ curves cannot be described satisfactorily with the 1-parameter Seth strain measure. This is due to the relatively high linearity for $\lambda \lesssim 1.5$-2. On the other hand, the increase of linearity with rising temperature and degree of crosslinking is qualitatively similar to results reported by Tschoegl et al.,[5-7] and to the well-known variation of the ratio of the Mooney-Rivlin constants, C_2/C_1, with these variables.[6]

REFERENCES

1. B.J.R. Scholtens, P.J.R. Leblans and H.C. Booij submitted to J. Rheol..

2. T.L. Smith Trans. Soc. Rheol., vol. 6, 61-80 (1962).

3. B.J.R. Scholtens J. Polym. Sci. Polym. Phys. Ed., vol. 22, 317-344 (1984).

4. B.J.R. Scholtens, E. Riande and J.E. Mark J. Polym. Sci. Polym. Phys. Ed., vol. 22, 1223-1238 (1984).

5. W.V. Chang, R. Bloch and N.W. Tschoegl in 'Chemistry and Properties of Crosslinked Polymers', S.S. Labana, ed., Academic Press, 1977, p. 431-451.

6. W.V. Chang, R. Bloch and N.W. Tschoegl J. Polym. Sci. Polym. Phys. Ed., vol. 15, 923-944 (1977).

7. R. Bloch, W.V. Chang and N.W. Tschoegl J. Rheol., vol. 22, 1-32 (1978).

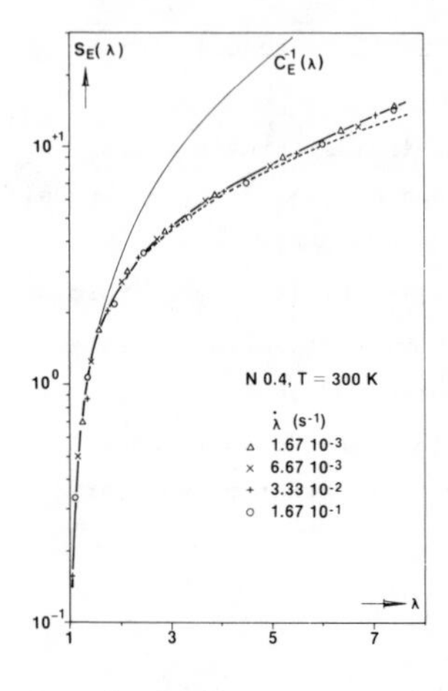

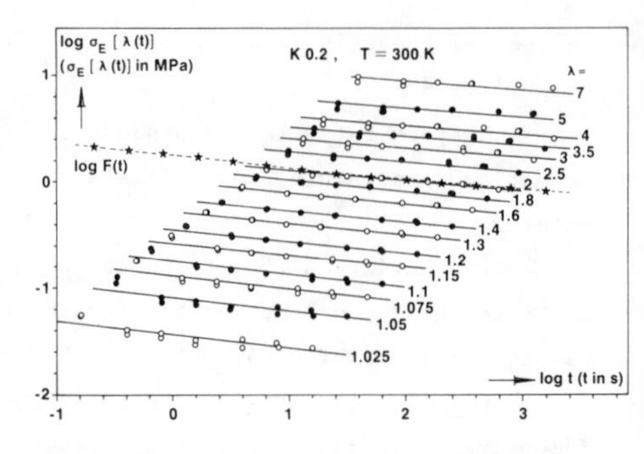

Fig. 2. Comparison of the variation of log $\sigma_E[\lambda(t)]$ and log F(t) with log t for K 0.2 at 300 K.

Fig. 1. Nonlinear strain measure for N 0.4 at 300 K as a function of the tensile velocity $\dot{\lambda}$; the solid curve and symbols refer to the numerical method, the broken curve refers to eq. [1].

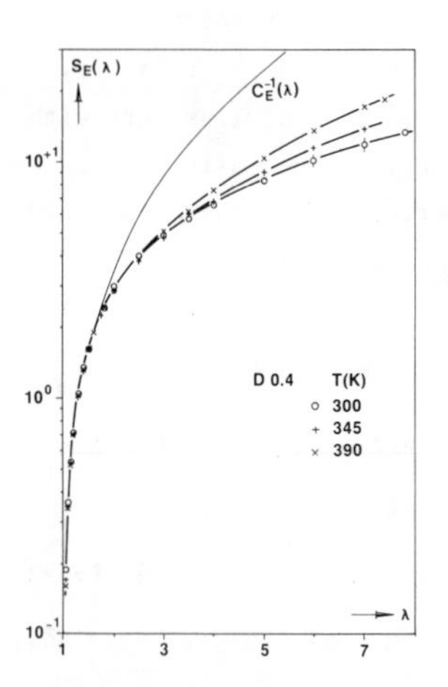

Fig. 3. Variation of $S_E(\lambda)$ with measuring temperature for D 0.4.

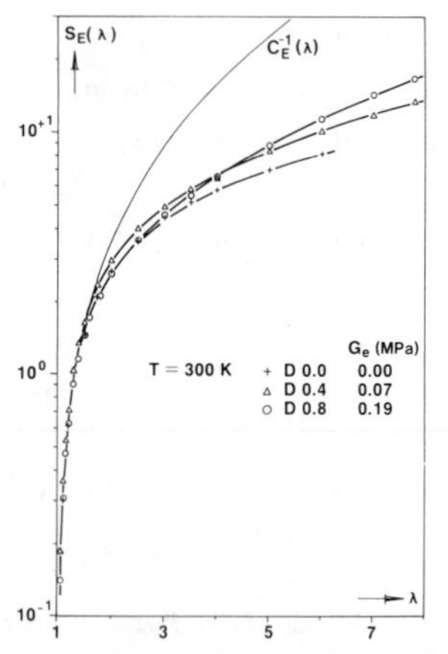

Fig. 4. Variation of $S_E(\lambda)$ with degree of crosslinking for D 0.0, 0.4 and 0.8 at 300 K (G_e = equilibrium shear modulus).

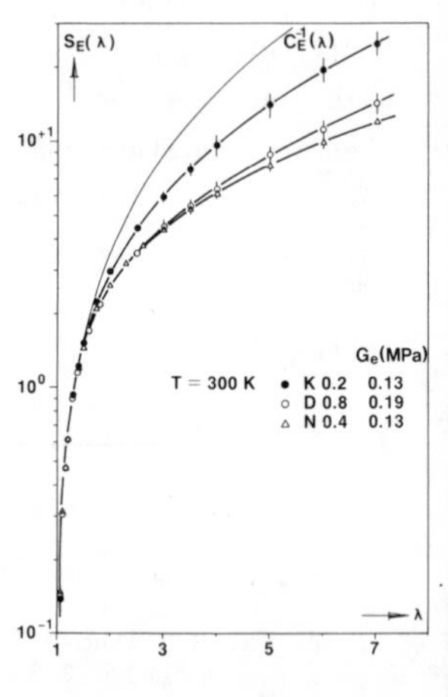

Fig. 5. Variation of $S_E(\lambda)$ with the type of the prepolymer at roughly equal degrees of crosslinking (G_e = equilibrium shear modulus).

SOME COMMENTS ON THE THERMODYNAMICS OF SWELLING

M. NAGY

Department of Colloid Science, Loránd Eötvös University,
Budapest

SYNOPSIS

The paper presents some new results on both experimental and theoretical aspects of swelling of network polymers with special attention to the problem of the reference state and homogeneity of gel structure.

1 INTRODUCTION

It was a hope in the earlier works on gels that swelling studies combining with stress-strain measurements provide us at least with two types of basic information, i.e., with the value of polymer-solvent interaction parameter, χ, and the concentration of the elastically active chains, ν, the latter being an important characteristic of the network structure[1,2]. However, on the basis of an extensive experimental and theoretical work performed in our laboratory it turned out that the equations proposed to evaluate experimental data might lead to erroneous values of both the polymer-solvent interaction parameter and the concentration of the elastically active elements.

2 THERMODYNAMICS OF SWELLING

The swelling equilibrium of gels in a pure solvent can be phenomenologically described as a sum of at least two solvent chemical potential terms[3]

$$\Delta\mu_{1,total} = \Delta\mu_{1,mix} + \Delta\mu_{1,c} \qquad (1)$$

where $\Delta\mu_{1,total}$ can be directly calculated from swelling pressure, vapor pressure or from deswelling measurements, $\Delta\mu_{1,mix}$ is related to the mixing of solvent molecules with the cross-linked polymer, and $\Delta\mu_{1,c}$ is a contribution due to the presence of the network structure. These two additive components of the overall change in the solvent chemical potential, $\Delta\mu_{1,total}$, cannot be measured separately, but they can be calculated using different equations based on different assumptions[4,5,6]. At present it is generally accepted in the literature that the application of the Flory-Huggins equation for calculation of $\Delta\mu_{1,mix}$ is a good approximation, thus from swelling measurements the χ can be unambigouosly determined.

In the following two parts of this paper first the effect of the inhomogeneities on $\Delta\mu_{1,total} = f(v_2)$ functions and also, their effect on χ as well as the role of the reference state in calculation of $\Delta\mu_{1,c}$ will be discussed.

3 THE EFFECT OF INHOMOGENEITIES ON THE SHAPE OF THE $\Delta\mu_{1,total} = f(v_2)$ FUNCTIONS

By using of a simple, additive, two-phase network modell it was shown[7] that the presence of inhomogeneities can strongly affect the shape of the $\Delta\mu_{1,total} = f(v_2)$ functions. Because this experimentally accessible functions serves as a basis for estimation of χ it is a reasonable assumption that the value of χ parameter and its concentration dependence should be somehow related to the topology dependence of the mixing term.

Table 1:

Concentration dependence of χ values for homogeneous (A) and inhomogeneous (B) gels

	v_2	0.10	0.12	0.14	0.16	0.18	0.20
χ	gel A	0.524	0.524	0.527	0.532	0.540	—
	gel B	—	0.542	0.549	0.557	0.565	0.573

In fig.1, as typical examples, two $\Delta\mu_{1,total} = f(v_2)$ functions are shown for poly (vinyl alcohol) gels swollen in water and differing considerably in homogeneity of their structure. As can be seen there is a great difference in the shape of the two curves though the volume degree of swelling ($q = v_2^{-1}$) in pure water for the two gels is practically the same. From these curves with the application of the Flory-Huggins equation χ values at different volume fractions were estimated. In these calculations the elastic contributions of the networks to $\Delta\mu_{1,total}$ were neglected. The Table 1 shows clearly that for the inhomogeneous structure the χ values are higher and their concentration dependence is relatively steeper than that for the homogeneous gel.

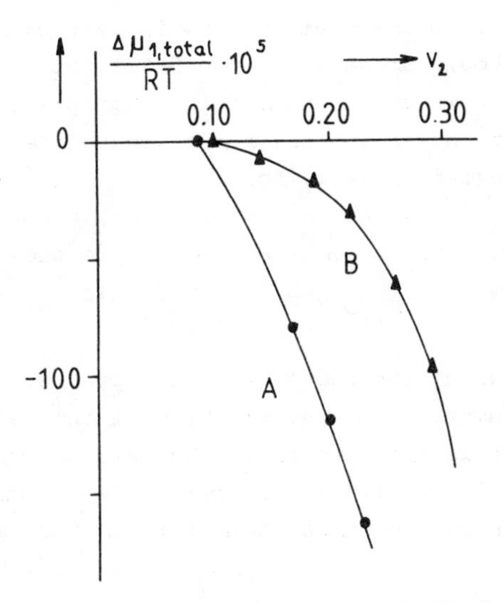

Fig.1. Comparison of $\Delta\mu_{1,total} = f(v_2)$ functions for homogeneous (A) and inhomogeneous (B) gels

4 THE REFERENCE VOLUME FRACTION

One of the essential differences between the osmotic and swelling phenomena is that for swollen networks it is always necessary to introduce a reference state at which the elastic contribution to $\Delta\mu_{1,total}$ is zero. Although some attempts have been made this problem cannot be considered as a solved one because most of the equations currently used in the literature predict incorrect values for $\Delta\mu_{1,c}$.

In order to illustrate briefly the task to be solved let us consider the cross-linking of a pure polymer in the rubber elastic state. At low degrees of cross-linking it is reasonable to assume that during the cross-linking reaction the average conformation of polymer chains does not change, i.e., such a network should be to a good approximation in its reference state.

This means that no elastic part of solvent chemical potential should arise at $v_2 = 1$. Still, at this volume fraction the well-known equations predict high positive value for $\Delta\mu_{1,c}$ if unity is chosen for the reference volume fraction or for reference degree of swelling.

Based on the classical theory of rubber elasticity it is possible to derive a new expression

$$\Delta\mu_{1,c} = K^{*}RT\rho_{\nu_{dry}} v_2(v_{2,0}^{1/3} - v_2^{1/3})\bar{V}_1 \qquad (2)$$

which, for ideal networks gives correctly the form of $\Delta\mu_{1,c} = f(v_2)$ function and the position of the reference volume fraction. As it is shown in fig.2 the maximum curves intersect the horizontal axis exactly at the reference volume fractions as indicated. Since the mixing term for miscible systems is always negativ the volume fractions at swelling equilibrium in pure solvent should be within the range where the $\Delta\mu_{1,c} = f(v_2)$ curves have positive values. At volume fractions higher than the reference one the network chains are to a more or less extent in a state of compressed conformation.

The maximum character of the $\Delta\mu_{1,c} = f(v_2)$ curves is supported by experimental data taken from the literature[8].

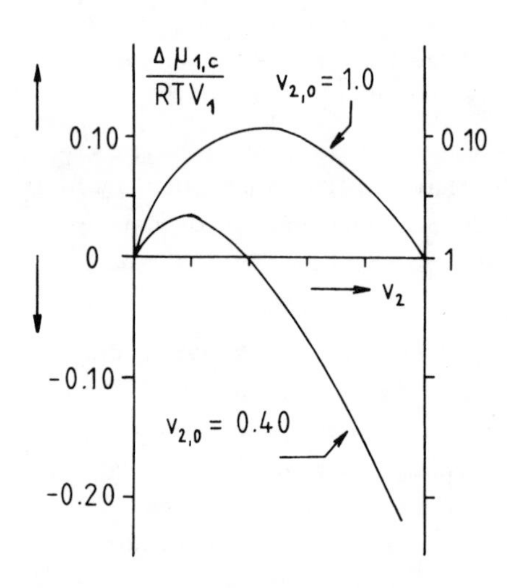

Fig.2. Graphical representation of eq.2 at different reference volume fractions. All the other parameters was taken to be one

REFERENCES

1. Flory, P.J. 'Principles of Polymer Chemistry', Cornell University Press, 1953
2. Dušek, K. and Prins, W. 'Adv. Polymer Sci.', Vol. 6, 1 (1969)
3. Nagy, M. and Horkay, F. 'Acta Chim. Acad. Sci. Hung.', 104, 49 (1980)
4. Boyer, R.F. 'J. Chem. Phys.', 13, 363 (1945)
5. Gee, G. and Orr, W.J.C. 'Trans. Faraday Soc.', 42, 585 (1966)
6. Kraats, E.J.van de, Winkeler, M.A.M., Potters, J.M., Prins, W. 'Res. Trav. Chim. Pay-Bas', 88, 449 (1969)
7. Nagy, M. 'Colloid and Polymer Sci.', in press
8. Yen, L.Y., Eichinger, B.E. 'J. Polymer Sci.', Polymer Physics Edition, 16, 121 (1978)

NOTATION

v_2 volume fraction of polymer

$v_{2,0}$ reference volume fraction of polymer

$\bar{V}_1$ partial molar volume of solvent

$\rho_{\nu_{dry}}$ the quantity of the elastically active chains referring to unit volume of dry polymer

K^* constant, which is independent of the gel volume

THERMOREVERSIBLE GELATION OF VINYLPOLYMERS

H. BERGHMANS and W. STOKS

Laboratorium voor Polymeeronderzoek, Celestijnenlaan 200 F, B-3030 Heverlee-Leuven Belgium

SYNOPSIS

The phenomenon of thermoreversible gelation of polymer solutions is discussed. The influence of crystallizability, solubility and chain structure is illustrated. A detailed analysis of the gelation of poly(vinyl alcohol) in ethylene glycol is presented. Calorimetric and optical measurements reveal the occurrence of liquid-liquid phase separation. Combined with X-ray scattering data, these observations lead to two different mechanisms when gelation is performed above and below the demixing temperature.

1 GENERAL ASPECTS OF THERMOREVERSIBLE GELATION

Polymer gels are macromolecular networks, swollen with solvent(1). The crosslinks, responsible for this network formation, can be of chemical or physical origin. Chemical crosslinking, performed during or after the polymerization, results in non-reversible networks, characterized by an absence of flow. When crosslinking is of physical nature, thermoreversible gels are formed. On cooling the polymer solutions, a sol to gel transition is observed. Heating results in the reverse phenomenon. The physical crosslinks originate from specific interaction between the polymer chains. They eventually lead to the formation of crystallites of different morphology.

When crystallization is at the origin of thermoreversible gelation, the generally accepted theories on polymer crystallization can be applied. It is well known that polymer crystallization is a nucleation controlled process which needs a certain degree of undercooling. A metastable crystalline structure is formed, the properties of which depend strongly on the crystallization conditions. The melting point (T_m) is determined by the degree of undercooling and consequently by the crystallization temperature (T_c). It can be altered by any thermal treatment given after the crystallization. It also depends on the morphology of the crystallites. Folded chain, lamellar crystallites and bundlelike, fringed micellar crystallites can be obtained at the same degree of undercooling. The melting point of lamellar crystallites is determined by their lamellar thickness and fold surface free energy. Because of their bundlelike nature, fringed micellar crystallites melt at much lower temperatures, close to their formation temperature. Calorimetric measurements are very well suited to reveal the presence of these different morphologies.

Both morphologies can also be present in crystalline, thermoreversible gels. This was concluded from calorimetric as well as X-ray scattering investigations[(2)(3)]. It was proposed that fringed micellar crystallites, mainly responsible for network formation, orient with their molecular axis parallel to the stretching direction. Folded chain crystallites orient with their molecular axis perpendicular to this direction[(1)(3)]. It was also suggested that their contribution to the network formation could be less important[(3)].

Because gelation occurs in a two component polymer-solvent system, the solubility of the polymer is very important. Many polymer solutions show liquid-liquid phase separation on cooling. Phase separation can be nucleation controlled or occur by a mechanism of spinodal decomposition[(4)]. Crystallization can complicate this behavior to a large extend. An increase in concentration in small domains will raise the rate of crystallization and "freeze" the initial morphology. Because T_m is far above the crystallization temperature, an imported hysteresis will be observed. Amorphous phase separation can also result in the formation of gel-like structures as long as the temperature of flow is above the experimental temperature[(5)].

Also important is chain composition and chain microstructure or distribution of monomer units along the chain. A high degree of tacticity is required for the crystallization of vinyl polymers. Stereoregular sequences of a certain length are needed to build up these crystallites. Consequently, industrial polymers, prepared by a radicalar mechanism, are not very well suited for crystallization. One important exception is poly(vinyl alcohol) (PVAL). The atactic isomer crystallizes very well because the hydrogen atom and the hydroxyl function can be exchanged in the lattice[(6)].

The length of the crystallizable sequences can also be reduced by copolymerization. The distribution of comonomers is of great importance. A supplementary complication arises with these polymers because of the difference in chemical nature of the segments. In presence of a solvent, a modified polymer-solvent interaction will be observed and the gelation-crystallization behavior will be modified[7].

Thermoreversible gelation has been observed with many industrially as well as academically interesting polymers. Because of the importance of crystallization, polymers can be classified according to this parameter.

Vinyl polymers can roughly be subdivided into two main categories : well crystallizable and poorly or not crystallizable chains.

The first group includes polymers as poly(ethene) (PE), isotactic poly (styrene) (IPS) and poly(vinyl alcohol) (PVAL). These polymers crystallize from the melt with a lamellar, folded chain morphology. This mode of crystallization can be encountered in the preparation of gels of PE and IPS. It results in the formation of turbid gels or suspensions because of the presence of these large crystallites. Fully transparent gels however can also be obtained with both polymers. At sufficiently high temperatures, PE solutions can be transformed in transparent solutions on mechanical solicitation[8][9]. The formation of bundle-like crystals was proposed[10]. On stretching, these gels, very strong fibres can be obtained. This gel spinning of PE has been developed for the production of ultra high modulus fibres[1][12].

Solutions of IPS in decaline also can be transformed into transparent gels on cooling to sufficiently low temperatures. From X-ray scattering experiments, it was concluded to the formation of a nearly all trans extended helical conformation[12]. It is different from the well known 3/1 helical conformation, encountered when crystallization is performed from the melt or from dilute solution.

Solutions of PVAL in e.g. ethyleneglycol have been reported to transform in gels on cooling. Crystalline, turbid gels are obtained[13]. The system seems to be complicated by liquid-liquid phase separation[14]. This system will be discussed in more detail in the second part of this paper.

The problem becomes more complex when less or non crystallizable systems are studied. A typical exemple is poly(vinylchloride) (PVC). The crystallinity of this polymer is far from the values obtained with the polymers cited before. It forms however very easily thermoreversible gels in many solvents. Because of its industrial importance, the gelation was thoroughly studied from both rheological and morphological point of view.

The crystalline nature of the gelation was recognized very early[15][16]. The presence of both lamellar folded chain and fringed micellar crystallites was revealed only recently[18]-[21]. These crystallites are build up by the syndiotactic sequences present in the atactic chains. While on stretching the gels the fringed micellar crystallites orient with the molecular axis parallel to the stretching direction, the folded chain ones orient perpendicular to it. The a-axis corresponding to the C-Cl resulting dipole[18] is then parallel to this direction. The presence of microcrystallites, already formed during the polymerization, was clearly demonstrated by studying the difference in gelation behavior of heterogeneously and homogeneously chlorinated PVC. In the first case, gelation is still observed at relatively high degree of chlorination. In the second case, the gelation ability disappears already at much lower chlorine concentrations[22].

A major problem, not yet fully solved up to now, is the possibility of finding enough sequences of the required length to build both types of morphologies. Accepting Bernouillian statistics, the fractions of these sequences are to small[21]. One way out of this impasse is to change the concept of the crystalline morphology of these gels. An alternative way is a different chain statistics, resulting in a different distribution of the syndiotactic placements.

Analogous problems were also encountered with copolyesters obtained by random copolycondensation of terephthalic and isophthalic acid with ethylene glycol[2]. Although DSC observations reveal only the presence of a fringed micellar morphology, two molecular orientations were observed on stretching. One of them corresponds to an orientation of the aromatic rings perpendicular to the stretching direction, corresponding to a direction of important intermolecular interaction in the lattice.

Recently, an alternative gelation of an atactic polymer was proposed. The thermoreversible gelation of atactic polystyrene was reported to be an equilibrium phenomenon[23][24]. This means that no difference between gelation temperature and gel melting temperature was observed. Complex phase diagrams were proposed and the formation of transparent as well as opaque gels was reported.

2 THERMOREVERSIBLE GELATION OF PVAL

A Experimental results

Moderately concentrated solutions of PVAL in ethyleneglycol gelify quickly on cooling. The exothermic phenomenon can easily be followed by calorimetric observations. Heating of the gels results in an endothermic melting. The thermal and mechanical properties of the gels depend strongly on polymer concentration, gelation temperature and time.

a. Influence of gelation time

The effect of gelation time at room temperature is illustrated in fig.1 (weight fraction of polymer (W_2) = 0,082). Melting of the gel, immediately after its formation, results in a broad, endothermic signal with an enthalpy change be-

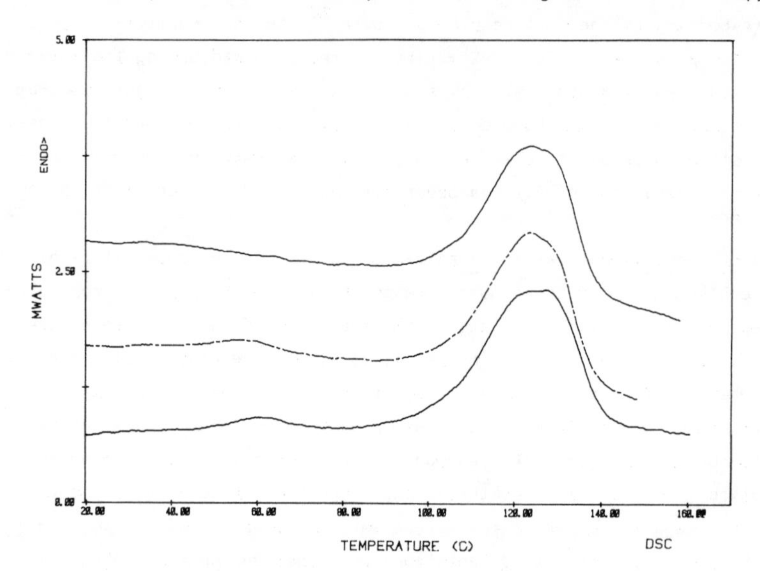

Figure 1 . DSC scans of PVAL-Ethylene Glycol (W_2 = 0,082) previously cooled to room temperature. Gelation time : A, 0 Hr; B, 118 Hr and C, 552 Hr.

tween 50 and 60 J g^{-1}. No change with gelation time was observed. The corresponding melting point, T_m^1 , increases with polymer concentration from 120° C (W_2= 0,082) to 137 °C (W_2 = 0.31). On standing, a small melting domain develops at much lower temperatures. The corresponding enthalpy change remains very small. Its melting point (T_m^2) increases with gelation time but seems not to be strongly influenced by polymer concentration. The corresponding induction period however decreases with increasing polymer concentration.

b. Influence of gelation temperature

The influence of this parameter was investigated for a solution of PVAL with W_2 = 0,082. The samples were heated to 150°C and cooled immediately to the gelation temperature. They were kept there long enough to develop both melting endotherms. The highest melting point, T_m^1, is constant and equal to 120°C up to T_{gel}

= 90°C. At higher T_{gel} it increases with this parameter (fig. 2). The corres-

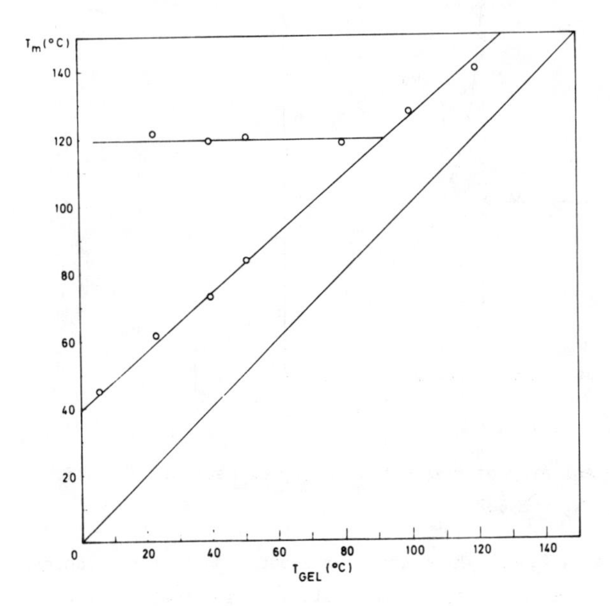

Figure 2. Dependence of gel melting point (T_m^1) on gelation temperature T_{gel} (W_2 = 0,082)

ponding change in enthalpy, ΔH^1, is also constant and equal to 52 J g^{-1} for T_{gel} 90°C. At 90°C it reaches a much higher value, corresponding to that of a dried gel, and then decreases as T_{gel} increases further (fig. 3) . On cooling, these partially crystallized gels crystallized further.

Above 90°C, the melting domain is much sharper and the corresponding gels are quite elastic while those formed at much lower temperatures are more pastlike.

c. Dynamic observations

From the previous observations, it is quite clear that a change in the gelation mechanism occurs at 90°C (W_2= 0,082). Because liquid-liquid phase separation can occur in this system, dynamic measurements were performed. Calorimetric data were compared with optical observations.

In figure 4, the onset of opalescence of solutions cooled from 150°C is reported. This curve resembles very much the phase diagrams reported in the literature. Their position on the temperature scale is independent on scanning rate

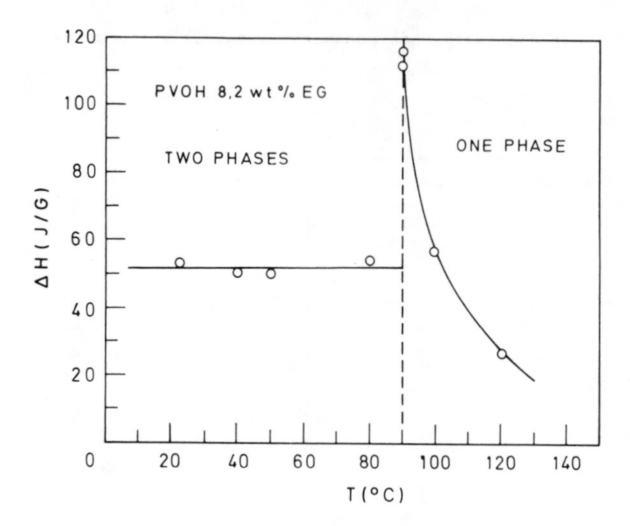

Figure 3. Melting enthalpy of PVAL-ethylene glycol gels as a function of T_{gel} (W_2 = 0,082)

but changes with molecular weight. The onset of the DSC signal on the contrary is very sensitive to scanning rate. It is interesting to see that for low molecular weight samples opalescence and DSC curves intersect, indicating that crystallization, as detected by DSC measurements, sets in before opalescence can be observed.

d. X-ray analysis

Crystallization of PVAL results in the formation of a crystalline structure with a monoclinic unit cell[(6)]. The most intense reflection is observed at 2θ = 19.3°, corresponding to a diffraction by both the 101 and the $10\bar{1}$ planes. They are parallel to the molecular axis.

Diffraction patterns of gels were obtained on stretched gels (5 to 6 times) from which the solvent was extracted during the stretching procedure. This results in an increase of the crystallinity.

The diffraction pattern of a gel prepared at room temperature (fig. 5) shows an intense ring at 2θ = 19.3 corresponding to a random orientation of the corresponding crystallites. This random scattering is superimposed by an equatorial scattering at the same diffraction angle. This means that at least a fraction of the crystallite is oriented with its molecular axis parallel to the stretching direction. This orientation is supported by the broad arcs corres-

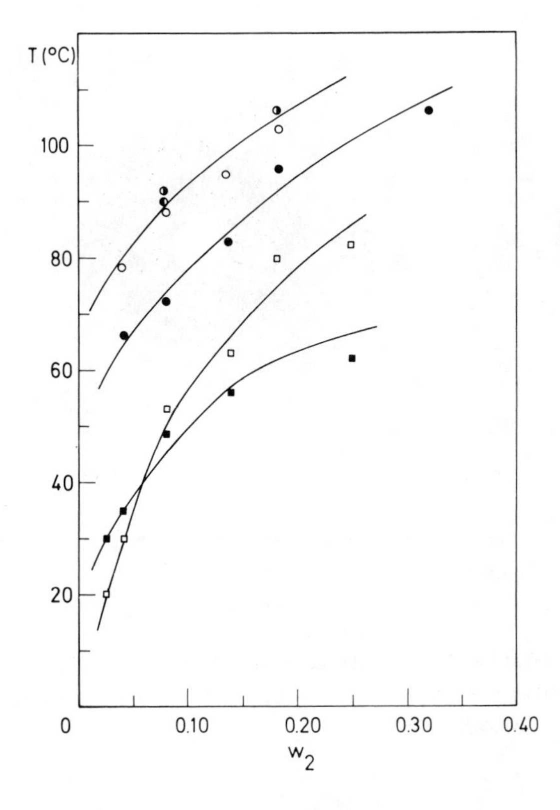

Figure 4. Onset of opalescence and DSC exotherm on cooling PVAL-Ethylene glycol solution as a function of W_2 and molecular weight ($\overline{M}_w$)

Onset of opalescence $\overline{M}_w$ = 115.000 ◑ : -15°/min, ◐ : -10°/min

○ : - 5°/min

$\overline{M}_w$ = 14.000 □ : - 5°/min

Onset of DSC exotherm (-5°/min) ● $\overline{M}_w$ = 115.000

■ $\overline{M}_w$ = 14.000

ponding to the 110 planes.

Gels prepared at 100°C give a much more complex diffraction pattern (fig. 6b). A well developed crystal structure orients with the 101 plane, and also with the molecular axis, perpendicular to the stretching direction. This is concluded from the meridional 101 reflection. In the unit cell this direction

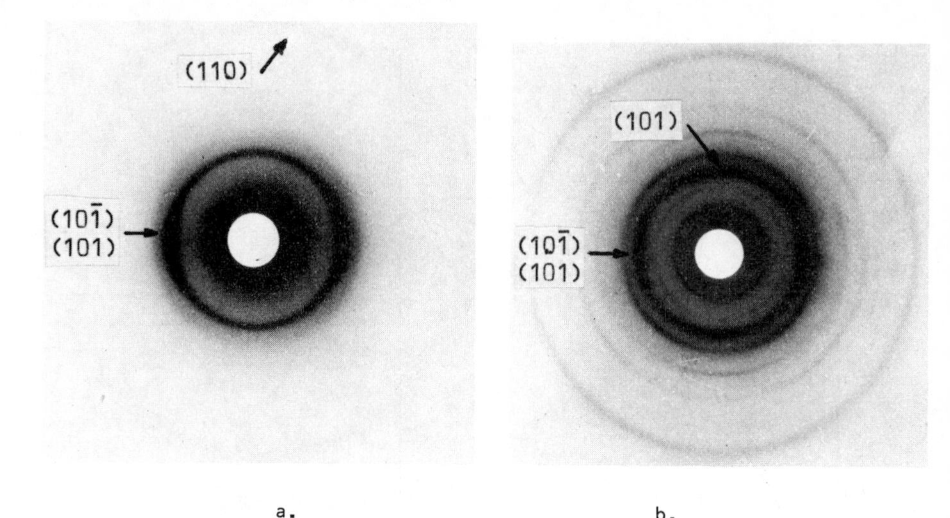

a. b.

Figure 5. X-ray scattering pattern of a stretched gel of PVAL
a. prepared at room temperature (extension 6 x) b. prepared at 100°C (extension 6 x)

coincides almost with the orientation of the intermolecular hydrogen bonds. A second type of orientation, quite similar to the one obtained with gels prepared at room temperature, is also observed.

B Discussion

From these experimental data, the following picture of gelation of PVAL can be obtained.

On cooling PVAL solutions in ethylene glycol phase separation occurs as manifested by the onset of opalescence. This is very rapidly followed by crystallization in the most concentrated phase. The onset of both phenomena occurs at different temperatures but this difference decreases as the scanning rate is decreased and is mainly due to the nucleation controlled nature of the crystallization phenomenon. This two step gelation mechanism results in the formation of poorly elastic, very opaque gels. The formation of the crystallites prevents any large scale phase separation. It is also responsible for an important hysteresis effect. On standing, secondary crystallization occurs and solvent is expelled. Stretching in methanol results in an increase of the crystallinity. The newly formed crystallites in the diluted phase orient with their molecular axis parallel to the stretching direction. The crystallites already present in the concentrated phase will not orient and contribute to the random diffraction (fig. 6a).

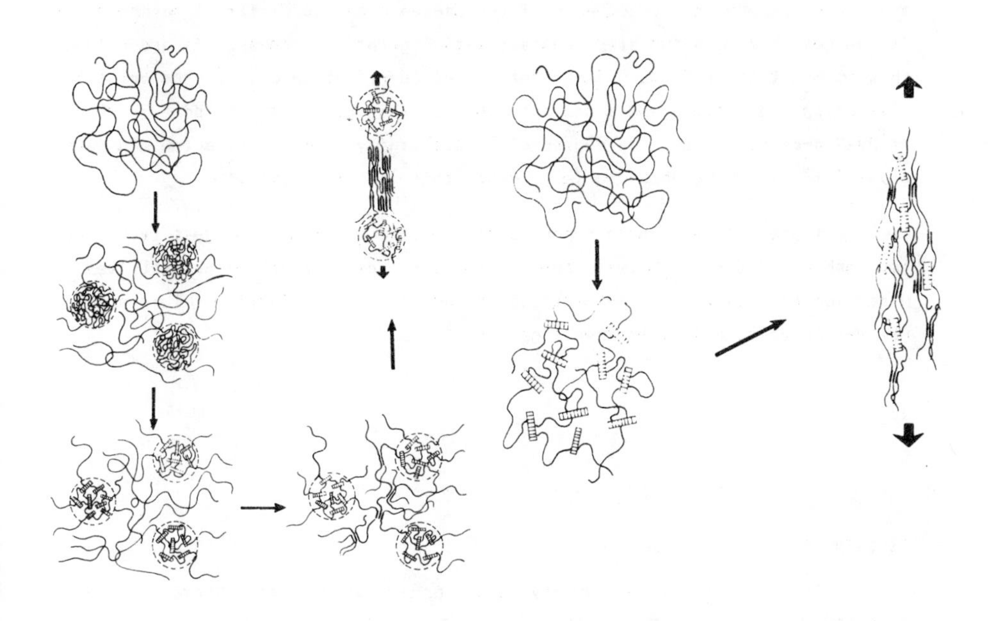

a. b.

Figure 6. Schematic representation of the behavior on stretching of PVAL gels prepared a) at room temperature b) at 100°C

Gelation above the demixing temperature results from the formation of folded chain crystallites. The induction period is concentration and temperature dependent. At the highest temperatures, only part of the polymer molecules participate in the gelation, the remaining fraction crystallizes on cooling. These folded chain crystallites provide the necessary cross links and introduce the opalescence in the gels. On stretching they orient with their molecular axis perpendicular and with the intermolecular hydrogen bonds nearly parallel to the stretching direction. Crystallites formed during the stretching procedure have their molecular axis parallel to the stretching direction. Consequently, the twofold orientation is the consequence of a two step crystallization (fig.6b).

These data clearly illustrate the importance of polymer solubility in the formation and final properties of a gel. They also show that twofold orientations are not necessary the consequence of the presence of two different morphologies. It can result from a two step crystallization-gelation process. In order to get this information it is very important to get quantitative data on the crystallinity of the gel before and after stretching. This can easily be done in the case of PVAL because of its high degree of crystallinity. When only a low degree of crystallinity can be obtained (e.g. PVC), this is problematic.

These data and those reported in the literature also suggest that a relationship could exist between the perpendicular orientation of crystallites and direction of preferential interaction in the crystalline lattice. This point however needs some further investigation.

REFERENCES

(1) A. Keller in "Structure-Property Relationships of Polymeric Solids", Anne Hiltner Ed., Plenum Press, 1983

(2) H. Berghmans, B. Overberg and F. Govaerts, J. Polymer Sci., Phys.Ed., 17, 1251 (1979)

(3) S.J. Guerrero and A. Keller, J. Macromol.Sci., Phys., B, 20, 167 (1981)

(4) O. Olabisi, L.M. Robeson and M.T. Shapo in "Polymer-Polymer Miscibility" chap. 2, Academic Press (1979)

(5) S.P. Papkov and S.G. Yefimova, Vysokomol.soyed., 8, 1984 (1966)

(6) C.W. Bunn, Nature, 161, 929 (1948)

(7) H. Berghmans, IUPAC Symposium on Macromolecules, Amherst, USA (1982)

(8) S. Zwijnenburg and A.J. Pennings, Colloïd and Polymer Sci., 259, 868 (1978)

(9) P.J. Barham, M.J. Hill and A. Keller, Colloïd and Polymer Sci., 258, 899 (1980)

(10) J.D. Hofmann, Polymer, 20, 1071 (1979)

(11) P. Smith and P.J. Lemstra, Polymer Bull., 1, 733 (1979)

(12) P. Smith and P.J. Lemstra, J. Material Sci., 15, 505 (1980)

(13) E.D.T. Atkins, M.J. Hill, D.A. Jarvis, A. Keller, E. Sarkene and J.S. Shapiro, Colloïd and Polymer Sci., 262, 22 (1984)

(14) G. Rehage, Kunststoffe 53, 605 (1963)

(15) E. Pines and W. Prins, Macromolecules, 6, 888 (1973)

(16) W. Aiken, T. Alfrey, A. Janssen and H. Mark, J. Polymer Sci., 2, 178 (1947)

(17) T. Alfrey, N. Widerkom, R. Stein and A.N. Toboloky, Znd. Eng. Chem., 41, 701 (1949)

(18) S.J. Guerrero, A. Keller, P.L. Sorri and P.H. Geil, J. Polymer Sci., Phys.Ed. 18, 1533 (1980)

(19) S.J. Guerrero, A. Keller, P.L. Sorri and P.H. Geil, J. Macromol.Sci., Phys. B 20, 161 (1981)

(20) S.J. Guerrero and A. Keller, J. Macromol. Sci., Phys., B 20, 167 (1981)

(21) P.J. Lemstra, A. Keller and M. Cudby, J. Polymer Sci., Phys.Ed., 16, 1567 (1978)

(22) A. Dorrestijn, P.J. Lemstra and H. Berghmans, Polymer Communications, 24, 226 (1983)

(23) S. Wellinghoff, J. Shaw and E. Bear, Macromolecules, 12, 932 (1979)

(24) Hwi-Min Tan, A. Moet, A. Hiltner and E. Bear, Macromolecules, 16, 28 (1983)

STATIC AND DYNAMIC LIGHTSCATTERING OF THERMOREVERSIBLE GELLING IOTA-CARRAGEENAN

H.-U. ter Meer *,W. Burchard
University of Freiburg

ABSTRACT : The static (ILS) and dynamic light scattering (QELS) of the ionic polysaccharide iota-carrageenan have been examined in the gel- and in solution.Differences in structure and dynamic behaviour are interpreted in terms of transient networks and/or entangled solutions.

INTRODUCTION :

The algal polysaccharide iota-carrageenan is known to form thermoreversibly gels on cooling at moderately low polymer concentrations (app. 0.5 % of weight). The gelation process is accompanied by a conformational disorder/order transition (1). From x-ray fiber diffraction of solid state samples it is deduced, that the ordered conformation is a double helix (2). Until now it was not known , whether double helices also exist in the gel-state and what the conformation is in solution. In the present work we have been able to elucidate the structure of iota-carrageenan in two salt solutions and we have gained some insight into the molecular dynamics within the thermoreversible gel.

EXPERIMENTAL:

A series of iota-carrageenan solutions (0.2 - 2.0 % by weight) have been studied in two different ionic solutions (0.1 N KCl and 0.25 N tetra-methyl-ammonium-chloride (TMACl)) . The light scattering measurements have been carried out on a self built instrument that measures static and dynamic lightscattering (3) simultaneously. Some static LS has also been carried out on a SOPHICA instrument . The intensity fluctuations were analysed by a 96 channel single-bit correlator (MALVERN INSTRUMENTS) . To measure the full correlation function , up to 7 measurements at different sample times had to be spliced together . Data analysis of the full correlation-funtion has been done by a numerical inversion method (programm CONTIN by S. Provencher (4)) .

* current address : Unilever Research ; Colworth Laboratory ; Sharnbrook , Bedford MK44 1LQ (GB)

RESULTS :

In 0.1 N KCl iota-carrageenan was found to be a worm like chain with a persistence length of about 180 nm at all temperatures . This allowed the mass per unit length (M/L) to be directly measured by light-scattering . It was found to decrease with increasing temperature and concentration . (fig. 1)

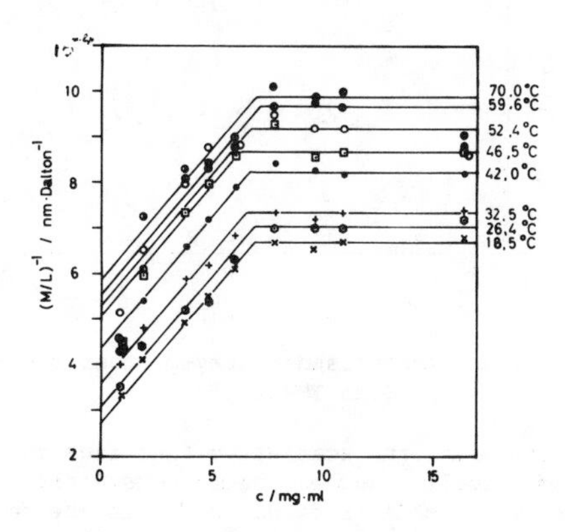

fig 1 : concentration dependence of reciprocal mass density 1/(M/L) of iota-carrageenan at various temperatures.

Even at the highest temperature and high concentrations the mass density was found to be close to the value estimated for a double strand (found M/L = 1100 Dalton/nm ; theor. value for double strand M/L = 1220 Dalton/nm) . This indicates , that even well above the gelling temperature no isolated strands exist. This is in contrast to a model proposed by Morris et. al (5) .
Knowing the mass density (M/L) , the molecular weight Mw and the radius of gyration <S> , a new method to calculate the Kuhn-length b (b is twice the persistence length) is shown.
From the definition of the Kuhn length

$$\langle S \rangle = n \cdot b \cdot b / 6 \quad \text{with}$$

$$L = n \cdot b = 1/((M/L) * 1/Mw)$$

one gets

$$b = (\langle S \rangle / M) * (M/L) * 6$$

The Kuhn/length estimated by this procedure (b~ 380 nm) agrees well with the values found by fitting the Kojama-theory (6) of worm-like chains to the data . Surprisingly there was only a limited change in persistence length with temperature and with concentration . This again is an indication that the structure of the macromolecular aggregates in the gel and the sol state are very similar.

The structure in TMACl differs from that found in KCl , as can be deduced from a closer examination of the Kratky-plot (fig 2) which suggests that the structure in this ionic form is highly , but not randomly branched .

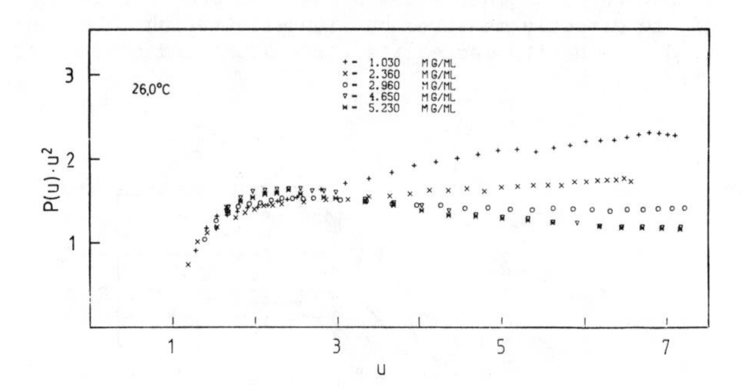

fig 2 : normalised Kratky-plot of iota-carrageenan in 0.25 TMACl

By comparing the scattering functions as measured with ILS and small angle neutron scattering (SANS) the Kuhn-length in TMACl is found to lie in the range 60 nm <= b <= 150 nm . The structure in TMACl is not affected by temperature.

Despite the constant structure at low and high temperatures , the mobility within these structures is changing . As was found by dynamical lightscattering (QELS) , at least two relaxation processes exist in both the carrageenan solution and the gel (fig 3) . The amount of each component is concentration dependent .

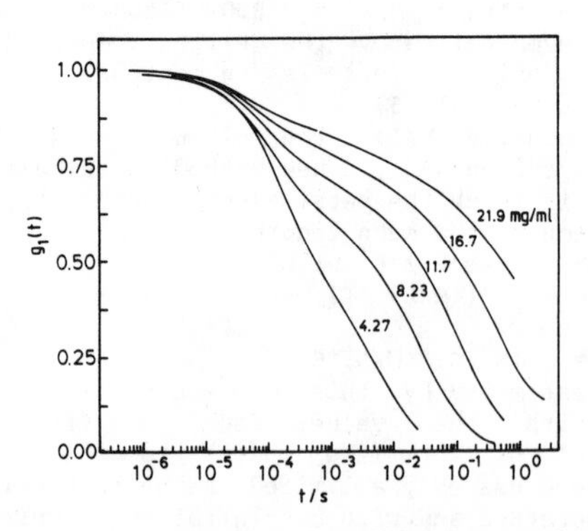

fig 3 : concentration dependence of the normalised field-time-correlation function

The mean relaxation time of the slower process depends on concentration . It also shows a dramatic change with temperature on going from the gel to the sol state . Whereas in the gel state the exponent of the concentration dependence of the relaxation time is v= -4 , it increases to a value of v=-2.4 in the sol state.
For an entangeld solution an exponent of v = 2 ... 3 is predicted by theory , and for a permanent network no concentration dependence of the slowest relaxation process is expected. From the fact that there is a concentration dependence of the slowest process in the gel state we suggest that iota-carrageenan is a transient network . Creep experiments on other thermoreversible gelling systems are consistent with this finding .

REFERENCES

1) A.A.McKinnon , D.A.Rees and F.B.Williams ; Chem. Comm. , 701 (1969)
2) S.Arnott , W.E.Scott , D.A.Rees , C.G.A.McNab and J.W.Samuel J.Mol.Biol. 90 , 235 (1974)
3) S.Bantle , M.Schmidt and W.Burchard ; Macromelecules 15 , 1604 (1982)
4) S.W.Provencher ; Computer Phys. Comm. 27 , 213 (1982)
5) E.R.Morris , D.A.Rees , G. Robinson ; J.Mol.Biol. 138 , 349 (1983)
6) R.Koyama ; J.Phys.Soc.Jap. 34 , 1029 (1973)

EFFECTS OF POLY(ACRYLAMIDE) ON THE SOLUTION AND GEL PROPERTIES OF WATER-GELATIN SYSTEM

A.TURTURRO*, A.SHOUSHTARIZADEHNASERI*, E.PEDEMONTE*, S.FRANCO† and A.VALLARINO†

*Istituto Chimica Industriale, Genoa University, Italy
†3M Italia Ricerche S.p.A., Ferrania, Italy

SYNOPSIS

The phase relationship in ternary mixtures water-gelatin-poly(acrylamide) has been studied performing measurements of cinematic viscosity, gel → sol transition temperature and heat and morphological observations.

1 INTRODUCTION

Gelatin is a natural polymer of fundamental importance in the technology and manufacture of photographic materials.

Since the physical properties of gelatin are not completely satisfactory, in the recent years the interest in gelatin synthetic polymer blends had grown. However, few studies on this system have been reported in literature[1-3].

The aim of our research is to investigate the factors affecting the compatibility or mixibility between gelatin and synthetic polymers in order to achieve binder compositions suitable for photographic film.

In the present work we report results regarding aqueous mixtures of gelatin/poly(acrylamide), Gel/PAM.

2 MATERIALS AND TECHNIQUES

2.1 Materials

A commercial gelatin of photographic grade of low isoelectric point was used ($\overline{M}_n = 1 \cdot 10^5$; $\overline{M}_w = 2 \cdot 10^5$). Two samples of poly(acrylamide), PAM, were synthesized using the free-radical solution polymerization technique, with conditions chosen to result in very different molecular weights. The viscosity measurements in $NaNO_3$ 1N[4] at 30°C have given the following molecular mass : PAM-1 : $\overline{M}_v = 1.45 \cdot 10^4$; PAM-2 : $\overline{M}_v = 14.4 \cdot 10^4$.

Both gelatin and PAM samples were used immediately after standard drying process.

2.2 Techniques

Aqueous blends of two polymers were prepared by mixing at 60°C the previously prepared water solutions of each polymer.

Investigations were carried out on water/gelatin/PAM mixtures containing 5 and 10 w/volume % of total polymer; the Gel/PAM polymer ratio covered the full range from 90/10 to 10/90.

The gels were prepared by fast cooling the solutions from 60°C to 8°C and examined after a standard maturation time of 18 hrs.

The effects of PAM on water/gelatin mixtures were analyzed by measurements of : cinematic viscosity (Cannon Fenske Viscometer) at 60°C; gel → sol transition heat, ΔH_m, and temperature, T_m, by differential scanning calorimetry (Perkin Elmer DSC-2 equipped with the model 3500 Data Station), using a heating rate of 2.5 K/min. The morphological observations on gel were performed by means of phase contrast microscopy (Reichert optical microscope).

The compositions of separated phases were determined by calorimetric measurements, using the microbiuret reagent[5] selective for gelatin.

3 RESULTS AND DISCUSSION

3.1 Viscosity behaviour

The dependence of ternary solution viscosities as a function of PAM molecular mass, PAM content and total polymer concentration is shown in fig.1 and fig.2.

The viscosity always resulted in lower than "ideal" additive value; the maximum deviation from the ideality has been found near the Gel/PAM ratio 70/30 and for solution of 4-5 wt.% of total polymer.

The specific interactions between different polar groups, present on the two polymer chains, could explain the complex behaviour of solution viscosity. However, it has been observed that the solutions water-gelatin-PAM 2, 10 w/volume %, are always not completely clear.

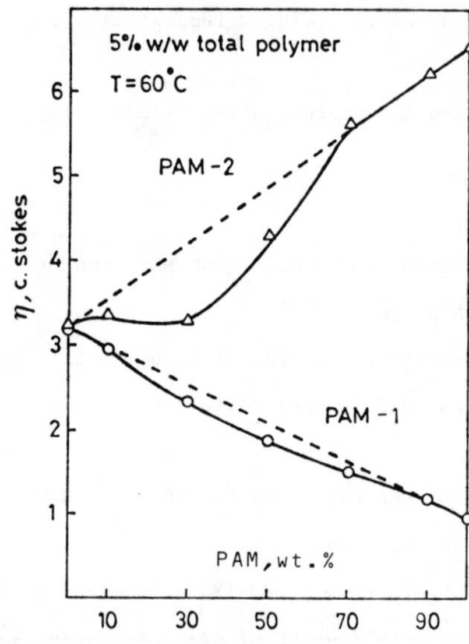

Fig.1-Cinematic viscosity of water/gelatin/PAM solutions vs PAM content.

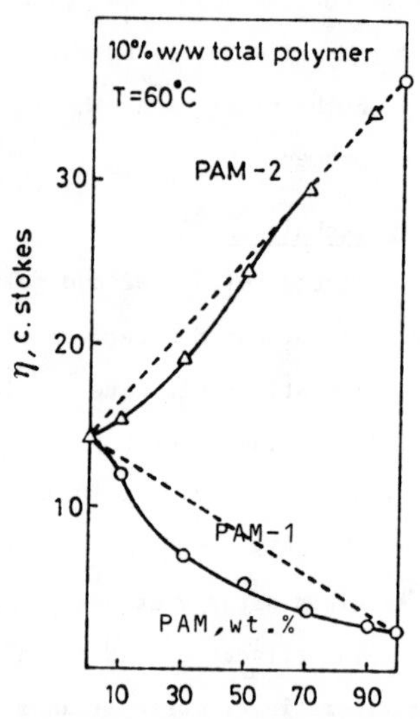

Fig.2-Cinematic viscosity of water/gelatin/PAM solutions vs PAM content.

3.2 Phase separation

In order to test the stability of mixtures, the 10 w/volume % solutions were stored at 60° undisturbed for 48 hrs.

The Gel/PAM-1 solutions do not present any segregation phenomenon; on the contrary, the Gel/PAM-2 solutions, ranging from 95/5 to 30/70 polymer ratio, segregate forming two well defined phases: the upper phase is always a ternary mixture of water-gelatin-PAM 2, up to about 20 w/w % of Gelatin respect to PAM-2; the lower one is a binary mixture of water and gelatin.

3.3 Thermal behaviour

In fig.3 the influence of PAM on ΔH_m is illustrated. As can be seen, the decrease of ΔH_m increases with lowering the PAM molecular mass.

As regarding T_m, no significant variation has been measured for all observed systems.

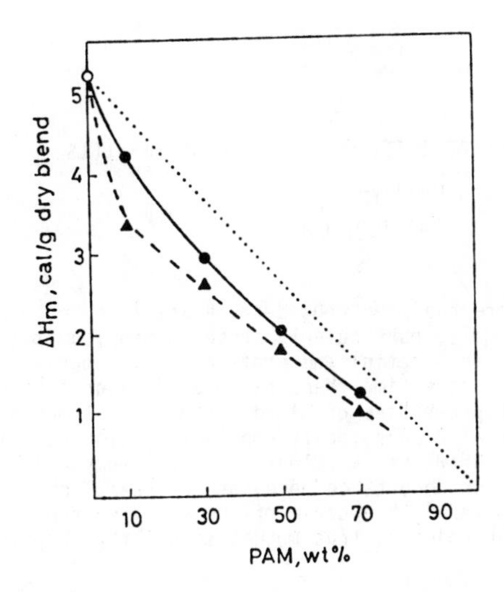

Fig. 3-Gel-sol transition heat vs PAM content. 10% total polymer.

These results would suggest that only the amount of gel is affected by the presence of PAM molecules.

3.4 Morphology of the gels

The microscopical observations show clear evidence of sharp phase segregation only when PAM-2 is present, throughout almost all of the composition range.

All these results suggest the presence of specific interactions between different polar groups on the two polymer chains; however, they are not enough to allow the formation of homogeneous gels with high molecular mass PAM.

REFERENCES

1. Croome, R.J. J.Phot.Sci., 30, 181, 1982

2. Clark, A.H., Richardson, R.K., Ross-Murphy, S.B. and Stubby, J.M. Macromolecules, 16, 1367, 1983

3. Petrak, K.L. J.Appl.Polym.Sci., 29, 555, 1984

4. Sorenson, W.R. and Campbell, T.W. "Preparative Methods of Polymer Chemistry", Interscience, N.Y., 1961

5. Gornall, A.G., Bardawill, C.S. and David, M.M. J.Biol.Chem., 177, 751, 1949

This work has been performed with the partial financial support of "Progetto Finalizzato Chimica Fine e Secondaria", C.N.R.

COMPATIBILITY AND VISCOELASTICITY OF MIXED BIOPOLYMER GELS

A. H. CLARK and S. B. ROSS-MURPHY*
Unilever Research, Bedford MK44 1LQ, U.K.

Over the last decade there has been considerable interest in, and growth of knowledge of biopolymer gels, particularly those formed from polysaccharides[1] (e.g. carrageenans, agarose, pectin) and proteins[2] (e.g. bovine serum albumin, gelatin, casein). At the same time there has been a major interest in the structure and mechanical properties of mixed synthetic polymer systems. Such mixtures include mechanical blends, graft copolymers, block copolymers and interpenetrating polymer networks[3] . Clearly the thermodynamics of the mixed polymer systems is of major importance here, as has been frequently emphasised by Koningsveld and co-workers[4,5] : because of the small entropy gain on mixing two high molecular weight species, true mutual solubility in polymer blends is quite rare.

The usual morphology of such systems is for one phase (the disperse phase) to be distributed throughout the other (matrix or continuous phase) as discrete domains or microphases. As the proportion of the two polymer species is changed, phase inversion can occur, and then the original disperse phase becomes the new matrix or supporting phase. If both polymeric species are network or gel forming systems, more complex morphologies including mutually interpenetrating polymer networks can be formed, and such systems have been studied by e.g. Frisch and Frisch[6] and Sperling and co-workers[7] .

The particular interest in this class of materials is, of course, in modifying the mechanical properties, studying the structure-property relationships, at both the molecular and supramolecular distance scales, and the thermodynamics of mixing. Recently some of these aspects were addressed by ourselves for mixed biopolymer gels[8] , of which one component was a protein and the other a polysaccharide. In particular 'cold set' gels are studied by cooling warm solutions of gelatin and agarose.

In both these materials, the "cross links" are formed when a coil-helix transition oocurs on cooling the solutions, and involve cooperative sequences of ordered chains ("junction zones") separated along the chain contour by disordered regions. In particular the junction zones for gelatin and agarose are believed to involve multiple helices; triple (collagen type) helices for gelatin and double helices for agarose; there is also evidence that stacking of these multiple helices can occur, and contribute to the size of the junction zone domains.

The mechanical properties of these single component gels can be well described by adapting the theory of Hermans[9] which treats the physical crosslinks as an equilibrium between 'reacted' and 'unreacted' site, and incorporating a more realistic count of the number of elastically active network chains[10] . The mechanical properties of the mixed gels can then be calculated by a suitable blending law, which takes into account the very high volume fraction of water solvent (~ 0.8), and the competition for this water between the polymer species.

Experimental

Mixed gels of agar and gelatin were prepared by adding a suitable amount of agar to water (1% w/w), autoclaving the mixtures to dissolve the agar, cooling to 45°C, and dissolving the appropriate amount of gelatin (0-25%). Mechanical measurements were performed by introducing the warm solutions between the plates of a Rheometrics mechanical spectrometer, and allowing gelation to occur in situ. Gelation was followed for 3 hours until the trace of log G' storage modulus vs time was effectively horizontal; then the frequency dependence of G' and G" was measured ($\omega = 10^{-2}$ -10^{2} rad sec^{-1}, strain, γ = 0.1). Optical microscopy was carried out by fixing and staining (toluidine blue), sectioning (10^{-6} m) and photographing with a Leitz Ortholux II microscope. For transmission electron microscopy, samples were stained with uranyl acetate and lead citrate, the micrographs were recorded with a JEOL 100 CX microscope.

Results and Discussion

Figure 1 shows light microscope pictures of the mixed gel systems (the dark areas indicate gelatin); as the concentration of this is increased the apparent volume fraction increases, there is an apparent phase inversion at 2.5% w/w gelatin, and at higher concentrations the system becomes gelatin continuous with light inclusions of agarose gel. In practice the morphology is much more complex than this, because the inclusions themselves contain subinclusions of agarose as can clearly be seen in electron micrographs.

Figure 2 shows the frequency dependence of G' and G" for a 1% agar/10% gelatin gel measured at 25°C; G' is almost independent of frequency and the trace suggests we can use this value as an estimate of the equilibrium shear modulus of the composite system, G_c. The slight dip in G" is due to the gelatin component since it is present in pure gelatin gels, but not in agarose.

Figure 3 shows the results for G_c plotted against % w/w gelatin at a fixed level of 1% w/w agarose. The theoretical curves 1, 2 and 3 represent the upper, lower and modifed upper bound behaviour for the mechanical behaviour of the composite system (see below). The latter is quite a reasonable fit, and this is confirmed by composite data at the other agarose concentrations[7].

The curves 1 and 2 represent the Takayanagi isostrain and isostress ("upper" and "lower" bounds) given by[3]

$$G_c = \phi_x G_x + \phi_y G_y \quad \text{(curve 1)}$$
$$1/G_c = \phi_x/G_x + \phi_y/G_y \quad \text{(curve 2)}$$

where G_x, G_y are the moduli of pure systems x and y, and ϕ_x, ϕ_y are their respective volume fractions. These equations are formulated originally for bulk polymer systems, so that G_x and G_y will have fixed values for the components and $\phi_x + \phi_y = 1$. For the present systems where $\phi_s \sim 1$, the volume fraction of solvent is high (~ 0.8), clearly $\phi_x + \phi_y \neq 1$, and further it is in the nature of such gelling systems that $G_x = f(\phi_x)$ and $G_y = f(\phi_y)$. The function f is in fact given by the modified Hermans treatment already mentioned, in terms of which at high ϕ, $G \propto \phi^2$, but as ϕ decreases there is a critical volume fraction, or critical concentration below which $G \approx 0$. In the intermediate range $G \propto \phi^n$, where the exponent $n \rightarrow \infty$ as $\phi \rightarrow \phi_{crit}$[10].

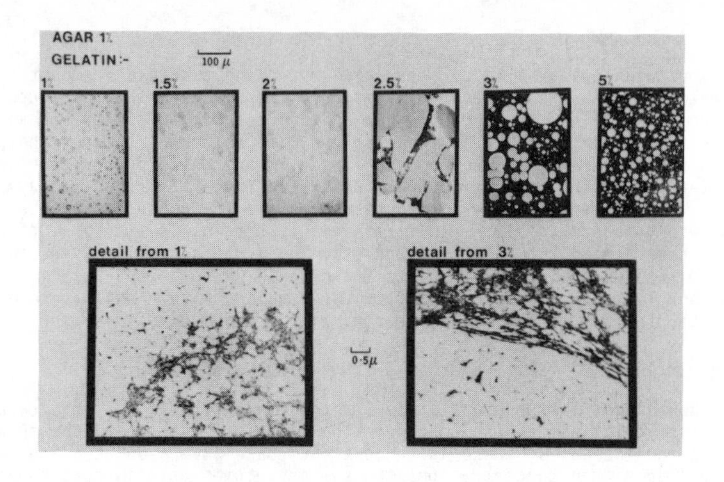

Figure 1. (a) Light microscope (top row) and electron microscope (bottom row) results for the 1% w/w agar series of agar/gelatin gels. Gelatin appears as the darkly staining material, the maximum concentration shown being 5% w/w.

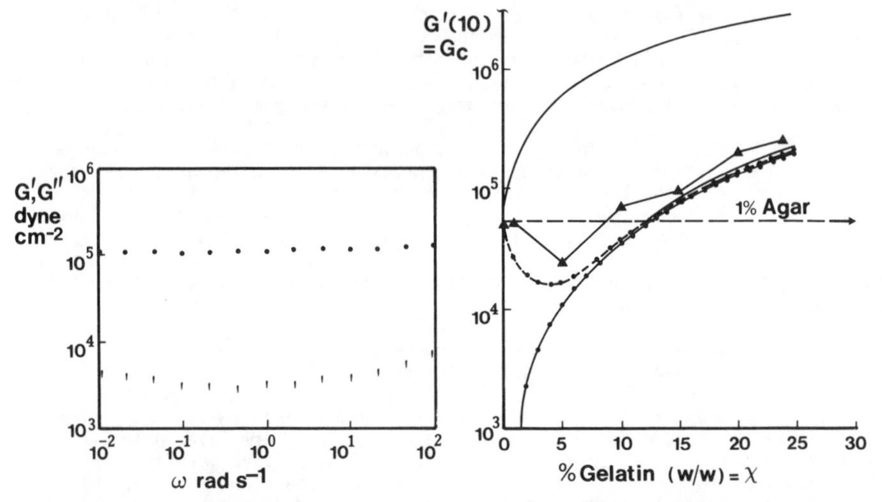

Figure 2. Frequency dependence of G' (...) and G" (""") for a 1% gelatin gel measured at 25°C.

Figure 3. Modified (p=1.0) upper and lower bounds (-·-) calculated using C_{agar}^{nom} instead of C_{agar}^{eff} are compared with the original estimates, only the upper bound is substantially changed.

However by measuring the $G = f(\phi)$ behaviour for pure systems x and y respectively, the theoretical curve 3 can be obtained, on the assumption that the agarose contribution to the overall modulus is not different from that which would apply if no gelatin molecules were present at all, consistent with the observed slightly higher temperature of gel formation on cooling (32°C compared to 27°C for gelatin) and the pronounced hysteresis on melting (agarose gels melt out at around 85°C, whereas gelatin gels melt at $\approx$ 29°C). It is interesting to note that curve 3 shows the pronounced minimum in G_c close to the phase inversion composition, and that at high gelatin the system approaches the lower bound behaviour of curve 2, as it should do as the system becomes more gelatin continuous.

The model above has been recently[11,12] extended to other gel network systems, and also to the regime of large deformation and failure properties, where a rewarding generality of response has been noted.

ACKNOWLEDGEMENTS

The authors are grateful to Bob Richardson, Lesley Linger and Helen Rose for discussions and technical assistance.

REFERENCES

1. Rees, D. A., Morris, E. R., Thom, D. and Madden, J. K. in 'The Polysaccharides - Vol. 1' Aspinall G. O. (ed) Academic Press, 1982, p196.
2. Clark, A. H. and Lee-Tuffnell, C. D. in 'Functional Properties of Food Macromolecules' Mitchell, J. R. and Ledward, D. A. (eds). Applied Science, in press.
3. Manson, J. A. and Sperling, L. H. 'Polymer Blends and Composites' Plenum Press, 1976.
4. Koningsveld, R. Adv. Polym. Sci. 7, 1, 1970.
5. Koningsveld, R. and Kleintjens, L. A. J. Polym. Sci. Polym. Symposia 61, 221, 1977.
6. Siegfried, D. L., Manson, J. A. and Sperling, L. H. J. Polym. Sci. Polym. Phys. Ed. 16, 583, 1978.
7. Kim, S. C., Klempner, D., Frisch, K. C. and Frisch, H. L. Macromolecules 10, 1187, 1977.
8. Clark, A. H., Richardson, R. K., Ross-Murphy, S. B., and Stubbs, J. M. Macromolecules 16, 1367, 1983.
9. Hermans, J. R. J. Polym. Sci. Part A 3, 1859, 1965.
10. Clark, A. H. and Ross-Murphy, S. B. Brit. Polym. J., 15, 000, 1985.
11. McEvoy, H., Ross-Murphy, S. B., and Clark, A. H. in 'Gums and Stabilizers for the Food Industry 2' Phillips, G. O. (ed.), Pergamon, 1984, p111.
12. McEvoy, H., Ross-Murphy, S. B. and Clark, A. H. Polymer, submitted.

HALATO-TELECHELIC POLYMERS AS MODELS OF ION-CONTAINING POLYMERS AND THERMOREVERSIBLE POLYMER NETWORKS

R. JEROME

Laboratory of Macromolecular Chemistry and Organic Catalysis. University of Liège, SART TILMAN, 4000 Liège, Belgium.

SYNOPSIS

Halato-telechelic polymers are interesting models of the more complex ionomers. A nonpolar environment triggers off the association of the ion-pair end-groups and the building-up of a tridimensional polymer network. This phenomenon is discussed in relation with the strength of the ion-pair interactions and the root mean square end-to-end distance between the ionic end-groups.

1 INTRODUCTION

The properties of hydrocarbon polymers are drastically modified by the random distribution of even modest amounts of ionic groups. The resulting ionomers might be considered as cross-linked by multiplets of a few ion-pairs and/or clusters of ionically concentrated material. The formation of microphase-separated ionic domains is however hindered to some extent by the random distribution of the ion-pairs onto entangled polymeric backbones. A better insight into the ion aggregation would be obtained from linear macromolecules carrying salt groups selectively attached at both ends and enjoying a large independence of each other. Halato-telechelic polymers (HTP) meet this requirement and provide a family of model compounds. An efficient synthetic pathway to well-defined carboxylato-telechelic polymers have been reported elsewhere[1]; this paper aims at discussing some of their original features.

2 GEL FORMATION

In nonpolar solvents, the metal carboxylates attached at both ends of hydrocarbon polymers are little or not at all ionized; as a consequence they exist as ion-pairs, the mutual interaction of which control the solution behavior. Gelation is observed at concentrations of a few wt % of polymer and higher, even though the molecular weight is sufficiently low to preclude any entanglement effect[1]. Upon subsequent dilution, either a viscous solution forms or a phase separation occurs. For weak ion-pair interactions and metal cations of low valency (<2), a critical concentration for gel formation (C_g) can be defined from the experimental viscosity-concentration relationships.

C_g strongly depends on ionicity and geometry of the ion-pairs, solvent polarity and temperature. It is worth noting that the thermal effect on C_g is perfectly reversible. Sufficiently strong and/or multiple (metal valency>2) ion-pair interactions efficiently prevent the disruption of the network, and beyond a given degree of dilution a demixing takes place at constant temperature. Demixing is also a reversible process as supported by the gel homogeneization upon distilling off the excess nonpolar solvent. All these experimental results convincingly support the electrostatic origin of gelation.

3 DEPENDENCE OF GELATION ON THE PREPOLYMER MOLECULAR WEIGHT ($\overline{M}_n$)

Equation 1 is experimentally observed for a series of α,ω-Mg dicarboxylato polydienes and poly(vinyl aromatics) in toluene at 25°C[1].

$$C_g = k\,\overline{M}_n^{-0.5} \qquad (1)$$

It means that C_g is directly proportional to $f_p^{0.5}$, where f_p is the mole fraction of repeat units carrying ionic groups. Using scaling concepts, Joanny predicted that C_g should be proportional to f_p^{-1} or $f_p^{-8/5}$ depending on the dimerization energy (Θ)[2]. The statistical mechanical analysis of dimerization of ion-pairs in ionomer solutions provide results in qualitative agreement with those of Joanny for Θ<2 Kcal/mol and with our observations for Θ>6 Kcal/mol[3]. A model wherein multiplets occupy well-defined positions within ideal space lattices enable us to derive eq.2 [4].

$$\bar{n}_g/(0.0074a_g) = (\bar{r}^2/M)^{3/2}.C_g.\bar{M}_n^{0.5} = K \qquad (2)$$

where $\bar{n}_g$ and a_g are the critical value of the mean number of cations per multiplet and of a parameter depending on the coordination number of the lattice, respectively.
As the conformational state of a polymer (or $\bar{r}^2/M$) is largely independent of its molecular weight M in a range as limited as that covered by our experiments [1,5], eq.1 is easily accounted for and the derivation of eq.3 is straightforward.

$$k^{-1/3} = K^{-1/3}\ (\bar{r}^2/M)^{1/2} \qquad (3)$$

For given ion-pair, solvent and temperature, the validity of eq.3 is well-supported by using the experimental values of k for chemically different polymers and that one of $(\bar{r}^2/M)^{1/2}$ as determined for the unneutralized material, i.e. assuming that the chain conformation is unmodified by the neutralization of the carboxylic acid end-groups.

4. MORPHOLOGICAL INVESTIGATION BY SAXS

At f_p>0.5, bulk carboxylato telechelic polydienes exhibit a broad maximum in the small angle X-ray scattering profile[6], which is usually taken as an evidence for ionic aggregation in ionomers. As supported by the slight dependence of the reciprocal of the Bragg spacing on the charge density of the metal cation, the nature of the cation is only of secondary importance in the overall organization of the aggregates. No evidence can be found for a change in the mechanism of the ion-pair aggregation and especially for cluster formation. This situation is deeply different from that reported by Eisenberg for ionomers[7]. Once formed, the ion-pair multiplets are very stable at least up to about 100°C. The $\bar{M}_n$ of the chain between charges is of prime importance in governing the organization of the multiplets. Up to now, only three different $\bar{M}_n$ have been considered and the variation law of the Bragg spacing with $\bar{M}_n$ (both $\bar{M}_n^{1}$ and $\bar{M}_n^{0.5}$ variations would be possible) cannot be ascertained and compared to eq.1. This is a key-point in view of the conformational effects due to the ion-pair association.

ACKNOWLEDGMENT

The author is grateful to Drs G. Broze, and J. Horrion (Liège), C.E. Williams (LURE, Paris) and T.P. Russell (IBM, San José) for their collaboration to this research.

REFERENCES

1. Broze, G., Jérôme, R. and Teyssié, Ph. Macromolecules, vol. 15, 920, 1982.
2. Joanny, J.F. Polymer, vol. 21, 71, 1980.
3. Forsman, W.C. and Hong, S. ACS Polymer Preprints, vol.25, n°2, 305, 1984.
4. Broze, G., Jérôme, R. and Teyssié, Ph. Macromolecules, vol.15, 1300, 1982.
5. Flory, P.J. "Principles of Polymer Chemistry", Cornell University Press, Ithaca, p.407, 1953.
6. Williams, C.E., Russell, T.P., Jérôme, R. and Horrion, J. Macromolecules, to be published.
7. Matsuura, H. and Eisenberg, A. J. Polym. Sci., Polym. Phys., vol. 14, 1201, 1976.

ION-CONTAINING NETWORKS: STRUCTURAL MODIFICATIONS INDUCED BY LITHIUM IONS

J.F. LeNEST, J.P. COHEN-ADDAD, A. GANDINI, F. DEFENDINI and H. CHERADAME

Laboratoire de Chimie Macromoléculaire et Papetière, Ecole Française de Papeterie (INPG), BP 65, Saint Martin d'Hères, 38402 France

SYNOPSIS

Polyether-urethane networks containing lithium perchlorate were submitted to swelling and proton spin-spin relaxation rate measurements to gain a better understanding of the interactions between the salt and the polyether chains. A model is proposed to account for the present results and previous data.

1 INTRODUCTION

Our laboratory has been engaged in research on ion-containing networks for several years and a thorough investigation has been made of the static and dynamic properties of polyether-urethane networks containing a variety of (mostly) alkali metal salts (1). One of the most interesting systems is certainly that involving polyethylene oxide chains and $LiClO_4$ because it offers the possibility of unravelling some basic features concerning the structure and transport properties of the materials.

The object of this study was to arrive at a full understanding of the interactions between lithium ions and ether groups, ie to carry out a series of experiments which would be complementary to those already carried out (1).

2 EXPERIMENTAL

2.1 Materials

2.1.1 Saltless networks: Polyethylene oxide glycols (PEO) with M_n= 200, 420, 640, 1050, 2100, 3200, 3800, 6000 and 10000 (Merck) were crosslinked with stoichiometric quantities of $HC(p\text{-}C_6H_4NCO)_3$ (Bayer), ie OH/NCO = 1. Similar networks were prepared from the same triisocyanate and polypropylene oxide glycols (POP) with M_n= 425, 1025 and 3000.

2.1.2 Networks containing salts: The following materials were prepared under the same conditions as above: (I) PEO-1000, (II) PEO-2000 and POP-1000 crosslinked with $HC(p\text{-}C_6H_4NCO)_3$ in the presence of variable amounts of $LiClO_4$.

2.2 Network Synthesis

Both saltless and salt-containing networks were prepared in methylene chloride using different amount of solvent, but always maintaining OH/NCO = 1. The details of these operations including the drying and conditioning of the products will be given elsewhere (2). All syntheses were carried out at 22°C.

2.3 Techniques and Measurements

Swelling of networks was conducted up to equilibration. The swelling ratio was then measured (2) and expressed as equilibrium volume to initial volume.

Proton spin-spin relaxation rates were determined by the spin-echo technique (2). All measurements were made at 22°C.

3. RESULTS AND DISCUSSION

3.1 Swelling Ratios

3.1.1 Saltless networks: Before proceeding to the systematic measurements, we verified that the volumes of various networks and solvents obeyed additivity rules. The actual study begun by fixing the volume ratio for the synthesis, V_c (volume of polymer to total volume) at 0.43 for all networks except that using PEO 200, for which V_c=0.65 because at V_c=0.43 syneresis would have occurred.

The equilibrium swelling of these networks provided the following qualitative informations: the swelling ratio, q, increased with PEO chain length as expected and for a given PEO size , it increased as a function of the type of solvent used following the order $CHCl_3 > CH_2Cl_2 > H_2O > C_6H_6$.

When log q was plotted against log M_n (of PEO chains), two straight lines were obtained for each solvent. The first, corresponding to the lower chain lengths, had a slope of 0.64±0.01, whilst the slope of that referring to the longer chains was 0.31± 0.04. The change of behaviour occurred at $M_n \sim 1500$ for all solvents. Since the highest M_n used was 10000, it seems unlikely that physical entangling among PEO chains would play an important role. Thus, the lower-slope region is probably due to chain crossover. For the swelling at low chain length, the slope found agrees closely with that expected from Flory-Huggings' theory rather than with the critical exponent of 4/5 expected from De Gennes' theory.

The effect of V_c was then studied for a series of networks and solvents, and with each series of measurements q was determined as a function of V_c, keeping M_n and the nature of solvent fixed; CCl_4 was added to the above solvents. Plots of q vs V_c were linear and extrapolated to a single crossing point corresponding to V_c=1.26, independent of solvent and PEO chain length. This striking feature has not been reported previously to our knowledge and probably illustrates the convergence towards a limit situation depicting a compact network, not only devoid of free volume, but also with its elements stacked to maximum density as in crystal structures.

Networks based on POP exhibited the same features as those described above for PEO networks.

3.1.2 Salt-containing networks: The systems I and II prepared with V_c=0.43 and in the presence of increasing quantities of $LiClO_4$ were swollen to equilibrium in four different solvents. Two different types of behaviour were encountered depending on the solvent polarity.

In the relatively polar solvents $CHCl_3$ and CH_2Cl_2 the swelling ratio was found to be linearly dependent upon the salt concentration as shown by the following experimental relationship:

I + $CHCl_3$, $q = 7.1 - 0.17\ c$; II + $CHCl_3$, $q = 8.8 - 0.16\ c$;

I + CH_2Cl_2 , $q = 5.8 - 0.08\ c$; II + CH_2Cl_2 , $q = 7.6 - 0.08\ c$;

where c is the concentration of $LiClO_4$ expressed in g of salt per hg of PEO in the network. Note that the slope was independent of the length of PEO chains but changed when changing solvent. The higher dielectric constant of CH_2Cl_2 favoured a higher degree of dissociation of the salt and thus a smaller dependence of swelling on its concentration.

In the non-polar solvents CCl_4 and benzene the swelling features were found to be quite different. The plots of q vs c were monotonically decreasing curves which could not be reduced to any simple relationship. The magnitude of the swelling ratio were much smaller than with the more polar solvents and the decrease of q as c increased much more pronounced.

In order to attempt a rationalisation of these results it is necessary to recall that previous work in our laboratory and particularly the study which accompanies this paper (1), has led to the formulation of a model in which the lithium cations coordinate with the lone pairs of ether-type oxygen atoms and perchlorate anions are "free" in the network below a certain critical concentration. No study had yet been carried out on the swelling of these systems and it turns out that such an investigation corroborates all previous findings and thus the model.

In non-polar solvents, physical crosslinking promoted by Li^+ or higher ionic aggregates from $LiClO_4$, adds onto the existing (irreversible) chemical crosslinks, ie the urethane bonds. Thus the lithium cations effectively increase the crosslink density by "subdividing" the PEO chains into subchains joined by electrostatic knots. Assuming that the role of the salt is restricted to that type of interaction, a calculation can be performed to establish the actual crosslink density (sum of physical and chemical crosslinks) and compare the swelling ratios corresponding to these systems to those of saltless networks. We found a very satisfactory agreement in that the slope of log q vs log M_n' (M_n' being the corrected length of PEO chains taking into account the physical crosslinks) was found to be 0.64, ie the same as for the saltless networks. Also, in a more quantitative comparison, it was found that q for a saltless network prepared with PEO 1000 had the same value as q for a network prepared with PEO 2000 and containing one Li^+ per PEO chain. The same correspondence was obtained when comparing the swelling ratios of a network from PEO 1000 containing one Li^+ per PEO chain with a network from PEO 2000 containing three Li^+ per PEO chain: the q values were the same. These results corroborate previous findings (1) in which two characteristic interactions between lithium ions and PEO chains were brought to light, namely the fact that Li^+ can coordinate more than one PEO chain and the "subdivision" into smaller chains down to a limiting value of about 12 EO units, ie a molecular weight of about 500. The present model is now supported by swelling measurements as well as by T_g, conductivity and transport number data (1).

In the more polar solvents, viz. chloroform and methylene chloride, the ionic associations as well as the binding of more than one PEO chain by the lithium cation tend to break down so that the linear relationships found between q and c reflect the behaviour of networks without physical crosslinks, but only single Li^+-$O(CH_2$-$CH_2)_2$ interactions.

The swelling of system I was also studied with $Mg(ClO_4)_2$ instead of the lithium salt. Using benzene as swelling agent, it was found that a given concentration of magnesium perchlorate gave the same quantitative behaviour as the same network containing twice that concentration of lithium perchlorate. It seems reasonable to interprete this observation as suggesting that the magnesium cation is twice as effective as the lithium one in complexing with oxygen atoms along the PEO chains. The markedly different charge density would account for this.

The study of the swelling behaviour of networks based on POP and containing lithium perchlorate gave results entirely similar to those expounded above for the PEO networks.

3.2 Magnetic Relaxation

The spin-echo technique was applied to the study of the transverse magnetic relaxation of hydrogen nuclei belonging to the methylene groups in PEO segments between crosslinks (3,4).

We followed the changes in relaxation rate δ in systems I and II with lithium perchlorate as a function of salt concentration. The results were treated using the recent theory developed by Cohen-Addad (3) which relates δ to the statistical chain length between crosslinks R_e and to the average number of monomer units in those chains N_e as

$$\delta \propto R_e^2/N_e^2 \quad .$$

Experimentally, it was found that δ was directly proportional to c^2. From this correlation and using the model discussed above (ie chain segmentation by lithium cations through complexation with PEO oxygen atoms up to one Li^+ per 12 EO units for physical crosslinking) we determined that

$$\delta \propto 1/N_e^2 \quad .$$

Once again the evidence is entirely compatible with the model proposed.

The fact that δ was found to be independent of R_e is not in conflict with the general picture. Indeed this observation suggests that the physical crosslinks introduced by the lithium ions are of a dynamic nature, ie they move continuosly along the PEO chains. Thus, the statistical value of R_e should be insensitive to their presence and remain that of the saltless network.

The topological model acquires thus a more detailed configuration after the information gathered by the relaxation measurements. The result of the lithium cation interactions with the oxygen atoms is a thwarting of PEO chain fluctuations.

4 CONCLUSIONS

The set of two papers presented in this book (1) provides the condensed version of the work carried out in our laboratory in last couple of years on the characterisation of polyether-based networks contaning ions. A thorough description and discussion of all these results will be submitted (2) together with further data which could not be included in these two manuscripts for lack of space. However, the proposed model appears to us a already adequately verified by the qualitative and quantitative observations given here and in ref.1.

REFERENCES

1. Gandini, A. et al., This Book, p.
2. Le Nest, J-F. et al., to be submitted.
3. Cohen-Addad, J-P., J. Physique, 43, 1509 (1982).
4. Cohen-Addad, J-P. et al., J. Physique, 45, 575 (1984).

ACKNOLEDGEMENTS

We are grateful to Direction des Recherches, Etudes et Techniques for financial help.

ION-CONTAINING NETWORKS: RECENT RESULTS CONCERNING TRANSPORT PROPERTIES

A. GANDINI, J.F. LeNEST, M. LEVEQUE and H. CHERADAME

Laboratoire de Chimie Macromoléculaire et Papetière, Ecole Française de Papeterie (INPG), BP 65, 38402 St. Martin d'Hères, France

SYNOPSIS

Various types of networks containing either an added salt or an ionisable function were thoroughly characterised in order to gain a better understanding of the interactions between ions and polymer segments and of the electrochemical properties displayed as a function of the network structure and the electrolyte.

1 INTRODUCTION

The steadily growing interest in ion-containing polymers in the last decade is mostly linked to the search of new solid-state electrochemical generators. The state of the art has been recently reviewed (1). Our laboratory has contributed to this field with several studies on (mostly) polyether networks crosslinked by urethane moieties and mixed with various alkali-metal salts (2-4).

This report deals with a similar global approach to the characterisation of new systems and attempts to rationalise succintly the diverse information gathered on the basis of reasonable static and dynamic models.

2 EXPERIMENTAL

2.1 Materials

2.1.1 Polyether networks without salts: Polyethylene oxide (PEO) glycols with M_n= 200,420, 640, 1050, 2100, 3200, 3800, 6000 and 10000 (Merck) were crosslinked with stoichiometric quantities of $HC(p-C_6H_4NCO)_3$ (MTI,Bayer), ie OH/NCO = 1.

2.1.2 Polyether networks containing salts: A variety of materials were synthesised of which the following are the most representative: (I) PEO 1000 + MTI; (II) PEO 1000 + $OCN(CH_2)_6N[CONH(CH_2)_6NCO]_2$ (HMTI, Bayer); (III) PEO 2000 + MTI; (IV) a b-copolymer glycol polyethylene oxide-polypropylene oxide-polyethylene oxide (Wyandotte, PEO-PPO-PEO) with block molecular weights of 2000-1000-2000 + MTI; (V) PEO-PPO-PEO 3350-1700-3350 + HMTI; (VI) polydimethylsiloxane (M=3500) randomly grafted with a diblock PPO-PEO-OH (600-500-OH) at one out of three siloxane units (PDMS-g-POP-b-PEO-OH, Rhône-Poulenc) + hexamethylenediisocyanate (HMDI, Merck). Again, crosslinking was carried out with OH/NCO = 1. These polyether-urethane networks were prepared in the presence of variable amounts of different salts (see below) but mostly lithium perchlorate.

2.1.3 Ionomers: A series of phosphate and thiophosphate anions were incorporated into polyether-urethane networks by first treating $POCl_3$ or $PSCl_3$ with PEO or PPO and then crosslinking with isocyanates. Conditions were set so as to substitute two chlorine atoms in the first reaction, whilst the third was removed by LiOH in order to generate a phosphate- or thiophosphate-type anion attached to the polyether chain and a lithium cation free to move within the resulting network. The latter was synthesised under stoichiometric OH/NCO ratios. The following systems are the most relevant examples: (VII-O, VII-S) lithium phosphate and thiophosphate with PEO 400 diol crosslinked with MTI; (VIII-O, VIII-S) lithium phosphate and thiophosphate with PEO 1500 diol crosslinked with MTI; (IX-O, IX-S) phosphate and thiophosphate with PPO 1500 triol (from glycerine, Merck) crosslinked with HMDI; (X-O, X-S) phosphate and thiophosphate with PPO 1500 triol crosslinked with 2,4-toluenediisocyanate (TDI, Merck).

2.2 Network Syntheses

The details concerning the preparation of gelled membranes without and with salts have been described (2-4). As for the ionomers, the subsequent reactions leading to the lithium phosphate and thiophosphate polyether-urethane networks will be described in full elsewhere (5). The same applies to the drying and conditioning of these materials before measurements.

2.3 Characterisation of networks

The techniques and procedures used to obtain the data discussed below are to be found in our previous publications (2-4) or in forthcoming ones (5).

3 RESULTS AND DISCUSSION

3.1 Specific Volume

The variation of specific volume V as a function of crosslinking density d (moles of urethane links per gram of network) for saltless networks at 22°C was found to be linear, viz., with MTI as crosslinking agent,

$$V = V_{\infty} - (106 \pm 5)\, d \quad ,$$

where V_{∞} represents the specific volume of a "linear" PEO of infinite molecular weight not bearing terminal hydroxyl group. Its extrapolated value was 0.896 cc/g. Extrapolation to the theoretical situation representing a purely urethane network (absence of PEO) gave a specific volume of 0.608 cc/g. No comparison is possible with experimental values for this limiting case since such a network has never been synthesised.

The effect of salt concentration upon the specific volume of a given network also gave a linear relationship, eg, for system I with $LiClO_4$ at 22°C,

$$V = V_{\circ} - 0.48\, c \quad ,$$

where c is the concentration of the salt expressed in w/w and $V_{\circ}$ is the specific volume of the saltless network I. Assuming simple additivity of densities for a mixture of network I and $LiClO_4$, one calculates the linear dependence

$$V = V_{\circ} - 0.43\, c \quad .$$

Since the measured values of V are smaller than those calculated for a non-interacting solid solution, it seems most likely that the contraction is due to strong electrostatic attraction between the salt and the network and more specifically between the lithium cation and the oxygen atoms forming ether linkages, ie to the formation of donor-accetor complexes. The contraction effect produced by these interactions is a function of the salt concentration, viz., $\Delta V = 0.05\, c$, ie a linear dependence.

3.2 Glass Transition Temperature

For PEO networks without salt obtained with MTI we obtained

$$1/T_g = (1/T_{g\infty}) - 0.76\ d' \quad ,$$

where d' is the crosslink density now expressed in moles of urethane links per cc (since we know V). In the absence of reticulation, the extrapolated value of $T_{g\infty}$ corresponding to a linear PEO without terminal OH groups and of infinite length was 209 K.

The variation of T_g as a function of salt concentration for a given network followed a similar trend, viz. with systems I and III containing $LiClO_4$,

$$1/T_g = (1/T_{g\circ}) - 0.27\ c' \quad ,$$

where $T_{g\circ}$ is the transition temperature of the saltless network and c' is now the $LiClO_4$ concentration expressed in moles of salt per cc. This relationship suggests that the added salt plays the role of a pseudocrosslinking agent. A plot of $1/T_g$ vs d'+ c' gave a straight line from which the quantitative effect of the salt as physical crosslinker could be evaluated. For the two systems above it turned out that about three salt molecules produced the same effect upon T_g as one urethane crosslink. Most likely, Li^+ is the "binding" entity.

Systems II, IV, V and VI with $LiClO_4$ behaved likewise, the coefficient of c' depending upon the nature of the polyether segments. In particular, it was found that the PEO chains interacted with the salt (Li^+) more strongly than POP chains. The PDMS elements showed virtually no interaction.

The above results can be rationalised in terms of the following model: lithium cations are solvated by oxygen atoms belonging to ether linkages and this donor-acceptor complexes involve more than one oxygen atom per Li^+. Thus, electrostatic crosslinks arise and their concentration increases with salt concentration and the corresponding increase in T_g follows the typical dependence encountered with chemical crosslink density.

These considerations also apply to most of the polyether networks-alkali metal salts previously studied in our laboratory (2-4).

3.3 Electrical Conductivity

We have already shown that free volume considerations adequately explain the dependence of ionic conductivity upon salt concentration for a number of polyether networks containing alkali metal salts (2-4). All the systems studied here showed the same behaviour and were amenable to a clear WLF treatment. Thus, for systems I to VI containing a whole variety of salts ($LiClO_4$, $LiSO_3CF_3$, $LiBF_4$, LiSCN, $LiOCOCF_3$, $NaClO_4$, $KClO_4$ and $Mg(ClO_4)_2$) the elaboration of the data using T_g as the reference temperature gave consistent results illustrated by the reproducible values of the WLF parameters, viz.,

$$c_1 = 10 \pm 2 \qquad \text{and} \qquad c_2 = 55 \pm 5\ \text{K} \ ,$$

independent of network structure and salt used.

The actual magnitude of the ionic conductivity measured with a given network at a given temperature and lithium salt concentration depended upon the nature of the anion. The trend followed roughly the relative basicity of the anion, ie the lower the basicity (the higher the strength of the parent acid), the higher the conductivity. Thus, for network II at 50°C and a concentration corresponding to 0.04 lithium cations per ether oxygen atom, the conductivities (10^6 S/cm) were 50 for ClO_4^-, 40 for $CF_3SO_3^-$, 19 for BF_4^-, 7 for SCN^- and 2 for CF_3COO^-. Given the tendency of the oxygen atoms belonging to ether functions to complex the lithium cation through the electron lone pair, these results can be interpreted as reflecting the competition between the anion and those oxygen atoms towards association with Li^+: as the basicity of the anion decreases, the proportion of the salt dissociating to give a lithium-oxygen complex and a "free" anion increases, as witnessed by a corresponding increase in conductivity.

No clear trend could be discerned when analysing the changes in conductivity as a function of the nature of the cation, all other conditions being kept constant.

With the ionomer networks VII to X we calculated $c_1 \sim 15$ and $c_2 \sim 50$ K, but much lower conductivities than with the polyether-salt systems under comparable conditions.

Among the many networks and salts tested, the material which gave the best performances in terms of conductivity and could therefore be a good candidate as solid electrolyte for electrochemical generators, was system V with lithium perchlorate. Values of about 10^{-4} S/cm were measured at room temperature. The presence of POP blocks in this network tend to minimize the propensity of PEO towards crystallization, although too high a percentage of POP in a polyether formulation was found to be detrimental to the conductivity. The latter observation corroborates the poorer solvating capacity of POP towards cations, compared with PEO, as already noticed in the study of T_g trends.

With systems I, II and III, plots of the logarithm of the conductivity vs the logarithm of $LiClO_4$ concentration were linear with unit slope up to a critical concentration, above which the conductivity did not increase further in an appreaciable manner. This critical concentration corresponded to one salt molecule per PEO segment for systems I and II and to three salt molecules per PEO segment for system III. Of course the above plots were made at constant $T-T_g$.

The first important comment about these results is that the combination PEO-$LiClO_4$ seems particularly favourable to the salt dissociation through Li^+ complexation with the ether oxygens. Indeed, the unit slope below the critical concentration suggests complete dissociation of the salt giving free perchlorate anions and $Li^+.O(CH_2-CH_2)_2$ complexes. With networks containing POP and/or salts derived from less strong acids, the extent of dissociation was lower, as expected from the various results already gathered (see above and 2-4).

The second relevant observation which requires an interpretation is the abrupt attainment of a levelling-off of the conductivity when certain specific ratios of salt-to-PEO-chains are attained. In fact, all the critical values correspond to the insertion of lithium cations along the PEO chains until sub-chains of about 12 ethylene oxide units are formed. These segments seem to represent the shortest PEO elements capable of solvating one Li^+. The model proposed is therefore one in which the lithium cations are accomodated within PEO chains: up to one entity per PEO-1000 network segment and up to three entities per PEO-2000 network segment. In other words the lithium cations can be seen as complexing moieties introducing sub-segments within the PEO chains bound between two urethane crosslinks: the length of these subsegments cannot be shorter than about 12 EO units, ie one lithium cation is only fully complexed by at least that length of PEO. These considerations must be placed in the context of the physical crosslinking role of Li^+ found from T_g measurements in order to attempt the construction of a complete and non-contraddictory model.

Above the critical concentrations observed the conductivity stops growing or at least suffers minor fluctuations as the concentration of lithium perchlorate is raised. It seems probable that in this domain,the fact that no more Li^+ can be accomodated individually into the PEO chains imposes the clustering of salt molecules or of their ions to give non-conducting aggregates. It remains to understand why this transition is so abrupt, ie why the critical ratio of one Li^+ to about 12 EO units is indeed critical to such a rigorous extent that below it all the salt appears to be dissociated and above it the additional salt molecules form ionic cluster and do not contribute to any further conductivity increase of substantial magnitude. The elucidation of this interesting problem is being pursued actively in our laboratory.

3.4 Transport Numbers

Two years ago, we were the first to show (6) that, contrary to previous suggestions and results, cationic contribution to conductivity in polyether networks containing a variety of monovalent metal perchlorates was much less important that the anionic counterpart.

The present study has confirmed this behaviour extending its range of validity to several new systems and salts (5). Thus, for networks I to VI and a variety of alkali metal salts the cationic transport number t^+ ranged from 0.02 to 0.3 depending on the system and the conditions. Of course, with the ionomeric networks VII to X t^+ was always close to unity since the anion is bound to the network.

It was found that with systems I, II and III, t^+ for $LiClO_4$ decreased as the salt concentration increased up to a critical concentration above which it remained constant. Interestingly, the critical concentrations corresponded to the same $Li^+/O(CH_2-CH_2)_2$ ratios encountered for the abrupt change in conductivity behaviour discussed above. These observations confirm the existence of "saturation" phenomena for the interaction of the lithium cations with the ether functions. In the present context, we interpret these trends as follows: the movements of lithium cations towards the cathode are the easier the higher the proportion of unoccupied ether groups, ie uncomplexed oxygen atoms along the PEO chains. At the critical saturation concentration the conductivity stops growing and the transport numbers remain constant with further salt additions.

With a given salt concentration, t^+ was foundto be the higher the shorter the PEO chain between crosslinks. This trend is quantitatively more important than that concerning the concentration dependence and can be rationalised assuming a direct correlation between crosslink density and number of favourable jumping sites for Li^+, ie the urethane moieties constitute bridging elements (through a non-complexing action or more local free volume) for the ease of migration of these cations towards the negative electrode. On the contrary, the pathway of the perchlorate anion towards the cathode is not sensitive to this factor because of the lack of specific interaction of this ion with PEO units and its larger size which makes it less sensitive to local changes in free volume density.

With systems I, II and III and magnesium perchlorate, t^+ was found to be zero within experimental error indicating that the magnesium cation is totally immobilised by the ether groups, whilst the perchlorate anion can move, as shown by conductivity measurements. Indeed, under the same experimental conditions, the conductivity of magnesium perchlorate (ie that due to the anion only) was only slightly lower than that of lithium perchlorate (of which about 20% was due to the cation).

3.5 Dynamic Mechanical Properties

Much work has been carried out in our laboratory on the viscoelastic behaviour of polyether networks containing different metal salts (2-4). Superposition laws were obtained by WLF treatment of the effect of temperature, excitation frequency and salt concentration.

In the present study these correlations were confirmed. Typically, with systems IV and VI and $LiClO_4$, good master curves were derived from the study of effect of the same variables, confirming that free volume is the decisive property governing the behaviour of these materials. The data from system IV with T_g as reference temperature gave c_1=10 and c_2=50 K, which clearly shows that the transport properties (conductivity, see above) and the viscoelastic response of these networks are governed by the same basic phenomena, namely the frequency of segmental motions. These conclusions had already been reached in the context of previous work on other materials (2-4).

3.6 Spin-Spin Relaxation

Line broadening studies related to the ^{7}Li nucleus in systems I-VI containing lithium perchlorate showed that the movements and the "freezing" of Li^{+} follow closely the extent of segmental motions allowed in these networks. These observations constitute an additional element favouring a general model based on free volume as the basic parameter influencing the mobility and transport properties of the ionic species, particularly the anion. Similar results had already been obtained and discussed (2-4).

With the ionomer networks VII-X, we looked at the spin-spin relaxation of three nuclei: ^{1}H, ^{7}Li and ^{31}P. We found the same temperature dependence of T_2 for all species, a fact which strongly suggests that both "free" and bound elements of these networks move according to the same laws, ie again those imposed by free volume considerations.

3.7 Electrochemical Stability

Cyclic voltammetry applied to systems III, IV and V with lithium perchlorate showed that the reduction pathway was characterised by a reversible barrier due to the passage of Li^{+} to Li°. The oxidation branch gave an irreversible barrier for the perchlorate anion, indicating that this ion is indeed a mobile species within these networks, as already clearly suggested by the transport number data. The potential difference between these two barriers, defining the redox stability window of the systems, was close to 4 V, which is wide enough to allow the use of a large variety of intercalation compounds for the assembly of electrochemical generators.

4 CONCLUSIONS

The different experimental techniques applied in this study have provided both complementary and confirmatory evidence concerning the structure and the dynamics of the solid polyether network electrolytes. The most relevant feature of these systems is the strong donor-acceptor interaction of Li^{+} with ether functions which translate into low t^{+} values, physical crosslinking and saturation levels above which the interaction is with ionic aggregates and non-conducting. Moreover, it was amply confirmed that the transport properties in these materials are subject predominantly to free volume constraints.

This work will shortly be described in full and more detailed models proposed, also on the basis of further studies described elsewhere in this book (7).

REFERENCES

1. Armand, M., Solid State Ionics, 9-10, 745 (1983).
2. Killis, A. et al., Macromolecules, 17, 63 (1984) and references therein.
3. Killis, A. et al., Solid State Ionics, 14, 231 (1984) and references therein.
4. Lévêque, M. et al., J. Power Sources, 14, 27 (1985) and references therein.
5. Le Nest, J-F., Gandini, A. and Cheradame, H., to be submitted.
6. Lévêque, M. et al., Makromol. Chem. Rapid Commun., 4, 497 (1983).
7. Le Nest, J-F., et al., This Book, p.

AKNOWLEDGEMENTS

We thank the Direction des Recherches Etudes et Techniques for financial support.

Part V

DIFFUSION/BARRIER PROPERTIES

DIFFUSION OF GASES AND LIQUIDS IN GLASSY AND SEMI-CRYSTALLINE POLYMERS

H. L. Frisch*

*Department of Chemistry, State University of New York, Albany, Albany, New York 12222.

SYNOPSIS

The diffusion of small molecules in polymers above the glass transition temperature appears to be dominated by those factors which affect the Rouse segment mobility, e.g. the diffusion coefficient is strongly concentration dependent in accord with predictions of free-volume theories. The major complication arises due to dispersed homogeneities such as crystallinity. The effect of such inhomogeneities are not well described by simple two phase models. We shall review previous attempts to deal with crystallinity in rubbery polymers. While rubbery polymers respond rapidly to changes in their condition this is not the case with glassy polymers. Diffusion in such polymers is often dominated by finite rate relaxation effects which produce a whole range of anomolous behavior. We will discuss some recent attempts to describe the coupling of mass transport with relaxation behavior whose origin can be mechanical or structural.

1 INTRODUCTION

The non-equilibrium sorption and molecular dispersal of small molecule penetrants in solid polymers has been extensively studied[1-5] usually by permeation cell and sorption balance experiments. The volume fraction of penetrant ϕ_1, rarely exceeds by much 15% in such measurements. The resulting data are basic to estimates of: barrier properties of polymers (e.g. in packaging), membrane separators, the efficiency of removal of solvent, unreacted monomer or polymer additives in processing applications, etc. Often in amorphous polymers (above the glass transition temperature, T_g)

the time scale of these processes determined solely by a single binary, (polymer fixed) diffusion coefficient, $D(\phi_1)$, which we describe in the next section under Fickian diffusion. Polymer crystallinity introduces complications which are briefly reviewed in the third section. Finally in the last section we describe the delayed relaxation effects which play an important role with the diffusion of organic penetrant fluids in glassy polymers.

2 FICKIAN DIFFUSION INAMORPHOUS POLYMERS

Above T_g the polymer-penetrant system can respond rapidly to a change in their condition, e.g. in a sorption balance. A change in the activity (relative vapor pressure) of the penetrant reservoir in which the polymer sample is immersed. Mechanical equilibrium can be easily maintained as Fickian diffusion progresses to a new, final, overall mechanical and chemical equilibrium state determined by the rapid sorption equilibrium achieved at the boundaries of the sample. Diffusion is mathematically described by the diffusion equation (Fick's 2° Law) with a concentration (ϕ_1) dependent diffusion coefficient $D(\phi_1)$ and constant boundary conditions. The characteristics of such Fickian sorption behavior are conveniently listed in references 1 and 2. The contribution of the intrinsic polymer center of mass motions (governed by reptation)are negligible and in excellent approximation the thermodynamic diffusion coefficient, $D_T(\phi_1)$, reflects only intrinsic penetrant center of mass motions and can be defined via

$$D_T\left(\frac{\partial \ell n a_1}{\partial \ell n \phi_1}\right)_T = \frac{D(\phi_1)}{(1-\phi_1)^3} \tag{2.1}$$

with a_1 the penetrant activity.

The penetrant motion is strongly associated with the Rouse mobility of the chain segments $m_R(\phi_1)$ modified by the presence of the penetrant and hence a function of ϕ_1. Following Fujita[6] this mobility m_R has a Doolittle free volume theory form

$$D_T = RTm_R(\phi_1) = RTA\exp[-B_d/f_{12}^a(\phi_1,T)] \tag{2.2}$$

with A and B_d phenomenological constants and $f^a_{12}(\phi_1,T)$ a suitably defined free volume fraction, satisfying[6]

$$f^a_{12}(\phi_1,T) = f^a_2(0,T) + \beta\phi_1$$
$$f^a_2(0,T) = f_2(T_g) + \alpha(T-T_g) \qquad (2.3)$$

In (2.3) $f_2(T_g)$ and α are close to their WLF values and β is the difference between the specific free volume of pure penetrant and polymer. Within the given number of parameters (A, B_d, $f_2(T_g)$, α,β) this theory has been remarkably successful in correlating Fickian diffusion data for both single and mixtures of penetrants in amorphous polymers[5].

3 DIFFUSION IN CRYSTALLINE AMORPHOUS POLYMERS

The simplest view of the dense, crystalline regions in amorphous polymers is that they act like fixed wholly impermeable domains ("particles") suspended in the otherwise unperturbed, amorphous state of the polymer. Semi-crystalline polymers are thus considered to be ideal two phase media; whose phases do not interact[1,2]. The crystalline phase is often further idealized into a set of dispersed "particles" of simple (e.g. spherical) shape. The random geometry of the dispersion is solely characterized by the volume fraction of crystallinity, ϕ_c. Different methods of determining ϕ_c (X-ray, density, etc.) rarely agree to better than 5-10%. Generally, for most semi-crystalline samples nothing more is known about the random geometry.

Departures from this model must occur because the presence of crystallinity reorganizes structurally the amorphous phase and stiffens it mechanically thus also affecting mass transport. At sufficiently high crystallinity, void or craze-like regions can appear. Sorption and desorption can cause recrystallization in some systems so that the crystalline domains cannot be thought to be fixed in space during diffusion. Early (say from the initial slope of the sorption curve) and late (steady-state permeation) time determinations of "diffusion coefficients" will differ (not only because of concentration dependence) but because late time determinations can take account of slower, more tortuous diffusion paths (One should thus compare D's obtained by similar diffusion measurements).

A convenient necessary condition that a penetrant-crystalline polymer system departs significantly from ideal two phase medium behavior is provided by the mass transport analog of the viscosity number (intrinsic viscosity) the diffusion numer [D]. If $D = D(\phi_1, \phi_c; T)$ approaches $D^a = D^a(\phi_1; T)$ as ϕ_c vanishes then [D] is defined by

$$[D] = \lim_{\phi_c \to 0} \left(\frac{D^a - D}{\phi_c D^a} \right) \tag{3.1}$$

If [D] varies significantly with ϕ_1 and T then the ideal two phase medium description fails. This can often be the case.

The simpliest theory[7], utilizing only ϕ_c to describe the crystallinity, which reduces to (2.2) when $\phi_c \to o$, is obtained by retaining only the bulk effect due to crystallinity on the free volume. Using additivity of densities of crystalline and amorphous domains f_{12}^a in (2.2) would then be replaced by

$$f_{12} = f_{12}(\phi_1, \phi_c; T) = f_{12}^a(\phi_1; T)(1-\phi_c) \tag{3.2}$$

since free volume mediated penetrant motion can only occur in the amorphous regions; the B_d having to reflect what method is used to determine D and thus D_T via (2.1). In this case

$$[D] = B_d / f_{12}^a(\phi_1; T) \tag{3.3}$$

good results are reported by some investigators[5] using (3.2) in (2.2).

4 RELAXATION EFFECTS IN DIFFUSION OF ORGANIC FLUIDS IN GLASSY POLYMERS

Glassy polymers respond noticeably more slowly to changes in their condition-often with a whole spectrum of relaxation times. Departures from Fickian diffusion behavior occurs whenever the diffusion rate is of the same order or more rapid than a characteristic relaxation rate of internal stress or rate of structural change resulting from mass transport[1-4]. We cannot review all mathematical models of this but will explore an old and new attempt.

Long and Richman[8] showed that even surface relaxation suffices, i.e. they found tnat the surface concentration, c_s for the sorption of organic vapors into glassy polymers exhibited experimentally a characteristic time dependence

$$c_s = c_i + (c_e - c_i)(1 - e^{-\beta t}) \tag{4.1}$$

with c_e the final sorption equilibrium and $c_i \ll c_e$ a small initial value and

β possibly a mechanical relaxation time of the glassy polymer. Eq. (4.1) could be a irreversible thermodynamic consequence of a relaxation of the chemical potential μ, of the penetrant[5]. Solution of Fick's 2° Law with constant D subject to (4.1) can account for many observed sorption anomalies such as "sigmoid" and "two stage" sorption[1,2,8]. Actually (4.1) can account for much more e.g. a linear in time uptake, M_t, of the organic penetrant in a sheet of thickness 2ℓ. For simplicity we set $c_i = 0$ and taking β very small so that for $t < 1/\beta$ we can replace (4.1) by

$$c_s = c_e \beta t \tag{4.2}$$

also observed by Long and Richman[8] using (4.2) one finds (cf. eq. (4.33) of reference 2)

$$\frac{DM_t}{c_e \beta \ell^2} = \frac{2Dt}{\ell^2} - \frac{2}{3} + \frac{64}{\pi^4} \sum_{n=o}^{\infty} \frac{\exp\{-D(2n+1)^2 \pi^2 t/4\ell^2\}}{(2n+1)^4} \tag{4.3}$$

which becomes linear in time (with a time lag).

To be consistent the relaxation behavior of μ must also be taken into account in formulating the transport equation replacing Fick's 2° Law. If one adopts an anology to stress relaxation in a viscoelastic medium one can suppose that the chemical potential μ can be decomposed into an ideal contribution $\mu_{id} = k_B T \ln c$ + constant, c the penetrant concentration, and an interactive part arising from retarded mechanical interactions, μ_{int}. Following a sudden change in Δc in concentration at $t = 0$, the interactive part of μ relaxes from the initial increase $(\partial\mu_{int}/\partial c)_\infty \cdot \Delta c$ corresponding to high frequency response to a smaller equilibrium increase $(\partial\mu_{int}/\partial c)_o \cdot \Delta c$, given by the static, zero-frequency derivative. The total response of the local chemical potential is given in terms of the relaxation function $\psi(t)$ by a Boltzmann superposition[9]

$$\mu = \mu_{id} + (\partial\mu_{int}/\partial c)_o c$$

$$+ \{(\partial\mu_{int}/\partial c)_\infty - (\partial\mu_{int}/\partial c)_o\} \int_{-\infty}^{t} \psi(t-t') \frac{\partial c}{\partial t'} dt' \tag{4.4}$$

Combining this with the usual expression for the flux, j, one obtains[9] (in say in one dimension, $j = -Bc\partial\mu/\partial x$, B the mobility)

$$\frac{\partial c}{\partial t} = Bk_BT \frac{\partial c^2}{\partial x^2} + B(\partial\mu_{int}/\partial c)_o \frac{\partial}{\partial x}(c \frac{\partial c}{\partial x})$$

$$- B\{(\partial\mu_{int}/\partial c)_o - (\partial\mu_{int}/\partial c)_\infty\} \int_{-\infty}^{t} dt'\psi(t-t') \frac{\partial}{\partial x}(c\frac{\partial^2 c}{\partial x \partial t'}) \qquad (4.5)$$

Many diffusion anomalies can be explained by this equation.

5 ACKNOWLEDGEMENT

This work was supported by grant NSF, DMR 8305716A01

This paper is dedicated to Ron Konigsveld.

6 REFERENCES

1. "Diffusion in Polymers" (ed. J. Crank and G. S. Park), Academic Press, London 1968.
2. J. Crank, "The Mathematics of Diffusion", 2° Edit., Clarendon Press, Oxford, 1975.
3. H.B. Hopfenberg and V. Stannett in "The Physics of Glassy Polymers" (ed. R. N. Haward), J. Wiley and Sons, New York, 1973.
4. "Sorption and Transport in Glassy Polymers" (ed. D.R. Paul) in Polymer Eng. and Science Volume 20, No. 1, 1980.
5. H. L. Frisch and S. A. Stern, CRC Critical Rev. in Solid State and Mat. Sci., 11, 123 (1983).
6. H. Fujita, Fortschr. Hoch. Polym. Forsch. 3, 1 (1960).
7. A. Kreituss and H. L. Frisch, J. Polymer Sci. (Phys. Ed.) 19, 889 (1981).
8. F. A. Long and D. Richman, J. Amer. Chem. Soc. 82, 513 (1960).
9. J. Jäckle and H. L. Frisch, J. Polymer Sci. (Phys. Ed.), in press.

TRANSPORT REGULATED ELECTROCHEMICAL REACTIONS IN POLYIMIDE FILMS

STEPHEN MAZUR and SHIMON REICH[11]

Central Research and Development Department
E.I. duPont de Nemours Inc.
Wilmington, De. 19898

SYNOPSIS

A film of ODA/PMDA polyimide can accept electrons from a cathode in a conventional electrochemical cell. When metal ions, such as Ag+, are included in the electrolyte solution, these can permeate into the film where they are reduced via the polymer to submicron metal particles. Under suitable conditions the deposition process can be controlled to produce a continuous metal interlayer embedded within the polymer. The thickness and position of this interlayer may be controlled with great precision and high resolution patterns can be made.

1 INTRODUCTION

Interest in the properties and uses of polymeric redox reagents originated with H. G. Cassidy's studies of quinone subtituted polystyrene resins (1). More recently, attention has been focused on a certain subclass of redox-active polymers, generally referred to as "electrochemically active". These materials are capable of exchanging electrons not only with small molecules in solution, but also with solid electrodes. A thin film of such a polymer supported on the surface of an electrode may be electrochemically driven between two or more stable oxidation states.

The first section of this paper describes electrochemistry of the polyimide derived from 4,4'-oxydianiline (ODA) and pyromellitic dianhydride (PMDA). While this commercially important material is best known as a high temperature dielectric with good barrier properties, it can be made to undergo two reversible electrochemical processes.

Beyond the requirement of appropriate chemical functionality, two other characteristics are essential for electrochemical activity. First, there must be some mechanism for electron transport within the polymer. This may involve simple, pairwise redox exchange reactions (hopping) between proximal functional groups, or else delocalization of charge over a system of extended pi-conjugation. Secondly, it must be possible for ions from solution to permeate the polymer in order to balance the charge associated with a change in electron population. These transport mechanisms are distinct. For example, only the latter involves mass transport of molecular entities.

There has been much interest in developing new kinds of polymer technology based on these properties. Recent work on polymer batteries (2) is a familiar example. The second part of this paper concerns an entirely different approach to exploiting electron and ion mobility in polymers. Namely, the reaction between electrons and metal ions permeating from opposite sides of a polyimide film can be controlled to grow electrically continuous metal interlayers embedded within the polymer. The dimensions and position of this interlayer depend upon the transport coefficients and the experimental conditions such that the structure can be controlled with great precision and in a predictable manner. Electrodeposited metal interlayers share many of the properties of metal vapor deposited films. There are features which may be uniquely well suited for certain kinds of thin film devices (4),(10).

2 ELECTROCHEMISTRY OF ODA/PMDA POLYIMIDE

The aromatic imide group is an electron acceptor comparable to a quinone. Haushalter and Krause (3) (hereafter referred to as HK) demonstrated that commercial polyimide film could be reacted with certain potent reducing agents (Zintl ions) from solution to produce either green or purple materials which they identified as the anion radical and dianion resulting from addition of one and two electrons respectively to a pyromellitic diimide repeat unit of the polymer. They also showed that these colored films could themselves be employed as reducing agents for electroless plating of various metals onto the surface or dispersed in the bulk of the polymer.

Attempts (5) to reduce commercial polyimide films electrochemically using a mercury cathode were totally unsucessful, regardless of the nature of the electrolyte. The cell resistance (more than a megaOhm for a 12 micron film) indicated that the ion permeability was far too low to support a significant faradaic current. Similar restrictions should also apply to the chemical reductions reproted by HK, it was therefore of interest to resolve the inconsistency. Experimental conditions described by HK include ethylenediamine as a constituent of the reaction solution. Ethylenediamine is known to react with polyimides resulting in partial degradation. Indeed, addition of small amounts of ethylenediamine to the electrolyte solutions was found to greatly enhance the salt permeability of these films to the extent that faradaic currents could be measured. This procedure is, however, unsatisfactory because the chemical state of the polymer is poorly defined and is quite nonuniform.

These problems were resolved by focusing on the conditions used for imidization of the polyamic acid precusor. ODA/PMDA polyimide is an infuseable, virtually insoluble material. Films are produced by casting a solution of the polyamic acid precursor followed by dehydration to the final imide structure, generally by heating to temperatures in excess of 200°C (6). Commercial films are characterized by a substantial degree of orientation and paracrystalline order (7). Either of these properties, particularly crystallinity, could severely limit the permeability. As an alternative to high temperatures, dehydration may be effected using a chemical dehydrating agent. Films of ODA/PMDA polyamic acid were prepared on a variety of electrodes (glassy Carbon, Pt, and tin oxide coated glass). These electrodes were then soaked in a solution of acetic anhydride in dry pyridine for several hours, rinsed with methanol and dried in a vacuum oven at 100°C. The resulting films remained clear and their optical and infra-red spectra were virtually identical to those of high-temperature processed polyimide. Fig. 1 shows the results of a cyclic voltammetry experiment performed with one of these electrodes.

The following details of the electrochemical response are noteworthy. The two independent pairs of cathodic and anodic current peaks represent discrete stages of reduction by one and two electrons per polymer repeat. Indeed the color changes (yellow to blue to rose) are just those observed for the chemical reductions, and this sequence is reversed as the film is reoxidized on the positive-going sweep of potential. The integrated current under each of these peaks (20 $mcoul/cm^2$) agrees within 10 percent with that required for complete conversion of the polymer film. The separation in voltage between the anodic and cathodic peaks for each of the two steps (200 mV) is believed to be caused by substantial ohmic resistance of the film. The apparent thermodynamic half-cell potential for each of the two processes may be estimated as the average between the two peak potentials , namely -0.73 and -1.13 V vs. SCE respectively. Similar responses were observed with other electrolyte cations (Na+, Cs+, and $(CH_4)_4N+$).

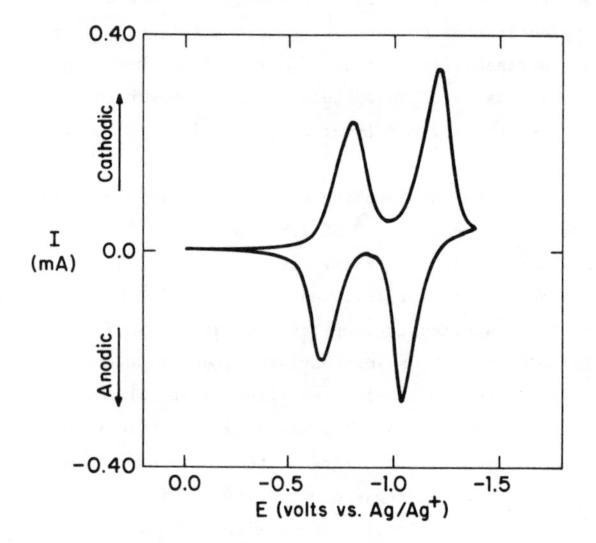

FIGURE 1: Cyclic voltammogram of polyimide film (0.24 micron) on glassy C electrode. Scan rate 2 mV/sec., 0.10 $\bar{M}$ KPF_6 in acetonitrile.

The kinetics of charging and discharging cylces, under conditions of fixed potential, are complex. The current transients deviate strongly from a simple Fick's law behavior (8). Undoubtedly electric field effects are important (9), particularly considering the high resistivity of the polymer in its neutral state.

3 ELECTRODEPOSITION OF METAL INTERLAYERS

The first reduction potential of ODA/PMDA polyimide is negative of the reduction potential of several monovalent metal ions, notably Ag+, Cu+, and Au+. It turns out that these ions can also permeate the polyimide film and may therefore be reduced either directly at the cathode/polymer interface (electroplating) or else indirectly via the polymer anion radicals or dianions. The latter process is especially interesting since it generates the reduced metal in an imaginary plane within the film whose location is governed directly by the experimental conditions.

In a typical experiment, an electrode coated with an ODA/PMDA film, 9 microns thick, was exposed to a solution of 0.10 M KPF_6 and 0.010 M $AgBF_4$ in N,N-dimethylformamide. The solution was stirred as the potential of the electrode was fixed at a value of -1.30 V vs. Ag/Ag+. The initial burst of cathodic current decayed within a few hundred seconds to a constant value (approximately 0.2 mA/cm^2). The process was allowed to continue for several minutes until 0.35 mg/cm^2 of Ag had been deposited. During this process the color of the film progressed from yellow to blue (the color of the radical anion) and gradually became very dark. Samples containing more than 0.10 mg Ag/cm^2 appeared highly reflective (gold in color) when viewed from the side facing the solution and dark black from the side facing the cathode. In transmitted light they where optically homogeneous.

FIGURE 2: TEM of polyimide film containing an electrodeposited silver interlayer. The upper surface of the film originally contacted solution. Calibration mark is 1.0 micron.

Fig. 2 is a transmission electron micrograph (TEM) of a microtomed cross-section of one such sample. Silver metal appears as dark microparticles (average diameter 150 Å) concentrated in a dense interlayer located roughly midway accross the film. Additional particles are also distributed more diffusely towards the side of the interlayer which originally faced the cathode. This morphology is consistent with the optical properties described above.

The development of the interlayer was followed in a series of experiments where the deposition process was interrupted at various stages. It was discovered that the diffuse particles are created early in the process, prior to establishment of a stable current (steady-state). The dense, optically reflective interlayer is established simultaneously with the onset of the steady-state and it grows in thickness linearly with the quantity of Silver incorported from that point on. The volume fraction of Silver in the interlayer was 0.71, independent of thickness. Electrical contact to the interlayer was made by etching holes through the film. Its resistivity was determined to be 58.8 microOhm cm (compared with 1.47 for bulk Silver).

A simple model was developed to describe the steady-state and, in particular, to predict dependence of the interlayer position upon experimental variables. It is assumed that electric field effects are negligible (by contrast with non-steady-state), that the particles form at a plane within the film where fluxes of metal ions and electrons are equal and opposite in sign, and finally it is assmumed that equilibrium prevails within the solution and at the two external surfaces of the polymer.. From Fick's first law and the Nernst equation the following relations are derived:

$$\frac{d}{t} = \left(1 + \frac{D_m K_s [M^+]_s}{D_e [e]_o}\right)^{-1} \qquad (1)$$

$$[e]_o = \frac{[P]\exp\left(\frac{nF}{RT}(E^\circ - E)\right)}{1 + \exp\left(\frac{nF}{RT}(E^\circ - E)\right)} \qquad (2)$$

where d is the distance from the cathode surface to the point of deposition and t is the total film thickness. De and Dm are the effective diffusion coefficients for electrons and metal ions respectively and Ks is an equilibrium constant governing the partitioning of metal ions between solution (at concentration $[M+]_s$) and polymer $[e]o$ is the concentration of electron carriers, ie. reduced polymer equivalents, at the cathode surface which is, in turn, determined by the applied potential according to equation 2. The dependence of d/t on applied potential and metal ion concentration is illustrated by the experimental points in Fig. 3 where the solid lines were calculated from the model employing parameters indicated in the figure caption. The rather good agreement shown here does not persist for higher metal ion concentrations. The deviation is likely due to the intrusion of migration effects as the transference numbers of these ions become comparable to those of the supporting electrolyte.

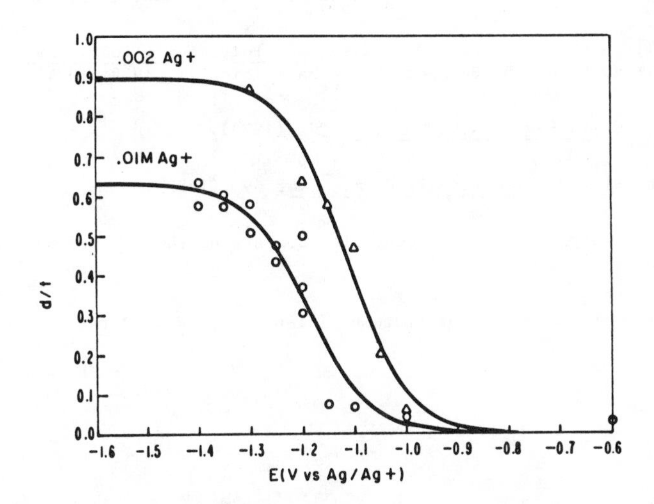

FIGURE 3: Effect of applied potential and metal ion concentration on interlayer position. Points are experimental. Lines were calculated from eqs. 1 & 2 where De = 8×10^{-9} cm^2/sec, DmKs = 1.7×10^{-7} cm^2/sec, $[P]_o$ = 3.67 $\bar{M}$ and E° = 1.25 V vs. Ag/Ag^+.

Our continuing efforts are directed at a more sophisticated understanding of the precipitation process, the resulting morphology, and at defining the generality of the phenomenon.

REFERENCES

1. H.G. Cassidy and K.A. Kun "Oxidation-Reduction Polymers", Interscience, 1965.

2. C.K. Chiang Polymer, 22, 1454 (1981).

3. R.C. Haushalter and L.J. Krause, Thin Solid Films, 102, 161 (1983).

4. K.L. Chopra and I. Kaur "Thin Film Device Applications", Plenum, N.Y., 1983.

5. C. Yarnitzky, unpublished results.

6. N.A. Adrova et.al., "Polyimides", Technomic Publ., Chap. I, 1970.

7. T.P. Russell, H. Gugger, and J.D. Swalen, J. Poly. Sci., Poly. Phys. 21, 1745 (1983); M. Kochi, H. Shimada, H. Kambe, Ibid, 22, 1979 (1984).

8. A. Hubbard, Crit. Rev. Anal. Chem., 3, 201 (1973).

9. W.T. Yap et. al. J. Electroanal. Chem., 144, 69, (1983).

10. A. U.S. Patent application has been filed concerning the subject matter of this paper.

11. On leave from the Weizmann Institute of Science, Rehovot, Israel.

PROCESSING OF BARRIER FILM BY COEXTRUSION

F. HENSEN

barmag Barmer Maschinenfabrik AG., P.O.Box 11 02 40
D-5630 Remscheid 11 (Western Germany)

IKV, RWTH Aachen

SYNOPSIS

This paper demonstrates that excellent barrier film constructions can be achieved by coextrusion and that economical advantages are possible with well selected polymers and film construction.

1. INTRODUCTION

Chemical structure and varying polymerization processes determine different specific properties of polymer film, such as gas permeability, water permeability, tenacity, elongation, adhesion, sealability, durability, appearance, and - finally - cost efficiency. There is no polymer resin which meets all of the requirements to produce and apply packaging film for goods that have to be protected against loss or absorption of moisture or gases, for instance food products. However, with coextrusion of different polymers the required properties can be achieved.

Today, the estimated annual growth rate of coextruded film is 20 - 30 per cent, which is considerably higher than the growth rate of single-layer film.

The main reasons for this fast growth rate are:

- Polymer producers have developed new barrier polymers with improved barrier properties, bonding polymers with improved adhesion, and supporting polymers with excellent mechanical properties and good processibility.
- Producers and converters of film are interested and prepared to manufacture barrier film tailored to specific end uses for packaging methods that ensure improved protection of the goods.
- Machine manufacturers offer advanced technology and production lines for coextruded multilayer film with up to 7 layers and 4 or 5 different polymer components. These lines include microprocessors for monitoring the film dimensions and the process itself.

2. POLYMER FOR COEXTRUDED BARRIER FILM

Figure 1 shows the differences in film polymers in regard to water and oxygen permeability, and it shows how the barrier properties can be improved by coextrusion. To meet all requirements of protection, processibility, and cost efficiency, supporting polymers, barrier polymers, and bonding polymers must be combined. There are as many combinations as there are applications.

Supporting polymers are used on the outside and/or the inside of a multilayer construction. Their properties should include weldability, transparency, and barrier against moisture, dyeability, printability, and others as it is shown in

Barrier polymers,Table-1,are special polymers with excellent gas impermeability. They are used in thin layers which are coextruded with supporting layers in order to improve the density of the film against gases as oxygen , aroma and flavour. As shown in Fig.1 and Table 1/2 barrier polymers are more or less hygroscopic, and special processing conditions must be considered. It is also important that the barrier layer is protected against crashes and other damages. In multilayer constructions the barrier layer is therefore extruded as inside layer between the supporting layers.

Supporting polymers and barrier polymers are so different in their

Table-1: Barrier Materials

Polymers	Properties	Particularities at coextrusion
barrier materials in general	polymers with high gas impermeability : for foods through aroma preservation criteria: gas impermeability of O_2 und CO_2	gas impermeability O_2 of approx. $1 \frac{cm^3}{m^2 \cdot 24h \cdot bar}$ is required for very delicate food products, such as wine, fruit juices for meat, sausages, cold cuts, etc. $20 - 80 \frac{cm^3}{m^2 \cdot 24h \cdot bar}$
copolyamides	: types developed specifically for coextrusion good transparency good tenacity trouble-free extrusion good gas impermeability good availability	processing at 220 - 250° C used preferably for 3-layer film; tendency of curling prevented by treatment in water bath film lay-flat with rolls required
polyamide 6	: better gas impermeability than copolyamides good transparency for special types less expensive than copolyamides good availability	because of strong curling tendency in 3-layer film, preferably used for 5-layer film. processing temperature 240 - 260° C limited blow-up ratio
polyester	: good transparency good gas impermeability as copolyester processible at lower temperatures	processing temperature of 260 - 280° C little use within blown film coextrusion
EVOH	: excellent gas impermeability (30 - 50 times better than PA) good transparency high rigidity increased brittleness hygroscopic	temperature - sensitive with thickness above 15 µm danger of breaks through bending preferred use for 5-layer film

Table-2 Supporting Materials

Polymers	Properties	Particularities at coextrusion
LDPE	good processibility good transparency good weldability easy availability good water barrier relatively low cost	MFI range 0.3 - 2 g/10 min. with increased density - increased rigidity used preferaby for layers which are to be welded
EVA	with VA portion in excess of 8 - 10 % excell.impact resistance good bonding to PP	temperature-sensitive; must be purged with LDPE befor line shut down. As outer layer to be laid flat wrinkle-free with rolls
MDPE	good rigidity high tenacity high melting point with narrower molecular weight distribution	preferable as outer layer to avoid sticking to the welding bar improved machinability
HDPE	considerable rigidity good tenacity low transparency	high tenacity - achieved by combination with LDPE in processing with long neck
LLDPE	good transparency good tenacity relatively low cost good weldability	no special tubular die head design needed can also be processed as a blend with LDPE
Ionomers	excellent weldability bonding to PA and LDPE good impact resistance special resins for increased shrinkage	because of aggressivity, special types of steel must be used for the melt channels preferable as inner layer; ionomers must be purged before line shutdown
PC	excellent transparency good temperature constancy high surface gloss good tenacity	high process temperature required limited to special applications

Table-3: Bonding agents

Polymers	Properties	Particularities at coextrusion
in general :	good availability for all common combinations	usual layer thickness around 3 - 5 µm
	large MFI range	must be purged before line shutdown
Ionomers :	good bonding to LDPE and PA, EVA, and LLDPE	with thicker bonding layers, also improved mechanical film properties
mod. EVA :	large variety of resin types	good processibility
	suitable for bonding with LDPE, LLDPE, EVA, PA, EVOH, PC, PETP	depending on combinations, bonding may still change after 2 weeks of extrusion

Table-4:Polymer bonding strength

Polymeres	LDPE	HDPE	EVA	Ionomers	LLDPE	PP	PACop	PA 6	PETP	EVOH
LDPE	+	+	+	0	+	-	-	-	-	-
HDPE	+	+	+	0	+	-	-	-	-	-
EVA	+	+	+	0	+	0	-	-	-	-
Ionomere	*	*	*	+	0	-	0	0	-	-
LLDPE	+	+	+	0	+	-	-	-	-	-
PP	*	*	*	*	*	+	-	-	-	-
PACop.	*	*	*	*	*	*	+	+	-	+
PA 6	*	*	*	*	*	*	+	+	-	+
PETP	*	*	*	*	*	*	*	*	+	-
EVOH	*	*	*	*	*	*	+	+	*	+

← with bonding agent → (lower left); ↑ without bonding agent ↓ (upper right)

+ good adhesion without bonding agent

- no adhesion

0 adhesion depending on the type of Polymers

* good adhesion with bonding agent, suitable bonding agents obtainable

chemical structure that they normally do not adhere during coextrusion. For the end use of multilayer films it is important that the individual layers are bonded so that they behave as if they were one film. This is the reason why a bonding agent must be extruded between the supporting and the barrier layer, as it is demonstrated in Table-3. Since this bonding agent has to fulfil bonding only, the thickness of its layer can be extremely thin. Good adhesion means that the force to separate these film layers is higher than 5 N/15 mm width of the film stripe.

Table-4 shows the bonding strength of different film polymers and its improvement with the aid of a bonding agent.

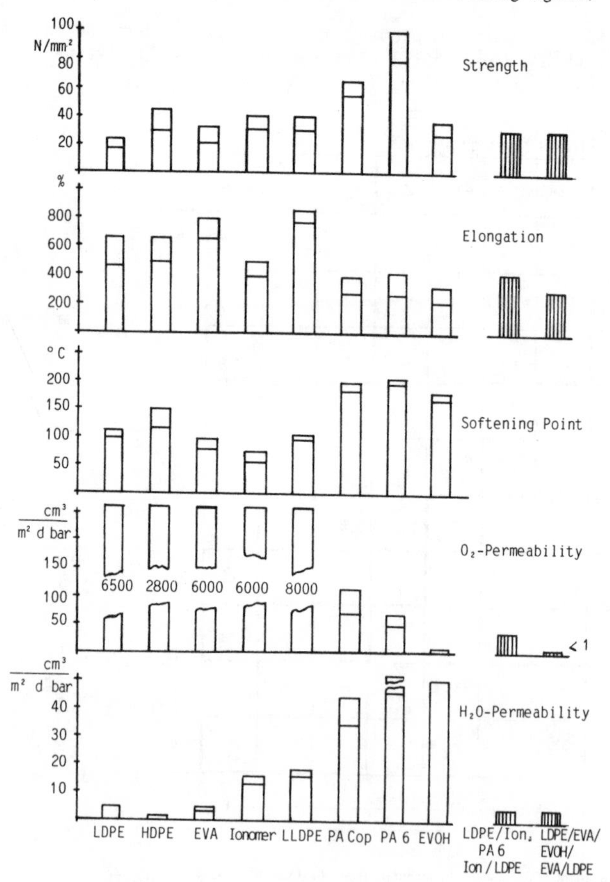

Fig. 1: Properties of Polymers in monolayer and multilayer films (Film thickness monolayer 25 µm; multilayer 70 µm)

3. ADVANTAGES OF COEXTRUSION

The main reason for the demand of coextrusion is its combination of physical film properties with the aim to produce high barrier film constructions, see Fig. 1. Apart from this big advantage concerning the film quality there are two other ones which can be reached by coextrusion:

- Improvement in converting the film
- Cost reduction

3.1. Improvement in converting the film means better weldability, better machinability and better handling of film packed goods.

Higher welding speeds and better welding quality can be obtained by the use of EVA or LLDPE at those layers that are to be welded.

Even in cases where contamination cannot be avoided on film layers which is originated by grease or oil , a good welding possibility is given if this layer is made out of Ionomer. In order to avoid bonding of the multilayer film onto the welding bar at the outside, polymers with higher density-as there are LDPE, MDPE, HDPE or PA-. should be used. In automatic welding and packaging machines a better machinability will be reached when polymers with higher stiffness are extruded into the middle- or outside layer, as par example MDPE or HDPE.

In addition of an exact width, forming, filling- and sealing machines need a web-sack-free film without any curling. Curling is caused by different physical properties of different polymer film layers. It can be avoided by symmetrical structures of multilayer films as shown in Fig.2. Such symmetrical structures are also possible at 5 or 7 layers.

Better handling of film packed goods in form of sacks or pouches can be achieved by using either slipfree, sticky film polymers as EVA at the outside or rough foamed LDPE film with open foam bubbles.

Smooth and good sliding HDPE layers at the outside are, sometimes, preferred for pouches.

3.2. Cost Reduction is obtained through coextrusion instead of making use of the lamination and through savings of raw material.

Coextrusion as a one step process is cheaper than laminating. Less storage place is needed for semi-products and less waste is produced by the edge trim.

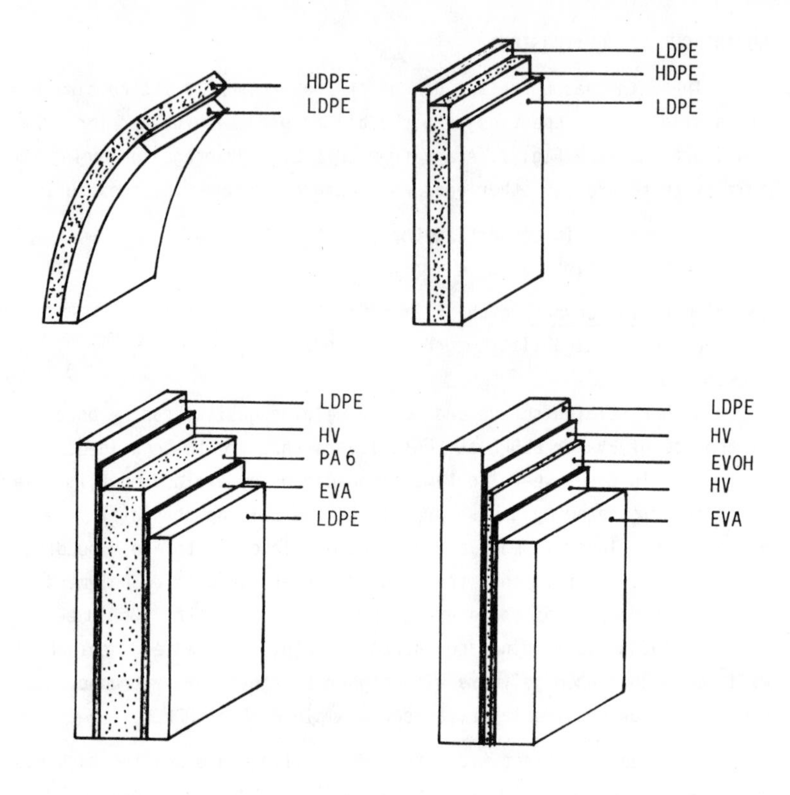

Fig.2: Structures of coextruded multilayer films

Even costs can be more reduced through the raw material savings. There are three possibilities to save costs when the same film quality is taken into consideration:

- Reduction of film thickness
- Use of low price polymers
- Use of better polymer qualities

Concerning the film thickness it should be mentioned that generally thinner thicknesses of film layers can be reached in coextrusion than they are attainable in monolayer-film extrusion.

The physical properties of the different polymers can be more different than their prices. If the tenacity of polymer B is higher than that of polymer A, the thickness of the film layer B can be thinner. For that reason the total thickness as well as the material costs for the film can be reduced.

$$P_M = \frac{d_A \varrho_A P_A + d_B \varrho_B P_B}{d_{ges} \varrho_m} \qquad \frac{DM}{m^2} \tag{1}$$

P_M = material costs [DM]
d = film layer thickness [μm]
ϱ = density of polymer [g/m^3]
P = price of Polymer DM/kg
Index A = polymer A (e.g. LDPE)
Index B = polymer B (e.g. HDPE)
Index ges = total
Index m = medium

This effect of cost reduction can be economically turned to advantage at the production of heavy duty bags, shopping bags, garbage bags and special packaging.

Since barrier polymers are normally much more expensive than supporting polymers, enormous cost reductions can be obtained when the permeability is taken into account. It is important to mention that the division of a given barrier layer into 2 separated layers of the same total thickness decreases drastically the permeability. This is another reason for choosing the five layer constructions which consist of more than one barrier layer.

Figure 3 shows how to improve the oxigen permeability through the protection of a barrier layer being placed in the centre of 5 layers instead of being arranged at the outside of 3 layers. Here PA 6 is used in place of PA Copolymer so that the barrier layer is divided up into two layers.

The O_2 barrier can be determined as below:

$$K = P \cdot d \qquad \frac{cm^3 \cdot \mu m}{24\,h \cdot m^2 \cdot bar} \tag{2}$$

$$d = \frac{K}{P} \qquad \mu m \tag{2a}$$

with P = Permeability $\frac{cm^3}{24\,h \cdot m^2 \cdot bar}$

d = layer thickness μm

K = specific measured value $\frac{cm^3 \cdot \mu m}{24\,h \cdot m^2 \cdot bar}$

$K_{3\text{-layer with PACop.}}$ = 1 500 $\frac{cm^3 \cdot \mu m}{m^2 \cdot 24\,h \cdot bar}$

$K_{5\text{-layer with PACop.}}$ = 1 000 - " -

$K_{5\text{-layer with PA 6}}$ = 700 - " -

$K_{5\text{-layer with EVOH}}$ = 15 - " -

If, for instance, a barrier of 10 $\frac{cm^3}{m^2 \cdot 24h \cdot bar}$ is wanted, the necessary thickness of the barrier layer is calculated as follows:

PA 6: $d = \frac{K}{P} = \frac{700}{10} = 70\ \mu m$

EVOH: $d = \frac{K}{P} = \frac{15}{10} = 1{,}5\ \mu m$ (less than 5 μm)

This leads to the following film construction:

	EVA	/	HV	/	PA 6	/	HV	/	LDPE	
	35		5		<u>70</u>		5		35	µm
or	EVA	/	HV	/	EVOH	/	HV	/	LDPE	
	65		5		<u>5</u>		5		65	

Comparing the film area/price potential of these film combinations

$$A_{PA6} = \frac{1000}{d_{EVA} \cdot \rho_{EVA} \cdot P_{EVA} + d_{HV} \cdot \rho_{HV} \cdot P_{HV} + d_{PA6} \cdot \rho_{PA6} \cdot P_{PA6} + d_{LDPE} \cdot \rho_{LDPE} \cdot P_{LDPE}} \frac{m^2}{DM} \quad (3)$$

with

$$A_{EVOH} = \frac{1000}{d_{EVA} \cdot \rho_{EVA} \cdot P_{EVA} + d_{HV} \cdot \rho_{HV} \cdot P_{HV} + d_{EVOH} \cdot \rho_{EVOH} \cdot P_{EVOH} + d_{LDPE} \cdot \rho_{LDPE} \cdot P_{LDPE}} \frac{m^2}{DM} \quad (3a)$$

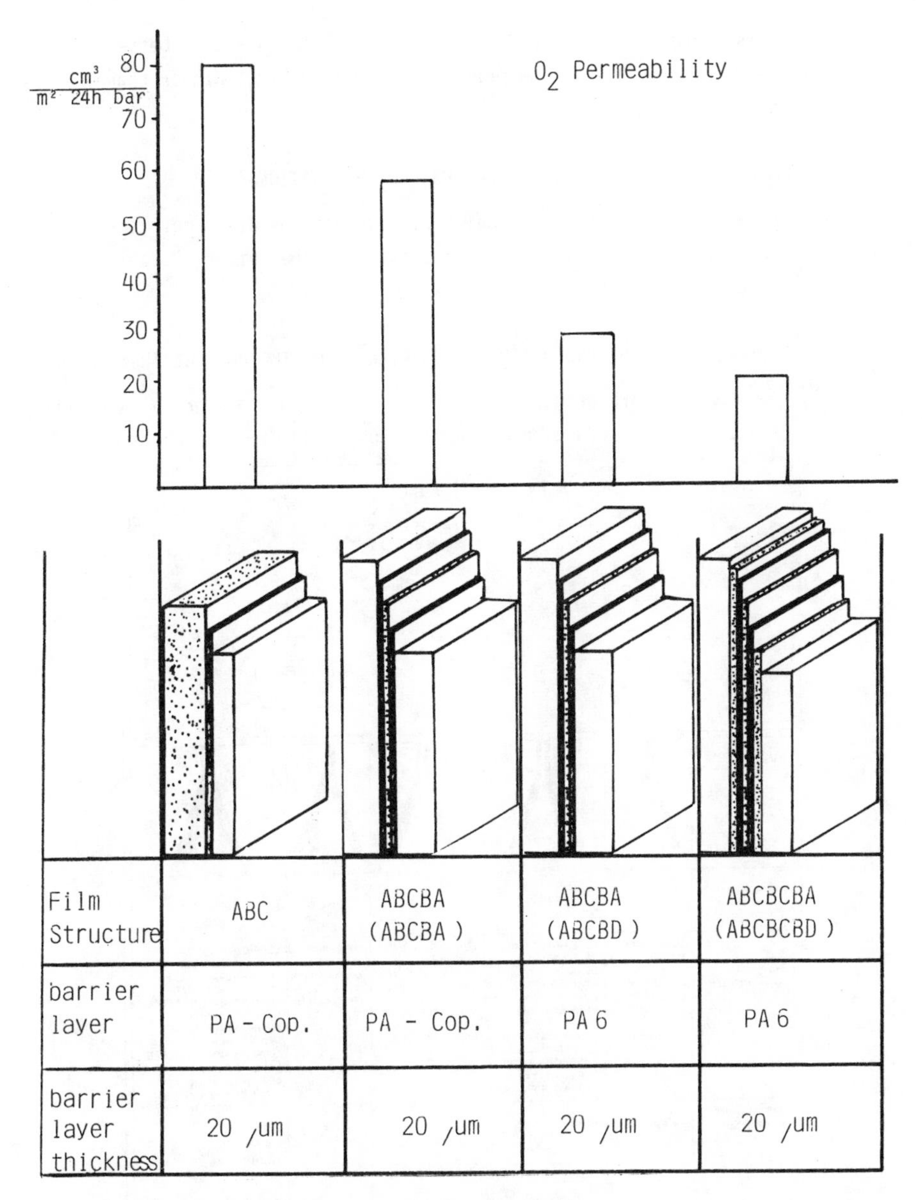

Fig.3: COMPARISON OF OXYGEN PERMEABILITY

it is found out that the EVOH multilayer film has a 20 % better O_2 - barrier, even when the price for EVOH is 3 times higher than that for PA 6.

4. MATERIAL COMBINATIONS, PROPERTIES AND APPLICATIONS

Table-5 shows 20 possible material combinations with their special properties and most important applications. The number of applications is steadily growing.

5. EQUIPMENT FOR THE PRODUCTION OF BARRIER FILMS THROUGH COEXTRUSION

Barrierfilms are produced either on blown film lines or on cast film lines. For the production of barrier tubes and pipes there is also a

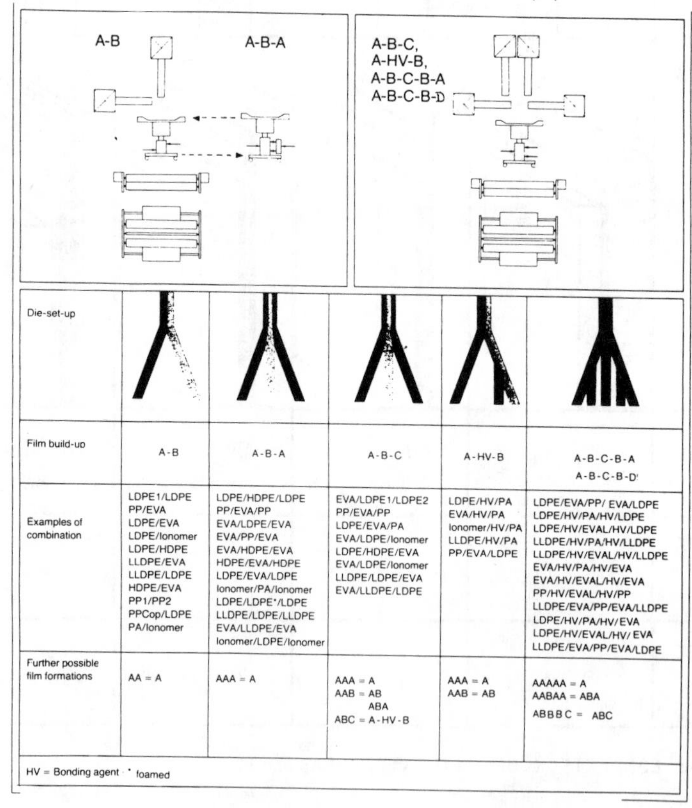

Die-set-up					
Film build-up	A-B	A-B-A	A-B-C	A-HV-B	A-B-C-B-A A-B-C-B-D'
Examples of combination	LDPE1/LDPE PP/EVA LDPE/EVA LDPE/Ionomer LDPE/HDPE LLDPE/EVA LLDPE/LDPE HDPE/EVA PP1/PP2 PPCop/LDPE PA/Ionomer	LDPE/HDPE/LDPE PP/EVA/PP EVA/LDPE/EVA EVA/PP/EVA EVA/HDPE/EVA HDPE/EVA/HDPE LDPE/EVA/LDPE Ionomer/PA/Ionomer LDPE/LDPE'/LDPE LLDPE/LDPE/LLDPE EVA/LLDPE/EVA Ionomer/LDPE/Ionomer	EVA/LDPE1/LDPE2 PP/EVA/PP LDPE/EVA/PA EVA/LDPE/Ionomer LDPE/HDPE/EVA EVA/LDPE/Ionomer LLDPE/LDPE/EVA EVA/LLDPE/LDPE	LDPE/HV/PA EVA/HV/PA Ionomer/HV/PA LLDPE/HV/PA PP/EVA/LDPE	LDPE/EVA/PP/ EVA/LDPE LDPE/HV/PA/HV/LDPE LDPE/HV/EVAL/HV/LDPE LLDPE/HV/PA/HV/LLDPE LLDPE/HV/EVAL/HV/LLDPE EVA/HV/PA/HV/EVA EVA/HV/EVAL/HV/EVA PP/HV/EVAL/HV/PP LLDPE/EVA/PP/EVA/LLDPE LDPE/HV/PA/HV/EVA LDPE/HV/EVAL/HV/ EVA LLDPE/EVA/PP/EVA/LDPE
Further possible film formations	AA = A	AAA = A	AAA = A AAB = AB ABA ABC = A-HV-B	AAA = A AAB = AB	AAAAA = A AABAA = ABA ABBBC = ABC

HV = Bonding agent · ' foamed

Fig.4: Coextrusion blown film

Material combination	Special properties	Most important application areas
Two layer sheet		
1. LDPE/LDPE	micropore sealed (multi-coloured)	milk sheet carrier bags general packaging
2. LDPE/EVA	easily welded can be sterilized	heavy duty bags stretch wrap medicinal articles
3. HDPE/EVA	can be sterilized	blood plasma baked goods food
4. HDPE/LDPE	high strength	baked goods foodstuffs tomato concentrate
5. LDPE/Ionomer	easily welded penetration resistant	milk products food medicinal instruments general packaging
6. LLDPE/LDPE LLDPE/EVA	high elasticity good surface adhesion	stretch film
7. Ionomer/EVA	greaseproof	coconuts cocktail snacks
8. Ionomer/PA	gas and aroma tight	meat, sausage, ham, fish, food, cheese
Three layer sheet, symmetrical		
9. LDPE/HDPE/LDPE	weldable on both sides reduced tendency to roll	as 4 pet food, cornflakes
10. EVA/PP/EVA	as 9	as 9
11. EVA/HDPE/EVA	as 3	as 9 cornflakes
Two layer + bonding agent		
12. LDPE/Bonding/PA	gas, water and aroma tight	foamed PS granulate meat, sausage, cheese, ham, fish, precooked foods, hops
13. EVA/Bonding/PA	as 12, in hot air channel, area welding	as 12
14. Mod. EVA/Bonding/PA		as 12 as vacuum packaging for ham (can be shrink packed)
Three layer sheet		
15. LDPE/HDPE/EVA	easily welded, good stiffness	backed goods, foodstuffs
16. LDPE/EVA/PP	as 15	as 15
Five layer sheet		
17. LDPE/Bonding/PA Bonding/LDPE rsp. LLDPE/Bonding/PA Bonding/LLDPE	no tendency to curl, improved sealing effect, since PA is protected against moisture absorption, improved layer bonding, weldable on all sides	as 12
18. EVA/Bonding/PA/Bonding/EVA	as 17	as 17
19. LDPE/Bonding/EVAL/Bonding/LDPE	as 17	as 17, fishmeal, wine packaging, milk powder (casein)
20. EVA/Bonding/EVAL/Bonding/EVA	as 19	as 19

Note: LDPE can also be replaced by LLDPE (linear low density PE)

Table-5: Polymer combinations, special properties and most applications for coextruded blown film recently developed coextrusion tube / pipe line available. All these lines are in operation with the same polymers as already mentioned before.

5.1. Multilayer Blown film lines

For the production of coextruded blown film a special line in modular design has been developed (Figure 4)

The concept allows the combination of any extruder size, adapted to process and fulfil the customers' requirements.

- Easily exchangeable breaker plate type filters with optimized melt flow design allow the processing even of sensitive materials.
- The rotating device allows the production of film rolls without piston rings by the continuous rotation of film die and cooling ring
- Blown film dies for AB, ABA, ABC, ABCD, ABCBA, ABCBD and even ABCBCBD films are available.

All types of dies can be mounted on the same rotating unit. The allocation of each extruder to a certain layer is feasible (without additional aggregates).

Aggregates for the automation of the equipment, as there are granule feeders, dosing stations, throughput control systems for each extruder incl. film thickness control, film width control, and automatic winders, can be incorporated in the line.

The equipment can be delivered in widths of 1 150 mm up to 2100 mm.

The output depends on the film combinations and widths, and is in a range of 100 up to 260 kg/h.

5.2. Multilayer cast film lines

As shown in Fig.5 coextrusion cast film lines are similar to monolayer cast film lines. They only differ in the number of extruders and the feed block system. In the feed block system the different polymers are formed into a multilayer strand with rectangular cross section which enters the monolayer die. Due to the rheology of thermoplastic polymers the individual layers keep their thickness relation when they are distributed inside the die over the total film width.

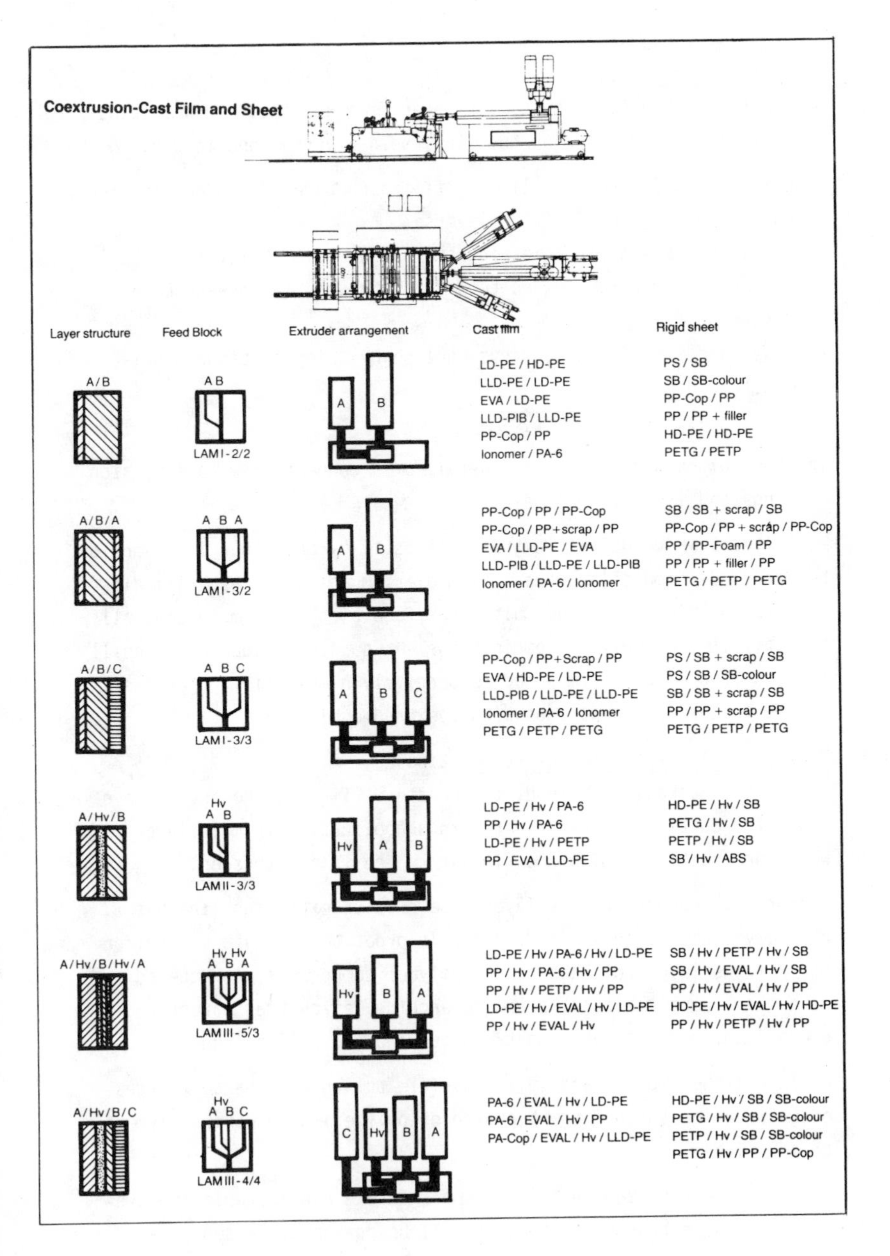

Fig.5: Coextrusion cast film and sheet

The Feed Block System

makes it possible to produce multilayer film and sheet in a thickness range from 20 - 2000 µm up to 7 layers in a very economic way.

- Quick variation and handling of film structure, resin combinations, number and thickness of the layers,
- good flow design for combining the different melt layers,
- minimum layer thickness up to 3 % for the top layers and core layers,
- easy handling of the coextrusion line by using a standard uni-manifold die

is guaranteed.

The feed block system is also installed in connection with extrusion coating- and laminating lines.

Polymers which differ in their viscosity by more than 1 : 3 should be extruded in multilayer cast film dies. Multilayer cast film dies can also be adapted to the multilayer feed block system so that all possibilities of polymer combinations are possible. Due to the chill roll cooling higher film speed respectively higher film thickness can be reached in comparison with coextruded blown film.

5.3. Multilayer Tube- or Pipe Extrusion Line

For the packaging of food, toothpaste or liquids the extrusion of tubes for bottles and containers finds application in multilayer wall constructions which are similar to the film extrusion.

For that purpose a multilayer tube or pipe extrusion line has been developed as shown in Fig.6 . In order to increase the uniformity of the layer thicknesses the polymer components are metered with gear pumps. With a special quenching device transparent tubes can be achieved even at high troughput and big wall thickness.

In all these processes it is possible to augment the advantages of coextrusion through the blending of polymers with other polymers or additives.

It can be foreseen that in the near future considerable improvements in the extrusion technology will be presented caused by a further optimation of polymers, their combinations and their processing in extrusion procedures.

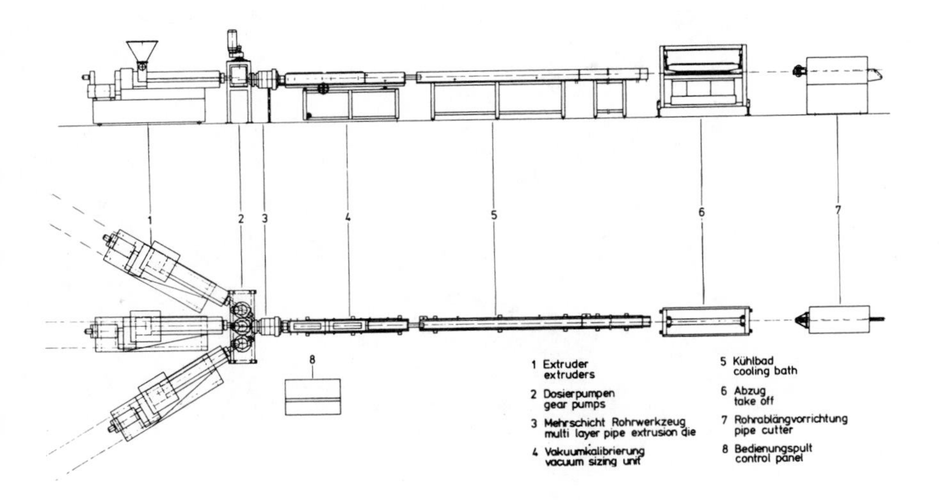

Fig.6: Coextrusion Pipes

References:

F. Hensen: Kunststoffe 59, 1969, S. 3 - 8

R. Hessenbruch: Plastverarbeiter 29, 1978, S. 68 - 70

F. Hensen, H. Bongaerts: Plastverarbeiter 30, 1979, S. 441 - 449

F. Hensen, R. Hessenbruch, H. Bongaerts: Kunststoffe 71, 1981, S. 530 - 538

H. Koslowski: Plastverarbeiter 7, 1984, S. 108 - 110

Part VI

CHAIN DYNAMICS

SINGLE-CHAIN DYNAMICS IN POLYMER CHARACTERIZATION

WALTER H. STOCKMAYER

Department of Chemistry, Dartmouth College, Hanover, New Hampshire 03755, U.S.A.

For several decades the researches of Koningsveld and his colleagues have largely centered on the effects of molecular weight and polydispersity on fluid-fluid phase quilibria in polymer/solvent and polymer/polymer systems. Effects of compositional heterogeneity and of varying topology (e.g., branching and ring formation) have also been considered. In the pursuit of these objectives, availability of molecularly well-characterized polymer samples as model systems has been and will continue to be essential.

The primary absolute methods of molecular weight determination are based on equilibrium behavior of dilute solutions. The thermodynamic properties of more concentrated solutions are essentially independent of molecular weight. In contrast, the transport properties of polymer systems (e.g., diffusion, viscosity) are highly chain-length dependent even up to the bulk state, indeed to a higher degree than in dilute solution; compare, for example the dependence of melt viscosity on a power of M exceeding 3 with the intrinsic viscosity (Mark-Houwink) exponent of less than 0.8. The theories of these bulk properties have recently undergone rapid development, thanks to the reptation or tube-flow concept, but they are not yet developed to a stage allowing an accurate description of polydispersity effects. When, through a combination of theoretical and experimental efforts, such a stage will have been reached, use of bulk transport measurements for molecular characterization may become desirable. At present, however, it remains safer to rely on the less sensitive but more easily measurable and theoretically better understood non-equilibrium behavior of dilute solutions--the "single chain" regime.

The transport coefficients of dilute polymer solutions--intrinsic viscosity, sedimentation coefficient, diffusion coefficient-have served as bulwarks of polymer characterization throughout the history of the discipline.

However, except in a minor way these coefficients do not really reveal anything about chain _dynamics_ as such. Because of the dominance of the hydrodynamic interactions among segments, the macromolecule behaves largely as an obstacle around which the solvent must go; the solvent molecules within the doman of the chain are almost completely immobilized. As a result, the aforementioned transport coefficients are essentially geometrical parameters, determined by the conformational _equilibrium_ behavior of the polymer molecule. This statement is illustrated by the well-known Kirkwood approximation (in the "non-draining" limit) to the translational diffusion coefficient,

$$D/k_B T = (3\pi\eta_o N^2)^{-1} \sum_{i<j}^{N} \sum^{N} \langle R_{ij}^{-1} \rangle \tag{1}$$

and the equally familiar "pre-averaged" (Kirkwood-Riseman) approximation to the intrinsic viscosity,

$$[\eta] = (N_A/6M\eta_o) \sum_{i<j}^{N} \sum^{N} \langle S_i (T^{-1})_{ij} S_j \rangle \tag{2}$$

Both of these expressions are known to give upper limits to the true values, with only about 10 to 15% deviations which may depend on more detailed aspects of dynamical behavior. In the above expressions, η_o is the viscosity coefficient of the solvent, $\underline{R}_{ij}$ is the distance between chain elements $\underline{i}$ and $\underline{j}$, and the total number of chain elements is $\underline{N}$. Further, $\underline{S}_i$ is the distance of element $\underline{i}$ from the center of resistance of the macromolecule, $(T^{-1})_{ij}$ is an element of the inverse of the (pre-averaged) hydrodynamic interaction tensor, which depends on the equilibrium averages $\langle R_{ij}^{-1} \rangle$, and N_A is Avogadro's number. The above expressions are convenient if not exact, and are still being exploited in calculations for systems with excluded volume effects, compositional heterogeneity or non-linearity (1,2,3).

The method of quasielastic light scattering yields a time-correlation function of the scattered intensity fluctuations, and for low scattering angles contains information about polydispersity of the translation diffusion coefficient. Meaningful inversion of the correlation function to extract this information is a desirable but formidable operation that often may not lead to unique results (4,5,6).

Quasielastic light scattering at higher angles does depend on internal dynamics, and can yield information about polydispersity, non-linearity and compositional heterogeneity; but usually the individual effects are hard to separate;

in particular, increased branching usually is accompanied by increased polydispersity, with largely cancelling contributions, so that great care is necessary in interpretation (7). At present, the theory of these effects is easily applicable only to the initial slope (first cumulant) of the time correlation function (8,9). The data for longer times contain information that might be useful in characterization but is not now readily accessible.

The real internal dynamics of the chain come into play in a number of other dilute solution properties. From the viewpoint of polymer characterization, attention should be directed to the larger-scale dynamical modes, with relaxation times that do depend on molecular size, being proportional to $M[\eta]$. Among suitable properties are dynamic viscoelasticity (e.g., in oscillating shear, as in the work of Ferry and his school) (10), dynamic flow birefringence (11), dynamic electrical birefringence (Kerr effect) and dielectric dispersion. The latter two phenomena show dependence on the slow modes only for chains with longitudinal components of electric dipole moment within repeat units (12). The viscoelastic and birefringence experiments are more universal, and with recent improvements of apparatus may become formidable candidates for use in characterization.

A new technique of great promise has been brought into view by the realization, at the hands of Keller and coworkers (13), of the observation of the "coil-stretch" transition in elongational flow predicted some years previously by Peterlin (14) and by DeGennes (15). The molecular weight dependence is again that of the terminal relaxation time, proportional to $M[\eta]$.

References

1. Barrett, A.J. Macromolecules, 1984, 17, 1566.
2. Garcia de la Torre, J.; López Martinez, M.C.; Tirado, M.M.; Freire, J.J. Macromolecules, 1984, 17, 2715.
3. Freire, J.J.; Prats, R.; Pla, J.; Garcia de la Torre, J. Macromolecules, 1984, 17, 1815.
4. Provencher, S.W. Makromol. Chem., 1979, 180, 201.
5. Gulari, Es.; Gulari, Er.; Tsunashima, Y.; Chu, B. J. Chem. Phys., 1979, 70, 3965.
6. Bertero, M.; Brianzi, P.; Pike, E.R.; deVilliers, G.; Lau, K.H.; Ostrowsky, N. J. Chem. Phys., 1985, 82, 1551.
7. Burchard, W.; Schmidt, M.; Stockmayer, W.H. Macromolecules, 1980, 13, 1265.
8. Akcasu, A.Z.; Gürol, H. J. Polym. Sci., Polym. Phys. Ed., 1976, 14, 1.
9. Burchard, W.; Schmidt, M.; Stockmayer, W.H. Macromolecules, 1980, 13, 580.
10. Ferry, J.D. Accts. Chem. Res., 1973, 6, 60.
11. Lodge, T.P.; Schrag, J.L. Macromolecules, 1982, 15, 1376.
12. Stockmayer, W.H.; Baur, M.E. J. Am. Chem. Soc., 1964, 86, 3485.
13. Farrell, C.J.; Keller, A.; Miles, M.J.; Pope, D.P. Polymer, 1980, 21, 1292.
14. Peterlin, A. Pure Appl. Chem., 1966, 12, 563.
15. DeGennes, P.G. J. Chem. Phys., 1974, 60, 5030.

NON-IDEAL STATISTICS AND POLYMER DYNAMICS

P.F. MIJNLIEFF and R.J.J. JONGSCHAAP
Rheology Group, Dept. Applied Physics, Twente University of Technology
Enschede, The Netherlands.

SYNOPSIS

The bead-and-spring representation of polymer molecules not displaying random-flight statistics is investigated on its merits to predict a relaxation time spectrum.

1 INTRODUCTION

The relaxation time spectrum of a solution of all-equal non-branched polymer molecules is a classical subject in polymer physics.

In most of the existing molecular theories the polymer molecules are (Zimm representation) modeled as a linear chain of N+1 "beads" connected by N equal submolecules ("springs"). A submolecule itself is a linear chain of many, say s, segments already.

For sufficiently long polymer molecules the semi-empirical relation between L^2, the mean-square polymer end-to-end distance, and M, the polymer molecular mass, reads

$$\overline{L^2} \sim M^{1+\varepsilon} \quad ,$$

where ε is the Peterlin parameter. As M is proportional to Ns and s is a constant, this may be rewritten as

$$\overline{L^2} \sim N^{1+\varepsilon} \quad . \qquad (1)$$

For sufficiently large s-values, this relation applies for N-values up from unity.

The dynamics of a solution of thus-modeled polymer molecules have been studied by Tschoegl[1] for the realistic case of "dominant" as well as for the academic case of "vanishing" hydrodynamic interaction. The relaxation spectrum was found to be that of N mechanisms, all with a strength nkT, where n is the number of polymer molecules per unit volume, and with relaxation times τ_p (p = 1, ..., N) which, for N>>1 and p < (about)N/5, are given by

$$\text{(dom.hydrod.interact., } \varepsilon \text{ arbitrary)} \quad \tau_p/\tau_1 = p^{-(3/2)(1+(1/3)\varepsilon)} \qquad (2)$$

$$\text{(vanish. hydrod. interact., } \varepsilon \text{ arbitrary)} \quad \tau_p/\tau_1 = p^{-2} \qquad (3)$$

The result (3) coincides with that of Rouse[2]:

$$\text{(vanish. hydrod. interact., } \varepsilon=0) \quad \tau_p/\tau_1 = \frac{\sin^2(\frac{\pi}{2(N+1)})}{\sin^2(\frac{p\pi}{2(N+1)})}, \tag{4}$$

and, thus, predicts that relaxation time spacing is affected by a non-zero ε only when hydrodynamic interaction is non-vanishing.

Recently, the τ_p-spacing was reconsidered[3] and, again for $N>>1$ and $p<$(about)$N/5$, predicted to be given, instead of by eq. 2, by

$$\tau_p/\tau_1 = p^{-(3/2)(1+\varepsilon)} \tag{5}$$

when hydrodynamic interaction is dominant and, instead of by eq. 3, by

$$\tau_p/\tau_1 = p^{-2(1+(1/2)\varepsilon)} \tag{6}$$

when hydrodynamic interaction vanishes.

The picture used for deriving eqs. 5 and 6 was based on the resemblance of the motions of a long ($N>>1$) polymer molecule to those of a continuous elastic bar immersed in a viscous liquid.

For $\varepsilon = 0.2$, the results 5 and 6 were, independently, derived by Hess et al.[4] along a quite different way.

Reported [5] and recent [6] experimental results are better described by eq. 5 than by eq. 2.

It should be noted that, in the derivation[2] of eqs 2 and 3, Zimm's represention was, as far as spring constants are concerned, maintained with the modification that all spring constants were multiplied by one and the same correction factor. This amounts to still representing a polymer molecule as a non-branched chain of N equal Gaussian submolecules and, at vanishing hydrodynamic interaction, naturally leads to the Rouse spacing 4 and,thus, 3.

In Section 2 we will see that, when ε differs from zero, non-branched polymer molecules can, at best, be respresented as a network of Gaussian chains.

2 NETWORK REPRESENTATION OF POLYMER MOLECULES

In a dilute solution of polymer molecules as modeled in Section 1, the probability, when bead 0 of a molecule is at $\underline{R}_0$, that the other beads are in $d^3\underline{R}_i$ about $\underline{R}_i$, is denoted by

$$W_N(\Delta\underline{R}_1, \ldots, \Delta\underline{R}_N)d^3\underline{R}_1 \ldots d^3\underline{R}_N \equiv W_N\{\Delta\underline{R}_i\}d^3\underline{R}_1 \ldots d^3\underline{R}_N \tag{7}$$

where:

$$\Delta\underline{R}_i \equiv \underline{R}_i - \underline{R}_{i-1}, \tag{8}$$

and where $\underline{R}_i$ are bead positions.

Consider $W_N\{\Delta\underline{R}_i\}$ to be continuous in $\Delta\underline{R}_i$, and to have a maximum, $W_{N,max}$, at $\Delta\underline{R}_i = \underline{0}$ (i = 1, ..., N). Expanding it about its maximum and applying its isotropy we then find

$$W_N\{\Delta\underline{R}_i\} = W_{N,max}\{1 - \sum_{l=1}^{N}\sum_{k=1}^{N} W_{kl}\Delta\underline{R}_k\cdot\Delta\underline{R}_l + ...\} \tag{9}$$

in which:

$$W_{lk} = W_{kl} \quad . \tag{10}$$

The double sum in the expression (9) can be rewritten as

$$\sum_{q=1}^{N}\sum_{r=1}^{N} \gamma_{q-1,r}\,(\underline{R}_r - \underline{R}_{q-1})^2 \tag{11}$$

The relation between the coefficients $\gamma_{q-1,r}$, of which there are (1/2)N(N+1) in total, and the coefficients W_{kl} follows from the equality

$$\underline{R}_r - \underline{R}_{q-1} = \sum_{t=q}^{r} (\underline{R}_t - \underline{R}_{t-1})$$

The expression 9, with the double sum replaced by (11), equals the expression

$$W_N\{\underline{R}_i\} = W_{N,max}\exp - \left[\sum_{q=1}^{N}\sum_{r=1}^{N} \gamma_{q-1,r}(\underline{R}_r - \underline{R}_{q-1})^2\right] \tag{12}$$

in the approximation: $\exp(x) = 1 + x$.

The distribution function (12) is that for N+1 beads, connected with each other by Gaussian springs with spring constants $2\,kT\gamma_{q-1,r}$. Only when the unbranched chain considered obeys Gaussian statistics all coefficients $\gamma_{q-1,r}$ for which $\{r-(q-1)\}>1$ are zero.

The relaxation behaviour of the network for which Eq(12) is the bead-position distribution function is that according to N mechanisms, all with a strength nkT. The relaxation times τ_p are inversionally proportional to the non-zero eigenvalues λ_p of the Zimm-matrix that can be constructed for this network.

3. SPECIAL CASES

According to Section 2, the relaxation time spectrum can be derived from the distribution function $W_N\{\Delta\underline{R}_i\}$. The distribution function leading to the result (1), however, is not known to us. We just know that, on integration over all $\underline{R}_i$ except for $\underline{R}_0$ and $\underline{R}_N$, it should result in a distribution function $W_N(\underline{R}_N - \underline{R}_0)$ with the general form

$$W_N(\underline{R}_N - \underline{R}_0) \backsim \exp - C(\varepsilon)N^{-(1+\varepsilon)}(\underline{R}_N - \underline{R}_0)^2 \ ,$$

leading to the result 1; in this expression $C(\varepsilon)$ is an ε-dependent constant.

From the various possibilities to account for a non-zero ε, we now choose one in which the ε-effect is brought in by factors added to the expression valid for $W\{\Delta\underline{R}_i\}$ at zero ε.

When N = 2 this leads to:

$$W_2\{\Delta\underline{R}_1,\Delta\underline{R}_2\} \sim \exp - \left[C(\varepsilon)\{\Delta\underline{R}_1{}^2 + \Delta\underline{R}_2{}^2 + (\frac{1}{2^{1+\varepsilon}} - \frac{1}{2})(\Delta\underline{R}_1 + \Delta\underline{R}_2)^2\}\right] ,$$

and, when N=3, to: $W_3\{\Delta\underline{R}_1, \Delta\underline{R}_2, \Delta\underline{R}_3\} \sim \exp - \Big[C(\varepsilon)\{\Delta\underline{R}_1{}^2 + \Delta\underline{R}_2{}^2 + \Delta\underline{R}_3{}^2\}$

$$+ C(\varepsilon)\{(\frac{1}{2^{1+\varepsilon}} - \frac{1}{2})((\Delta\underline{R}_1+\Delta\underline{R}_2)^2+(\Delta\underline{R}_2+\Delta\underline{R}_3)^2)+(\frac{1}{3^{1+\varepsilon}} - \frac{3-2^{\varepsilon}}{1+5\ 2^{\varepsilon}})(\Delta\underline{R}_1+\Delta\underline{R}_2+\Delta\underline{R}_3)^2\}\Big]$$

Results for relaxation time spacing, when hydrodynamic interaction is absent, are collected in the Table.

Table

Relaxation time spacings at vanishing hydrodynamic interaction

N = 2			N = 3		
	$\varepsilon = 0$	$\varepsilon = 0.2$		$\varepsilon = 0$	$\varepsilon = 0.2$
τ_2/τ_1	$\frac{1}{3} = 0.333$	$\frac{1}{3}\ 2^{-\varepsilon} = 0.290$	τ_2/τ_1	$\frac{2-\sqrt{2}}{2} = 0.293$	0.247
			τ_3/τ_1	$\frac{2-\sqrt{2}}{2+\sqrt{2}} = 0.172$	0.136

It is seen that a positive ε widens the τ-spacing.

ACKNOWLEDGEMENT

Thanks are due to Olaf Bouschè for the eigenvalue calculations.

REFERENCES

1. Tschoegl, N.W., J. Chem. Phys. 40, 473-479, 1964.

2. Rouse, P.E., J. Chem. Phys. 21, 1272-1280, 1953.

3. Mijnlieff, P.F., Advances in Rheology (IX-th International Congress on Rheology, Acapulco, Mexico) Vol. I, 459-465, 1984.

4. Hess, W., Jilge, W. and Klein, L., J. Pol. Sci: Pol. Phys. Edtn, 19, 849-862, 1981.

5. Ferry, J.D., "Viscoelastic properties of polymers", Third Edtn (John Wiley, 1980) Chapter 9

6. Mijnlieff, P.F. and Bouschè, O. unpublished results.

COMPUTATION AND DISPLAY OF POLYMER CHAIN BEHAVIOUR

R.F.T. Stepto.

Department of Polymer Science and Technology, The University of Manchester Institute of Science and Technology, Sackville Street, Manchester, M60 1QD. England.

INTRODUCTION

Traditionally, the only type of molecular model available to the polymer scientist has been the conventional, static type which is constructed by hand or nowadays by computer. Such models are satisfactory for small molecules in which only small changes in size and shape can occur. However, one of the important characteristics of polymer molecules is their ability to change size and shape over wide extremes. It is impossible using static models to give a correct description of the configurational behaviour of polymer chains. Only in cases where configurations remain fixed, as in the crystalline state, are static models really adequate.

Using fast, large computers it is possible to generate sequences of polymer chain configurations calculated according to given theoretical models. By means of Monte Carlo calculations, such sequences can be used to describe the equilibrium configurational behaviour of polymer chains in given situations. The configurations can then be displayed directly using a subsidiary display computer and recorded directly on video-tape, or graph-plots of configurations can be obtained and made into a movie film.

The present contribution surveys the use of Monte Carlo calculations and both types of computer display to depict macromolecular adsorption, molecular size and shape in solution, differences between cyclic and linear macromolecules, and centre-of-mass diffusion. The Monte Carlo calculations use Metropolis sampling[1], according to which sequences of configurations are generated by moving parts of a polymer chain. The method is more efficient than random sampling. For non-lattice calculations, algorithms have been devised[2] to reduce the rounding errors resulting from the numerous matrix multiplications involved in moving sections of chains. For the numerical work, samples of the order of 10^6 configurations were used to evaluate configurational averages of the particular properties being investigated. The displays of configurations were constructed from screen- or graph-plots of about 50 configurations, each separated by 10^4 elementary chain movements, so that the samples of configurations spanned (5×10^5) were sufficient to represent equilibrium behaviour.

ADSORPTION FROM SOLUTION

Tetrahedral-lattice and rotational-isomeric-state chains with excluded volume have been studied. The numerical work[1,3] shows the dominance of polymer surface interactions over polymer-solvent interactions in defining configurations of adsorbed molecules. Terminal sequences of segments (tails) are shown to be important in determining adsorbed-layer thickness under conditions of weak adsorption, as illustrated in fig. 1. The display[4] shows a 50-bond lattice chain under various strengths of adsorption.

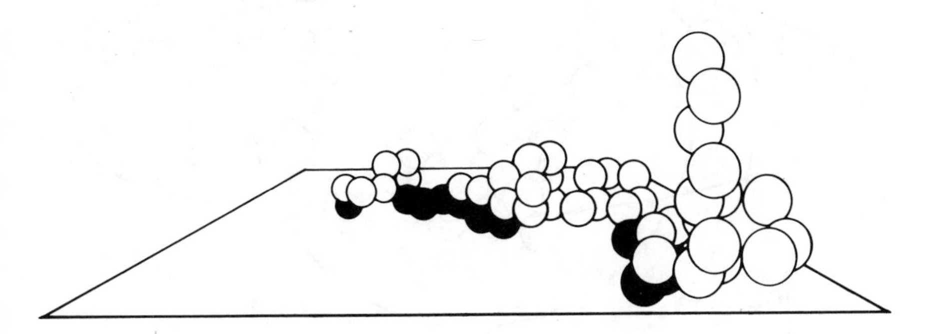

Fig.1 Configuration of a 50-bond polymethylene chain under weak adsorption. Filled circles represent adsorbed segments.

MOLECULAR SIZE AND SHAPE IN SOLUTION

The effects of chain flexibility and segment-solvent interaction have been investigated. In the numerical work, comparisons between lattice and non-lattice models have been made[2], and particle scattering functions of different chains have been calculated[5]. In the display work, the uses of end-to-end distance, radius of gyration and equivalent ellipsoid for describing molecular size and shape have been compared[6].

CYCLIC AND LINEAR MACROMOLECULES

Algorithms for generating configurations of cyclic molecules have been devised and calculations made comparing properties of cyclic and linear, unperturbed poly(dimethyl siloxane)(PDMS) chains[7,8], including comparisons with experimental particle scattering functions[5,9] and diffusion coefficients.[10,11] The display work compares cyclic and linear PDMS in terms of the equivalent ellipsoid model[12]. Fig. 2 shows a configuration of a linear PDMS chain together with a section of its average equivalent ellipsoid[7].

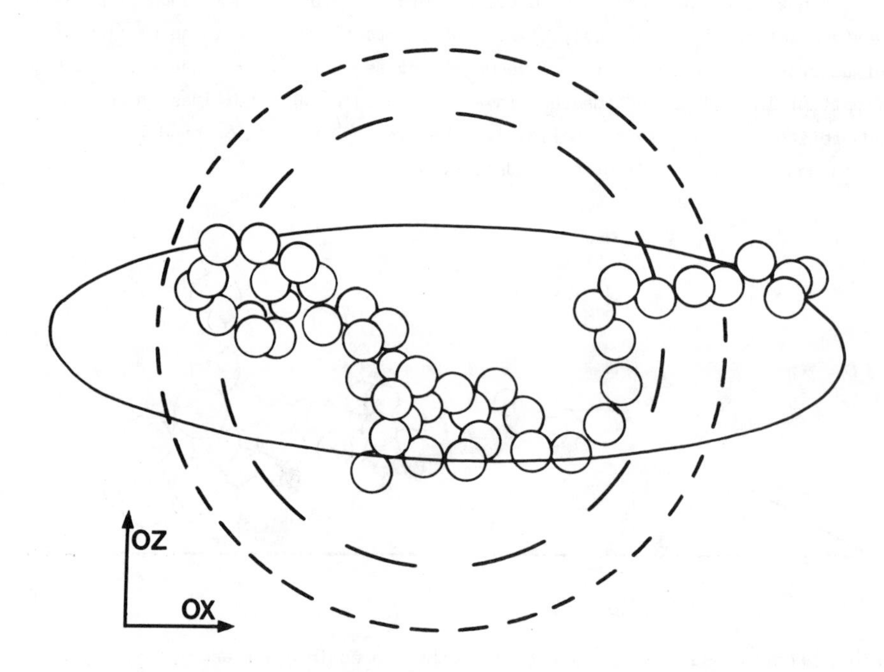

Fig. 2. Configuration of a linear PDMS chain of 100 skeletal atoms. View along OY axis of average equivalent ellipsoid of constant segment density. Ellipse shown (——) has axes along OX and OZ which are the longest and the shortest axes, respectively, of the average equivalent ellipsoid. ---, circle of radius corresponding to equivalent sphere of constant segment density. ––, circle of radius equal to the usual root-mean-square radius of gyration. Circles represent $-Si(CH_3)_2O-$ segments.

TRANSLATIONAL DIFFUSION

The numerical calculations have involved detailed interpretations of experimental diffusion coefficients (D) in terms of Kirkwood-Riseman[10,13] and full Oseen hydrodynamics[14]. The display work[6] illustrates the change of D with chain length by using Einstein's jump-model and experimental values of D to estimate the real time intervals required for computer moves of parts of chains.

ACKNOWLEDGEMENTS

Thanks are due to Unilever Research and the Science and Engineering Research Council for their support.

REFERENCES

1. M. Lal and R.F.T. Stepto, J. Polymer Sci., Polymer Symposia 1977 61 401.
2. M.A. Winnik, D. Rigby, R.F.T. Stepto and B. Lemaire, Macromolecules 1980 13 699.
3. A. Higuchi, D. Rigby and R.F.T. Stepto in Adsorption from Solution, eds. R.H. Ottewill, C.H. Rochester and A.L. Smith, Academic Press, 1983, p.273.
4. M. Lal and R.F.T. Stepto, Comparative Adsorption of Macromolecules (film and video recording), University of Manchester Audio Visual Services, 1976.
5. C.J.C. Etwards, R.W. Richards and R.F.T. Stepto, Macromolecules 1984 17 2147.
6. R.F.T. Stepto, Chain Configuration and Molecular Movement in Solution (film and video cassette), University of Manchester Audio Visual Service, 1978.
7. C.J.C. Edwards, D. Rigby, R.F.T. Stepto, K. Dodgson and J.A. Semlyen, Polymer 1983 24 391
8. C.J.C. Edwards, D. Rigby, R.F.T. Stepto and J.A. Semlyen, Polymer 1983 24 395.
9. C.J.C. Edwards, R.W. Richards, R.F.T. Stepto, K. Dodgson, J.S. Higgins and J.A. Semlyen, Polymer 1984 25 365.
10. C.J.C. Edwards, D. Rigby and R.F.T. Stepto, Macromolecules 1981 14 1808.
11. K. Dodgson, C.J.C. Edwards, and R.F.T. Stepto, Br.Polymer J. 1985, in press.
12. D. Rigby and R.F.T. Stepto, Comparative Behaviour of Linear and Cyclic Poly(dimethyl siloxane)(video recording), University of Manchester Audio Visual Services, 1981.
13. I.J. Mokrys, D. Rigby and R.F.T. Stepto, Ber. Bunsenges. Phys. Chem. 1979 83 446.
14. C.J.C. Edwards, A. Kaye and R.F.T. Stepto, Macromolecules, 1984 17 773.

DEUTERON-NMR STUDIES OF MOLECULAR MOTIONS IN SOLID POLYMERS

H.W. Spiess

Max-Planck-Institut für Polymerforschung
Jakob-Welder-Weg 11, Postfach 3148, D-6500 Mainz, FRG

SYNOPSIS

Pulsed deuteron NMR yields clear-cut information about the molecular dynamics in solid polymers. The technique is illustrated by two experimental examples characterizing ultraslow chain motions in the vicinity of the glass transition in polystyrene and the determination of the distribution of correlation times for local motions in polycarbonate.

1 INTRODUCTION

Knowledge of type and timescale of molecular motions in solid polymers should provide a much better understanding of their mechanical properties. Although many techniques are used in this area[1] the relationship between the macroscopic behaviour and the molecular dynamics is not yet fully understood. One of the reasons for this may be, that most of the techniques employed to study molecular motions probe the dynamics rather unselectively and on the wrong time scale.

Deuteron-NMR, on the other hand, has a number of advantages which make it a particularly powerful tool for elucidating molecular dynamics[2]. Rotations involving individual C-H bond directions are detected separately and by analysis of the solid echo and the spin-alignment spectra molecular motions can be followed over an extraordinary wide range of characteristic frequencies (10 MHz - 1 Hz). The solid echo is generated[2] by applying two rf pulses delayed by the evolution time τ_1 and the spin-alignment echo is generated via the Jeener Broekaert sequence[3], where a third pulse is applied after a mixing time τ_2. The corresponding spectra are obtained by Fourier Transform (FT) of the echo signals.

2. EXAMPLES

2.1 Chain motion at the glass transition in polystyrene

The ultraslow motions associated with the glass transition can be characterized by deuteron spin-alignment[2]. It can detect motions which change the NMR frequency during the rather long mixing time τ_2, which typically is in the range 1ms-10s. The onset of chain motion above the glass transition temperature T_g first of all manifests itself in a rapid decay of the alignment echo[2,4]. The time constant of the initial decay strongly depends on τ_1. This indicates diffusive motions by small but not well-defined angles. In close proximity to T_g this diffusive motion is highly restricted, however, involving small angle rotations only, even after long times. This is illustrated in figs.1 and 2, where experimental and calculated spin-alignment spectra[5] are plotted for polystyrene. For τ_2=2s and 0.2s at 378 K and 388 K, respectively, the alignment echo has decayed far below 10% of its initial intensity, cf. also the corresponding curves shown in fig.19 of ref.2b. Nevertheless at 378 K the line shape changes associated with this decay are rather minor indicating that after 2s each chain segment has rotated by approximately $\pm$ 4° only. At 388 K, on the other hand, not

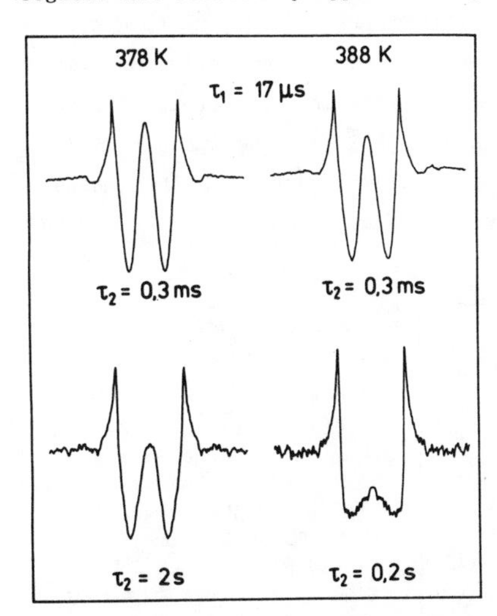

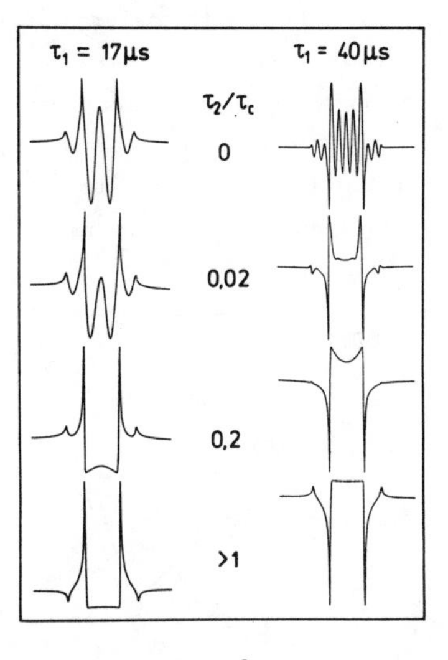

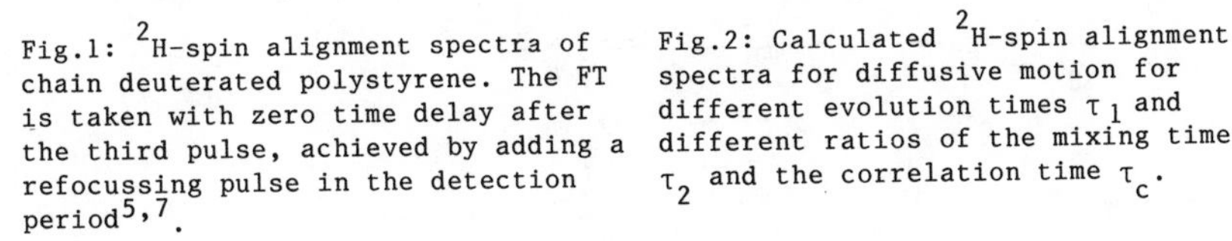

Fig.1: ^{2}H-spin alignment spectra of chain deuterated polystyrene. The FT is taken with zero time delay after the third pulse, achieved by adding a refocussing pulse in the detection period[5,7].

Fig.2: Calculated ^{2}H-spin alignment spectra for diffusive motion for different evolution times τ_1 and different ratios of the mixing time τ_2 and the correlation time τ_c.

only does the decay occur on a much shorter time scale, but also is associated with a major change in line shape. In fig.2 theoretical line shapes are plotted, calculated for diffusive motion. Note in particular that for an unrestricted motion leading to a complete randomization of the molecular orientation for $\tau_2/\tau_c > 1$ the line shape becomes independent of τ_1 apart from the sign. The experimental spin-alignment line shapes at 388 K and τ_2=0.2s, in fact, are largely independent of τ_1. This clearly shows that at the higher temperature due to the chain motion a given C-H bond direction can explore a considerably larger angle, i.e. approximately $\pm$ 15° within 0.2s, details will be published at a later date[5]. Our experiments thus allow us to follow the increasing cooperativity of the chain motion with increasing temperature.

2.2 Distribution of correlation times for motions in polycarbonate

It is well-known that the molecular motion in polymers often is not uniform but must, instead be described by a distribution of correlation times[1]. Although the phenomenon as such is well established, little is known about the nature of this distribution. In particular, most techniques employed in this area do not allow a distinction of a *heterogeneous* distribution, where spatially separated groups move with different time constants, and a *homogeneous* distribution, where each monomer unit shows essentially the same non-exponential relaxation. Even worse, relaxation processes resulting from different motional mechanisms often cannot be separated. Deuteron NMR offers new possibilities in this area. The total spectrum is a weighted superposition of line shapes for the different correlation times, allowing a direct determination of the

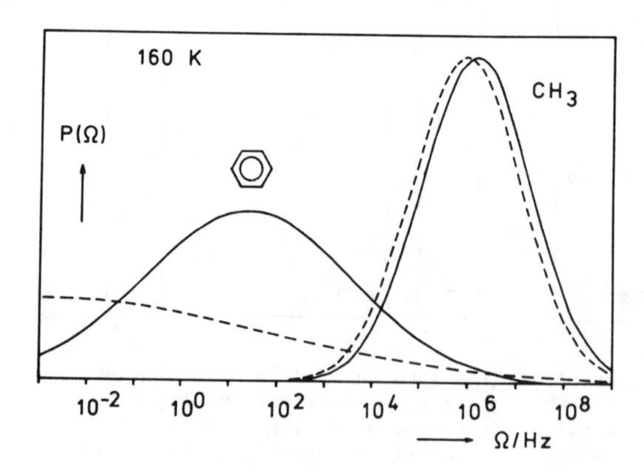

Fig.3: Distribution of correlation frequencies $\Omega=1/\tau_c$ for the methyl- and the phenyl groups in PC (solid lines) and mixtures of PC with 25% polychlorinated biphenyls[8,9] (dotted lines).

corresponding distribution function[2,6,8]. Moreover, by combination with spin-lattice relaxation techniques homogeneous and heterogeneous distributions can be dinstinguished[2b]. As an example we determined the distribution of correlation times both for the methyl- and the phenyl motions in bisphenol-A-polycarbonate (PC), as plotted in fig.3. In both cases the motion is highly restricted again: the phenyl groups exhibit 180° jumps[2] and the methyl groups three-fold jumps about their axis of symmetry. The corresponding spectra are published in refs.2b and 9. Clearly at 160 K the widths of the distributions for the methyl- and phenyl motions are similar and rather narrow (2.3 and 4.3 orders of magnitude, respectively). The distributions do overlap, however. Thus unselective techniques might suggest a much broader distribution for the same polymer. By recording partially relaxed spectra we could prove that the distribution is *heterogeneous* in nature for both methyl- and phenyl groups[2], presumably due to differences in packing. This is confirmed by our experiments on mixtures of PC and low molecular mass additives which suppress the low temperature mechanical relaxation[9]. These additives hinder the *phenyl* mobility and in particular drastically increase the width of the corresponding distribution of correlation times but leave the *methyl* mobility essentially unchanged, cf. fig.3. Our selective experiments thus strongly suggest a relation between phenyl mobility and mechanical properties in this system[9].

ACKNOWLEDGEMENTS

The results on polystyrene presented in figs.1 and 2 were obtained in collaboration with Dr. F. Fujara and S. Wefing; their work is highly appreciated. We would also like to thank the Deutsche Forschungsgemeinschaft (SFB 41) for the continuous financial support.

REFERENCES

1. Bailey, R.T., North, A.M., and Pethrick, R.A. Molecular motion in polymers, Clarendon Press, Oxford, 1981.
2. Spiess, H.W., a) Coll. & Polym. Sci. 261, 193 (1983);
   b) Adv. Polym. Sci. 66, 24 (1985)
3. Jeener, J. and Broekaert, P., Phys. Rev. 157, 232 (1967)
4. Rössler, E., Sillescu, H. and Spiess, H.W. Polymer 26, 203 (1985)
5. Wefing, S., Fujara, F., Sillescu, H. and Spiess, H.W. to be published.
6. Hentschel, D., Sillescu, H. and Spiess, H.W. Polymer 25, 1078 (1984)
7. Jeffrey, K.R., Bull. Magn. Res. 3, 69 (1981)
8. Schmidt, C., Kuhn, K. and Spiess, H.W. to be submitted to Colloid & Polym.Sci.
9. Fischer, E.W. Hellmann, G.P., Spiess, H.W., Hörth, F.J., Ecarius, U., and Wehrle, M. Makromol. Chem. 186 (1985), in press.

A TWO-DIMENSIONAL NMR STUDY OF VERY SLOW MOLECULAR MOTIONS IN POLYMERS

A.P.M. KENTGENS, A.F. de JONG, E. de BOER and W.S. VEEMAN

Department of Molecular Spectroscopy, Faculty of Science, University of Nijmegen, Toernooiveld, 6525 ED Nijmegen

SYNOPSIS

A combined use of the 2D-exchange NMR experiment with cross polarization and magic angle spinning (CP-MAS) is described. This enables us to study very slow molecular motions (1 kHz - 0.1 Hz) via natural abundance ^{13}C NMR. Precise information about the molecular motion involved can be obtained from the characteristic two-dimensional spinning sideband patterns. This is illustrated with spectra of dimethylsulfone. Furthermore some results on polyoxymethylene are presented.

1 INTRODUCTION

It needs no further explanation that molecular reorientations of macromolecules exert great influence on relevant physical properties of polymers. Among the different techniques used to study molecular motions NMR has become a very valuable one, capable of determining motions over a wide range of frequencies from 10^8 Hz to 0.1 Hz. Especially for the study of very slow molecular motions (1 kHz - 0.1 Hz) deuterium NMR can be successfully used but then deuterated samples are required[1]. Here we report a new 2D exchange NMR technique to study such slow molecular motions via natural abundance ^{13}C NMR.

2 TECHNIQUE

A detailed description of 2D exchange NMR can be found elsewhere[2]. The pulse scheme for the experiment is given in fig. 1. First we consider a static, powdered sample[3]. In this case each spin packet has a different resonance frequency, depending on the position of its chemical shielding tensor with respect to the

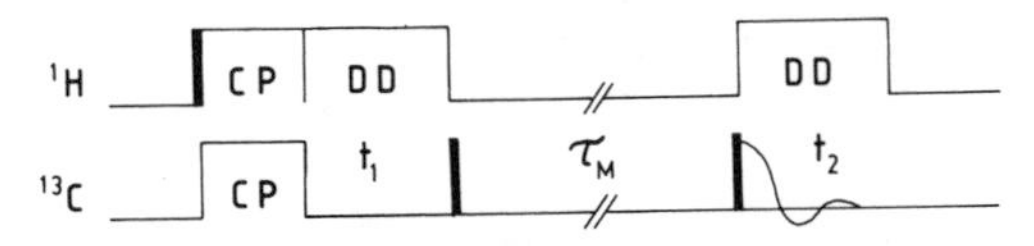

Fig. 1: Pulse-sequence for the 2D-exchange experiment with cross-polarization (CP) and high power dipolar decoupling (DD). 90° pulses are in black.

external magnetic field. If we apply the 2D exchange technique to such a sample without any molecular motions present, we will get the normal 1D powder pattern along the diagonal. If, however, molecular motions occur during the mixing time τ_m, these spins will have a different resonance frequency in the detection period t_2 than they had during the evolution time t_1. This will manifest itself as off-diagonal intensity. The resulting 2D powder pattern enables one to characterize the motion. The sensitivity, however, is rather low. In order to overcome the sensitivity problem and to resolve possibly overlapping resonances we would like to apply MAS, at a rotation frequency smaller than the chemical shielding anisotropy so that we get spinning side bands which contain information about the shielding tensor[4]. The effect of sample rotation is that the resonance frequency of each spin packet becomes time dependent, so that we can no longer speak of spin *isochromates* during the evolution time and the detection period. This results in a 2D exchange spectrum that contains off-diagonal lines even if there are no molecular motions. However, if we choose the mixing period an integer number of spinner rotations the off-diagonal signals disappear (if there is no molecular motion) and we get the 1D spinning side band pattern along the diagonal[5]. Molecular motions ($t_1 < \tau_c < T_1$) cause off-diagonal cross-peaks between the different spinning side bands. The intensities of these cross-peaks enable us to relate the position of the shielding tensor during t_2 to that during t_1 and thus make it possible to characterize the motion.

3 RESULTS

Dimethylsulfone serves as a testcase for this experiment. The molecule rotates about the C_2 axis (fig. 2c) at a rate of 314 s^{-1}[6]. Figs. 2a and 2b show the 2D exchange spectra of dimethylsulfone with synchronous mixing times of 0 and 6.4 ms, respectively. As expected fig. 2a shows the normal 1D spinning side band pattern along the diagonal while in fig. 2b the effect of molecular motion is clearly present. The principal axes systems of the shielding tensor of the two interchanging methyl groups can be related by a rotation over an angle of 108° about the

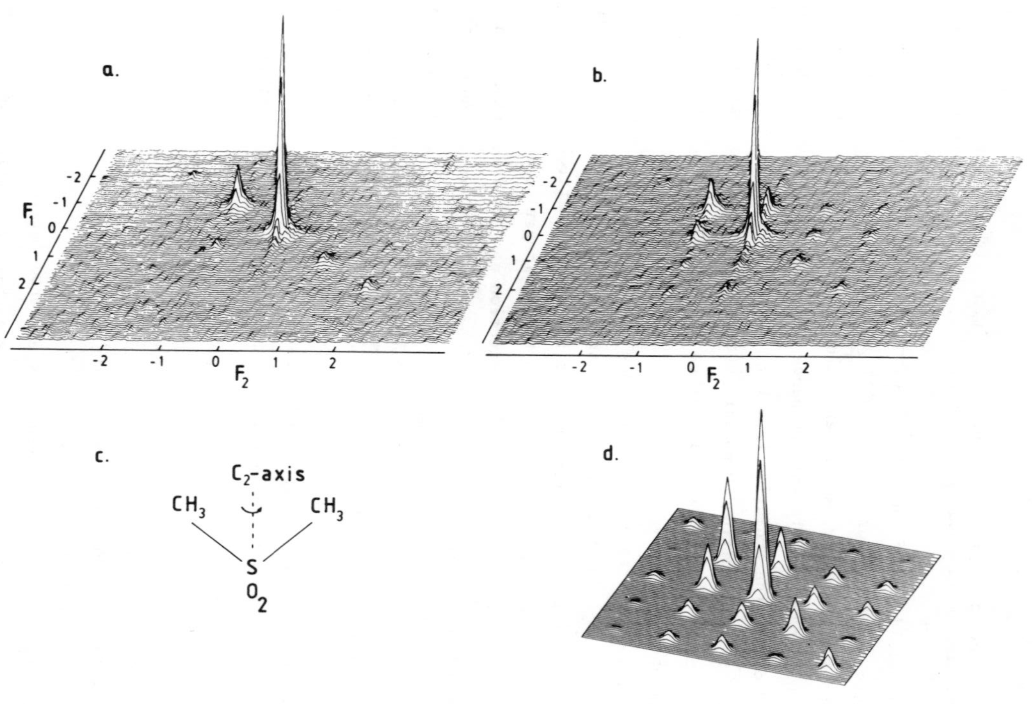

Fig. 2: 2D exchange spectra of dimethylsulfone, the experimental spectra were obtained at a spinner speed of 1250 Hz. Numbers indicate the positions of the spinning side bands. a) Mixing time of 0 ms. b) Synchronous mixing time of 8 rotor periods (~ 6.4 ms). c) Structure of DMS indicating the C_2-axis. d) Computer simulation of spectrum b).

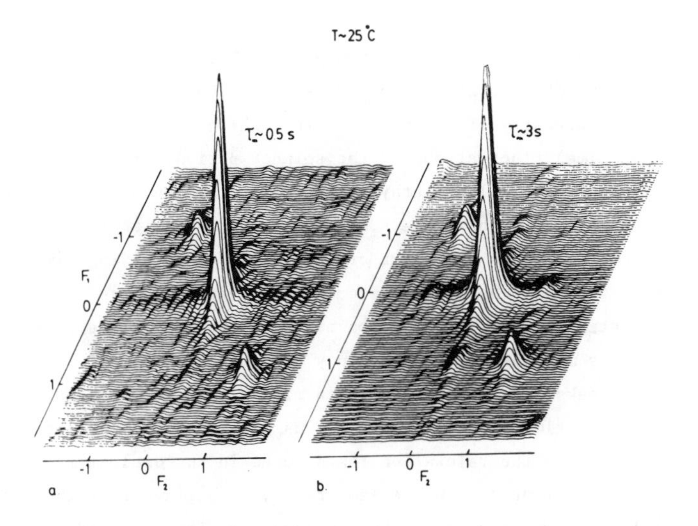

Fig. 3: 2D exchange NMR spectra of polyoxymethylene at room temperature at a spinning frequency of ~ 1450 Hz. a) Mixing time 0.5 s. b) Mixing time 3 s.

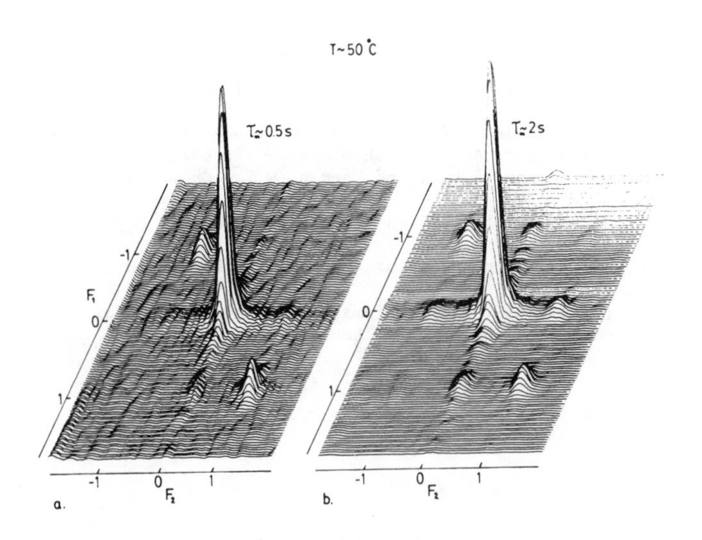

Fig. 4: POM spectra obtained at 50°C. a) Mixing time 0.5 s. b) Mixing time 2 s.

σ_{22} axis which is perpendicular to the C-S-C plane[6]. This information results in the simulated spectrum of fig. 2d, which agrees well with the experimental spectrum except for the intensity of the central line, which is the only line not broadened by (long time) fluctuations of the spinner period. Figs. 3 and 4 show some preliminary results on polyoxymethylene[7] (commercially available Hostaform-C from Hoechst). The resonance shown stems exclusively from the crystalline part of the polymer. We see that there are clear cross-peaks after a mixing time of 3 s at room temperature (fig. 3b). The effect becomes even more pronounced at 50°C with τ_m = 2 s. Extrapolation of the α-relaxation map of polyoxymethylene from dielectrical loss data[8] yields motions with a frequency of ~ 0.1 Hz at 50°C. In fig. 4 we see that the cross peaks are very weak after a mixing time of 0.5 s but become more pronounced at a mixing time of 2 s. This tells us that the observed motion must be slower than ~ 1 Hz. Since the relaxation times found in the dielectric loss measurements and the NMR experiment are in the same range, we believe that the motion observed here should be associated with the α-relaxation. This then proves that the crystalline part of POM is involved in the α-relaxation as has been assumed before[8]. More investigations are needed to elucidate the exact nature of these motions.

REFERENCES

1. Spiess, H.W., J. Chem. Phys. 72, 6755, 1980.
2. Jeener, J.; Meier, B.H.; Bachmann, P. and Ernst, R.R., J. Chem. Phys. 71, 4546, 1979.
3. Edzes, H.T. and Bernards, J.P.C., J. Am. Chem. Soc. 106, 1515, 1984.
4. Herzfeld, J. and Berger, A.E., J. Chem. Phys. 73, 6021, 1980.
5. De Jong, A.F.; Kentgens, A.P.M. and Veeman, W.S., Chem. Phys. Lett. 109, 337, 1984.
6. Solumn, M.S.; Zilm, K.W.; Michl, J. and Grant, D.M., J. Phys. Chem. 87, 2940, 1983.
7. Kentgens, A.P.M.; de Jong, A.F.; de Boer, E. and Veeman, W.S., to be published in Macromolecules.
8. McCrum, N.C.; Read, B.E. and Williams, G. 'Anelastic and dielectric effects in polymeric solids' John Wily and Sons: London, 1967.

TRANSITIONS AND MOBILE PHASES BY NMR NORMAL ALKANES AND POLYETHYLENE

Martin Möller, Detlef Emeis and Hans-Joachim Cantow

Institut für Makromolekulare Chemie der Universität Freiburg
Hermann-Staudinger-Haus, Stefan-Meier-Str. 31, D-7800 Freiburg
Federal Republic of Germany

SYNOPSIS

Variable Temperature magic angle sample spinning ^{13}C-NMR experiments on nonadecane, hexatriacontane and low molecular weight polyethylene allow the observation of the arrangement of the chain ends within the lamella crystallites. Conformational order-disorder and intermolecular packing effects are reflected within the spectra. The effects are discussed with regard to the peculiar phase behaviour of the normal alkanes and the fringed micelle model for low molecular weight polyethylene.

INTRODUCTION

A question extensively studied but yet not fully understood is the arrangement of endgroups and folds within the lamella surface of normal alkanes and polyethylene crystallites. High resolution solid state NMR by means of heteronuclear high power decoupling and magic angle sample spinning (MAS) provides structure related information with regard to individual atoms in the amorphous and in the crystalline state. MAS NMR studies on cyclic alkanes can give insight regarding the transition between crystalline and amorphous phases with respect to the conformational characteristics of loops[1)]. Linear alkanes may yield corresponding information for the situation of the chain ends.

RESULTS AND DISCUSSION

Fig. 1 presents the ^{13}C-NMR spectrum of high molecular weight polyethylene at 300 K. The 'crystalline' signal at 33.6 ppm is separated by 1.8 ppm from the 'amorphous' one at 31.8 ppm[2)]. Additional signals are shown in the spectra of low molecular weight polyethylene and normal alkanes which are due to the terminal methyl and methylene carbons. Fig. 2 shows the MAS ^{13}C-NMR spectrum of hexatriacontane at 300 K. The chain is arranged in the all-anti conformation like it is the case for crystalline polyethylene. Besides the resonance at 33.6 ppm for the inner methylene

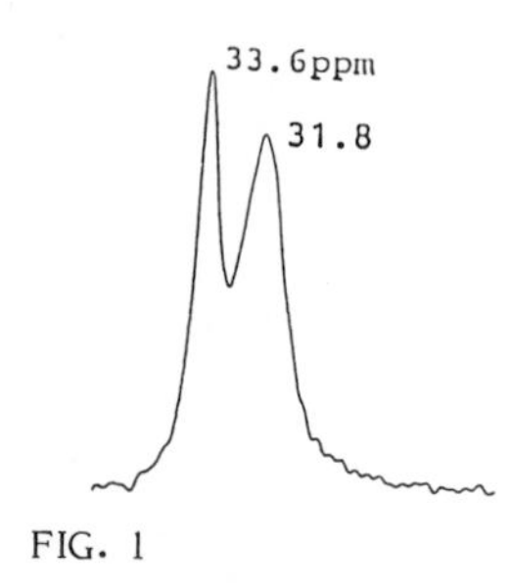

FIG. 1

carbons additional resonances appear at 15.7 ppm for the $\underline{CH}_3$, at 25.4 ppm for the $-\underline{CH}_2-CH_3$, and about 35 ppm for the $-\underline{CH}_2-CH_2-CH_3$ carbons[3].

Fig. 3 shows MAS ^{13}C-NMR spectra of nonadecane at various temperatures cooled down from the melt. The changes detected in the NMR spectra have to be discussed with respect to the polymorphism known for this compound. Beside the melting transition at 305.2 K nonadecane shows a solid-solid transition at 295.2 K below which the unit cell is orthorhombic and above which it approaches hexagonal symmetry. Due to the pressure gradient in the rotating sample holder the corresponding changes appear at slightly higher temperatures in the NMR spectra. Also in some cases the molecules can be observed at the same temperature in different modifications. Thus the spectrum at 310 K shows the molecules in the melt. The carbon-carbon bonds exchange fast between rotational isomeric states (anti and gauche). The NMR shifts are averaged for the diastereotopic positions the carbons can occupy. Below the melting point the carbon-carbon bonds are predominantly in the anti conformation and all signals are significantly shifted downfield in the spectrum at 308 K due to the γ-gauche effect[4]. However, IR studies have shown that a considerable number of gauche defects still exists within the rotator phase. It decreases slowly when the solid-solid transition is approached[5]. This is indicated by a small continuous downfield shift of all signals within the spectra between 302 K and 308 K. The spectrum at 300 K shows the compound when the major part is already in the orthorhombic modification. Now gauche defects are neglegible in terms of the NMR sensitivity. The chains may be considered to be entirely in the planar all anti conformation. However the resonances are slightly shifted upfield compared to those of the rotator phase. We explain this effect by the intermolecular packing of the chains[3]. Lowering the temperature further results in additional changes, most obviously seen at the methyl carbon resonance. At 298 K a second downfield shifted signal appears, which becomes the only methyl signal at even lower temperatures. In Fig. 4 the solid state spectra of a melt crystallized hexatriacontane sample is shown at different temperatures. Coming from lower temperatures a second upfield shifted CH_3 resonance is observed also. Raising the temperature further resulted in a spectrum at 344 K in which the downfield CH_3 signal had disappeared. A similar behaviour could be observed for the $-\underline{CH}_2-CH_3$ signal. However the splitting occured at higher temperature (334 K) as in case of the methyl carbons. At 344 K both the $-\underline{CH}_3$ and the $-\underline{CH}_2-CH_3$ carbons gave solely the upfield resonances. Lowering the temperature resulted in the inverse effect, although it has to be noted that the temperature dependence of the intensity ratios of the corresponding signals varies to some extend with the cooling rate. All the effects described appear well below the melt and the solid-solid transitions reported for hexatriacontane: 349 K, 347 K, and 345 K. Fig. 5 shows the 335 K spectrum of hexatriacontane in comparison with the room temperature spectrum of a polydisperse low molecular weight polyethylene sample. Due to the polydispersity the polyethylene sample contains a high fraction of amorphous material (30%). Hence we have to consider crystallites consisting of lamellas with a crystalline interior and two amorphous surface layers which incorporate excluded chain ends. "Amorphous" chain ends are indicated by the -$\underline{CH_2}$-CH_3 resonance at 24.3 ppm which coincides with the -$\underline{CH_2}$-CH_3 signal of nonadecane in the melt. A splitting of the "crystalline" -$\underline{CH_2}$-CH_3 similar to that observed for hexatriacontane might be indicated in the polyethylene spectrum, also this is not beyond doubt because of the unsufficient signal to noise ratio. The methyl carbon signals do not allow the discrimination of chain ends in the amorphous phase and those as observed in case of hexatriacontane.

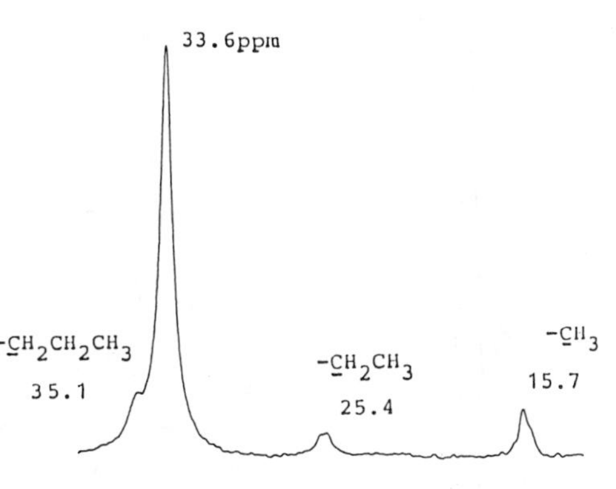

FIG. 2

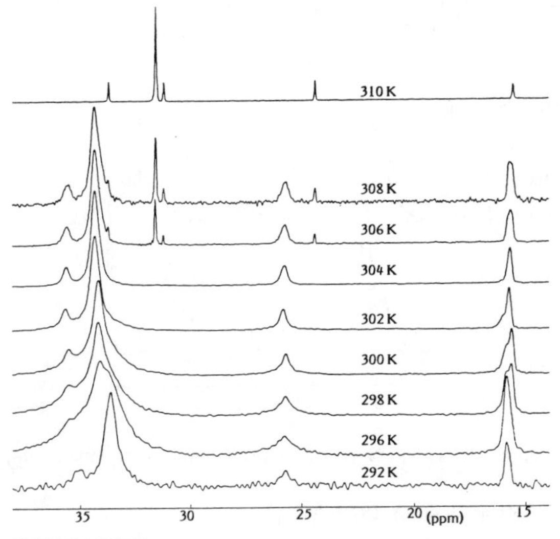

FIG. 3

CONCLUSION

The NMR spectra allow the observation of chain ends in the amorphous phase as well as in the crystalline phase. Additionally two types of chain ends were observed for hexatriacontane at temperatures significantly lower than the thermal transitions. This has to be explained by the occurence of disorder within the lamella surface

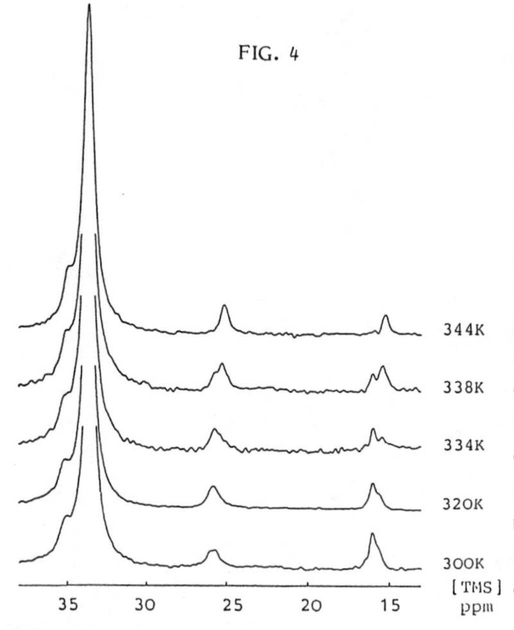

FIG. 4

under conditions when kinks and rotational disorder can be excluded as a major contribution. At the example of hexatriacontane it was shown that the effect occurs first for the methyl groups and then with increasing temperature for the outer methylene groups. This may be explained by longitudinal displacements of the chains. In order to preserve the crystalline order as far as possible the displacement should be correlated with a corresponding displacement in the adjacent lamella. Thus a roughness within the lamella surface might be generated to different extends.

ACKNOWLEDGMENT

Cordially we want to thank Mr. Alfred Hasenhindl for technical assistance. Generous financial support by the Deutsche Forschungsgemeinschaft (SFB 60) is gratefully acknowledged.

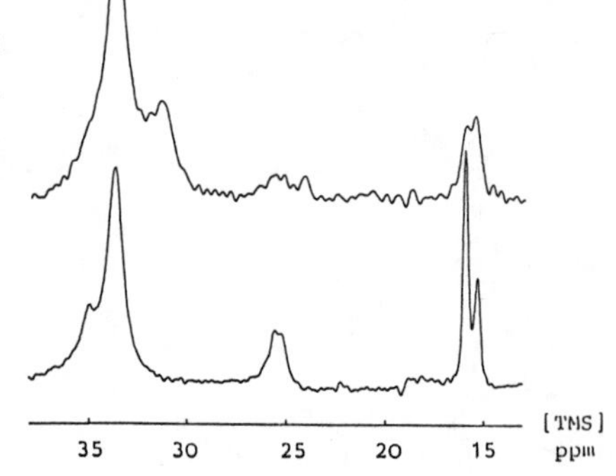

FIG. 5

LITERATURE

1). Möller, M.; Gronski, W.; Cantow, H.-J.; Höcker, H.; J. Am. Chem. Soc. **1984**, 106, 5093

2). Earl, W. L.; VanderHart, D. L.; Macromolecules **1979**, 12, 782

3). VanderHart, D. L.; J. Magn. Reson. **1981**, 44, 117

4). Möller, M.; Advances in Polym. Sci. **1985**, 66, 59

5). Snyder, R. G.; Marconelli, M.; Strauss, H. L.; J. Am. Chem. Soc. **1983**, 105, 133

MORPHOLOGY AND CHAIN DYNAMICS OF POLYMERS AS REFLECTED FROM POLYMER-DYE INTERACTIONS

C.D. EISENBACH*

*Polymer-Institute, University of Karlsruhe, P.O. Box 6380,
D-7500 Karlsruhe, FRG

SYNOPSIS

Dyes such as photochromes or fluorophores were used as labels or probes to study structure, morphology and chain dynamics of bulk polymers on the basis of the spectral behaviour of the chromophores. In multiblockcopolymers, e.g. segmented polyurethane (PU) elastomers, photochromic labels detect chain segmental motions in different morphological regions. In homo- or copolymers, e.g. acrylates, fluorescence probes directly reflect the accessibility of polymer elements to additives and relate it to the polymer constitution.

1 INTRODUCTION

A number of molecules exhibiting quite different physical phenomena have been employed in recent years as labels or probes to characterize polymers; generally known are stable free radicals in the ESR technique and fluorophores in fluorescence decay or quenching studies. However, the shape, intensity and wave length position of the absorption or emission bands of chromophores, which change from one matrix to the other and also exhibit time-dependent phenomena, can directly be analysed and allow conclusions on polymer properties[1-3].

In general, significant differences exist in the photoreactivity, reactivity and spectral behaviour of a dye molecule dissolved in a low molecular weight solvent as against embeded in a polymer matrix. This is mainly due to differences in the fluctuation of the free volume as well as the local free volume around the dye molecule in the different media[1], and to differences in the possibilities of interactions between the dye molecule and the surrounding medium[3] (small molecules vs. polymer chain elements). The analysis of these deviations does not only lead to a better understanding of polymer-dye interactions on a molecular scale, but also allows to obtain particular informations on the morphology and dynamics of a polymer system by using the dye molecule as a probe.

The practical aspects of this work are that the results might be used to improve the efficiency of fluorescence solar energy collectors[3,4] or to stimulate the development of devices based on the photomechanical effect of photochromes[5,6].

2 PHOTOCHROMES AND FLUOROPHORES AS LABELS OR PROBES

Two types of molecules were used: azophenolether derivatives, which undergo a reversible cis-trans isomerization around the azo-linkage, as a photochromic label, and 4-dimethylamino-4-nitrostilbene (DMANS) as a fluorescence probe; in the case of the aromatic azo compound, the trans → cis photoisomerization and the thermal cis → trans isomerization was followed by measuring the change in the optical density of the trans band, whereas for the stilbene primarily the shift in the absorption and emission band was investigated.

As a common feature, irrespective of the type of the dye and the (photo-) process investigated (photochromism, absorption, emission), a more or less distinct change in the observed parameter is found not only on the glass-rubber transition but also in the temperature range of secundary relaxation processes.

2.1 Photochromism in Segmented Polyether Polyurethane Elastomers

Azophenolether units were incorporated in the hard segment based on 4,4'-diphenylmethanediisocyanate and 1,4-butanediol, and in the poly(oxytetramethylene) soft segment at various distinct distances (oxytetramethylene units) apart from the hard segment. The change in the absorption spectrum of the polymer films upon trans-cis photoisomerization until the photostationary state (curve 5) has been reached, or the thermal back reaction to the all trans isomer (curve 1) is illustrated in fig. 1.

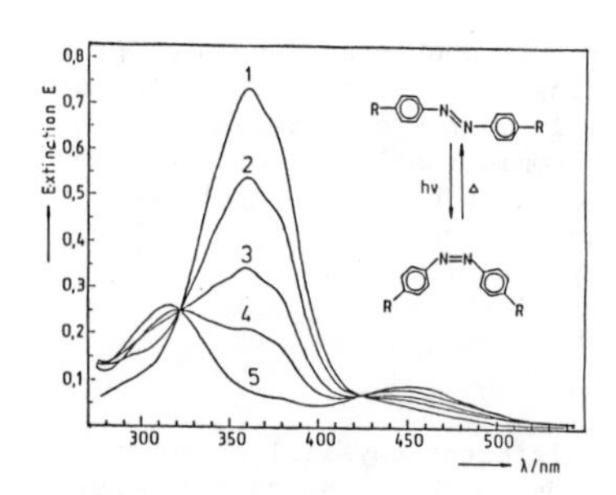

Fig. 1: UV spectra of photochromic PU elastomer films after different periods of irradiation (λ = 360nm)

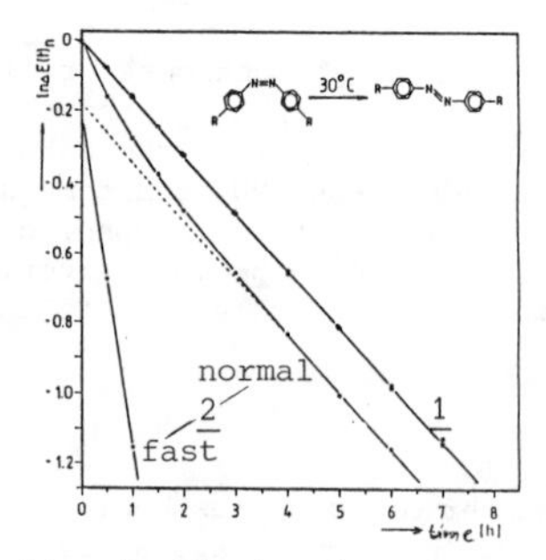

Fig. 2: First order plots of the thermal cis-trans isomerization of chromophores in the soft phase (1) and the interphase (2)

For chromophores in the hard segment, single first order kinetics for the thermal back reaction (increase of the optical density at the maximum trans absorption) are only observed at temperatures above the glass transition temperature Tg of the hard phase (~70°C); for chromophores in the rubbery soft segment, this process can be described by one exponential only if the chromophore is separated by more than four oxytetramethylene units from the hard segment. In all the other cases more or less pronounced deviations from first order kinetics occur, which normally can be resolved by two simultaneous first order processes (fig. 2); one of these processes is anomalously fast as compared to the bleaching reaction in solution, and also characterized by a very low energy of activation. As was already shown in earlier work (ref. [1,2]), this phenomenon can be correlated with the free volume theory [7] and has to be attributed to particular segmental motions detected by the chromophore.

One of the most interesting conclusions, which can be drawn from the decrease of the fraction of anomalously fast reacting cis-isomers with increasing distance from the hard segment (fig. 3), is that the dimension of the interface between the hard and the soft phase has to be extended up to four soft segment repeat units at most.

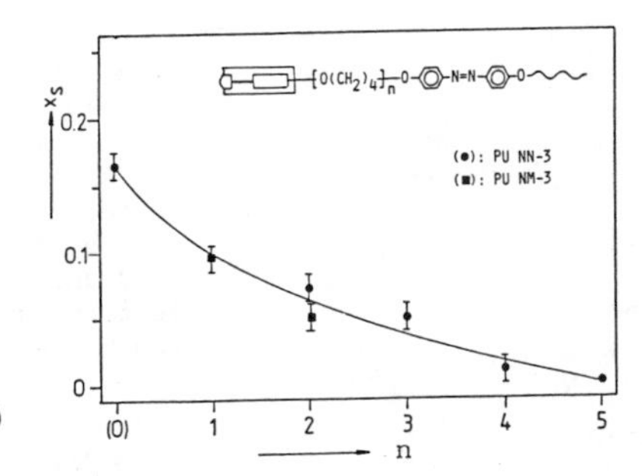

Fig. 3: Change of fraction x_S of anomalously fast isomerizing cis-isomers with distance n (in oxytetramethylene units) from hard segment

2.2 Fluorescence in Acrylate Homo- and Copolymers

DMANS is a fluorophore with the highest difference in the dipole moment between the ground state and the excited state ($\Delta\mu \sim 19D$); therefore this probe responds sensitively to changes in the matrix polarity and interaction possibilities with matrix elements. This is primarily reflected in the position of the absorption and emission band, and the magnitude of the difference between the maxima of these bands, the so-called Stokes' shift, is a rough measure for the extent of the polymer-dye interactions.

As a thumb rule, steric and polar effects known from low molecular weight systems (solvents), e.g. solvatochromism[8], can be applied to polymeric systems, but they are sometimes smeared or even masked by structural and dynamic properties of the particular polymer, or by additives[3].

The polarity of a polymer matrix is reflected by the Stokes' shift, which - at a given supercooling relative to Tg - increases in a series of methylacrylate (MA) styrene (St) copolymers with increasing MA-content, primarily due to a low frequency shift of the emission band (fig. 4,5); the opposite is observed for butylacrylate (BA))/St copolymers, where St is the more polar comonomer because of the shielding effect of the butyl residue on the ester group.

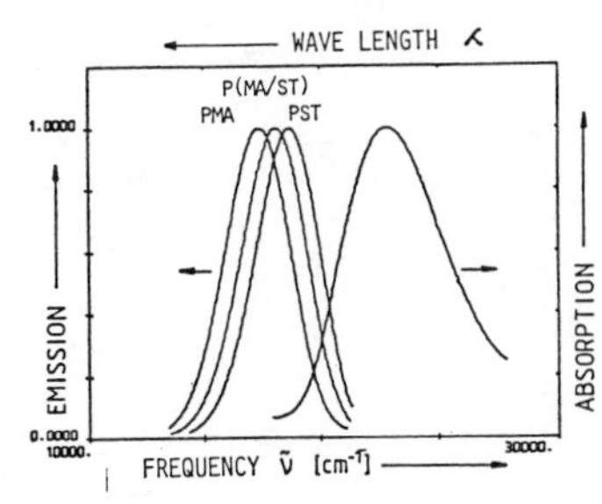

Fig. 4: Absorption and emission spectra of DMANS in pure PMA, PST and 50/50 MA/ST copolymer at $T-T_g = 50K$

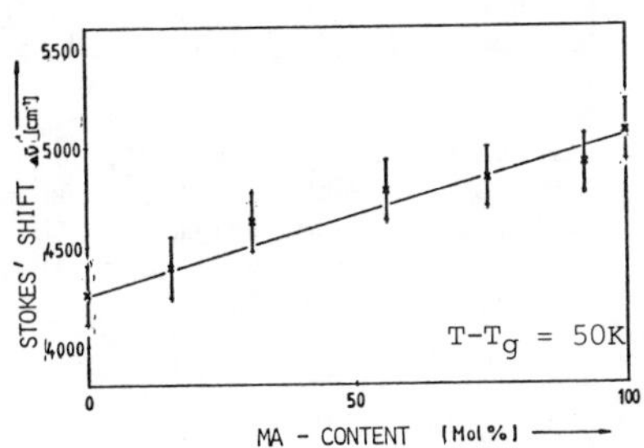

Fig. 5: Change of the Sokes' shift of DMANS in MA/ST copolymers with MA content

As already mentioned above, the temperature dependency of the photoreactivity of a dye in a polymer matrix changes when going through the glass-rubber transition. In addition to this, a second break in the Stokes' shift vs. reduced temperature plot is observed at temperatures far below Tg for a family of acry-

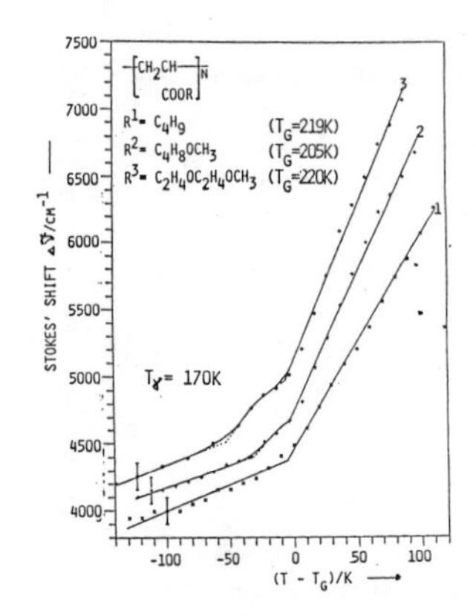

late polymers with highly polar ether as ester residues (fig. 6); this effect occurs in the temperature range of secundary relaxations in the glass and increases with the number of ether oxygens in the side chain. The fluorescence probe does not only detect changes in the mobility of the main chain but also of the side chain, since only minor conformational changes of the ether side group greatly effect the stabilization of the excited fluorophore through interaction with the ether oxygen.

Fig. 6: Temperature dependency of the Stokes' shift of DMANS in polyacrylates

ACKNOWLEGEMENTS

Most of the work presented here was carried out by H.King (photochromism), M.Vogel and K.Fischer (fluorescence) as part of their diploma and doctoral thesis, respectively. Financial support of this work by the Deutsche Forschungsgemeinschaft and the Deutsche Kautschukgesellschaft is gratefully acknowleged.

REFERENCES

1. Eisenbach, C.D., Ber. Bunsenges. Phys. Chem. 84, 680 (1980)
2. Eisenbach, C.D., in "Urethane Chemistry and Application", ACS Symp. Ser. Vol. 172, ACS, Washington D.C., 1981, p.219
3. Eisenbach, C.D., Sah, R.E. and Baur, G., J. Appl. Polym. Sci. 28, 1819 (1983)
4. Goetzberger, A. and Greubel, W., Appl. Phys. 14, 123 (1977)
5. Eisenbach, C.D., Polymer 21, 1175 (1980); King, H. and Eisenbach, C.D., Proc. German Rubber Conf., Wiesbaden, 1983, p.89
6. Smets, G., Advanc. Polym. Sci. 50, 17 (1983)
7. Schatzki, T.F., Polym. Prepr. ACS Polym. Div., Vol. 6, 646 (1965)
8. Liptay, W., Angew. Chem. 81, 195 (1969)

EMISSION SPECTROSCOPY AND THE MOLECULAR MOBILITY OF POLYEPOXIDE NETWORKS

E.W. Meijer and R.J.M. Zwiers

Philips Research Laboratories, P.O.Box 80 000, 5600 JA Eindhoven, The Netherlands.

SYNOPSIS

The paper presents the use of phosphorescence spectroscopy to reveal the structure of polyepoxide networks. The emission properties have been studied as a function of the degree of polymerization. The results strongly indicate that the increase in crosslink density proceeds via an inhomogeneous pathway. This statement is confirmed by ESR spectroscopy. The phosphorescence intensity and half-life were measured as a function of temperature for several polyepoxide networks at maximum conversion. Activation energies and transition temperatures have been estimated. These results are compared with data obtained by Dynamic Mechanical Thermal Analysis, showing that both techniques offer supplementary information concerning the mobility and structure characterization of polyepoxide networks.

1. INTRODUCTION

In recent years the application of emission spectroscopy to the study of a variety of phenomena in linear polymers has become widespread (1). The luminescence of a chromophore incorporated in a polymer is in many cases very sensitive to properties of the local environment. Important parameters are intensity, lifetime data, and degree of depolarization of both fluorescence and phosphorescence. Using the appropriate luminescent probe and monitoring the parameters to match, one sensitively can measure properties like polymer compatibility, ordering of polymer chains, viscosity, glass transition temperatures, micromobility, sub-group motion, or micropolarity (2-6). The large difference in rate for the $S_1 \rightarrow S_0$ spin-allowed transition ($10^6/10^9$ sec^{-1}) and the $T_1 \rightarrow S_0$ spin-forbidden transition ($10^{-1}/10^2$ sec^{-1}) opens the possibility to study molecular motion at a great diversity of frequencies.

Our interest is focused on three-dimensional polymer networks, densely crosslinked materials with a complicated structure and to which, to the best of our knowledge, phosphorescence spectroscopy has never been performed to gain insight in problems related to micromobility. We have investigated polyepoxides, obtained by photoinitiated cationic polymerization. Several factors governing the photoinitiated homopolymerization of epoxides have been investigated intensively (7,8). However, information about the structure of the produced polymers is scarce. The polymer networks under investigation are formed by homopolymerization of (cyclo)aliphatic bisepoxides using diaryliodonium salts as photoinitiators.

2. EMISSION CHARACTERISTICS

The photolysis of di(4-*tert*-butylphenyl)iodonium hexafluoroarsenate by UV light affords a Brønstedt acid which in turn is the catalyst for epoxide polymerization. It was found that intermediates in this photolysis exhibit remarkable luminescent properties in a polymeric matrix (9). Networks have been obtained using 2 mol % photoinitiator and a light intensity of about 1 mW/cm^2 for a period of 120 min at the wavelength of $\lambda_{max} = 254$ nm. The emission spectrum of a network from 1,2-epoxyethyl-3,4-epoxycyclohexane is depicted in Fig. 1. Both fluorescence and phosphorescence are observed at $\lambda_{max} = 340$ nm and $\lambda_{max} = 480$ nm, respectively. The latter shows a nearly exponential

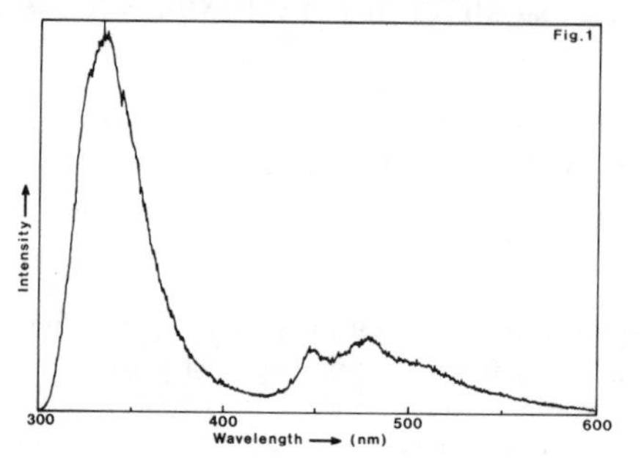

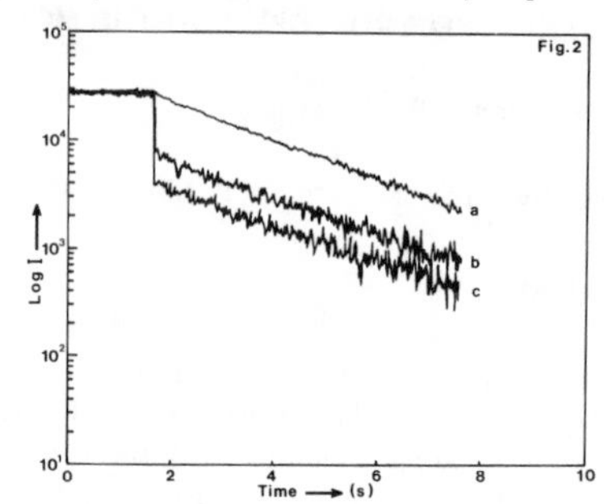

Fig. 1 and 2. The emission spectrum and decay curves of the triplet excited state for the network from 1,2-epoxyethyl--3,4-epoxycyclohexane upon excitation with light of $\lambda = 254$ nm. The halflife (Fig. 2) is detected at $\lambda = 475$ nm (a), 435 nm (b), and 425 nm (c).

decay with a halflife of 1.6 sec at ambient temperature (Fig. 2). The halflife, τ, is given by $\tau^{-1} = \tau_r^{-1} + \tau_{nr}^{-1}$, where τ_r and τ_{nr} are the radiative and nonradiative lifetimes of the triplet state, respectively. The nonradiative lifetime is determined by inter and intramolecular deactivation processes. The intermolecular quenching of the triplet state with molecular oxygen seems in our case the most plausible pathway (1) and can be attributed to a high local mobility. Excitation spectra reveal that maximum intensity for both fluorescence and phosphorescence is obtained by irradiation at $\lambda = 290$ nm.

Using diphenyliodonium hexafluoroarsenate as photoinitiator the fluorescence and phosphorescence intensity are decreased by a factor of 10 with respect to the initiator with the two *tert*-butyl groups. This is in contrast with the direct emission of aromatic species, in which the so-called "loose--bolt effect" is operative (10).

All the results are in accordance with the proposal that either the luminescent species consists of a charge-transfer complex of iodine and an aromatic moiety (iodine formed by the photolysis of iodobenzene, one of the photoproducts of the initiator) or that the luminescence is the result of an electron transfer process.

3. LUMINESCENCE DURING POLYMERIZATION AND CROSSLINKING

In order to investigate the photopolymerization of several bisepoxides with di(4-*tert*-butylphenyl)iodonium hexafluoroarsenate, we monitored the intensity of fluorescence and phosphorescence, as well as the lifetime τ of the triplet excited state of bisepoxides throughout conversion. Right from the start a rapid increase in fluorescence intensity was observed, in many cases followed by a build--up of the phosphorescence after variable periods of time (dependent on the structure of the bisepoxide). In Fig. 3 the results are presented of the bisepoxides 1,2-epoxyethyl-3,4-epoxycyclohexane, 1,4-butanedioldiglycidyl ether, and mixtures thereof. The ratio $I_{ph}(\lambda = 480\ nm)/I_{fl}(\lambda = 340\ nm)$ is plotted against the photopolymerization time. The onset of phosphorescence is very sensitive to the fraction of 1,4-butanediol-diglycidyl ether in the mixture. Even differences in composition up to a few percent give rise to a detectable change in onset time. The halflife of phosphorescence for a particular polyepoxide network at ambient temperature appears to be independent of degree of polymerization. At first glance this observation is remarkable. Since the observed lifetime is the result of radiative and non-radiative transitions, it is expected that at a gradual increase in crosslink density and hence in immobility, the observed lifetime will also increase gradually. However the observation of a constant lifetime leads to the following interpretation. The polymer network is thought to be composed of domains with low internal mobility (in which radiative phosphorescence predominates τ) and domains with higher internal

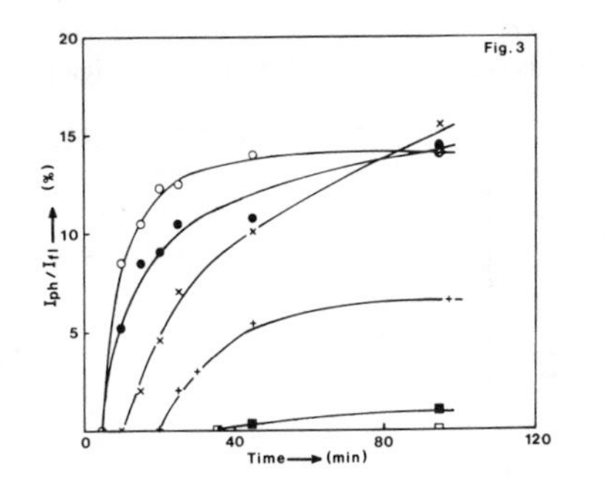

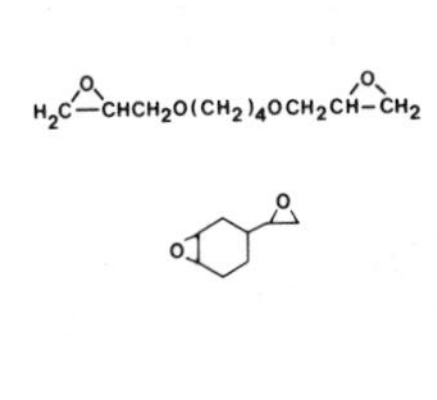

Fig. 3. The relative increase in phosphorescence intensity as a function of polymerization time for the mixture 1,2-epoxy-ethyl-3,4-epoxycyclohexane/1,4-butanediol-diglycidyl ether (ww%). (o) = 100/0; (●) = 80/20; (x) = 60/40; (+) = 40/60; (■) = 20/80; (□) = 0/100.

mobility (in which radiationless deactivation of the triplet state predominates τ). The volume fraction of domains with low internal mobility will increase as polymerization and crosslinking proceeds. This volume fraction is expressed by the ratio I_{ph}/I_{fl} (as depicted in Fig. 3).

The "two-domain hypothesis" outlined above is supported by data obtained with ESR spectroscopy. Since most intermediates of the photolysis of the iodonium salts are radicals, we performed ESR to detect trapped intermediates with unpaired electrons. Evaluation of the ESR signal intensity in time shows that the concentration of trapped radicals increases during polymerization, while the shape of the signal is unaffected. Moreover, the stability of the radicals is only slightly dependent on degree of polymerization. Again no gradual increase in decay time and hence in stability and immobility has been observed.

On the basis of these mobility studies we conclude that the photoinitiated cationic polymerization of bisepoxides is inhomogeneous in nature, leading to networks with domains of lower and higher internal mobility at room temperature.

4. MOLECULAR MOBILITY AS A FUNCTION OF TEMPERATURE

Detailed information about the structure of polyepoxide networks at maximum conversion was obtained by temperature dependent measurements (11). Since no marked differences in the mechanism of absorption, intersystem-crossing, and emission are to be expected between linear polymers and networks, we use in our study assumptions that are generally accepted for linear polymers (2). The following expressions for both phosphorescence intensity and decay time are well established,

$$(I_{fl}/I_{ph}) \cdot (I_{ph_0}/I_{fl_0}) - 1 = A \exp(-E/RT)$$

$$\tau^{-1} - \tau_0{}^{-1} = B \exp(-E/RT)$$

where I_{fl}, I_{ph}, I_{fl_0}, I_{ph_0} are the fluorescence and phosphorescence intensities at temperature T and the low temperature limit respectively, τ and τ_0 are the halflife of the triplet state at temperature T and the low temperature limit, A and B are constants, and E is the energy term governing the temperature dependence of the phosphorescence. Arrhenius plots of $\ln(\tau^{-1} - \tau_0^{-1})$ and $\ln((I_{fl}/I_{ph}) \cdot (I_{ph_0}/I_{fl_0}) - 1)$ versus 1/T should give straight lines of slope $-$ E/RT. Figure 4 and 5 show such Arrhenius plots for networks of 1,2-epoxyethyl-3,4-epoxycyclohexane and 1,4-butanediol-diglycidyl ether, respectively. It is clear that all curves show two distinct linear parts with a discontinuity within a narrow temperature region centered at T = 257K. Below the transition temperature the estimated activation energies for both polyepoxides differ only slightly. However, above this temperature the differences are significant.

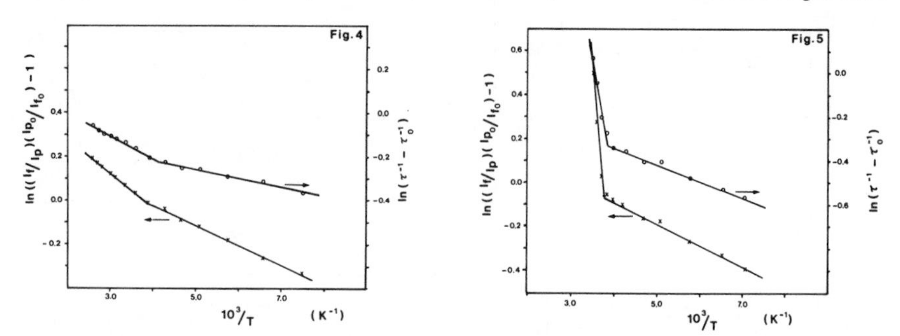

Fig. 4 and 5. The temperature dependence of triplet state lifetimes and phosphorescence intensities of 1,2-epoxyethyl- -3,4-epoxycyclohexane (Fig. 4) and 1,4-butanediol-diglycidyl ether (Fig. 5) networks.

A very rapid increase in molecular mobility within a narrow temperature region is observed for the network obtained from 1,4-butanediol-diglycidyl ether. The activation energy in this case is $>$ 150 kJ/mol, while for 1,2-epoxyethyl-3,4-epoxycyclohexane a value of 14 kJ/mol is estimated above the transition temperature from the intensity data. We have also investigated several other bisepoxides, they all show different activation energies. However, the transition temperature in all these cases is located around T = 257 (± 10) K.

Detailed analysis of the estimated activation energies of both intensity and halflife for all polyepoxide networks show that the values based on halflife data are always smaller than those found for intensity. The observed differences can be attributed to concentration effects, since there is only very minor temperature dependence of absorption and intersystem-crossing (10). This means that by increasing the temperature the concentration of emitting species is decreased, which in turn is determined by the volume fraction of domains with low internal mobility. These results clearly confirm the "two-domain hypothesis" outlined in the previous section.

In order to relate the local molecular mobility with mechanical properties of the individual polyepoxides, we performed Dynamic Mechanical Thermal Analysis on these polyepoxide networks. The frequencies of the DMTA measurements are matched with the halflives of the phosphorescence (10, 1 and 0.1 Hz versus 1-3 sec, respectively). The E-modulus for the two polyepoxide networks as function of temperature is depicted in Fig. 6. Again both show a significant difference in temperature range in which the E-modulus drops. This difference in DMTA data is in accordance with the luminescence results with respect to the activation energies. However, significant discrepancy is found for the values of the transition temperatures obtained by the two techniques. Although differences in technique and sample preparation cannot be ignored, we propose that both measurements demonstrate different transitions, to wit alpha and beta transitions for DMTA and emission spectroscopy, respectively.

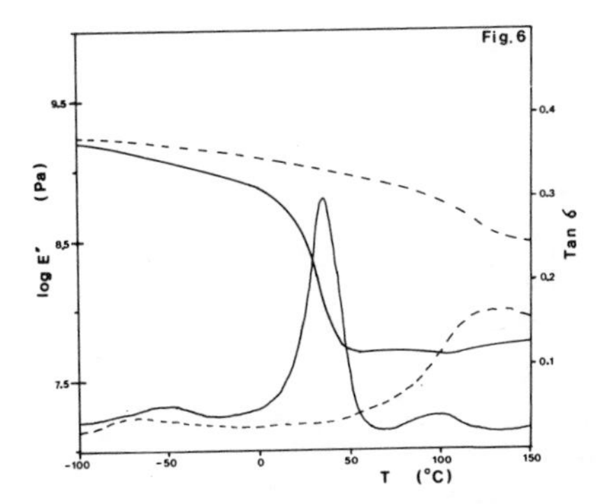

Fig. 6. The temperature dependence of the E-modulus at 1 Hz for 1,2-epoxyethyl-3,4-epoxycyclohexane (dotted line) and 1,4-butanediol-diglycidyl ether (drawn line) networks.

6. CONCLUSION

Through the years the use of luminescence from aromatic species incorporated in linear polymers has proved to be very useful in studying molecular mobility (1). This study shows that also polymer networks with a much more complicated structure can be subjected to this technique. In the photoinitiated cationic polymerization of bisepoxides the probe molecule orgins from the photoinitiator. The results can only be explained when we assume the polymerization being inhomogeneous in nature. A "two-domain hypothesis" is proposed. Activation energies for the local internal mobility could be estimated from the temperature dependence of the luminescence. These results appeared to be supplementary with data obtained for dynamic mechanical properties.

ACKNOWLEDGMENT

The authors gratefully acknowledge Dr. D.M. de Leeuw for discussions, Mr. C.J. de Boon for technical assistance, and Dr. F.J.A.M. Greidanus for the ESR spectroscopy.

REFERENCES

1. A.C. Somersall, E. Dan, and J.E. Guillet, Macromolecules 1974, **7**, 233.
2. K.J. Smit, R. Sakurovs, and K.P. Ghiggino, Eur. Polym. J. 1983, **19**, 49.
4. H. Rutherford and I. Soutar, J. Polym. Sci., Polym. Phys. Ed., 1980, **18**, 1021.
5. H. Morawetz, Science, 1979, **203**, 405.
6. L. Monnerie in "Static and dynamic properties of the polymeric solid state", (Reidel, The Netherlands) 1982, 383.
7. J.V. Crivello, J.H.W. Lam, Macromolecules 1977, **10**, 1307.
8. T. Saegusa, Topics in Current Chemistry 1982, **100**, 75.
9. E.W. Meijer, D.M. de Leeuw, F.J.A.M. Greidanus, and R.J.M. Zwiers, Polymer 1985, **26** (Commun.) 45.
10. N.J. Turro, "Modern Molecular Photochemistry", (Benjamin, Menlo Park) 1978.
11. E.W. Meijer and R.J.M. Zwiers, Macromolecules, submitted.

MOBILITY OF SIDEGROUPS IN POLYDIMETHYLSILOXANE

H.-H. GRAPENGETER[+], B. ALEFELD[++] and R. KOSFELD[+++]

\+ Institut für Physikalische Chemie, Universität Hamburg, Hamburg
++ Institut für Festkörperforschung, KFA Jülich, Jülich
+++ FB 6 - Physikalische Chemie, Universität -GH- Duisburg, Duisburg

SYNOPSIS

In this paper a report is given on investigations of the temperature and frequency dependence of sidegroup, i. e. methyl group, and segmental motions in polydimethylsiloxane by neutron scattering and proton spin-lattice relaxation experiments. It is concluded that besides the classical types of motional processes a quantum mechanical mechanism, the methyl group rotational tunneling, occurs in this polymeric system.

1 INTRODUCTION

The micro-brownian motions a linear polydimethylsiloxane (PDMS) molecule is capable of include methyl group rotation about the C_3-axis, segmental rotation about the Si-O-bond and cooperative segmental motion. Methyl group rotation which is the main objective of this paper comprises three different motional processes: thermally activated rotational hopping, librational transitions and rotational tunneling. Model parameters describing the dynamics of the methyl group are activation energies and correlation times characteristic of the classical thermally activated hindered rotation whereas the quantum mechanical types of motion, i. e. rotational tunneling and librational transitions, are sensitive to the hight and shape of the hindering potential.

2 NUCLEAR MAGNETIC RELAXATION EXPERIMENTS

The temperature dependence of the proton spin-lattice relaxation time T_1 reveals at least three motional processes a polydimethylsiloxane molecule performs in the temperature range 3 K to 300 K. By inspection of fig. 1a (insert) pronounced effects of the thermally activated hindered types of motion can be realized which result in the two T_1-minima I and II. The low temperature minimum is generated by methyl group rotational jumping whereas the high temperature minimum is due to segmental rotation. A further motional process, related to a first order phase transition, is responsible for the T_1-discontinuity in the segmental rotation regime. Residual molecular oxygen still considerably reduces the T_1-values below 35 K. Thus a possible effect of methyl group rotational tunneling on spin-lattice relaxation which is expected to occur in this low temperature region remains latent.

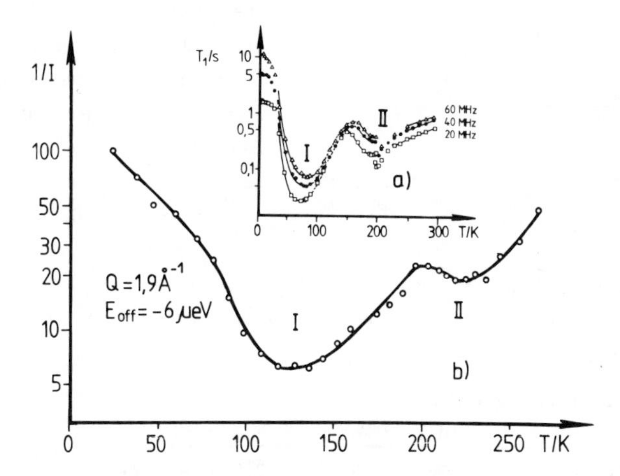

Fig. 1. NMR proton spin-lattice relaxation experiment a), and neutron scattering fixed-window experiment b), on polydimethylsiloxane (weight average polymerization degree: $\langle r \rangle_w = 1780$)

The temperature and frequency dependence of the proton spin-lattice relaxation controlled by thermally activated hindered motional processes is quite well understood by means of the theory of Bloembergen, Purcell and Pound (BPP):

$$T_1^{-1} = C\,[\tau_c/(1+\omega^2\tau_c^2) + 4\tau_c/(1+4\omega^2\tau_c^2)] \qquad \ldots (1)$$

The mean correlation time τ_c is given by an Arrhenius ansatz

$$\tau_c = \tau_c(\infty)\,\exp(\Delta E/RT) \qquad \ldots (2)$$

From these equations an expression is derived which allows to determine the model parameters appearing in eq. 2, the activation energy ΔE and the mean correlation time at infinite temperature $\tau_c(\infty)$, from the frequency dependent shift of the T_1-minima temperature positions (T_E).

$$\ln\omega = -(\Delta E/R)/T_E - \ln[\tau_c(\infty)/0.62] \qquad \ldots (3)$$

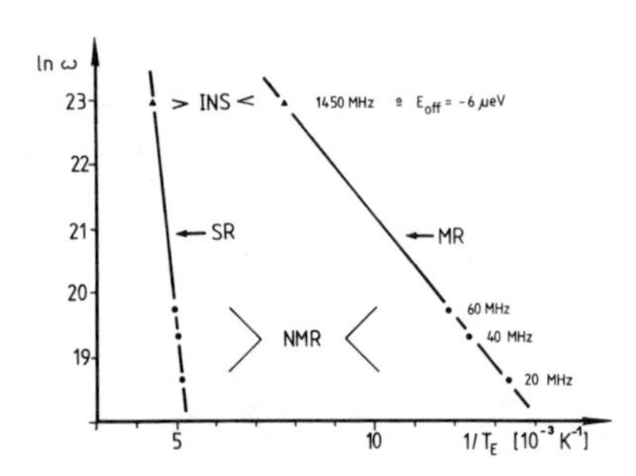

Fig. 2. Arrhenius plot from incoherent neutron scattering (INS) and nuclear magnetic resonance (NMR) data

Figure 2 shows an Arrhenius plot of the measuring radian frequency vs. $1/T_E$. The lower part represents the NMR (ω, T_E)-data drawn from fig. 1a. According to eq. 3 the experimental data lie on straight lines labelled as MR and SR referring to methyl group and segmental rotation respectively. Hence the model parameters follow to be $\Delta E^{MR} = 6.4$ $kJ \cdot mol^{-1}$ and $\tau_c^{MR}(\infty) = 1.7 \cdot 10^{-13}$ s for the methyl group hopping process and $\Delta E^{SR} = 50.2$ $kJ \cdot mol^{-1}$ and $\tau_c^{SR}(\infty) = 1.4 \cdot 10^{-22}$ s for segmental rotation. The incoherent neutron scattering (INS) data entering the upper part of fig. 2 will be discussed in the following section.

3 NEUTRON SCATTERING EXPERIMENTS

3.1 Thermally Activated Hindered Motional Processes

Supplementary investigations on this classical type of motions going beyond the NMR-frequency region are feasible by a neutron scattering fixed-window technique characterized by a definite spectrometer energy-offset $E_{off} \neq 0$ and variable temperature intensity scanning mode. The result of this experiment being complementary to the NMR T_1-measurements is shown in fig. 1b. Again methyl group and segmental rotation are manifested by the two distinct reciprocal scattering intensity minima I and II respectively. An effect analogous to the T_1-discontinuity is not recognizable in this experiment.

The analogy between the NMR and fixed-window experiment is evidenced by the simplified incoherent scattering law which is proportional to the scattering intensity I. Omitting the elastic part of the scattering function and taking into account only the quasielastic term a truncated expression for the scattering intensity follows

$$I = B[1 - A(Qr)]\,\tau/(1 + \omega^2\tau^2) \qquad \ldots (4)$$

which is essentially given by a single Lorentzian. The average jump time τ

again obeys an Arrhenius type equation

$$\tau = \tau(\infty)\, \exp(\Delta E/RT) \qquad ...(5)$$

From these equations an expression analogous to eq. 3 follows by the same arguments used in the NMR case.

$$\ln\omega = -(\Delta E/R)/T_E - \ln\tau(\infty) \qquad ...(6)$$

The (ω, T_E)-values taken from fig. 1b fit well to the NMR data in fig. 2 thus confirming that the motional processes under investigation obey an Arrhenius law over a wide frequency range. Furthermore the fixed-window experiment in contrast to other neutron scattering experiments[1,2] yields the same value of activation energy like the NMR spin-lattice relaxation experiment.

3.2 Methyl Group Rotational Tunneling

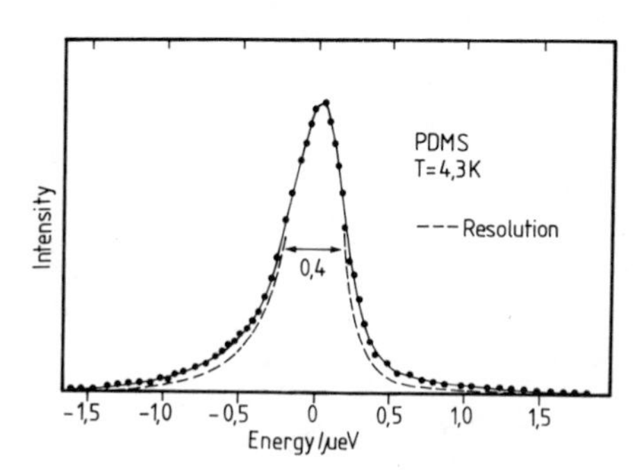

Fig. 3. Quasielastic neutron scattering spectrum of polydimethylsiloxane (weight average polymerization degree: $\langle r \rangle_w = 1780$)

First evidence of this quantum mechanical type of motion is given by a quasielastic neutron scattering experiment performed at the ILL, Grenoble. The result is given in fig. 3. The deviations of the experimental data from the resolution curve indicate a low temperature motional process which is interpreted to be methyl group rotational tunneling. From the extended spreading of the experimental points over the wings of the spectrum a broad distribution of the rotational barrier hights is deduced. This fact is confirmed by inelastic neutron scattering experiments done only recently on oligomeric dimethylsiloxanes[3] at the Jülich reactor.

ACKNOWLEDGEMENTS

The authors wish to thank Dr. A. Heidemann for his assistance in performing the neutron scattering experiment at the ILL, Grenoble. Thanks are also due to the Bundesministerium für Forschung und Technologie of the Federal Republic of Germany for the financial support to carry out this work.

REFERENCES

1. Allen, G. and Higgins, J.S., Rep. Progr. Phys., vol. 36, 1073-1133, June 1973

2. Amaral, L.Q., Vinhas, Z.A. and Herdade, S.B., J. Polym. Sci. - Polym. Phys. Ed., vol. 14, 1077-1085, June 1976

3. Alefeld, B., Grapengeter, H.-H., Kosfeld, R. and Prager, M., Report on the Workshop on Nuclear Solid State Research, Maria Laach, FR Germany, Oct. 1984

NOTATION

C,B	constants
ω	measuring radian frequency
R	gas constant
A(Qr)	structure factor
Q	momentum transfer
r	jump distance

GLASS TRANSITIONS IN UNSYMMETRICALLY SUBSTITUTED SILOXANES

R. KOSFELD, M. HEß, Th. UHLENBROICH

University of Duisburg, FB-6, Physical Chemnistry, D-4100 Duisburg, FRG

SYNOPSIS

A series of unsymmetrically substituted polysiloxanes has been synthesized and the glassy transition region has been investigated by means of calorimetry and dynamic-mechanical torsional analysis. Slope parameters of the glass transition process were calculated. Qualitative estimations on the energy of activation of transition process were achieved.

1 INTRODUCTION

The polymers synthesized were of the following general structure:

$$HO\left[\begin{array}{c} CH_3 \\ | \\ Si \\ | \\ R \end{array} - O\right]_n H \qquad R = -C_2H_5,\ -C_6H_5,\ -(CH_2)_2-C_6H_5$$

The empirical expression basing upon the corresponding temperature concept allows some closer insight into temperature depending behaviour of the processes taking place in the vicinity of glassy transition if applied to results which are obtained from dynamic-mechanical experiments[1].

From the original equation

$$\log G' = \frac{1}{2}\left\{\log G_1' \ G_2' + \log (G'/G_2') \ \mathrm{erf}\left[h_w \ \log (w/k)\right]\right\} \quad . . . (1)$$

the more useful eq. 2 may be derived by means of the well known frequency-temperature correspondence[1-4]:

$$\frac{\log \frac{G'}{G'_2}}{\log \frac{G'_1}{G'_2}} = \frac{1}{2}\left\{1 + \text{erf}\left[h_T\left(\frac{1}{T} - \frac{1}{T_o}\right)\right]\right\} \quad \ldots (2)$$

T_o is the temperature where $G' = \sqrt{G'_1 \; G'_2}$ is valid.
Plotting the left side of eq. 2 against $\frac{1}{T} - \frac{1}{T_o}$ on normal probability paper a straight line may be expected. The slope of the function should be represented by one slope parameter h_T of the temperature dispersion.
Comparison of the slope parameters of the different polymers synthesized gives information about the width of distribution of the chain segment motion setting in.
In practice it has been found that two slope parameters are necessary for the description of several systems, e. g. in the case of some epoxides[3].

Usually the change of the slope occurs at the corresponding temperature T_o where the ordinate has a value of approximately 50 %. A decrease in the steepness of the slope corresponds to an increase in motional dispersion distribution.
Further information on the beginning motions in the region of the glassy transition may be revealed from activation energy E_A of the process. According to Read, Williams[5] and Müller[6].

$$E_A = \frac{\pi R}{2} \frac{\Delta G'}{\int_o^\infty G''.\, d(\frac{1}{T})} \quad \ldots (3)$$

is valid if temperature depending measurements of free torsion experiments at a constant frequency are under the scope.
Application of eq. 3 is, however, restricted by some limitations[7]. The most important point is that E_A is considered to be temperature independent within the region under survey. Another point is the precision the integral can be evaluated with. In this work this evaluation was performed by means of a planimeter. As a consequence of the limitations mentioned above the calculated values of E_A cannot be looked upon as exact values but as relativ quantities which nevertheless are capable of giving ideas on the behaviour of the polymers under survey relativ to eachother.

RESULTS

The measurements were performed using a convenient torsional pendulum (Fa. Myrenne, D-5106 Roetgen) working at a constant frequency of 1 Hz and a

Perkin-Elmer DSC-2.

The results of the measurements are given in the following table 1 and in figure 1. All polymers expect polydimethylsiloxane showed no effects of crystalinity.

Table 1

R	$T_g/^{o}C$	ε_A/kJ mol^{-1}	glass-transition region/^{o}C
$-CH_3$	- 124	22	45
$-C_6H_5$	- 20	6	40
$-CH_2-C_6H_5$	- 40	18	30
$-CH_2-CH_2-C_6H_5$	- 52	16	45
$-CH_2-CH_3$	- 110	8	60

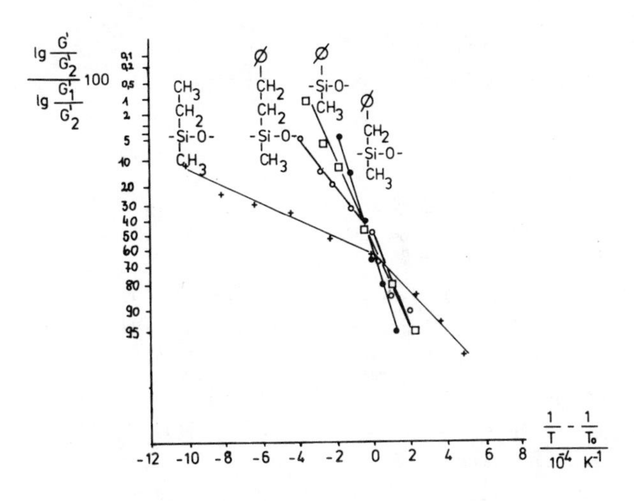

Fig. 1. Plot of the temperature dependence of modulus data according to eq. 2

The outer shape of the modul functions seem to be independent of chain length and degree of crosslinking just as it is known from natural rubber.

Additional dispersion areas observable in phenylsubstituted siloxanes. A flexible spacer as e. g. an ethyl group leads to a more pronounced effect than in the case of the directly linked phenyl group. These processes seem to be related to the ß-process found by Illers and Jenckel[8,9] in polystyrene.

The steepness of the glassy transition as revealed in fig. 1 is almost equal below the corresponding temperature T_o in the case of ring substituted siloxanes. At values above other values of the slope parameter become governing than below, especially in the case of the ethyl spacer.
The ethyl substituent shows a quite different behaviour in general. The flat slope above and below T_o indicates broader dispersion of beginning chain motion. Temperature dependence of main chain motion is influenced by the flexible sidechain even in systems of high main chain mobility.
The results from the analysis of activation energy E_A in the vicinity of crystalline polydimethylsiloxane which lies in the order of magnitude of a partially crystalline polyethylene[10]. It seems that the interactive ring system is able to enhance chain stiffness if it is in some sense decoupled from primary chain motions.

ACKNOWLEDGEMENTS

The authors want to thank Prof. W. Borchard and Prof. E.A. Hemmer for helpfull discussions and the Deutsche Forschungsgemeinschaft for financial support.

REFERENCES

1. Tobolski,A.V., Hoffmann, M.,"Mechanische Eigenschaften und Struktur von Polymeren", Stuttgart, 1967, 185 ff,219 ff
   Tobolski, A.V., McLaoughlin, J.R., J. Polymer Sci. 8, 1952, 543
2. Shibayama, K., Suzuki, Y., J. Polym. Sci. A, 3, 1965, 267
3. Kwei,T.K., J. Polym. Sci. A2, 4, 1966, 943
4. Theocaris, P.S., J. Appl. Polym. Sci. 8, 1964, 399
5. Read, B.E., Williams, G.,Trans. Faraday Soc. 57, 1961, 1973
6. Müller, F.H., Z. Elektrochem. 65, 1961, 152
7. Posselt, K., Kolloid Zt., Z. Polymere 223, 1968, 104
8. Illers, K.H., Jenckel, E., Kolloid Zt. 165(1), 1959, 73
9. Illers, K.H., Jenckel, E., Rheol. Acta 1, 1958, 322
10. Illers, K.H., Kolloid Zt., Z. Polymere 231, 1961, 622

Notation:

G' - storage modulus
G'' - loss modulus
h_w, h_T - slope parameters
ω - frequency

Part VII

PROCESSING/RHEOLOGY

FROM MOLECULAR MODELS TO THE SOLUTION OF FLOW PROBLEMS

L. E. WEDGEWOOD[*] and R. BYRON BIRD[*]

[*]Chemical Engineering Department and Rheology Research Center,
University of Wisconsin, Madison, WI U.S.A. 53706

SYNOPSIS

One of the goals in polymer fluid dynamics is to use kinetic theory to derive a constitutive equation for a polymeric liquid starting from a molecular model, and then to use the constitutive equation to solve flow problems. We illustrate this procedure by using a finitely extensible nonlinear elastic (FENE) dumbbell as a crude model of a polymer molecule and then derive an approximate constitutive equation. The latter can be used, along with the equations of continuity and motion, to solve fluid dynamics problems. By "interfacing" the constitutive equation with appropriate continuum-mechanics expressions, some problems can be solved at once by taking over published solutions. This technique is illustrated for squeezing flow between circular disks, the Weissenberg rod-climbing effect, and the torque on a rotating sphere.

1 MOLECULAR MODEL TO CONSTITUTIVE EQUATION

We model the polymer molecule by a "FENE dumbbell" consisting of two "beads" (each of which has a friction coefficient ζ) connected by a nonlinear spring with a connecting force $\tilde{F}^{(c)}$ given by:[1,2]

$$\tilde{F}^{(c)} = H\,\tilde{R}/[1 - (R/R_0)^2] \tag{1}$$

Here $\tilde{R}$ is the bead-to-bead vector, H is a spring constant, and R_0 is the maximum extension. For this force law it is not easy to solve the kinetic-theory equations; however, if one uses a modification of the "Peterlin approximation"[2], an approximate constitutive equation can be derived:[3]

$$\underline{\underline{\tau}} = - \eta_s \; \underline{\underline{\gamma}}_{(1)} + \underline{\underline{\tau}}_p \tag{2a}$$

$$Z\underline{\underline{\tau}}_p + \lambda_H \underline{\underline{\tau}}_{p(1)} - \lambda_H\{\underline{\underline{\tau}}_p - [b/(b+2)] \; a \; \underline{\underline{\delta}}\} \; D \; \ell n \; Z/Dt$$

$$= - [b/(b+2)] \; a \; \lambda_H \underline{\underline{\gamma}}_{(1)} \tag{2b}$$

$$Z = 1 + (3/b)\{[b/(b+2)] - [(\mathrm{tr} \; \underline{\underline{\tau}}_p)/3a]\} \tag{2c}$$

Here $\underline{\underline{\tau}}$ is the stress tensor of the polymer solution, $\underline{\underline{\tau}}_p$ is the polymer contribution, $\underline{\underline{\gamma}}_{(1)} = \nabla\underset{\sim}{v} + (\nabla\underset{\sim}{v})^\dagger$ is the rate of strain tensor, $\underline{\underline{\tau}}_{p(1)}$ is the "convected derivative" of the stress tensor,[1] D/Dt is the "substantial derivative",[1] and $\underline{\underline{\delta}}$ is the unit tensor.

The constitutive equation contains four parameters: η_s, the solvent viscosity; $\lambda_H = \zeta/4H$, a characteristic relaxation time; $b = HR_0^{\,2}/kT$, a ratio of the elastic spring energy to the thermal energy; and a = nkT, where n is the number density of polymer molecules. Although these four quantities do have molecular interpretations, we treat all of them as adjustable constants to be determined from rheometric data. That is, the molecular theory is used to suggest a useful form for the constitutive equation, and experiments determine the values of the constants.

2 DIRECT SOLUTIONS TO FLOW PROBLEMS

An earlier version of eq. 2 based on the original Peterlin approximation (this corresponds to replacing [b/(b+2)] by 1 in three places in eqs. 2b and 2c) has been solved along with the equations of continuity and motion by Mochimaru[4,5] for two flow problems: (i) fast squeezing flow between circular disks; and (ii) start up of Couette flow. The latter problem is of interest because a "velocity-overshoot effect" is predicted; however, no experimental data are available to verify the correctness of the prediction.

3 "INTERFACING" THE CONSTITUTIVE EQUATION WITH RESTRICTED CONTINUUM-MECHANICS EXPRESSIONS

Continuum mechanics provides several expressions for $\underline{\underline{\tau}}$ that are general in that they apply to a very wide class of viscoelastic fluids, but only within certain restricted flow regimes. Two examples are:

(1) The Criminale-Ericksen-Filbey (CEF) equation[1] for steady shear flow:

$$\underline{\underline{\tau}} = - \eta(\dot{\gamma})\underline{\underline{\gamma}}_{(1)} + \frac{1}{2}\Psi_1(\dot{\gamma})\underline{\underline{\gamma}}_{(2)} - \Psi_2(\dot{\gamma})\{\underline{\underline{\gamma}}_{(1)} \cdot \underline{\underline{\gamma}}_{(1)}\} \tag{3}$$

where η, Ψ_1, and Ψ_2 are the viscosity, first-normal-stress coefficient, and second-normal-stress coefficient, all of which are functions of the shear rate $\dot{\gamma} = \sqrt{(1/2)(\underline{\underline{\gamma}}_{(1)} : \underline{\underline{\gamma}}_{(1)})}$.

(2) The retarded-motion expansion (RME) for slow flows, slowly varying in space and time:

$$\underline{\underline{\tau}} = -[b_1\underline{\underline{\gamma}}_{(1)} + b_2\underline{\underline{\gamma}}_{(2)} + b_{11}\{\underline{\underline{\gamma}}_{(1)} \cdot \underline{\underline{\gamma}}_{(1)}\}$$
$$+ b_3\underline{\underline{\gamma}}_{(3)} + b_{12}\{\underline{\underline{\gamma}}_{(1)} \cdot \underline{\underline{\gamma}}_{(2)} + \underline{\underline{\gamma}}_{(2)} \cdot \underline{\underline{\gamma}}_{(1)}\} + b_{1:11}(\underline{\underline{\gamma}}_{(1)} : \underline{\underline{\gamma}}_{(1)})\underline{\underline{\gamma}}_{(1)} + \ldots] \quad (4)$$

in which the b's are constants.

From eq. 2 it can be shown that the functions in the CEF equation for the FENE dumbbell model are:

$$\eta - \eta_s = 3a\lambda_H[b/(b+5)](\lambda'\dot{\gamma})^{-1} \sinh[(1/3) \operatorname{arcsinh} \lambda'\dot{\gamma}] \xrightarrow{\dot{\gamma} \to \infty} m\,\dot{\gamma}^{n-1} \quad (5)$$

$$\Psi_1 = 2(\eta - \eta_s)^2/[ab/(b+2)] \xrightarrow{\dot{\gamma} \to \infty} m'\,\dot{\gamma}^{n'-2} \quad (6)$$

in which $\lambda' = \sqrt{27/2}\,[(b+2)/(b+5)^{3/2}]\lambda_H$ and

$$m = a\lambda_H[b\sqrt{b/2} \,/\, [(b+2)\lambda_H]]^{2/3} \quad ; \quad n = 1/3 \quad (7)$$

$$m' = a\,\lambda^2_H[(b+2)/b][b\sqrt{b/2} \,/[(b+2)\lambda_H]]^{4/3} \quad ; \quad n' = 2/3 \quad (8)$$

and that $\Psi_2 = 0$.

Also, from eq. 2 the constants in the RME corresponding to the FENE dumbbell model are:

$$b_1 = \eta_s + a\lambda_H\,[b/(b+5)] \quad (9)$$

$$b_2 = -\,a\lambda_H^{\;2}\,[b(b+2)/(b+5)^2] \quad , \quad b_{11} = 0 \quad (10)$$

$$b_3 = a\,\lambda_H^{\;3}\,[b(b+2)^2/(b+5)^3] = -\,(b+5)b_{1:11} \quad , \quad b_{12} = 0 \quad (11)$$

Note that $b_2 < b_{11}$ and that $b_3 > b_{12} > b_{1:11}$; these inequalities have been found for almost all molecular models and kinetic theories studied so far.

4 FAST SQUEEZING FLOW BETWEEN CIRCULAR DISKS

Figure 1 shows Leider's[6] data on η and $\tau_{xx} - \tau_{yy} = -\Psi_1\dot{\gamma}^2$ for a 5.2% PIB solution along with the curves calculated from eq. 2 for the parameters: $\eta_s = 0$, $\lambda_H = 1370_s$, $a = 133$ Pa , and $b = 0.24$.

Figure 2 shows Leider's data for squeeze flow between parallel disks[6] given as the time $t_{1/2}$ for squeezing out one-half of the liquid under an imposed force F; the squeezing time is made dimensionless by dividing by a time constant $\lambda = (m'/2m)^{[1/(n'-n)]}$. The initial dimensions of the cylindrical fluid sample are R(radius) and $2h_0$ (height). The dashed lines, labelled "Scott Equation", are obtained from eq. 3 with $\Psi_1 = \Psi_2 = 0$ (generalized Newtonian fluid), and the solid curves result from the CEF equation with η and Ψ_1 obtained from eqs. 7 and 8 with the parameters given above; the latter curves are obtained from eqs. 35 and 36 of McClelland and Finlayson[7], which are based on the CEF equation (although the authors did not state that they used this equation). The agreement between experimental data and molecular-theory curves is satisfactory.

5 THE WEISSENBERG ROD-CLIMBING EXPERIMENT

When a circular rod of radius R rotates with angular velocity W in a beaker of polymeric liquid of density ρ, the height of the liquid surface at a distance r from the rod axis is[8]:

$$h(r,W) = h_0(r) - \frac{R^4}{\rho g}\left[\frac{2}{r^4}(b_2 - 2b_{11}) + \frac{\rho}{2r^2}\right]W^2 + \ldots \qquad (12)$$

where $h_0(r)$ is the "static climb" (including meniscus effect), and g is the gravitational acceleration. Using the η and Ψ_1 data on PMMA solutions of Joseph et al.[8] (figs. 3 and 4), we determined the FENE-dumbbell parameters and also $(b_2 - 2b_{11})$ from eq. 10; from their rod-climbing data, they found the values of $(b_2 - 2b_{11})$ from eq. 12. The results are given in table 1. The agreement between theory and experiment is satisfactory for the higher concentrations.

6 THE TORQUE ON A SLOWLY ROTATING SPHERE

When a sphere of radius R rotates with an angular velocity W in a fluid of density ρ, the torque $\mathcal{T}$ required is:

$$\frac{\mathcal{T}}{8\pi R^3} = b_1 W + \left[\frac{(\rho R^2)^2}{1200} - \frac{(b_2 + b_{11})\rho R^2}{140} - \frac{2(b_2 + b_{11})^2}{15} + \frac{24 b_1 (b_{1:11} - b_{12})}{5}\right]\frac{W^3}{b_1} + \ldots \qquad (13)$$

Using the viscosity data of Walters and Savins[10] for a 1.5% aqueous solution of polyacrylamide (fig. 5), we determined two groupings of molecular parameters,

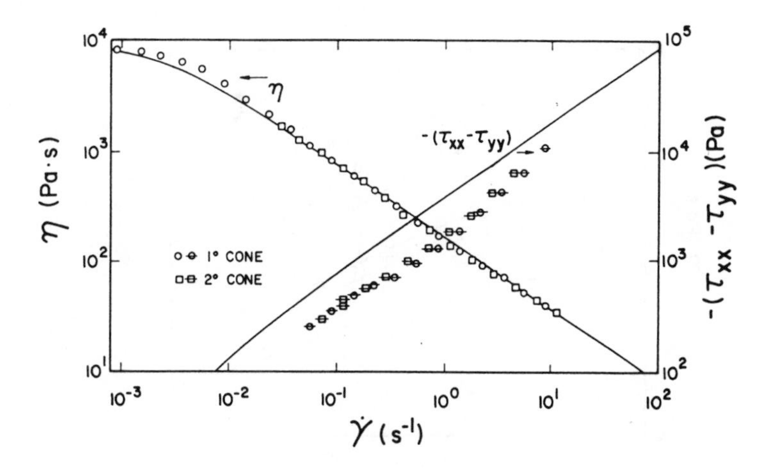

Fig. 1 Rheometric data for a PIB solution[6] compared with FENE-dumbbell functions.

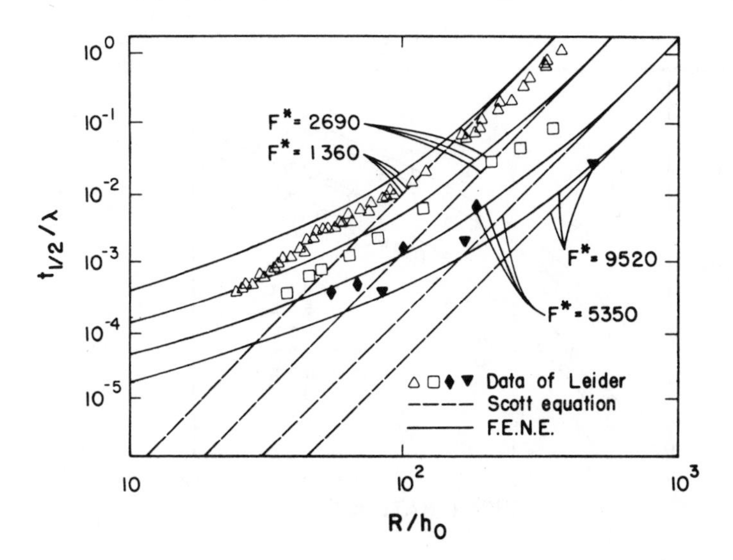

Fig. 2 Squeezing flow data[6] along with FENE-dumbbell predictions based on data from fig. 1. The dimensionless force F^* is defined as $F^* = F\,\lambda^n/\pi R^2\,m$ in which F is the applied force.

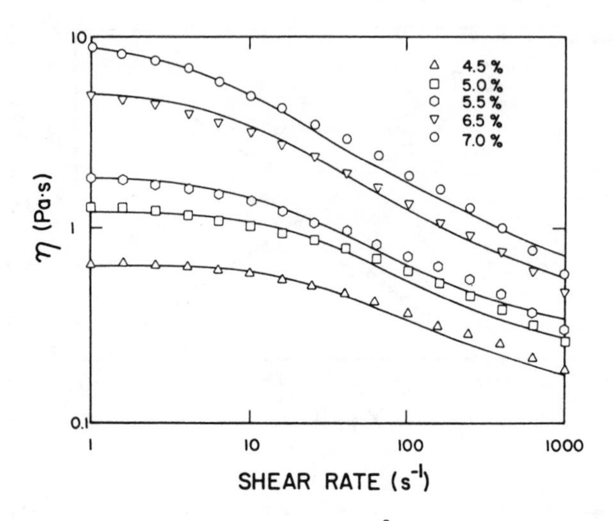

Fig. 3 Viscosity of PMMA solutions[8] along with FENE-dumbbell curves.

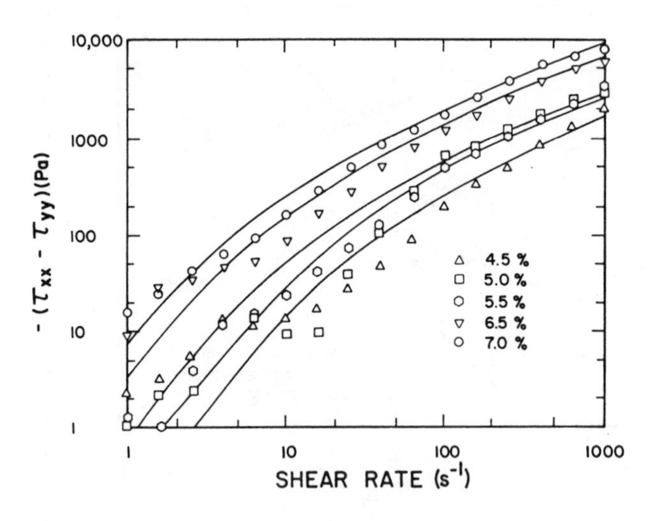

Fig. 4 First normal stress difference for PMMA solutions[8] along with FENE-dumbbell curves.

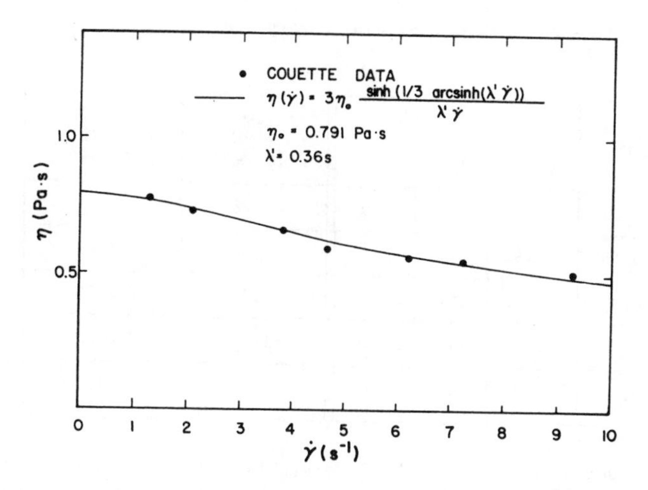

Fig. 5 Viscosity data for a polyacrylamide solution[10] along with FENE-dumbbell curve. The zero shear rate viscosity η_0 is defined as $\eta_0 = a\ \lambda_H b/(b+5)$.

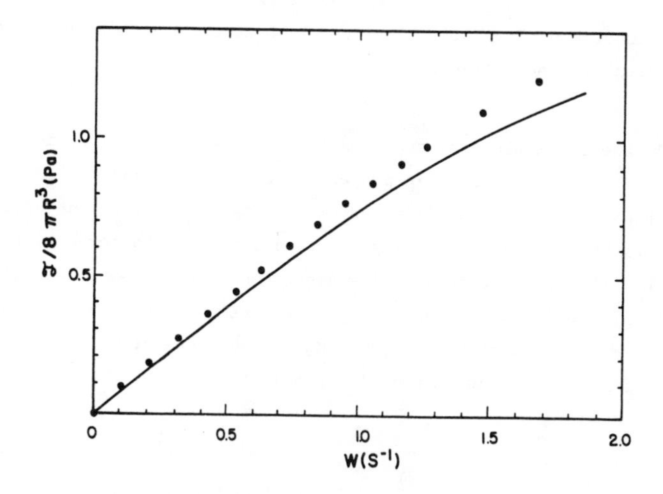

Fig. 6 Torque on rotating sphere in the polyacrylamide solution[10] of fig. 5, along with the FENE-dumbbell prediction.

Table 1:

The grouping $(b_2 - 2b_{11})$ from molecular theory and from rod-climbing

Polymer concentration (%)	η_0 (Pa·s)	λ_H (s)	a (Pa)	b (-)	Eq.10 $(b_2 - 2b_{11})$ (g/cm)	Eq. 12 $(b_2 - 2b_{11})$ (g/cm)
4.5	0.13	0.157	3.58	47.9	0.76	0.13
5.0	0.19	0.192	5.94	38.3	1.80	0.76
5.5	0.25	0.302	5.98	30.6	4.31	1.71
6.5	0.38	0.447	11.8	25.0	17.8	14.7
7.0	0.45	0.553	19.1	16.0	38.1	36.2

namely $\lambda' = 0.36$ s and $a\lambda_H[b/(b+5)] = 0.791$ Pa·s, taking η_s to be zero. Since no Ψ_1 data were available the values of η_s , λ_H , b , and a could not be determined uniquely. From the grouping of constants above, we determined b_1 and $b_{1:11}$ (b_{11} and b_{12} are zero for FENE dumbbells and b_2 was assumed to be negligible from information on the streamlines[10]). Then the torque was calculated from eq. 13 as a function of W; the resulting curve, from molecular theory, is shown in fig. 6 along with experimental values. The agreement is regarded as satisfactory in view of the inadequate rheometric data.

7 CONCLUSIONS

The FENE-dumbbell, although a crude model of a polymer molecule, seems to be capable of describing some rheometric properties fairly well and, in addition, is compatible with data from several flow systems. Complete evaluations of the FENE-dumbbell constitutive equation (or any other equation) is hampered by the lack of experimental rheometric data and flow data, both for the same test fluid. The discussion given here is not intended to promote the FENE-dumbbell model, but rather to illustrate the importance of making the connection:
molecular model → constitutive equation → flow problems.

ACKNOWLEDGMENTS

We thank the National Science Foundation (grant No.: CPE-8104705), the Vilas Trust Fund of the University of Wisconsin, and a MacArthur Professorship for financial support. Also we acknowledge Mr. X. J. Fan of Zhejiang University for assistance in the early part of this work.

REFERENCES

1. Bird, R. B., Hassager, O., Armstrong, R. C., and Curtiss, C. F., "Dynamics of Polymeric Liquids, Vol. 2, Kinetic Theory", Wiley, 1977; section 10.5 and Appendix C.
2. Bird, R. B., Dotson, P. D., and Johnson, N. L., J. Non-Newtonian Fluid Mech., Vol. 7, 213-235, 1980. [Because of an error in eq. 58, pp. 224-234 are invalidated.]
3. Fan, X. J., J. Non-Newtonian Fluid Mech., 1985 (accepted for publication).
4. Mochimaru, Y., J. Non-Newtonian Fluid Mech., Vol. 9, 157-178, 1981.
5. Mochimaru, Y., J. Non-Newtonian Fluid Mech., Vol. 12, 135-152, 1983.
6. Leider, P. J. and Bird, R. B., Univ. Wisc. Rheol. Res. Center Report No. 22, 1973.
7. McClelland, M. A. and Finlayson, B. A., J. Non-Newtonian Fluid Mech., Vol. 13, 181-201, 1983.
8. Joseph, D. D., Beavers, G. S., Cers, A., Dewald, C., Hoger, A., and Than, P. T., J. Rheol., Vol. 28, 325-345, 1984.
9. Giesekus, H., Rheol. Acta, Vol. 3, 59-71, 1963.
10. Walters, K. and Savins, J. G., Trans. Soc. Rheol., Vol. 9, No. 1, 407-416, 1965.

APPENDIX

The FENE-dumbbell model was used here for illustrative purposes because of its simplicity. It should be pointed out that much more complex molecular models have been investigated. For example, for polymer melts the Kramers freely-jointed bead-rod chain has been used (for a short summary of the method and further literature references see R. B. Bird and C. F. Curtiss, Physics Today, Vol. 37, No. 1, 36-43 (1983) and Nederlands Tijdschrift voor Natuurkunde, Vol. A47, No. 4, 133-136 (1981)). For polymer melts it is necessary to take into account the anisotropic hydrodynamic drag forces as well as the anisotropic Brownian motion forces, in order to account for the fact that a polymer molecule in a melt is constrained by the presence of neighboring molecules. This theory, which gives an integral constitutive equation containing four constants can describe many rheological properties reasonably well. It is currently being used in the solution of flow problems (see D. S. Malkus and B. Bernstein, J. Non-Newtonian Fluid Mechanics, Vol. 16, Nos. 1 and 2, 77-116 (1984)).

TRANSIENT - NETWORK THEORIES: NEW DEVELOPMENTS AND APPLICATIONS

R.J.J. Jongschaap* and H. Kamphuis†

*Twente University, Enschede, The Netherlands
†K.S.E.P.L.-Laboratory, Rijswijk, The Netherlands

SYNOPSIS

A generalized formulation is presented of the so-called transient network model for the rheological behaviour of polymeric liquids. This formulation is shown to be more flexible and applicable to other types of systems as well. Furthermore this formulation is shown to be closely related to the theory of Doi and Edwards.

1 INTRODUCTION

In the transient-network model[1,2], the polymer molecules are described by a network of segments which may be created and lost during the flow.
In the present paper a generalized formulation of this model is proposed, applied to various types of systems and also discussed in relation with the theory of Doi and Edwards[3].

2 GENERAL FORMULATION

A segment, present (or created) at time t with an end-to-end vector $\underset{\sim}{q}$ will be indentified as a $(\underset{\sim}{q}, t)$- segment. The density of such segments at time t is given by the distribution function $\Psi(\underset{\sim}{q}, t)$.
The macroscopical stress tensor is given by the average

$$\underline{T} = \int \Psi(\underset{\sim}{q}, t)\ \underline{f}\,\underset{\sim}{q}\ d^3\underset{\sim}{q}\,, \tag{1}$$

in which $\underline{f} = \underline{f}(\underset{\sim}{q})$ denotes the force in a segment. Instead of eq. 1 we will use the expression

$$\underline{T} = \int_{-\infty}^{t}\int \tilde{\Psi}_t^{t'}(\underline{q}')\ \underline{f}_t^{t'}(\underline{q}')\ \underline{q}_t^{t'}(\underline{q}')\ d^3\underline{q}'\ dt' \ , \tag{2}$$

in which $\tilde{\Psi}_t^{t'}(\underline{q}')\,dt'$ is the density of ($\underline{q}'$, t')- segments, created in the interval [t', t'+dt'] and still present at time t. The functions

$$\underline{f} = \underline{f}_t^{t'}(\underline{q}') \ ; \ \underline{q} = \underline{q}_t^{t'}(\underline{q}') \tag{3}$$

will be specified later on.
We introduce a creation function g(t') by assuming that

$$\tilde{\Psi}_t^{t'}(\underline{q}') = N_0\ g(t')\ \tilde{\psi}(\underline{q}') \ , \tag{4}$$

in which N_0 is the segment density at rest and $\tilde{\psi}(\underline{q}')$ a normalized creation distribution function (i.e. $\int\tilde{\psi}(\underline{q}')d^3\underline{q}' = 1$). The annihilation of segments is described by an annihilation function h($\underline{q}'$, t', t), defined by the equation

$$\frac{\partial\tilde{\Psi}_t^{t'}(\underline{q}')}{\partial t} = -\,h(\underline{q}', t', t)\ \tilde{\Psi}_t^{t'}(\underline{q}') \ . \tag{5}$$

In the simplest case the motion of the segments is "affine", which means that

$$\underline{q}_t^{t'}(\underline{q}') = \underline{F}_t^{-1}(t')\cdot\underline{q}' \ , \tag{6}$$

where $\underline{F}_t(t') = \partial\underline{x}(t')/\partial\underline{x}(t)$ is the macroscopical deformation-gradient-tensor. A special kind of nonaffine deformation is the so-called[4] "partially extending nonaffine convection". Instead of eq. 6 we then have:

$$\underline{q}_t^{t'}(\underline{q}')\ (1 + \Lambda) = \underline{F}_t^{-1}(t')\cdot\underline{q}' \ , \tag{7}$$

where Λ is a function, to be specified later on, which determines a "contraction" of the segments with respect to affine motion.
In colloidal systems (e.g. concentrated dispersions) the forces $\underline{f}$ may be highly nonlinear functions of the segment vector $\underline{q}$. In general we may write:

$$\underline{f} = \underline{f}(\underline{q}) = \underline{f}(\underline{q}_t^{t'}(\underline{q}')) \equiv \underline{f}_t^{t'}(\underline{q}') \ . \tag{8}$$

In some cases, an example of which will be discussed below, $\underline{f}$ may even depend upon the deformation history of the segment.

3 INCEPTION OF STATIONARY FLOW

We consider a stationary flow starting at time $t' = 0$. The integration in eq. 2 can then be split up into two parts: the interval $-\infty<t'<0$ in which the system is in a state of rest and the interval $0<t'<t$ in which stationary flow occurs. From the equations 4 and 5 and a few additional assumptions it can be shown[5] that in this flow

$$\mathring{\Psi}_t^{t'}(\underline{q}') = g(t')\, \Psi_{t-t'}^{0}(\underline{q}') , \tag{9}$$

where $\Psi_t^0(\underline{q}') = \int_{-\infty}^{0} \mathring{\Psi}_t^{t'}(\underline{q}')\, dt'$ is the density of $(\underline{q}', 0)$-segments that are still present at time t. By using eq. 2 the following expression for the stress response is obtained[5]:

$$\underline{T} = \underline{T}_t^0 + \int_0^t g(t')\, \underline{T}_{t-t'}^0\, dt' , \tag{10}$$

in which

$$\underline{T}_t^0 = \int \Psi_t^0(\underline{q}')\, \underline{f}_t^0(\underline{q}')\, \underline{q}_t^0(\underline{q}')\, d^3\underline{q}' . \tag{11}$$

These expressions may be used in numerical calculations on the stress response upon the inception of steady shear flow. Some results of such calculations will be discussed in the next section.

4 APPLICATIONS

In dispersions a transient network of dispersed particles may exist. In this case we will assume that $|\underline{f}|=0$ for $|\underline{q}|<q_0$ and $|\underline{f}|=k|\underline{q}-\underline{q}_0|$ for $|\underline{q}|>q_0$, where k is a constant and q_0 the stretched length of a chain consisting of dispersed particles. The creation of segments is assumed to take place at a constant rate with $g = g_0 = h_0 = \text{const.}$ and a spherical symmetric creation distribution $\mathring{\psi}(\underline{q}') = (1/4\pi q_0^2)\, \delta(q - q_0)$. The loss of segments is described by an annihilation function of the Arrhenius-type: $h = h_0 \exp(\Delta E/kT)$ in which $\Delta E = 0$ for $q < q_0$ and $\Delta E = \frac{1}{2} k (\underline{q} - \underline{q}_0)^2$ for $q>q_0$, the energy stored in the chains.
The segmental motion is assumed to be described by eq. 7 with

$$\Lambda = \lambda \frac{q_0}{q} \left(\frac{N_0 - N_t^0}{N_0}\right) f \quad , \tag{12}$$

in which λ is a constant and $N_t^0 = \int \Psi_t^0 (\underline{q}')\, d^3\underline{q}'$. Although the assumptions introduced so far of course are crude simplifications, it turns out[5] that the numerical results obtained from eqs 10 and 12, in this case, are qualitatively in accordance with experimental results.

Systems like protein gels sometimes are built up of a network of filaments consisting of a continuous viscoelastic medium. Such systems can also be treated with the present model. In this case no creation and loss of segments is assumed to take place, so $g = h = 0$. The motion is assumed to be affine (eq. 6) and the medium in the segments is modelled by a generalized Maxwell model. The flow in the segments is an uniaxial elongational flow with an elongational rate $\dot{\varepsilon}(t) = \dot{q}(t)/q(t)$. The results[6] of a calculation of the stress response upon the inception of steady shear flow show a significant overshoot. This is interesting, since a homogeneous Maxwell fluid does not show any overshoot at all in this experiment.

As a third application of our model we consider the Doi and Edwards theory[3] of dense polymeric liquids. In this theory, at a relatively long time scale, the motion of a polymeric chain is modelled by a "reptation motion" in a tube along its center line, the so-called primitive chain. At the ends of the tube the chain may create new tube elements of length "a" by disengaging from the tube or cause a loss of tube elements by moving into the opposite direction. By considering the tube elements as network segments this process fits the transient network model very well.

In this case the density function $\tilde{\Psi}_t^{t'}(\underline{q}')$ is of the form $\tilde{\Psi}_t^{t'}(\underline{q}') = \tilde{N}_t^{t'}\tilde{\psi}(\underline{q}')$, where $\tilde{\psi}(\underline{q}') = (1/4\pi a^2)\,\delta(q' - a)$ and $\tilde{N}_t^{t'}$, determined by the one-dimensional diffusion process in the tube, is given by

$$\tilde{N}_t^{t'} = \sum_{j(\mathrm{odd})} \frac{8cL}{\pi^2 a \tau_d} \exp\left(\frac{-(t-t')\,j^2}{\tau_d}\right) , \tag{13}$$

in which L is the primitive chain length, c the polymer concentration and τ_d the "disengagement time". The motion is a nonextending convection[4], which is a special case of eq. 7 with $(1 + \Lambda) = |\underline{F}_t^{-1}(t').\underline{q}'|/|\underline{q}'|$.. The segmental force adopts its equilibrium value $f = kT/a$ [this expression differs by a factor of 3 from the corresponding one of Doi and Edwards since it was derived by considering a one-dimensional Gaussian chain in accordance with the tube concept. Doi and Edwards, on the other hand, use an expression based upon three-dimensional Gaussian chain statistics]. From eq. 2 the following constitutive equation is obtained:

$$\underline{T} = kT \int_{-\infty}^{t} \tilde{N}_t^{t'} < \frac{\underline{F}_t^{-1}(t') \cdot \underline{q}' \; \underline{F}_t^{-1}(t') \cdot \underline{q}'}{|\underline{F}_t^{-1}(t') \cdot \underline{q}'|^2} > dt' \tag{14}$$

which, apart from a factor 3, is the one originally derived by Doi and Edwards[3].

REFERENCES

1. Lodge, A.S., Trans. Faraday Soc., 52, 120 (1956)
2. Lodge, A.S., R.C. Armstrong, M.H. Wagner and H.H. Winter, Pure Appl. Chem. 54, 1349 (1983)
3. Doi, M. and S.F. Edwards, J. Chem. Soc. Faraday Trans. II, 74, 1789, 1802, 1818 (1978); 75, 38 (1979)
4. Larson, R.G., J. Non Newtonian Fluid Mech. 13, 279 (1983)
5. Kamphuis, H., R.J.J. Jongschaap and P.F. Mijnlieff, Rheol. Acta 23, 329 (1984)
6. Kamphuis, H. and R.J.J. Jongschaap , to be published in Journal of Rheology.

RHEOLOGICAL PROPERTIES OF A LDPE MELT IN TRANSIENT UNIAXIAL ELONGATIONAL FLOW, DESCRIBED WITH A SPECIAL TYPE OF CONSTITUTIVE EQUATION

P.J.R. LEBLANS*, J. SAMPERS** and H.C. BOOIJ**

* Department of Chemistry, University of Antwerp, Belgium
** DSM, Central Research and Patents, Geleen, Netherlands

SYNOPSIS

It is shown that a single integral constitutive equation with an integrand factorable into a time-dependent and a strain-dependent function gives a satisfactory description of the rheological properties of a LDPE melt in uniaxial extensional flow. The time-dependent memory function obtained in simple shear has been inserted into the equation and the tensorial strain functions deduced from different types of tensile experiments coincide reasonably.

1 INTRODUCTION

A single integral constitutive equation with an integrand factorable into a memory function only dependent on time and a tensorial functional of the strain history has proven to be succesful in describing transient and steady-state rheological properties in unidirectional shear flow of a class of polymer melts[1,2]. A low-density polyethylene (LDPE) melt belonging to this class[3] has been subjected to a number of experiments in simple extension to investigate the applicability of the said constitutive equation to this type of flow.

2 EXPERIMENTAL

The LDPE material chosen is the DSM commercial grade, Stamylan® 2800. The linear viscoelastic properties have been investigated[3] by executing oscillation measurements in shear on a Rheometrics Mechanical Spectrometer over the temperature range from 140 to 250 °C and the frequency range from 10^{-1} to 10^{2} rad s^{-1}. Time-temperature superposition provided master curves from which the relaxation time spectrum $H(\tau)$ has been calculated (fig. 1).

A uniaxial extension apparatus of the conventional clamp type, the Göttfert Rheostrain, has been used. Great care was bestowed on sample preparation and performance of the measurements, which were all carried out in duplicate. Reproducibility was better than 10 % for all measuring results shown.

3 CONSTITUTIVE EQUATION

The data will be analysed on the basis of the stress equation:

$$\underline{P}(t) = -p_o \underline{1} + \int_{-\infty}^{t} m(t - t') \underline{S}[\varepsilon(t) - \varepsilon(t')] \, dt' \tag{1}$$

where the memory function is defined as:

$$m(t - t') = \int_{-\infty}^{+\infty} \frac{H(\tau)}{\tau} e^{-(t - t')/\tau} \, d\ln \tau \tag{2}$$

and the tensorial strain measure is uniquely dependent on the difference between the Hencky strain ε at present time t and previous time t'. $\underline{S}$ equals the Finger tensor for small strains. This equation is similar in form to the one derived by Doi and Edwards[4] and Curtiss and Bird[5] on molecular kinetic grounds.

4 PRESENTATION AND ANALYSIS OF THE MEASURING RESULTS

For a step strain relaxation experiment where a Hencky strain ε has been imposed instantaneously at time t' = 0, eq. 1 simplifies to:

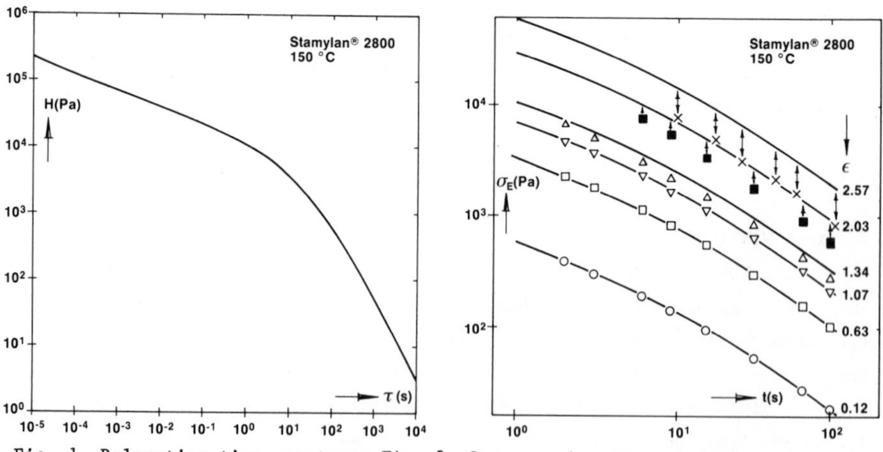

Fig. 1. Relaxation time spectrum Fig. 2. Stress relaxation at indicated strains

$$\sigma_E(t,\varepsilon) = S_E(\varepsilon) \int_{-\infty}^{+\infty} H(\tau)\, e^{-t/\tau}\, d\ln\tau = S_E(\varepsilon)\, G(t) \tag{3}$$

Experimental results are shown in fig. 2. Calculated values of $S_E(\varepsilon)$ appear not to be dependent on time and average values are represented in fig. 3.

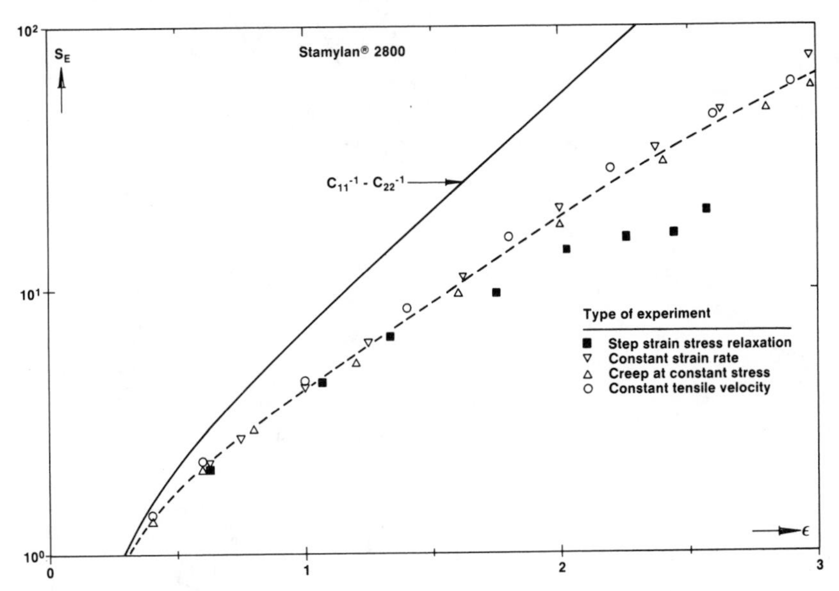

Fig. 3. Strain measures obtained from various types of experiment

The stress build-up under a constant stretching rate $\dot{\varepsilon}$ started at $t' = 0$ is described by the equation:

$$\sigma_E^+(t,\dot{\varepsilon}) = \eta_E^+(t,\dot{\varepsilon})\, \dot{\varepsilon} = S_E(\dot{\varepsilon}t)\, G(t) + \int_o^t m(s)\, S_E(\dot{\varepsilon}s)\, ds \tag{4}$$

from which the strain measure $S_E(\varepsilon)$ can be made explicit[6]:

$$S_E(\varepsilon) = \sigma_E(t,\dot{\varepsilon})/G(\varepsilon/\dot{\varepsilon}) - {}_o\int^{\varepsilon} m(\varepsilon'/\dot{\varepsilon})\, G^{-2}(\varepsilon'/\dot{\varepsilon})\, \sigma_E(\varepsilon')\, d\varepsilon' \tag{5}$$

Measured curves are shown in fig. 4, the average $S_E(\varepsilon)$ function in fig. 3.

The build-up of strain in a creep experiment started at $t' = 0$ at a constant stress σ_E is given by the general equation:

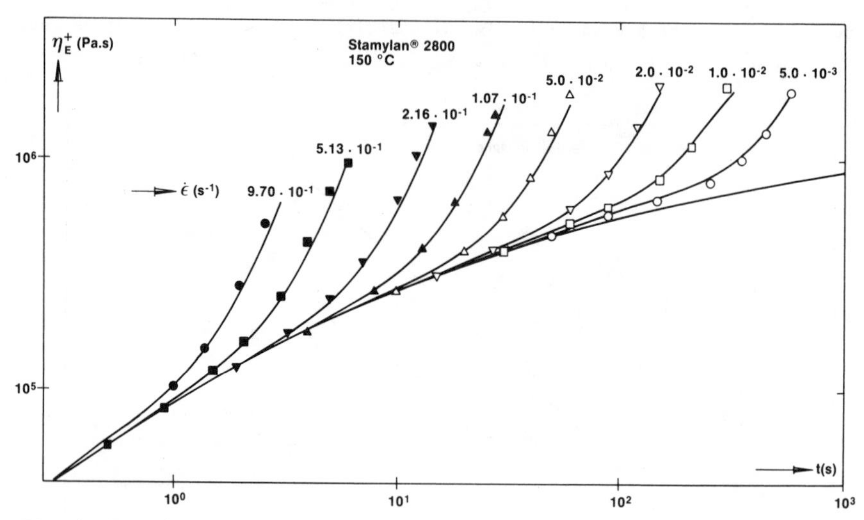

Fig. 4. Tensile viscosity at indicated constant strain rates

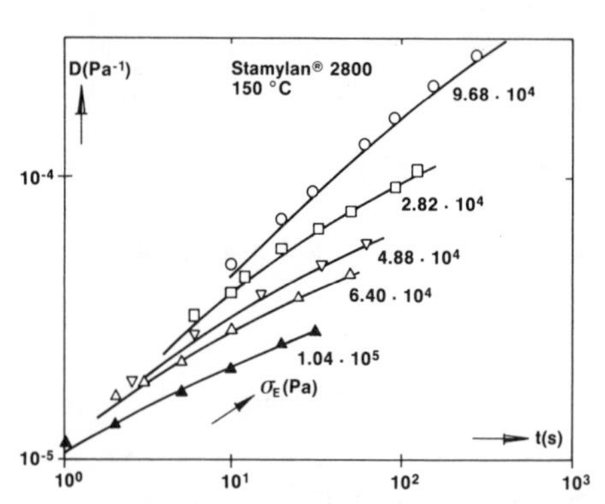

Fig. 5. Creep compliance at indicated tensile stresses

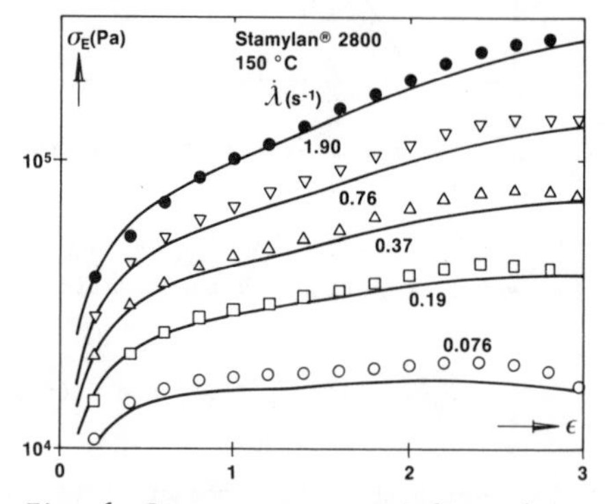

Fig. 6. Stress curves at indicated tensile velocities

$$\sigma_E = S_E(\varepsilon)\, G(t) + \int_0^t m(t - t')\, S_E[\varepsilon(t) - \varepsilon(t')]\, dt' \qquad (6)$$

Compliance curves, defined as $D(t) = \varepsilon(t)/\sigma_E$ are shown in fig. 5. The average $S_E(\varepsilon)$ curve obtained by numerical methods is shown in fig. 3.

The connection between the stress under a constant tensile velocity λ, defined as the ratio of the constant draw speed to the initial sample length, started at $t' = 0$, and the strain is also given by the general eq. 6. The stress curves are given in fig. 6, while fig. 3 shows the average $S_E(\varepsilon)$ curve that was calculated from those results, also by numerical methods.

CONCLUSION

The nonlinear strain measures $S_E(\varepsilon)$ obtained from the different experiments are approximately identical as displayed by fig. 3. The largest deviations occur for the step strain stress relaxation results, which may be due to experimental imperfections. The potentiality of the factorable single integral constitutive eq. 1 to describe the nonlinear behaviour in simple elongational flow is demonstrated by the agreement between the experimental points and the curves calculated (full lines in figs. 2, 4, 5 and 6), inserting the memory function (fig. 1), one single strain measure (broken line in fig. 3) and the specific strain histories into eq. 1.

REFERENCES

1. Booij, H.C. and Palmen, J.H.M., Rheol. Acta, vol.21, 376-387, 1982.

2. Larson, R.G., Rheol. Acta, vol.23, 10-13, 1984.

3. Booij, H.C., Palmen, J.H.M. and Leblans, P.J.R., in: B. Mena, A. Garcia - Rejón and C. Rangel Nafaile (eds.), 'Advances in Rheology', vol.3 (Proc. IX Intl. Congress on Rheology), 367-374, México 1984.

4. Doi, M., and Edwards, S.F., J. Chem. Soc. Faraday Trans. II, vol.74, 1789-1832, 1978; vol.75, 38-54, 1979.

5. Bird, R.B., Saab, H.H. and Curtiss, C.F., J. Phys. Chem., vol.86, 1102-1106, 1982.

6. Wagner, M.H., Rheol. Acta, vol.18, 33-50, 1979.

NOTATION

p_o	isotropic stress
σ_E	$p_{11} - p_{22}$ in uniaxial extension
S_E	$S_{11} - S_{22}$
G	shear relaxation modulus
η_E^+	tensile viscosity

PHYSICAL BACKGROUND OF MOULD FILLING WITH AND WITHOUT CRYSTALLIZATION

By H. JANESCHITZ-KRIEGL, G. KROBATH and S. LIEDAUER

Linz University, A-4040 Linz, Austria

SYNOPSIS

In pursuing a previously started research program on the physical processes occurring during mould filling of thermoplastic materials[1], one can show that the layer structure in a cross sectional cut of a moulded article of a crystallizable polymer is the consequence of temperature dependent crystallization kinetics. In contrast to the cited publication, however, a sharp crystallization front cannot be expected if the mould wall temperature is below the temperature of the maximum rate of crystal growth. In the course of these considerations a tentative explanation is given for the occurrence of layers of alternating degrees of orientation.

1 INTRODUCTION

Only little is known about the details of processes accompanying a "quench". The situation becomes very complicated if a phase transition is involved and/or shearing is applied. Results of mainly one recently designed model experiment will be reviewed.

The role of theory should not be overestimated in this connection. But proper experiments must certainly be guided by theoretical considerations in avoiding irrelevant experimentation or misinterpretations. In fact, heat transfer has always been a domain of applied mathematics, in particular, if a phase transition (crystallization) is involved (Stefan problem). However, as a typical example for the consequences of the absence of an experimental check the important phenomenon of supercooling has completely been overlooked, as shown by Janeschitz-Kriegl and coworkers[1][2], where further literature is cited. So far, however, the classical assumption of the occurrence of a crystallization front was maintained also in the recent work[2]. Since the speed of propagation of this front (linear growth speed) is a function of crystallization temperature $T_c < T_m$ (with T_m being the equilibrium melting point), one can solve this problem only by an iterative process of calculation, because T_c (as a function of time and location) is - itself - the solution of the problem.

However, as soon as the temperature T_w of the quenching contact surface becomes lower than the temperature T_p of maximum linear growth rate of nuclei, in general no crystallization front will depart from this surface. (It may be that some preoriented material, as occurring in mould filling, Tadmor[3], can cause temporal advantage for such a front also at $T_w < T_p$.) These aspects are related to the experiences gathered in zone crystallization[4] (for further references see loc.cit.1): In fact, at temperatures T_c close to the glass transition temperature T_g (with $T_g < T_w < T_c < T_p << T_m$) the linear growth rate becomes very small (as near T_m). As a consequence, the speed of propagation of any crystallization front formed at the contact surface is slowed down considerably. But this does not mean that the penetration of the "cold" into the material is retarded. As a consequence, however, a great number of potential growth centres ("embryos", athermal nuclei, heterogeneities) is stabilized before the crystallization front. As these centres can grow at temperatures higher than that of the contact surface or of any initially formed crystallization front, such a front must get lost.

Whereas we have a good idea about linear growth rate as a function of temperature for many polymers (maximum speed related to T_g/T_m, zero speed at T_g and at T_m [5]), little is known for $T_g < T_c < T_p$ with respect to the overall crystallization rate (i.e. the combined consequences of the number of growth centres and the linear growth rate). This holds, in particular, if polymers of industrially relevant overall crystallization speeds (like polypropylene) are considered. For the latter polymer one has $T_p \simeq 110°C$, whereas mould temperatures are usually set between 30° and 60°C. This makes quantitative predictions of the progress of solidification very difficult. An additional problem is introduced by the action of shear. So, at the present state of the art, even qualitative features - as the well known layer structure of the sample wall - can be deduced safely only from systematic experimentation.

2 APPARATUS

The apparatus resembles a parallel plate rheometer, but can be used also without rotation. The lower plate is formed by the bottom of a shallow trough, in which a tabloid of the polymer under investigation is melted at T_i well above T_m (subscript "i" for initial). The upper plate is formed by the face of a thermostated copper cylinder, kept at a temperature T_w chosen somewhere between T_m and T_g. The copper cylinder can quickly be lowered until it touches the upper surface of the molten tabloid at time t = 0. After various effective "contact times" t the trough is quenched by tap water led through channels in its body.

3 RESULTS AND DISCUSSION

In Fig. 1 a microphotograph, as made with the aid of a polarizing microscope, is presented. It shows the whole spectrum of textures obtained in the temperature range of $T_w < T < T_m$ for a polypropylene (Daplen KS 10) under the following condition: Zero shear stress, $T_i = 210°C$, $T_w \simeq 40°C$, t = 120s, gap with $D \simeq 1$ mm.

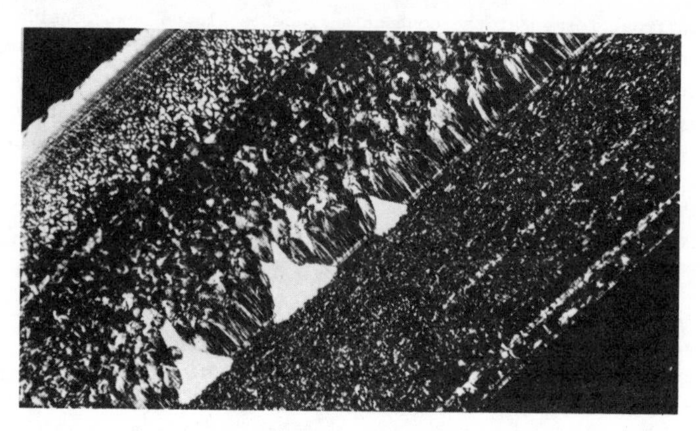

Fig.1 Texture of polypropylene after unilateral quench at zero shear stress (for details see text).

Near the contact surface the material is very fine grained. At the crystallization front, which is apparently obtained after such a long contact time, very large structures of the spherulitic type are found. In fact, these "spherulites" show the features of unidirectional unimpeded growth. These features are well known from zone crystallization at $T_p < T_c < T_m$. With the assumption of linear temperature profiles in the amorphous and crystalline phases and the use of the pertinent thermal data of polypropylene a value of 140°C was estimated for T_c at the nearly stagnant crystallization front. Apparently, the bright areas are caused by the β-modification. It should be noted that at much shorter effective contact times no sharp crystallization front was found, as could be expected from the considerations in the introduction.

In Fig. 2 a microphotograph is shown of a cut through a sample obtained under torsion. The shear stress was of the order of $0{,}5.10^5$ Pa. Most conditions were similar to those of the sample presented in Fig. 1 (t = 30s). Also in Fig. 2 a fine grained structure is observed near the contact surface. Apparently, this fine grained structure is not sensitive to the applied shear stress. In contrast, the pseudo-spherulitic structures near the stagnant crystallization front, as shown in Fig. 1, appear to be highly sensitive to the shear stress. In fact, with the application of the shear stress these structures are replaced by a less structured band of uniform high birefringence. The larger refractive index of this band is in a direction parallel to the

Fig. 2 Texture of polypropylene after unilateral quench at finite shear stress (for details see text).

crystallization front. The conclusion is that this new feature is a consequence of the influence of shearing on secondary growth of already existing growth centres. Where these growth centres are too closely packed, such a (secondary) growth does not seem to be possible.

In Fig. 3 a microphotograph of a cut obtained on an injection moulded strip of the same polypropylene is shown. (Dimensions of the strip 15x8x0,2 cm, cut taken at 8 cm from the line gate in the flow direction, injection time t = 3,6 s, mould and mass temperatures T_w = 50°C and T_i = 200°C, respectively.)

Apparently, the narrow, practically non-birefringent zones close to the mould walls can be explained in terms of the above mentioned qualitative picture. With shear stresses, as occurring in injection moulding, being more than one decade higher than in the above mentioned experiments, one may expect that (secondary) growth speeds are much more enhanced so that only a narrow zone of

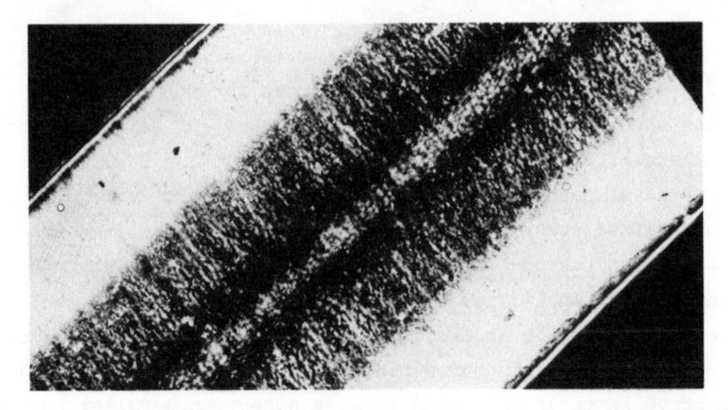

Fig. 3 Texture of injection moulded strip of polypropylene.

quenched material contains enough primary growth centres to remain uninfluenced by shear.

Only very recently, theoretical considerations as carried out by Berger and Schneider[6] in the course of the work on program S 33 (see acknowledgements), seem to support the general picture sketched.

ACKNOWLEDGEMENTS

The authors are indebted to the Austrian "Fonds zur Förderung der wissenschaftlichen Forschung" for its financial support under grant S 33/02 as part of the national research program on injection mouldings of plastics. They are also very grateful to Chemie Linz (Dr. Altendorfer, Mr. Seitl) for providing the injection moulded strip.

REFERENCES

1. Janeschitz-Kriegl, H. and Kügler F., "Heat transfer and kinetics of flow induced crystallization as intertwined problems in polymer melt processing", Polymer Processing and Properties, Astarita, G. and Nicolais, L., Eds. Plenum Publ. Co. NY.1984.
2. Eder, G. and Janeschitz-Kriegl, H., "Stefan problem and polymer processing", Polym.Bull. 11 , 93(1984).
3. Tadmor, Z., "Molecular Orientation in Injection Molding", J.Appl.Polym.Sci. 18 , 1753(1974).
4. Lovinger, A.J., Chua, J.O. and C.C. Gryte, "Studies on the α- and β-forms of isotactic polypropylene by crystallization in a temperature gradient". J.Polym.Sci., Phys.Ed., 15 , 641(1977).
5. Van Krevelen, D.W., "Crystallinity of polymers and the means to influence the crystallization process", Chimia 32 , 279(1978).
6. Berger, J. and Schneider, W., Vienna Univ.Tech., personal communication.

ON THE MATHEMATICAL MODELLING OF THE INJECTION MOULDING PROCESS

C.W.M. SITTERS[*] and J.F. DIJKSMAN[+]

*Eindhoven University of Technology, Eindhoven (The Netherlands)
+Philips' Research Laboratories, Eindhoven (The Netherlands)

SYNOPSIS

A numerical method will be presented for the computation of the behaviour of a molten thermoplastic material during the filling of a narrow cavity. The viscosity of the molten polymer is temperature, pressure and shear rate dependent. Effects such as convection, conduction and viscous heating as well as solidification at the cooled walls will be taken into account.

1 INTRODUCTION AND STATEMENT OF THE PROBLEM

Injection moulding is a widely used process for the manufacturing of small or thin walled products. The raw granulated material is melted and homogenized in a reciprocating screw extruder. As soon as the material is sufficiently deformable it is injected into the mould at high speed. Thermoplastic materials are poor heat conductors. To keep the cycle time short, the walls of the cavity have to be cooled far below the glass transition temperature. The temperature distribution development in the flowing melt is rather complex. In the middle of the flow domain heat is transported mainly by convection. Close to the walls heat flow by conduction dominates. Because of the high viscosity η and shear rate $\dot{\gamma}$, viscous heating will be important. During solidification at the walls the transition heat has to be removed by conduction through the solidified layer. In this paper we confine ourselves to the analysis of the non-isothermal flow of an incompressible generalized Newtonian liquid in a narrow cavity with a viscosity function of the Carreau type, depending on temperature T, pressure p and shear rate $\dot{\gamma}$ (ref. 1):

$$\eta = e^{p/p_0}\,\eta^* \quad ; \quad \eta^* = B_1 e^{A_1/T}\left[1 + (B_2 e^{A_2/T}\dot{\gamma})^2\right]^{(n-1)/2} \tag{1}$$

where A_1, A_2, B_1, B_2, p_0 and n are constants. We only regard amorphous polymers with a glass transition temperature T_m and a constant density ϱ in both the solid and liquid phase.

The geometry of the cavity can be composed of a series connection of rather simple flow elements describing plane, radial and conical flow (Fig. 1). We assume that the channel is symmetric with respect to the local coordinate z=0. Essential is that the height is much smaller then the length of the cavity. The leading terms of the equations governing the problem become (ref. 1):

$$\frac{\partial p}{\partial x} = \frac{\partial}{\partial z}\left(\eta \frac{\partial v_x}{\partial z}\right) \quad ; \quad \frac{\partial p}{\partial z} = 0 \qquad \text{momentum equations} \tag{2}$$

$$\varrho c_{pl}\dot{T} = \lambda_l \frac{\partial^2 T}{\partial z^2} + \eta\dot{\gamma}^2 \quad ; \quad \dot{\gamma} = \frac{\partial v_x}{\partial z} \qquad \text{energy equation in fluid} \tag{3}$$

$$\varrho c_{ps}\dot{T} = \lambda_s \frac{\partial^2 T}{\partial z^2} \qquad \text{energy equation in solid} \tag{4}$$

with $\dot{T}$ the material derivative of the temperature. The continuity equation $\vec{\nabla}.\vec{v}=0$ will be discussed later. These equations are subjected to the following boundary conditions. At the flow front the atmospheric pressure is prescribed. At z=0 we have the symmetry conditions. At the gate (x=0) the injection temperature T_i and either the volume flux or the pressure are prescribed. At $z=\pm h(x)$, i.e. at the walls, the transport of heat depends on the temperature gradient in the wall. Assuming that this gradient is constant we have (ref. 2,3):

$$\lambda_s \frac{\partial T_s}{\partial z}(x,h) + H(T_w - T_c) = 0 \tag{5}$$

The subscripts s, w and c refer to solid, wall and coolant, respectively. H is the heat transfer coefficient. In most cases H is very large, so $T_w \approx T_c$. At the solid-liquid interfaces $z=\pm d(x)$, the glass transition temperature and the no-slip condition are imposed.

The velocity $(\hat{v}_x, \hat{v}_z)$ at which the solid-liquid interface penetrates the flowing material, depends on the temperature gradients in the liquid and solid phase according to (ref. 4):

$$\hat{v}_z \approx \frac{1}{K\varrho}(\lambda_s \frac{\partial T_s}{\partial z} - \lambda_l \frac{\partial T_l}{\partial z}) \quad ; \quad \hat{v}_x \approx 0 \tag{6}$$

Integration of the momentum equation into the z-direction yields:

$$\frac{\partial}{\partial x}(p_o e^{-p/p_o}) = Q/(2L \int_0^d \frac{z^2}{\eta^*} dz) \tag{7}$$

$$v_x = \frac{Q}{2L} (\int_z^d \frac{z}{\eta^*} dz)/(\int_0^d \frac{z^2}{\eta^*} dz) \tag{8}$$

where Q is the total volume flux and L the length of the flow front. The viscosity η^* depends on the temperature and the shear rate, and therefore these relations have to be solved iteratively.

2 NUMERICAL APPROXIMATION

The mould is filled in a number of time steps. After each time step the position of the flow front is calculated and a new grid line is added there. All the other grid lines remain fixed. Suppose that at a certain time all equations and boundary conditions are satisfied. After a time step the whole domain is calculated line by line starting with the first line at the gate, using the following iterative computational scheme.

1) The equation governing the velocity of the solid-liquid interface. The velocity of this interface can be calculated with two different methods. One method uses the present position and the position at the previous time (backward difference), the other utilizes eq. 6 in an implicit manner using the present temperature distribution. The difference of these two velocities has to converge to zero during the iteration process.

2) The continuity equation. On the first grid line we define a number of grid points. On this line it is rather easy to calculate the velocity profile because the temperature is constant (T_i). When the velocity distribution has been approximated on any other grid line, we arrange the grid points on that line in such a way that the flux through two adjacent grid points equals the flux through the two corresponding points on the first grid line. After convergence the continuity equation will be satisfied automatically.

3) The energy equation. The temperature equations are solved by a finite difference technique. The material derivative of the temperature can be

approximated in two ways. It is possible to compute the track of a particle with its temperature history and give a direct approximation of $\dot{T}$ by a backward difference. Also a combination of the spatial time derivative $\frac{\partial T}{\partial t}$ by a backward difference and an implicit approximation of the convection term $v_s \frac{\partial T}{\partial s}$ (s is the coordinate along a streamline), can be used. The conduction term (central differences) and the viscous heating term will be approximated implicitly. Together with the boundary conditions this scheme leads to a tridiagonal matrix equation. In the solidified layer the problem is analog (without convection and viscous heating). After solution of the equations we have a new approximation of the temperatures at the grid points.

4) The momentum equation. After computation of the viscosity distribution η^* from the approximated temperatures and velocity gradients, the integrals of eqs. 7 and 8 can be solved numerically. From the results we obtain the pressure drop and the velocity profile.

The iteration loop consisting of the four steps listed, will be terminated as soon as convergence has been reached.

At the flow front this scheme has to be adapted with respect to the calculations of the temperature and solidified layer, because of the two-dimensional character of the flow (fountain effect, ref. 4). At the centre the temperature distribution is convected from the upstream. The thickness of the solidified layer and the temperature distribution therein will be calculated by the penetration theory with an initial temperature difference equal to the core temperature minus the wall temperature. These two temperature distributions are matched in a suitable way.

After the computations of the velocities and temperatures on all the grid points according to the iteration scheme above, it is possible to calculate the pressure distribution along the axis of the channel by integration of the pressure drop with respect to the x-direction. When the maximum machine pressure is exceeded, another iteration procedure will be started which adapts the volume flux Q until the calculated pressure at the gate equals the maximum pressure.

3 RESULTS

As a numerical experiment we have filled the cavity of Fig. 1 with a polycarbonate. The physical properties of the polymer and the process data are given in Table 1. The calculated pressure at the entrance at the end of the filling stage is 65.7 Mpa. In Figs. 2, 3 and 4 some results are plotted. Figures 2 and 4 refer to the instant that the mould is nearly filled.

A_1	$= 1.444\ 10^4$	K	A_2	$= 1.312\ 10^4$	K
B_1	$= 6.845\ 10^{-9}$	Pas	B_2	$= 5.050\ 10^{-13}$	s^{-1}
n	$= 0.39$		p_o	$= 1000$	MPa
$\lambda_l = \lambda_s$	$= 0.20$	$J(smK)^{-1}$	ϱ	$= 1176$	kgm^{-3}
c_{ps}	$= 2000$	$J(kgK)^{-1}$	c_{pl}	$= 1700$	$J(kgK)^{-1}$
T_m	$= 145$	^oC	K	$= 2250$	Jkg^{-1}
Q	$= 1.5\ 10^{-5}$	m^3s^{-1}	H	$= 10^6$	$Jm^{-2}(sK)^{-1}$
T_i	$= 320$	^oC	T_c	$= 80$	^oC

Table 1. Physical properties of Makrolon 6560 and process data

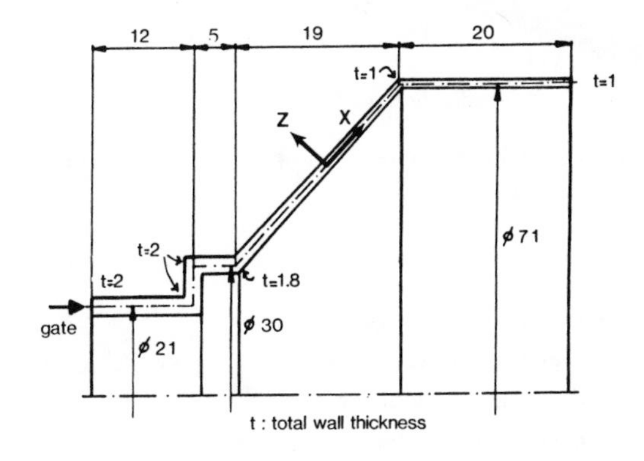

Fig. 1. Axial symmetric geometry

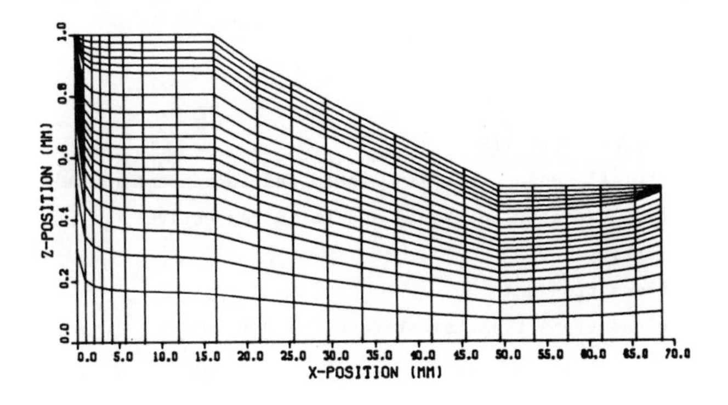

Fig. 2. Computational grid and instanteneous stream lines

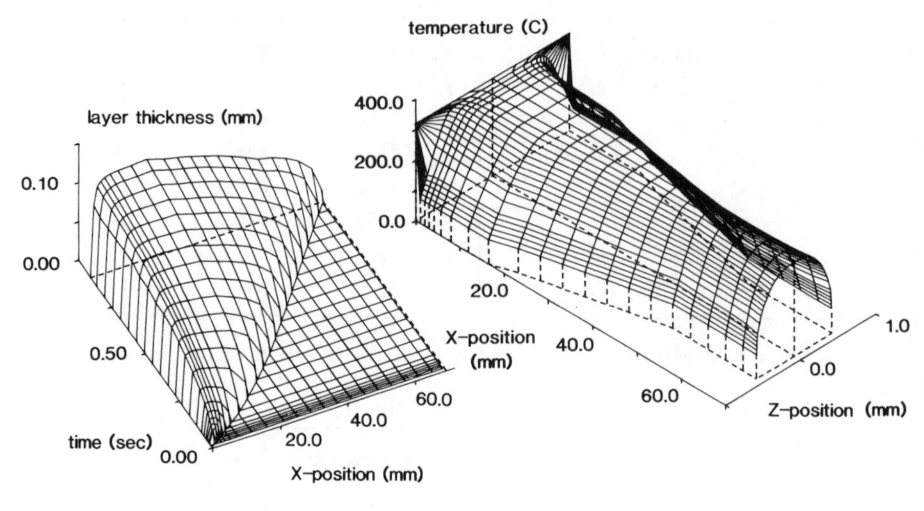

Fig. 3. Built-up of solidified layer

Fig. 4. Temperature distribution

REFERENCES

1. Bird R.B., Armstrong R.C., Hassager O., 'Dynamics of Polymeric Liquids', Wiley New York, 1977.
2. Bird R.B., Stewart W.E., Lightfoot E.N., 'Transport Phenomena', Wiley New York, 1960.
3. Wijngaarden H. van, Dijksman J.F., Wesseling P., J. of Non-Newtonian Fluid Mechanics 11 (1982) pp 175-199.
4. Tadmor Z., J. of Appl. polymer Sci., 18 (1974) 1753-1772.

SYMBOLS FOR PHYSICAL PROPERTIES

λ_l, λ_s	thermal conductivity in liquid and solid respectively
c_{pl}, c_{ps}	heat capacity in liquid and solid respectively
ρ	density
K	transition heat
T_m	glass transition temperature

MIXING PROCESSES IN POLYMER PROCESSING

Z. TADMOR

Department of Chemical Engineering, Stevens Institute of Technology, N.J.

SYNOPSIS

Pure polymers and copolymers are made processable and useful by compounding and mixing them with a broad variety of solid, liquid and gaseous additives. Moreover, as the number of chemically new, and commercially viable polymers diminish, increasing efforts are being made to meet new demands with existing polymers, by their mixing compounding and alloying with new additives in novel ways. In this paper, following a brief review of mixing technology and machinery, recent views on mixing mechanisms are discussed and a theoretical formulation of mixing processes is presented.

1. INTRODUCTION

Synthetic polymers, comprising of plastics, elastomers, fibers and coatings, are the exclusive products of the 20th century. They have a profound influence on our lives and technologies. Space technology would be inconceivable without them, as would be the electronic and computer technologies. The scale of the polymer industry can be well appreciated by recalling that world production of polymers (by volume) seems to have surpassed that of metals. Indeed the 20th century can well be coined the "polymer age." The phenomenal success of polymers is due perhaps as much to the relative ease and diversity of their processing into useful products, as to their unique combination of properties they possess.

The scientific understanding of polymers started with Herman Staudinger's macromolecular hypothesis in 1920. His work was followed by that of many other brilliant scientists to establish the polymer sciences as accepted and respected branches of chemistry and physics. The "engineering" of polymers, on the other hand, that is their conversion into useful products, though preceeding the scientific understanding of polymers (a phenomenon not uncharacteristic in the history of technology), is evolving into a well defined engineering discipline only in recent years. This evolutionary process of polymer engineering could only follow that of the polymer sciences as well as that of the science of rheology.

The structural breakdown of polymer processing (1) comprises of five elementary steps (solids handling, melting, mixing, pumping and stripping) and of five shaping steps (die forming, casting & molding, coating & calendering, mold coating, and secondary shaping). The elementary steps prepare the raw materials for shaping. Among the elementary steps, devolatilization (stripping) and mixing have been recently enjoying a great deal of attention. The interest in the former is driven by increasingly stringent health and environmental regulations as well as by the desire to improve properties by removing low molecular weight components; whereas, the interest in the latter stems from the realization that new property demands can no longer be economically met by chemically new polymers, but they can be frequently met by compounding, blending, alloying, foaming, reinforcing and reaction. Not only do these primarily mixing operations meet the new demands, but they also provide the best route for commercial competitiveness.

2. MIXING PROCESSES

Mixing is an ancient human activity, and consequently we have developed a great deal of intuition into its nature. Yet mixing processes of polymers are so varied (2) and demanding that they can tax the best designers. They include mixing of solids into a viscous liquid matrix, mixing of two viscoelastic melts which may be rheologically homogeneous or nonhomogeneous as well as thermodynamically compatible or non compatible, mixing of low viscosity liquids into high viscosity melts, and finally mixing of gases into viscous polymeric melts. All these operations are designed to ensure that the very many chemical components of a commercial polymer are sufficiently comingled. The additives convert a pure polymer into a useful one. Such mixing operations are generally called 'compounding,' which is a somewhat loosely defined term denoting in addition to mixing also a melting some-times devolatilizing as well as pelletizing.

The mixing process are not only varied in nature but highly constrained by the sensitivity of polymers to high temperature, oxidation and shearing. Thus, the desire for a thorough mixing must always be tempered by the need for the gentle handling of the polymer, and therefore, temperature, time and shear histories must be closely controlled. Yet in spite of all precaution the proper-ties of polymer do change each time they go through a processing machine. A crude measure of this change is the Melt Index shift, which is always followed up by a battery of physical and mechanical property tests.

Mixing processes of polymers can be divided into two categories involving different physical phenomena. One category is associated with the reduction in size of a segregated component which has a cohesive nature such as cohesive granular solids, liquid regions with surface tension as well as vapor or gas bubbles. This type of mixing is called dispersive or intensive mixing. It is dominated by the stress level within the deforming liquid matrix. A critical stress level must be exceeded to break up cohesive agglomerated solids (3). For breaking up a visco-elastic 'blob' or droplets in additon to stress the stress history becomes important (4). This type of mixing is relevant to blending and alloying of noncompatible polymeric melts, where both viscous and elastic properties of the melt play key roles (5). Vapor and gas bubble mechanics in viscoelastic liquids is important not only in classical foaming processes, but also in melt devolatilization where vapor and gas bubble deformation, breakup, and coalescence within the sheared liquid and bubble rupture at the liquid gas interface, determine to a large extent the outcome of the process (6).

In the absence of cohesive barriers, as in the mixing of thermodynamically compatible melts, or the mixing of melt regions at different temperatures or composition (e.g. containing non dispersive pigment or other additive which do not affect significantly rheological properties), the mixing is determined by the strain history imparted to the liquid. This type of mixing is called extensive or laminar mixing or simply blending. The primary mechanism of this mixing process is convection associated to the deformation of the liquid. The deformation is mostly by shear with some elongation. Thus, terms such as 'kneading' denoting squeezing type flow (shear and elongation) followed by folding, and 'milling' denoting smearing and wiping (combined shear and elongation with some dispersion), are frequently used to describe the nature of the mixing in the machine.

Mixing of polymers, in particular dispersive mixing is energy intensive.

For example preparing a LDPE color masterbatch requires a specific energy input of the order of 1.5 MJ/Kg (2). For comparison, heating the polymer from room temperature to 200 ^{o}C requires only 0.6 MJ/Kg. Specific energy inputs for rubbers range up to 10 MJ/Kg. Clearly, the high energy inputs in the mixing processes, focuses attention on the need for effective heat removal to avoid polymer damage. In dispersive mixing, excessive temperature rise will not only damage the polymer, but due to decreasing viscosity may bring the dispersion process to a complete halt.

The quantitative analysis of the mixing processes must deal with the characterization of the mixture and that of the mixing process. Mixture characterization was reviewed elsewhere (1,7) and will not be discussed here. The quantitative analysis of the mixing processes involves the mathematical modeling and formulation of the associated physical mechanisms, and that of the processing machines. Certain aspects of these for both laminar and dispersive mixing will be discussed, subsequent to a brief review of mixing machinery.

3. MIXING MACHINERY

Mixing machinery and technologies have been recently reviewed in detail by Matthews (2). There is a broad variety of batch mixing machines mainly for low and medium range viscosities. For melt mixing, however, and primarily for dispersion the heated roll-mill and the Banbury type internal mixer (fig. 1) are the most important ones. Both were developed for rubber mixing and adopted by the plastics industry. The fundamentally different continuous mixers and compounders are schematically shown in fig. 1. Most common among these is the single screw extruder, with dozens of specialized 'mixing' screws (e.g. barrier type screws) and mixing sections (e.g. pins, torpedos, planetary gears, reverse flights, interupted flights etc.). Motionless mixers are attached sometimes at the discharge end of single screw extruders. The single screw extruder with all these modification becomes a good extensive mixer and mild compounder, but cannot perform as an intensive mixer. The reciprocating single screw variation which provides for fully wiped surfaces and improved compounding capability, is a very imaginative extension of the single screw concept. Twin screw extruders-compounders are subdivided into four groups. The continuous mixer has two non-intermeshing rotors, much like those of the internal mixer, extended with short screw like elements to feed the rotors. It is, in fact, a conversion of the batch internal machine to a continuous one. It is considered a medium intensity compounder capable of high production rates, but it has no

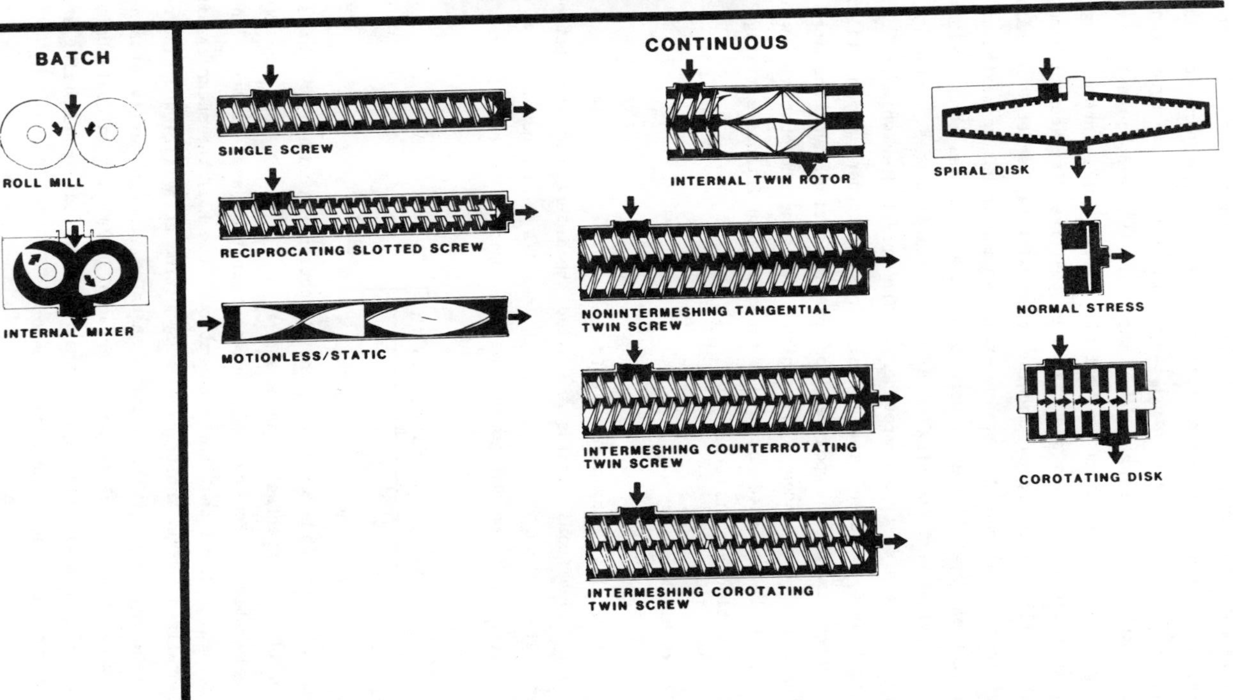

Fig. 1 Schematic view of batch and continuous mixers

devolatilization or pumping capabilities. In order to resolve the latter deficiency a screw extruder or more recently even a gear pump is added to the machine. The non intermeshing tangential twin screw type machine are basically similar to the single screw machine, with improved extensive mixing (8) and good devolatilization capabilities. The fully intermeshing counter rotating twin screw machines are basically positive displacement pumps. The intermeshing corotating screw machine come with a variety of mixing and kneading elements tailored to provide extensive and intensive mixing as well as devolatilization capabilities. Gear pumps are sometimes connected in tandem for energy savings and lowering of extrudate temper-atures. Finally, fig. 1 schematizes three disk type compounders. The spiral disk which is tantamount to a flat screw extruder, the normal stress extruder, which utilizes the elastic properties of the polymer for pressurization, and the more recent corotating disk compounder (9), which like the corotating twin screw compounder, can be designed to provide extensive and intensive mixing, as well as devolatilization capabilities.

4. EXTENSIVE MIXING

4.1 Rheologically Homogeneous Liquids

When two viscous liquids are mixed the interfacial area between them increases. The ratio of final to initial area A/A_o, is a quantitative measure of laminar mixing and for homogeneous deformation is given by (1,10):

$$\frac{A}{A_o} = \left[\frac{\cos^2\alpha'}{\lambda_x^2} + \frac{\cos^2\beta'}{\lambda_y^2} + \frac{\cos^2\gamma'}{\lambda_z^2} \right]^{1/2} \qquad (1)$$

where $\cos\alpha'$, $\cos\beta'$ and $\cos\gamma'$ are the directional cosines of the initital orienation of the interfacial area and λ_x , λ_y, and λ_z are the principal elongational ratios. For randomly oriented initial area elements in simple shear flow, eq. 1 shows that the area ratio is one half the shear strain. Eq. 1 was generalized for an arbitrary flow by Ottino et al (12) who also reviewed recently laminar mixing (13). Simple shearing, though easily attainable in polymer mixers, tends to align the interfacial area in a uniquely unfavorable direction (11,12). Erwin (14) showed that if N periodic randomizing steps are introduced, the interfacial area ratio becomes

$$\frac{A}{A_0} = \frac{1}{2}\left(\frac{\gamma}{N}\right)^N \qquad (2)$$

which explains the very positive effect of even the simplest mixing section in screw extruders. If optimal orientation of 45° to the direction of shear is maintained at all times, that is the deformation becomes pure shear or planar extension, the inter-facial area will increase exponentially with deformation (1,15).

The first attempt to deal with a realistic mixer is McKelvey's (16) derivation of an average shear strain in single screw extruders, modified by Tadmor et al. (17), who also defined a strain distribution function $f(\gamma)d\gamma$, derived it for the screw extruder(1) and suggested the mean strain as a quantitative measure of continuous mixers. However, if significant reorientation occurs along the mixer, the total strain by itself cannot be a useful measure of mixing anymore, and the exact flow path has to be followed (12,13). But, calculating mixing performance from basic principles is limited to relatively simple configurations such as helical annular mixer, certain motionless mixers and constant depth single screw mixer with numerous approximations. The computational effort for real mixers is enormous and experimental techniques using tracer studies (7) are being applied to elucidate laminar mixing mechanism. Such techniques were used very effectively by Erwin et al in studying single screw, twin screw and motionless mixers as well as by David (18) for studying laminar mixing in co-rotating disk processor. The complexity of the flow pattern induced in real mixers, reflects the necessity to randomize orientation and to distribute composition and interfacial area elements throughout the volume.

4.2 RHEOLOGICALLY NON-HOMOGENEOUS LIQUIDS

In mixing rheologically nonhomogeneous systems, viscosity ratio, elasticity and surface tension are of interest. Mixing is no longer measured just by interfacial area and its distribution throughout the volume, but also by the morphology of the blend. Van Oene (19) reviewed many of the phenomena associated with both dispersed and stratified flows of two liquids. It is generally asserted that it is more difficult to mix a low viscosity liquid into high viscosity one, than the other way around. Simple layered analysis, (1,20) shows that the more viscous the liquid the less will it tend to deform.

But, such analysis is oversimpli-fied. For example, the low viscosity liquid tends to encapsulate the other liquid (19), and the interface between two viscoelastic liquids will be distorted in complex ways (21). Big and Middleman (22) investigated the effect of viscosity ratio on interface evolution in cross channel drag flow, and though it appears that equal viscosity are to be preferred results cannot be generalized. Arimond and Erwin (23) explored the mixing of a randomly distributed molten pellets in realistic shear flows. They claim that, in systems of viscous pellets in less viscous continuous phase, wall effects and pellet-pellet interaction play a key role in the mixing process.

One of the most important areas of mixing of different polymer melts is alloying (19,24,25). The morphology of the 'blend,' hence its physical properties, depends to a great extent on the mixing process (25). Van Oene (5) derived a thermodynamic criterion based on the primary normal stress function of the components and the interfacial tension, to predict the basic morphology of the blend, that is which will be the dispersed phase and which will be the continuous one. However, much further work is neded in this important area of mixing, both in analyzing realistic systems as well as devising design criteria for mixer design.

5. DISPERSIVE MIXING

5.1 Dispersive Mechanisms

Dispersive mixing is perhaps the most demanding type of mixing. The mixing of carbon black into rubber is a prime example, and the most carefully investigated one. Carbon black aggregate of a typical size of 150 nm, cluster into large agglomerates of sizes up to 100,000 nm. They are held together by Van-der-Waals forces, and the objective of dispersive mixng is to rupture the agglomerate into its constituents, and distribute them throughout he volume. Rumpf (26) modelled a randomly packed cohesive agglomerate and derived an expression for the tensile strength σ,

$$\sigma = \frac{9}{8}\left(\frac{1-\varepsilon}{\varepsilon}\right)\frac{F}{d^2} \qquad (3)$$

where ε is the porosity, d is the diameter of the particles forming the agglomerate and F is the cohesive force between two individual particles which can be obtained from the Bradley-Hamaker theory. Thus, for equal sized particles $F = C_o d$, and for carbon in polystyrene environment C_o is of the order $4.0 - 4.8 \times 10^{-11}$ N/nm (27). Rumpf's model is an approximate one, and perhaps the approach developed recently by Thornton (28), of micromechanical examination of particulate matter using numerical simulation, may yield a much more realistic method to deal with agglomorates in dispersive mixing.

Mixing of carbon black into rubber, mostly a batch operation carried out in inernal mixers, has four stages: incorporation, dispersion, distribution and plasticization (29). During incorporation encapsulation and wetting of the solids takes place, and it is probably during this stage that the nature, strength, and initial size distribution of agglomerates are determined. Hess (29) distinguishes between 'soft' and 'hard' agglomerates based on the level of rubber penetration into the agglomerate, but there is probably a whole spectrum of agglomerate varying in size, strength and nature.

In the course of the dispersion stage, the agglomerate are successively broken apart and reduced in size. In this stage the agglomerates are freely suspended in the liquid, which is repeat-edly passed over some narrow gap region. In this high shear field, the hydrodynamic forces acting on the agglomerate surface generate internal stresses, and when these exceed the cohesive strength of the agglomerate (eq. 3), rupture occurs. The hydrodynamic separating force within a freely suspended axisymmetric particle in shear flow is (30):

$$F_h = \chi \pi \mu \dot{\gamma} c^2 \sin^2\theta \sin\phi \cos\phi \qquad (4)$$

where χ is a numerical constant dependent on particle shape (27), c is characteristic radius μ is the Newtonian viscosity, $\dot{\gamma}$ the local shear rate, and θ and ϕ are instantaneous orientation angles. The condition for rupture is $F_h/F_c > 1$, where F_c is the cohesive strength of the agglomerate evaluated from eq. 3. The ratio F_h/F_c is given by

$$\frac{F_h}{F_c} = Z \sin^2\theta \sin\phi \cos\phi \qquad (5)$$

where

$$Z = \frac{8}{9}\chi(\mu\dot{\gamma})\left(\frac{\varepsilon}{1-\varepsilon}\right)\frac{d}{C_0} \quad (6)$$

This dispersion model, proposed by Manas-Zloczower et al (27), predicts, therefore, that rupture depends on shear stress in the liquid, agglomerate porosity and the size and cohesive forces between the particles forming the agglomerate. The size of the agglomerate is conspicuosly absent, implying that it is equally likely to break large or small agglomerates, a prediction which seem to have received some experimental support (31). The model also correctly predicts that it is easier to disperse 'high structure' blacks (i.e., larger aggregate) than low structure blacks. Thus, for example the estimate cohesive force F_c of a 100,000 nm agglomerate is 10.6×10^{-3}N for 150 nm aggreage size, and it is only 3.2×10^{-3} N for a 500 nm aggregate size ($C_o = 4.5 \times 10^{-11}$ N/nm). The former will rupture in a shear field of 0.22 MPa (32psi) and the latter requires only 0.66 MPa (9.6 psi). These shear stress values are well within the practical range considering that the shear rates in the high shear zone are in the range of 200-500 s^{-1}.

Upstream the high shear narrow gap zone, there is a tapered entrance region where strong elongational flow components exist. The exact role of this region is not clear. It may have a critical function in separating closely placed agglomerate fragments, in generating significant hydrodynamic pressure to eliminate slip at the wall in the high shear zone, in pulling apart 'soft' agglomerate, and in provididng extensive mixing. Another point of view expressed by Funt (32) is that dispersion is altogether controlled by separation and, therefore, the elongational flow in the entrance zone will play a central role in the process. Yet another dispersion model is the 'onion peeling' model proposed by Shiga and Furuta (33). They suggest a dispersion mechanism based on gradual peeling of the external layers of the agglomorates. Additional detailed experimental and theoretical work is needed to critically test the validity of the various models, to account for the many non-Newtonian phenomena that may exist, and to account for particle-particle interactions including possible reagglomeration. Some of these effects are discussed by Manas-Zloczower et al (27,34).

5.2 Pass Distribution Functions

A basic characteristic of dispersive mixing is that the material experiences repeated passes over high shear zones in narrow gaps. In order to reduce agglomerates to acceptable sizes (e.g. a typical criterion is 99% below 9μ), virtually all agglomerates must experience a sufficient number of passes. In between pases the material is transported to other regions of the mixer where, by and large, extensive mixing and a randomization of composition takes place. Hence different material elements experience different number of passes, and passes over the high shear zones are bests characterized by a function that accounts for this effect termed as the Pass Distribution Functions (PDF). Thus, for a batch mixer g_k is defined as the fraction of material volume that experienced k passes over the high shear zones. The mean number of passes is

$$\bar{k} = \sum_{k=0}^{\infty} k g_k \tag{7}$$

The volume fraction of material that experienced k passes or less, is

$$G_k = \sum_{j=0}^{k} g_j \tag{8}$$

G_k, g_k and $\bar{k}$ are functions of time and by definitiona $G_\infty = 1$. Similarly, for a steady continuous mixer, we define f_k as the volume fraction of exiting flow rate Q, that experienced k passes over the high shear zone. The mean number of passes $\bar{k}$ is

$$\bar{k} = \sum_{k=0}^{\infty} k\, f_k \tag{9}$$

The volume fraction of exiting flow rate that experienced k passes is

$$F_k = \sum_{j=0}^{k} f_k \tag{10}$$

and by definition $F_\infty = 1$. The function F_k and f_k as well as $\bar{k}$ are functions of the mean residence time in the mixer. Next we define an internal PDF of a continuous mixer i_k as the fraction of volume of the material in the mixer that experienced k passes. This function is related to F_k via

$$i_k = \frac{Q}{q}(1 - F_k) \tag{11}$$

where q is the total flow rate over the high shear zone. Finally I_k is defined as the fraction of material volume in he mixer that experienced k passes or less

$$I_k = \sum_{j=0}^{k} i_j \tag{12}$$

5.3 Dispersion Model of the Batch Internal Mixer

The dispersion model of section 5.1 can be combined with the PDF concept to derive a dispersion model for example of a batch internal mixer (27). For this purpose the mixer is considered as a well mixed tank of volume, V (representing the region beetween the rotors), from which a steady stream q passes over the high shear zone and recycles to the tank. The fraction of agglomerates that rupture per pass, X, is given by

$$X = \int_0^1 W(\xi)f(\xi)d\xi \tag{13}$$

where $W(\xi)$ is the fraction of agglomerates that break at dimensionless location ξ, and $f(\xi)d\xi$ is the fraction of flow rate between ξ and $\xi + d\xi$. Assuming that the axisymmetric agglomerates are randomly oriented at the entrance to the high shear zone X can be computed via Eq. 5 and equations describing the rotation of the particles in the shear flow (35). The computed value of X, in simple shear flow, is a function of only the dimensionless groups Z and L/H where L is the length of the high shear zone, and H is the gap size. The PDF is given by

$$g_k = \frac{1}{k!}\left(\frac{t}{\bar{t}}\right)^k e^{-t/\bar{t}} \quad (14)$$

where t is the mixing time, and $\bar{t} = V/q$ is the mean residence time in the well mixed region. The mean number of passes, from eq. 14, is simply $\bar{k} = t/\bar{t}$. The fraction of agglomerates at any time t, that ruptured j times is given by

$$Y_j = \sum_{k=j}^{\infty} g_k\, u_{jk} = \frac{(Xt^*)^j}{j!} e^{-Xt^*} \quad (15)$$

where $t^* = t/\bar{t}$ and u_{jk} is the fraction of agglomerate that experienced j ruptured after κ passes over the high shear zone. Agglomerate size can be approximately related to the number of ruptures via

$$\frac{D_i}{D_{i+1}} = 2^{1/3} \quad (16)$$

Eq. 15 and 16 permit to calculate agglomerate size distribution as a function of mixing time, and a comparison gave good agreement with experimental data (27). Finally, Eq. 15 suggest Xt^* as a dimensionless scale up criterion (36).

REFERENCES

1. Tadmor, Z. and Gogos, C.G., 'Principles of Polymer Processing,' Wiley-Interscience, 1979.
2. Matthews, G., 'Polymer Mixing Technology' Applied Science Publishers, 1982.
3. Bolen, W.R. and Colwell, R.E., Soc. Plast. Eng. J., vol. 14, no. 8, 24-28, 1958.
4. Flumerfeld, R.W., Drop breakup in simple shear field of viscoelastic fluid,' Ind. Eng. Chem. Funcdam., vol. 11, 312-318, 1972.
5. Van Oene, H., 'Modes of dispersion of viscoelastic fluids in flow,' J. Colloid and Interface Sci., vol. 40, no.3, Sept. 1972.
6. Mehta, P.S., Valsamis, L.N. and Tadmor, Z., 'Foam devolatili-zation in multichannel corotating disk processors,' Polym. Process Eng., vol. 2, 103-128, 1984.
7. Strasser, R.A. and Erwin, L., 'Experimental techniques in analyzing distributive laminar mixing in continuous flows,' Advances in Polym. Tech., vol. 4, 17-32, 1984.
8. Howland, C., and Erwin, L., 'Mixing in counter rotating tangential twin

screw extruders', Dept. Mech. Eng. M.I.T. Cambridge, Mass., 1984.
9. Tadmor, Z. Method and apparatus for processing polymeric materials, U.S. Patents 4, 142, 805, 1979 and 4, 194, 841, 1980.
10. Spencer, R.S. and wiley, R.N., 'The mixing of very viscous liquids,' J. Colloid Sci., vol. 6, 133-145, 1957.
11. Erwin, L., 'Theory of laminar mixing,' Polym. Eng. Sci., vol. 18, no. 13, 1044-1048, 1978.
12. Ottino, J.M., Ranz, W.E., and Macosko, C.W., A lamellar model for analysis of liquid-liquid mixing,' Chem. Eng. Sci., vol. 14, 877, 1979.
13. Ottino, J.M. nd Chella, R., 'Laminar mixing of polymeric liquids. A brief review and report on theoretical develop-ments,' Polym. Eng. Sci., vol. 23, 357-379, 198314. Erwin, L., 'Theory of mixing sections in single screw extruders,' Polym. Eng. Sci., vol. 18, 512-516, 1978.
15. Erwin, L., An upper bound on the performance of plane strain mixers,' Ibid, 738-740, 1978.
16. McKevey, J.L., 'Polymer Processing', Wiley, 1962.
17. Tadmor, Z. and Klein, I., 'Engineering Principles of Plastic-ating Extrusion,' Van Nostrand, Reinhold, 1970.
18. David, B. 'Extensive mixing in corotating disk processors', M.S. Thesis, Dept. of Chem. Eng., Technion, Haifa.
19. Van Oene, H.,' Rheology of polymers blends and dispersions, in Paul, D.R. and Newman, S., edts. 'Polymer Blends,' vol. 1, Academic Press, 1978.
20. Middleman, S., 'Fundamentals of Poloymer Processing,' McGraw Hill, 1977.
21. Han, C.D., 'Rheology in Polymer Processing,' Chapter 3, Academic Press, 1976.
22. Bigg, D.M. and Middleman, S., Laminar mixing of pair of fluids in rectangular cavity,' Ind. Eng. Chem. Fundam., vol. 13, 184, 1974.
23. Arimond,J. and Erwin, L., 'An approach to mixing of highly viscous mixtures having components of moderately different viscosities', Am. Inst. Chem. Eng. Annual Summer Conf., 1982.
24. Utracki, L.A., Economics of polymer blends, Polym. Eng. Sci., vol. 22, 1166, 1982.
25. Plochocki, A.P., 'Melt rheology of polymer blends, the morpho-logy feedback,' ibid vol. 3, 618, 1983.
26. Rumpf, H., 'The strength of granules and agglomerates,' Ch. 15, Knepper, W.A., (ed), 'Agglomeration ,' Wiley Interscience, 1962.
27. Manas-Zloczower, I., Nir, A. and Tadmor, Z., 'Dispersive mixing in internal mixers - A theoretical model based on agglomerate rupture,' Rubber Chem. Tech., vol. 55, no. 5, 1250-1285, 1982.
28. Thornton, C., 'A micromechanical examination of particulate material using numerical simulation,' 16th IUTAM Int. Cong. on Theor. & Appl. Mech. Lyngby August, 1984.
29. Hess, W.M., Swor, R.A and Micek, E.J. 'The influence of carbon black, mixing and compounding variables on dispersion.' 124th Meeting, Rubber Div. ACS, October 25-28, 1983, Houston.
30. Nir. A. and Acrivos, A., J. Fl. Mech., vol. 59, 209, 1973.
31. Cotten, I. Rubb. Chem. Tech., vol. 57, 118, 32.Funt, J.M., 'Rubber mixing,' ibid, vol. 53, 772-779, 1980.
33. Shiga, S. and Furuta, M., Nippon Gomu Kyokaishi, vol. 55, 491, 1982.
34. Manas-Zloczower, I., Nir, A. and Z. Tadmor, 'Dispersive mixing in rubber and plastics,' Rubber Chem. Tech., vol. 57, 583-620, 1984.
35. Zia, I.Y.Z., Cox, R.G. and Mason, S.G., Proc. Roy. Soc., vol. A300, 427, 1967.
36. Manas-Zloczower, I. and Z. Tadmor, 'Scale up of internal mixers,' Rubber Chem. Tech., vol. 57, no. 1, 48-54, 1984.

BLENDING OF INCOMPATIBLE POLYMERS

A.K. VAN DER VEGT and J.J. ELMENDORP

Laboratory of Polymer Technology, Delft University of Technology.

SYNOPSIS

The morphology of a polymer blend, composed of incompatible components, is governed by deformation and break-up of the dispersed phase but also by capillary instabilities of thread-like particles and by coalescence of the dispersed phase. Additional effects may be created by the presence of a yield stress and by orientation-induced crystallization.

1 INTRODUCTION

Most polymer pairs are incompatible, so that blending operations generally result in a dispersion. During blending the scale of the dispersion is gradually reduced until an equilibrium is reached, depending on polymer properties and process conditions. Although several aspects of this process are fairly well understood, a reliable prediction of the type and the scale of the dispersion for a given system is, as yet,beyond reach. In our laboratory some aspects of the blending mechanism have, over the past few years, been studied in some detail. In this paper a survey will be given of the various stages of a blending process, partly based on literature and partly on some recent findings in our research group.

2 DEFORMATION AND BREAK-UP OF FLUID ELEMENTS

In the initial stage of a blending process the system consists of a mixture of fluid elements of the same size as the original solid particles (e.g. nibs or powder). If the volume fractions of the components do not differ too much, isolated elements of both components may be present within a local matrix of the other component. When the viscosities of the components differ significantly, the most probably situation will be that the component with the lowest viscosity acts as the continuous phase.

In a mixing operation this two-phase system is subjected to a flow field which in most cases is a combination of shear flow and elongational flow. The first effect of this flow is a deformation of the dispersed elements from e.g. a sphere into a rotation ellipsoid. For shear flow Cox[1] has derived an expression which, for Newtonian fluids at not too high deformations, has been proven to be valid:

$$D = \frac{5\ (19\lambda + 16)}{4\ (\lambda+1)\ [(19\lambda)^2 + (20\ k)^2]^{\frac{1}{2}}} \qquad (1)$$

in which $D = (L-B)/(L+B)$, (L is the long and B the short axis),
$\lambda = \eta_d/\eta_c$ (viscosity ratio of dispersed and continuous phase),
$k = \frac{\sigma}{G.\eta_c.R}$ (σ = interfacial tension, G = shear rate, R = initial sphere radius).

In the first stage of the process, when R is still large, no stable deformation occurs, since pressure differences due to interfacial tensions are neglegible compared to the shear stresses in the matrix; threads are being formed. With decreasing length scale the cohesive action of the interfacial tension increases until stable droplets are originated, which can only be deformed to a limited extent. Deformation of such droplets has been studied in our laboratory using Newtonian liquids in shear flow[2]. The results were in agreement with Cox' equation. For non-Newtonian liquids deviations were observed, which could not be straightened out by substituting appropriate values for the apparent viscosities at the pertinent shear rates. Shear rate distributions inside and outside the drop probably invalidate this simplified approach.

Deformation of the droplets increases up to a level where they break up. According to Taylor[3] the shear rate, G_b, at which this happens, can be written as

$$G_b = \frac{\sigma}{2\ \eta_c\ R} \cdot \frac{16\lambda + 16}{19\lambda + 16} \qquad (2)$$

Smaller droplets deform less and have a higher value for G_b. In a shear field with constant shear rate, G, an equilibrium drop size will be reached, which, by approximation, can be written as

$$R = \frac{\sigma}{2\ \eta_c\ G} \qquad (3)$$

Here again, the non-Newtonian behaviour causes a complication. In addition fluid elasticity may effect the result. As to the latter, VanOene[4] suggested that the interfacial tension could be corrected with a term containing the first normal stress differences of both components:

$$\sigma_{eff} = \sigma + \frac{R}{6}\left[(\tau_{11}-\tau_{22})_d - (\tau_{11}-\tau_{22})_c\right] = \sigma + \frac{R}{6}\Delta\tau \qquad (4)$$

Combination with eq. 3 results in

$$R = \frac{\sigma}{2\,\eta_c\,G - \frac{\Delta\tau}{6}} \qquad (5)$$

This means that elasticity of the matrix tends to reduce the scale of the dispersion, whereas elasticity of the dispersed phase leads to a coarser dispersion. These effects have been confirmed qualitatively by our experiments on single drops, using elastic model liquids of various nature. Quantitative agreement was, however, not obtained; e.g. in cases where $\Delta\tau/6$ is of the same order of magnitude as $2\,\eta_c\,G$, the effect on droplet size appeared to be much smaller than predicted by eq. 5.

A number of trials have been carried out on molten polymers of known visco-elastic properties and interfacial tension in a simple Couette-type shear field. It was found in all trials with concentrations higher than 10% that the equilibrium particle size of the dispersed phase was several times greater than expected. Because of the higher deformability of larger particles this finding may be related to the observation that, in some blending experiments, next to dispersed particles several long strands were found to be embedded in the matrix phase. These two phenomena will be discussed in the next two sections.

3 COALESCENCE

It would be an oversimplification to assume that droplets formed by break-up of bigger drops (e.g. according to Taylor's formula) or by break-up of a cylindrical thread due to capillary instabilities (see section 4) will retain their individuality in the course of a blending process. Small drops may recombine into bigger ones by coalescence. When in the flow field two particles collide they will deform so that a layer of the matrix polymer is formed between the two drops. This layer will gradually be squeezed out until the distance between the interfaces reaches a critical value (of the order of 50 nm) at which the matrix fluid breaks and the drops flow together. The rate at which this squeezing-out occurs, or the speed at which the interfaces approach each other, can, with some generalization, be expressed as[5]

$$\frac{dh}{dt} \; (:) \; \frac{h^{\alpha}}{\eta_c\, R^{\gamma}} \qquad (6)$$

The values of α and γ depend on the nature of the interface: $\gamma = 1$, $\alpha = 1$ for a fully mobile interface, and $\gamma = 5$, $\alpha = 3$ for a rigid interface which can withstand tangential stresses without interfacial flow occurring. The latter condition holds in most cases as a result of the presence of surfactants even in extremely low concentrations (e.g. 10^{-10} mol/l).

The time required for coalescence is very strongly dependent on γ; therefore we have investigated which of the two cases occurs in polymer blending. From the measurement of coalescence time under the influence of gravity of a number of droplets of various sizes on a flat interface, it appeared that this time was proportional to the radius of the droplet (see fig. 1).

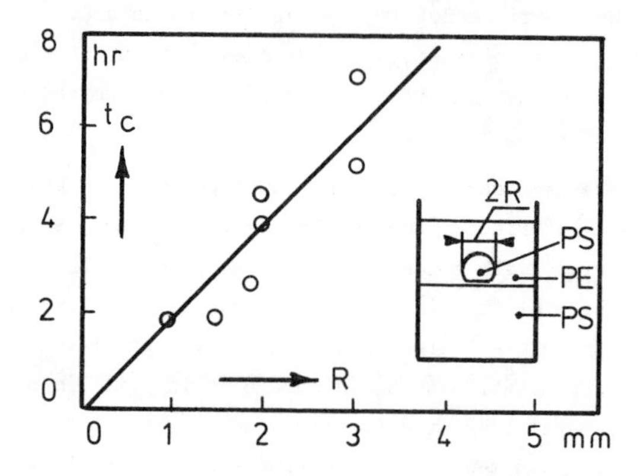

Fig.1. Coalescence time as a function of drop size

This means that the interface is fully mobile; apparently the diffusion of surfactant material, when present, is too slow to effect the interface mobility. Coalescence is, therefore, much more probable to occur in polymer blending and the distribution of particle sizes will be governed by the dynamical balance between break-up events and coalescence events. Proper modelling of this balance is extremely complicated and has only scarcely been applied to predict the resulting scale of dispersion for molten polymer systems. An obvious first conclusion is that the equilibrium particle size will be strongly dependent on the volume fraction of the dispersed phase: the less particles present, the less the probability of collision and coalescence will be. In our laboratory we have carried out blending experiments of PS and PP in a single-screw extruder with a broad range of PP concentrations. The resulting average particle size as a function of volume fraction is presented in fig. 2, which clearly shows the strong effect of concentration.

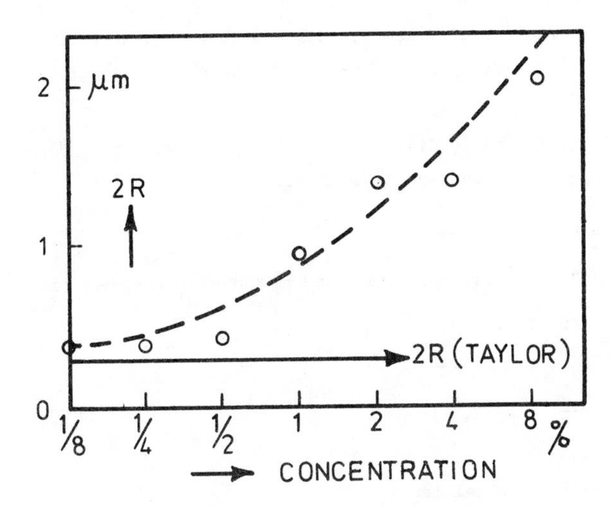

Fig.2. Equilibrium drop size vs. concentration

Extrapolation to zero concentration yields a particle size which is in good agreement with the value from Taylor's equation for Newtonian fluids. Other series of measurements have confirmed this result. A first-order approximation of the average particle size as a function of concentration can be expressed as

$$\overline{R} = R_{Taylor}\ (1 + \beta\ \phi) \tag{7}$$

with β, dependent on conditions, varying between 50 and 200. β appears to be much higher than values reported in literature on low-molecular liquids (viz. 5 to 10), apparently as a result of the mobility of the interface.

4 FORMATION AND STABILITY OF THREADS

In Section 2 the breaking-up of fluid elements under influence of shear and elongational stresses was mentioned. Closer consideration of eq. 4 shows that the break-up is practically controlled by surface tension, geometry and matrix viscosity only, whereas the viscosity of the dispersed phase has a negligible effect. It seems, therefore, logical to assume that eqs. 4, 5 and 6 can only be applied to quasi-static conditions, at which the relative displacement of volume elements within the drop, necessary for breaking-up, are not delayed by the high viscosity of the dispersed phase. At deformation rates exceeding G_b manyfold, this is, however, no longer the case so that highly elongated drops may be formed, which then break up after a considerably longer time. This fragmentation of filaments can be treated quantitatively on the basis of Rayleigh's description of capillary instabilities[6], later extended by Tomotika[7].

If a thread is sinusoidally distorted, so that its radius, R, depends on the coordinate along the thread, z, as follows:

$$R(z) = \overline{R} + \alpha \sin \frac{2\pi z}{\Lambda} \tag{8}$$

then for $\Lambda > 2\pi R$ the distortion will grow exponentially with time:

$$\alpha = \alpha_o \exp qt \quad , \tag{9}$$

in which

$$q = \frac{\sigma}{2\eta_c R_o} \cdot \Omega(\Lambda, \lambda)$$

$$\lambda = \eta_c/\eta_d$$

σ is the interfacial tension.

$\Omega(\Lambda, \lambda)$ is a function showing a maximum at a certain value of Λ, Λ_m, which depends on the viscosity ratio λ. Initially the thread will contain random distortions with a broad spectrum of wavelengths; after some time the distortion with wavelength Λ_m will become dominant. By imposing initial disturbances on a thread, we were able to study its break-up due to distortions with non-dominant wavelengths. Figure 3 represents the wavelength dependence of the growth rate of distortions imposed on threads of silicon oil in castor oil.

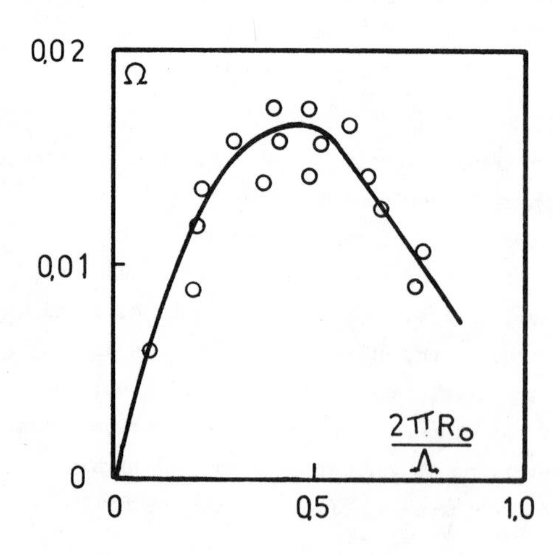

Fig.3. Theoretical and experimental distortion growth rate vs. wavelength

The points in this graph are the experimentally determined values of the $\Omega\ (\Lambda,\ \lambda)$ function; the drawn curve has been calculated from Tomotika's original equations. In spite of some scatter good agreement is obtained.

The time needed to convert the thread-like particles to a string of small droplets can be calculated by extrapolating eq. 9 to the point of breakage:

$$t_b = \frac{1}{q}\,\ell n\,\left(\frac{0.8\ R_o}{\alpha}\right) \qquad (10)$$

Experiments carried out on molten polymer threads, embedded in another polymer, lead to results which were in quantitative agreement with the above equation. Depending on the conditions the break-up time varied between a few seconds and a few hours[8].

Apparently, thread-like particles will be present in a polymer blend if the time needed for full development of instability exceeds the time of blending and solidification. Such threads have actually been observed in blends prepared on a two-roll mill.

The driving force in the breaking-up of a cylindrical thread is the interfacial tension; the viscosities counteract the growth of distortions but they can only retard the process. If, however, the thread material exhibits a finite yield stress which exceeds the pressure difference due to the interfacial tension in a distorted thread, a distortion will be unable to grow and the thread will be stable. A quantitative criterion for this phenomenon can be derived by calculating the pressure difference between the thickest and the thinnest part of a sinusoidally distorted cylinder. This criterion has been checked experimentally by using polymer solutions with finite yield stresses and investigating the stability of threads with various thicknesses. Figure 4 shows some of the results; the drawn curve is the theoretically predicted boundary between stable and unstable behaviour, the + and - signs each represent an observation on stability.

This phenomenon apparently forms the basis for the fact that some polymers, as e.g. the three-block copolymer SBS, tend to form a continuous network in blends, even at small concentrations, since it is known[9] that SBS exhibits a finite yield stress.

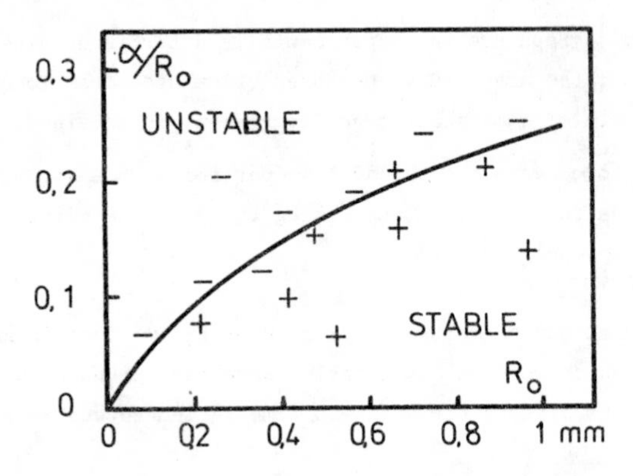

Fig.4. Stability observations on fluid threads with a yield stress; the drawn line is the calculated boundary.

Another possibility for filaments to be present in a blend is rapid crystallization of the dispersed phase. If a crystallizable polymer is dispersed in another polymer and the system is subjected to a very rapid elongational deformation, the high molecular orientation in the fibres formed by the dispersed phase, may lead to spontaneous crystallization if the temperature level has been adjusted to a value not too far above the melting point[10]. The condition for this to happen is proper balance between four times scales, viz. of the deformation, of the formation of instabilities, of the relaxation of the orientation and of the crystallization.

CONCLUSIONS

The morphology of a blend, formed from incompatible polymers, is governed by a number of factors. Among these are the viscosities and elastic properties of both components, and the magnitude of the shear and elongation gradients present in the flow field of the mixer. Also the time scale of mixing and further processing plays an important role, in particular with respect to the formation and growth of instabilities in elongated strands and in the coalescence of small drops into bigger ones.

Moreover, special features of the components, as e.g. the presence of a small yield stress, or orientation-induced crystallization, may result in blends with deviating morphologies and, therefore, with unusual properties.

Further research on the mechanism of blending is expected to open ways to the manufacture of novel types of blends.

REFERENCES

1. Cox, R.G. J. Fluid Mech., 37 (3), 601-623, 1969.

2. Elmendorp, J.J. and Maalcke, R.J. Pol. Eng. Science, in preparation.

3. Taylor, G.I. Proc. Roy. Soc., A 14 b, 501-523, 1934.

4. VanOene, H.J. J. Coll. and Int. Sci., 40 (3), 448-467, 1972.

5. Elmendorp, J.J. to be published in "Mixing in Polymer Processing", ed. C. Rauwendaal, Marcel Dekker Inc., 1986.

6. Lord Rayleigh, Proc. London Math. Soc., 10, pp. 4, 1878.

7. Tomotika, S. Proc. Roy. Soc. (London), A 150, pp. 322, 1935.

8. Elmendorp, J.J. Pol. Eng. Science, in preparation.

9. Ghijsels, A. and Raadsen, J. Pure and Appl. Chem., 52, pp. 1359, 1980.

10. Van der Vegt, A.K. and Verbraak, C.L.J.A. in "Interrelations between Processing, Structure and Properties of Polymeric Materials", ed. J.C. Seferis and P.S. Theocaris, Elsevier, Amsterdam, pp. 57, 1984.

POLYMER REACTIONS DURING MELT-PROCESSING

Shaul M. AHARONI

Chemical Sector Research Laboratories, Allied Corporation, Morristown, New Jersey 07960, U.S.A.

SYNOPSIS

It is shown that when melt-processed in the presence of organic phosphites, chain extension of polyamides and polyesters, and block or graft copolymers can be obtained from polymers containing carboxyl end groups and primary amines or hydroxyl residues. The reaction mechanism is described.

We have recently discovered[1,2,3,4] that amide or ester bonds between carboxyl chain ends and primary amine or hydroxyl chain ends can be effected during melt processing in the presence of certain organic phosphites. This allowed us to obtain three kinds of polymeric systems: (a) high molecular weight linear homopolymers obtained from their difunctional low molecular weight analogs; (b) linear block copolymers obtained from two difunctional polymers different from each other, and (c) graft copolymers obtained from, preferably, monofunctional polymers reacted with polyfunctional polymers of different nature.

The formation of amide or ester bonds during melt-processing in the presence of organic phosphites was found to be applicable to various polymers and polymer-pairs, provided that (a) they contain available carboxylic groups and aliphatic amine or hydroxyl groups, (b) they are stable under the processing conditions, and (c) their melt viscosities are sufficiently close to allow for thorough mixing in the molten state to facilitate the reaction.

A convenient method to gauge the changes in chain length is to compare the intrinsic or relative viscosity of the sample with the viscosity of a blank subjected to the same processing conditions. Typical results, obtained while monitoring the progress of chain extension reactions of homopolymers in the presence of triphenyl phosphite (TPP) and tributyl phosphite (TBP) are shown in Table 1.

<u>Table 1:</u>

Molecular weight enhancement of single polymers.

Polymer	Phosphite Weight %	Reaction time	Reaction temp.	Viscosity
Nylon-6	0	10	265	1.85
Nylon-6	0	5	300	1.84
Nylon-6	1% TPP	15	200	1.84
Nylon-6	1% TPP	10	245	2.00
Nylon-6	1% TPP	7.5	265	2.24
Nylon-6	1% TPP	3.75	285	2.57
Nylon-6	1% TPP	3.75	305	3.98
Nylon-66	0	7.5	300	1.66
Nylon-66	0.6% TBP	7.5	300	2.41
Nylon-11	0	5	300	1.38
Nylon-11	1% TPP	5	300	2.17
Nylon-12	0	5	300	1.36
Nylon-12	1% TPP	5	300	2.06
PET	0	12.5	265	0.68
PET	1.5% TPP	1.5	265	0.79
PET	1.5% TPP	7.5	265	0.95
PET	1.5% TPP	30	265	1.07

Note: Time in minutes, temperature in ^{o}C, relative viscosity of polyamides at 0.5% concentration in m-cresol, intrinsic viscosity of PET in mixed solvent.

Qualitative tests of the efficiency of various phosphites in promoting the formation of amide bonds at elevated temperatures were conducted by evaluating the increase in intrinsic viscosity of nylon-6 brought about by extrusion at 300°C for identical residence time. In all cases the same grade polyamide was used. The various phosphites were introduced in amounts corresponding to about one phosphorus atom per each nylon-6 chain. The efficiency of organic phosphites in increasing the chain length of nylon-6 decreases in the series: triphenyl phosphite $>$ diphenyl phosphite $>$ tris(nonylphenyl)phosphite $\gtrsim$ tri(2,4-di-t-butylphenyl)phosphite $\gg$ tributyl phosphite $\gtrsim$ tri(chloroethyl)phosphite $>$ diphenyl isodecyl phosphite $\gtrsim$ poly(4,4'-isopropylidenediphenol "neodol 25" alcohol)phosphite $\gtrsim$ bis(2,4-di-t-butylphenyl)pentaerythritol diphosphite $>$ tri(isodecyl)phosphite $\gtrsim$ di(isodecyl) phosphite $\gg$ triethyl phosphite $>$ distearyl pentaerythritol diphosphite. Inorganic phosphites were found to be ineffective. Less extensive tests indicated that the above hierarchy of phosphite efficiency is not limited to nylon-6 alone, but carries over to chain extension of other melt-processable polyamides and to block or graft formations typified by the systems listed in Table 2 below. Esterification reactions require, in general, the use of the most reactive, aryl and substituted aryl, phosphites in order for the reaction to progress in an acceptable rate during the residence in the extruder.

Table 2:

Block or graft copolymers formed within 5 minutes at $300^{\circ}C$ in presence of 1% TPP.

Family	Polymer pair
Block copolyamides	Nylon-6/nylon-66
	Nylon-6/nylon-12
	Nylon-6/nylon-11
	Nylon-6/nylon-6T
	Nylon-6/Trogamid T
Block poly(ester-amide)	Nylon-6/PET
	Nylon-6/PBT
	Nylon-6/Kodar A-150
Graft poly(amide-olefin)	Nylon-6/EAA-455
	Nylon-6/EAA-452
	Nylon-6/Surlyn 1855
	Nylon-6/Surlyn 1856

Studies on model compounds were conducted at $280^{\circ}C$ in the absence of any solvent, and at $100^{\circ}C$ in the presence of solvent and pyridine; the latter serving as a hydrogen scavenger. The analysis of the reaction end-products, by-products and intermediates, mostly by phosphorus NMR, showed[5,6] that in the amide-formation reaction, the first step was the formation of diaryloxy aminophosphine by the reaction of the primary amine with the arylphosphite. This intermediate slowly reacts with the available carboxylic acid to produce an amide group and a diarylphosphite. In the esterification reactions, the triarylphosphite first reacts with the available aliphatic hydroxyl to produce an alkoxyaryloxy phosphite. This intermediate slowly reacts with the available carboxylic acid to form an ester group and a diarylphosphite. In the absence of either primary amine or aliphatic hydroxyl, the phosphite slowly reacts with carboxylic acid to yield an aryl ester and diarylphosphite. The rate of reaction of the three species with TPP is primary amine>> aliphatic hydroxyl>> carboxy acid.

REFERENCES.

1. Aharoni, S.M. and Largman, T. U.S. Patent 4,417,031 of 22/11/83.
2. Largman, T. and Aharoni, S.M. U.S. Patent 4,433,116 of 21/2/84.
3. Aharoni, S.M. Polym. Bull., 10, 210 (1983).
4. Aharoni, S.M. U.S. Patent, submitted.
5. Aharoni, S.M. Hammond, W.B. Szobota, J.S. and Masilamani, D. J. Polymer Sci. Polym. Chem. Ed., 22, 2567 (1984).
6. Aharoni, S.M. Hammond, W.B. Szobota, J.S. and Masilamani, D. J. Polymer Sci. Polym. Chem. Ed., 22, 2579 (1984).

ASSESSING RUBBER PROCESSING AIDS EFFECTIVENESS

Jean L. LEBLANC

Monsanto Europe S.A., Brussels, BELGIUM

1 INTRODUCTION

Processing aids are quite an important class of rubber compounding ingredients whilst their actual mode of action is as yet by no means clearly understood. Therefore their choice and use remain more a question of pragmatical experience than of rubber science.

As experienced on the factory floor by most people in industry, rubber processing aids do work whilst laboratory assessment often leads to disappointing results. Although shear differences between laboratory and factory equipments are key factors, a clearer understanding of processing aid modes of action would lead to more pertinent laboratory test procedures.

This paper intends (i) to discuss the possible modes of action for rubber processing aids, (ii) to draw preliminary conclusions about most appropriate test procedures and (iii) to report assessment results with selected processing aids.

2 MODES OF ACTION

Various natural products, such as long chain fatty acids, their soaps and esters, petroleum and petrolatum blends and mixtures, natural and synthetic resins, waxes and paraffins are amongst the number of materials which have been marketed for years as rubber processing aids. The variety and diversity of these rubber compounding ingredients is challenging any classification and most reviews deal with their chemical nature (1, 2).

Completely different products frequently offer similar processing effects in such an extent that nearly all classes of processing aids overlap each other with respect to their benefits.

With the progress made in recent years in studying the particular flow phenomena involved in processing rubber compounds, it is likely that the mode of action of processing aids can be somewhat cleared out, which would lead to a better selection of these compounding ingredients.

Basically there are two extreme possible modes of action for rubber processing aids, either as bulk viscosity modifiers, or as external lubricants at rubber-metal interface. As schematically illustrated in Figure 1, the simple lubrication mechanism would result rather in (increased) slippage of the rubber compound at the wall of processing equipment, whilst bulk viscosity modification would be obtained through intermacromolecular or intramacromolecular action.

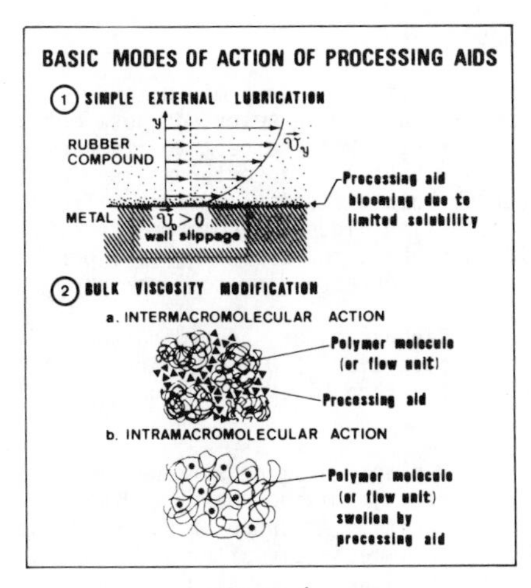

Figure 1

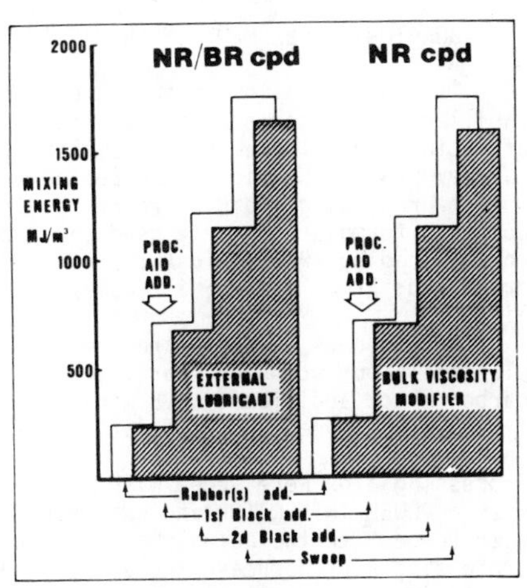

Figure 2

In the simple lubrication mode, a certain uncompatibility is required between the processing aid and the elastomer. For bulk viscosity modification, the processing aid must be either soluble (intramacromolecular action) or at least compatible with the rubber (intermacromolecular action). Such extreme modes of action are obviously model mechanisms which have to be further refined to take into account interactions with fillers and other compounding ingredients. Actual mechanisms are most probably more complex than the simplistic models of Figure 1 and, depending upon the chemical considered, somewhere in between these extreme modes of action. This model approach allows, however, laboratory test procedures to be selected in a more appropriate manner, with respect to expected benefits in industrial operations.

3 SELECTION OF LABORATORY TEST PROCEDURES

Further to their likely modes of action previously discussed, most of the processing aids are expected to play some rôle in rubber-filler interactions. In a first approach, processing aids must consequently be evaluated through mixing experiments. Whilst similar temperature and shear levels can be achieved with both laboratory and factory mixers, there remain, however, some differences not the least being the actual batch size. The scale-up problem has somewhat been clarified with the mixing energy unit concept (3, 4), which allow pertinent information to be obtained from laboratory mixing experiments, providing the mixer is equipped with the appropriate power integrator.

The capillary rheometer allows the high shear levels of rubber processing to be investigated, in carefully controlled conditions. In addition, with respect to the actual flow situations in the barrel-and-die system, pertinent information can be obtained about the rôle of processing aid not only in shear flow but also in elongational flow conditions (5).

In summary, a suitable combination of mixing experiments and capillary rheometer tests allows an adequate preliminary assessment of processing aid to be made.

4 LABORATORY ASSESSMENT OF PROCESSING AIDS

In order to investigate the significance of laboratory mixing and capillary rheometer experiments, two processing aids were selected whose modes of action are believed to be completely different. With respect to its chemical nature, processing aid A is expected to mainly behave as an external lubricant, whilst processing aid B is rather an internal bulk viscosity modifier. Tread compound formulations were used (*) and processing aids were compared at similar volume loading (Proc. aid A : 3.45 cm^3/100 g rubber, corresponding to 2 phr proc. aid; Proc. aid B : 3.26 cm^3/100 g rubber, corresponding to 3 phr proc. aid).

Compounds were prepared in a 1.5 l BR Banbury mixer, according to a classical four step procedure, i.e. rubber mastication (1.0 min), first half carbon black addition (+ 1.5 min), second half carbon black addition (+ 1.5 min), sweep and dump (at 5.5 min).

Processing aids were added with the first half black. Mixing work was measured with a POWER INTEGRATOR and kWh reading converted in mixing energy (MJ/m^3) with respect to the actual batch weight and density. As shown in Figure 2 both processing aids decrease the energy consumption immediately after addition but, due to increased wall slippage, the external lubricant gives slightly larger differences. However, as the external lubricant is somewhat compatible with the compound, it apparently allows adequate filler dispersion to be obtained since no undispersed carbon black agglomerate was observed. These results suggest that the type of processing aid - and its associated mode of action - is critically influencing the mixing process. Further experiments are nevertheless needed in order to understand the complex interactions between the rubber, the filler and the processing aid.

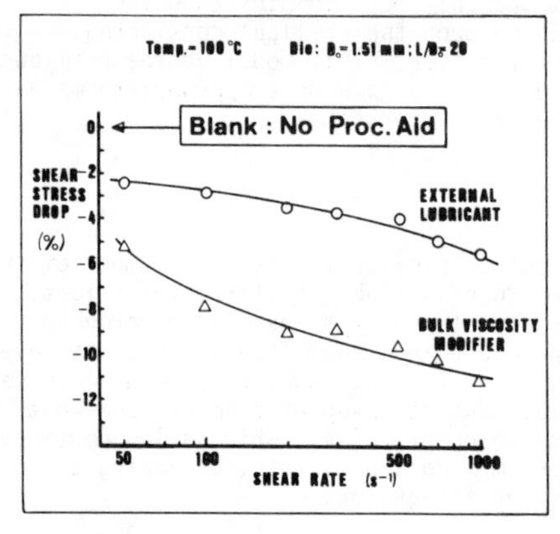

Figure 3

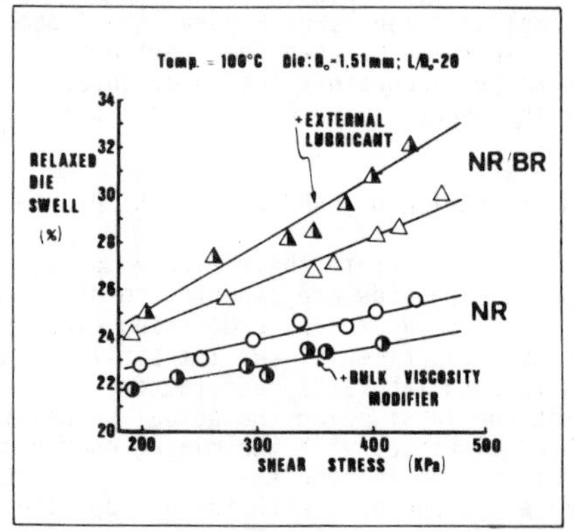

Figure 4

(*) NR/BR CPD : SMR 10 : 75 / Cariflex BR : 25 / N242 black : 45 / ZnO : 5 / Stearic Acid : 2 / Oil : 6

NR CPD : SMR 10 : 100 / N330 : 50 / ZnO : 5 / Stearic Acid : 2 / Oil : 3

Capillary rheometer experiments were performed with the MONSANTO PROCESSABILITY TESTER at 100°C, using a 1.51 mm diameter die (L/D_o = 20). Several tests were made in order to cover a 50 - 1000 s^{-1} shear rate range and processing aid effects were analyzed in terms of shear stress drop (%) with respect to the blank compound without processing aid. Figure 3 shows the large difference due to the type of processing aid. Both products give significantly lower shear stress (and hence higher output) and the benefit increases with the shear rate, but the bulk viscosity modifier (Proc. aid B) appears largely more efficient, despite the fact that the volume loading of processing aid A was slightly higher.

Figure 4 shows the effect of processing aids on relaxed die swell. As can be seen, the external lubricant increases the die swell whilst the bulk viscosity modifier decreases it. These observations correspond very well with the assumed mode of action of both products. Processing aid A increases slippage at the entrance of the die and therefore higher elastic stresses are stored in the elongational flow region, which give higher extrudate swell. Processing aid B decreases the bulk viscosity of the compound and consequently relaxation processes are favoured and it results in lower die swell values.

5 CONCLUSIONS

A better approach of rubber processing aids selection is likely to arise from a basic understanding of their mode of action. Processing aids modify the flow properties of rubber compounds in a complex manner, somewhere between the two extremes modes of external lubrication and bulk viscosity modification.

Mixing and capillary rheometer experiments allow processing aid effects to be studied in a suitable manner and, when interpreting test results with respect to the assumed mode of action of processing aids, key information can be obtained.

REFERENCES

(1) O'Connor F.M. and Slinger J.L. 'Rubber World', 187 (1), 19 (Oct. 1982)

(2) Crowther B.G. 'Plast. Rubber International', 9 (5), 14 (1984)

(3) Van Buskirk P.R., Turetzky S.B. and Gunberg P.F. 'Rubber Chem. Technol.', 48, 577 (1975)

(4) Dizon E.S. and Papazian L.A. 'Rubb. Chem. Technol.', 50, 765 (1977)

(5) Leblanc J.L. 'Rubb. Chem. Technol.', 54, 905 (1981)

PLASTICS PROCESSING

G.Menges
IKV-Aachen/FRG

1. Introduction

The tasks involved in plastics processing are extremely varied and hence the manufacturing methods differ widely as well. There is, however, a uniform development trend in evidence, in that all the sectors are working on easier and error-free setting of production plant and on maximum possible automation. This trend is still being backed up to a large extent by the huge increase in wage costs registered in all industrialised nations and also by the reluctance of many employees to work night shifts. Quality requirements have also increased to a very large extent, particularly since industrial components have become a big growth area for plastics. If plastics are to compete with moulded parts in other materials here then it is essential for dimensions and properties to be reliably observed. A wave of new investment is thus to be seen in a large number of countries - particularly in injection moulding machines equipped for fully automatic operation.

Another area which is attracting a great deal of interest in parallel to industrial components is the manufacture of moulded parts with long fibre reinforcement. It is thus fairly certain that the next generation of military aircraft in the first instance, and then civil aeroplanes, will be extensively equipped with structural components in fibre composite materials with a plastic matrix. The corresponding production units are currently being prepared.

The car industry is similarly equipping itself to switch bodywork parts over to plastic and load-bearing components in the drive and chassis over to fibre composite materials.

This is being prompted in both cases first and foremost by weight savings and the attendant economies in design.

High precision and reproducible properties are required in the finished parts for all these cases, and hence there is a call not only for totally reliable production but also for virtual 100% monitoring by means of supervision facilities and testing incorporated into production. It also has to be ensured, however, that the parts used will attain the required service life, and thus very extensive practical trials are currently being undertaken, some of which are to run for years. Test components have already been fitted in aircraft for 10 years or so undergoing flight trials. The same applies to the car industry, which can likewise ill afford failures which could entail expensive recall campaigns and damage its image. Practical trials of this type are not only very time-consuming, however, but are also very expensive, and hence a remedy is urgently required. The only remedy is to be seen in improved planning, incorporating scientific findings. It is for this reason that raw materials manufacturers and research institutes have been working on the determination of long-term properties under a wide range of loads for many years.

It unfortunately has to be ascertained, however, that the wide range of knowledge already acquired on plastics is scarcely known or readily accessible to the experts, let alone to the designer. We thus feel it ought to be possible to bring these scientific findings to the workplace through a type of expert program. These expert programs must be capable of running on mini-computers, such as those installed in CAD systems. The usual designer is then guided from decision to decision by the order specifications right through to the end of the planning work, backed up by calculations and in parallel to his graphic design (1,2). A first expert program of this type for the design of injection moulded parts is currently being developed at our Institute. The initial sections of this will be brought on to the market towards the end of this year (1,2). We hope that this will serve to shorten the duration of trials for moulded parts. Experience has shown that virtually none of the comprehensive scientific knowledge has been applied in practice to date because this knowledge is located in a large number of individual sources and the designer finds it too much trouble to first have to look for this knowledge and assimilate it.

One factor which is still largely lacking for the compilation of this type of program today is knowledge of the way in which moulded part properties are influenced by the production process. We thus regard the establishment of these causal relationships as a prime task. We are actively engaged on this work together with six other institutes from our University in a special research program which is being given long-term funding by the German Research Association (Deutsche Forschungsgemeinschaft) (19).

2. Automated processes and in-line control

Injection Moulding

Fully automatic injection moulding production units which make use of the opportunities offered by computer process control have been on the market for two years (3,4,5). At the command of the central master computer, production is switched automatically from one particular part to another part in a different material (cf Figs. 1 and 2). This not only requires a change of mould but also a change of plasticising unit. On the Netstal unit, for example, a crane, guided into the correct position by a laser, removes the mould that is no longer required along with the plasticising cylinder and conveys these to a holding station. It then takes up the mould, which has already been prepared for use, and the plasticising cylinder and inserts these into the machine. Molten plastic is injected through the plasticising cylinder to clean it, the nozzle is cleared and then

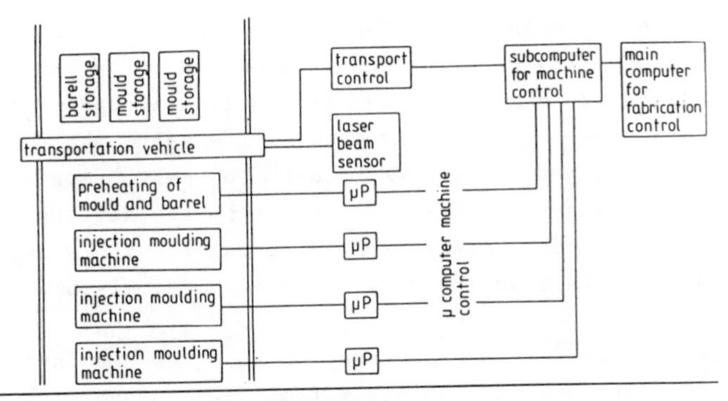

Fig. 1: Diagram of the Netstal unit
(Source: Netstal)

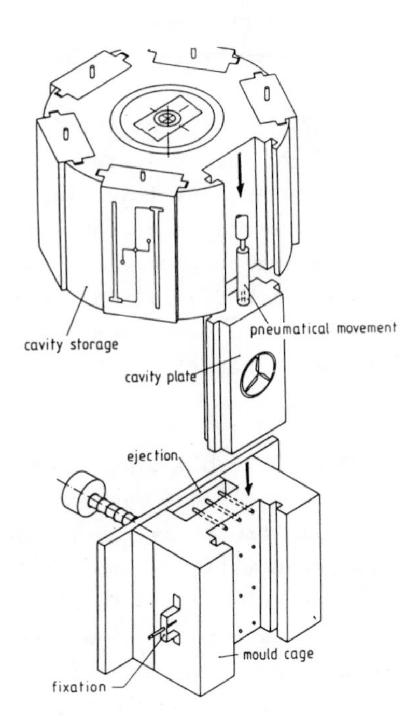

Fig. 2: Changeover of injection units

production of the new part can begin. The point at which the first quality parts begin to be manufactured can be seen from the built-in monitoring system which incorporates appropriate sensors. Only then will the parts start to be conveyed to the packing station with the rest being automatically discarded. Units of this type are already being used now to produce highly complex products with a recurring demand, such as image-storing discs and medical components.

Figure 3 shows a less expensive and also less comfortable solution for a rapid changeover from one moulded part to another. This allows the parts manufacturer to upgrade a single modern machine into a flexible production centre tailored to his individual requirements. This system comprises a standard commercial robot which serves a likewise standard commercial, computer-controlled injection moulding machine and assembles the parts immediately after removal or at a later stage. The machine incorporates a mould carrier which can be supplied in a matter of seconds with the interchangeable plates that carry the mould cavities from the plate container where they are stored and pre-heated. All three system components of injection moulding machine, robot and mould carrier are controlled and supervised by a higher-ranking master computer. It is of particular advantage here that small and very small series can also be manufactured on call - at a very much lower cost. The manufacturer is, however, restricted to a single material in this case.

It goes without saying that systems of this type cannot get by without facilities for monitoring all the machine functions and quality. It must, for instance, be guaranteed that any unserviceable part produced will be immediately discarded automatically so as to prevent any defective parts from reaching dispatch. Further production must not be disturbed either. The monitoring of machine functions is perhaps even more important, since system components could otherwise be easily damaged.

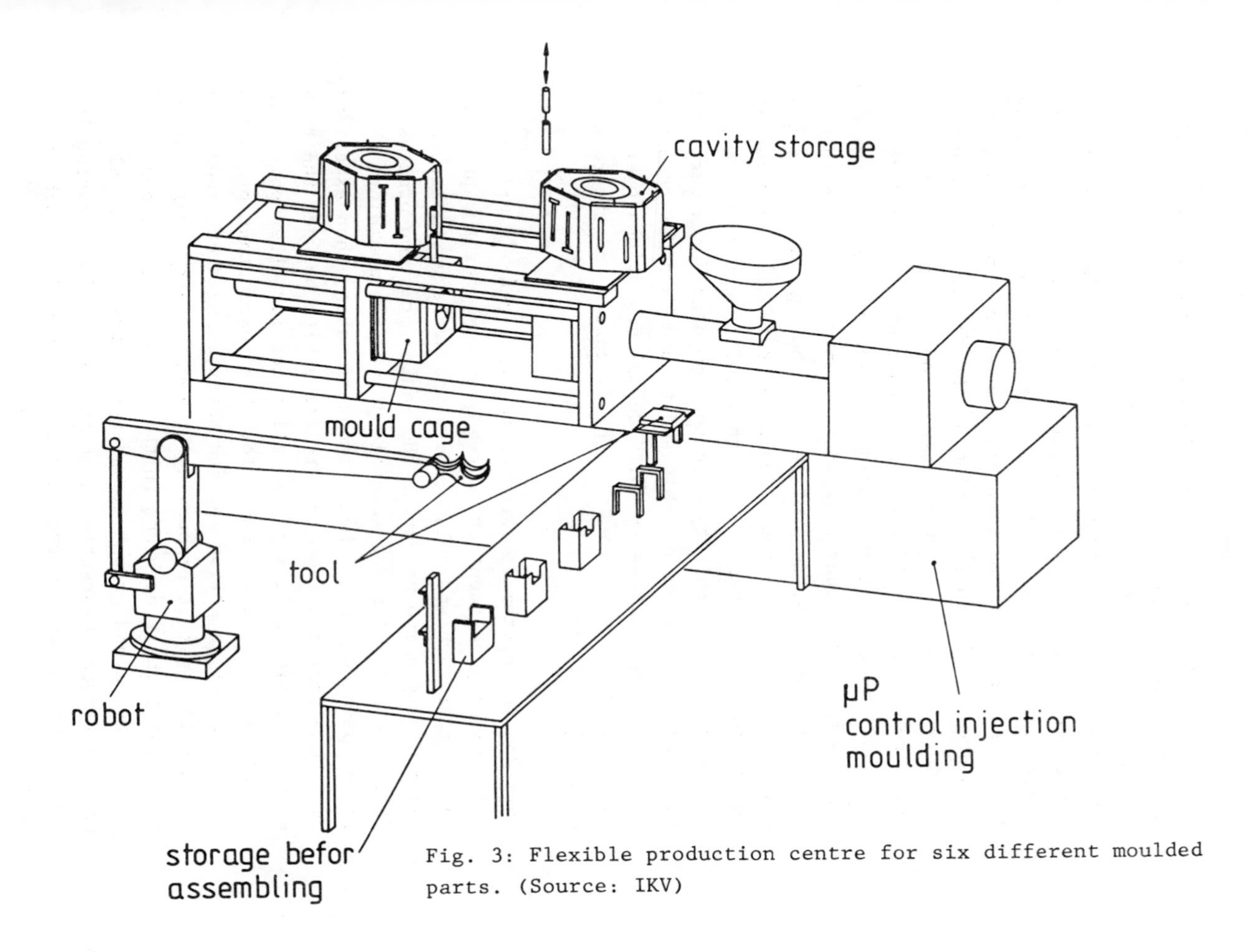

Fig. 3: Flexible production centre for six different moulded parts. (Source: IKV)

It has been seen in a large number of tests based on methods for temperature and, more especially, pressure control in the mould cavity, worked out at the IKV (6), that it is sufficient to monitor moulding quality on the basis of these two process parameters to ensure uniform production (7,8). A pressure sensor close to the gate in the moulding cavity - the point where the melt enters the mould - measures the pressure for the full duration of the filling and holding pressure stages until such time as cooling in the mould is complete (Fig. 4). If the temperature of the moulding compound and that of the mould are kept constant at the same time by means of separate feedback control circuits then the course of the pressure profile will show up any change that comes about in the process sequence or in the behaviour of the moulding compound. It is thus sufficient to store a sample profile of this type in the computer for a part that has been found to be suitable and to use this for comparison. Poor quality parts which reveal deviations over and above an admissible level will be discarded and the rest automatically packed.

This control system is based on the fact that identical parts display an identical curve over time for the pressure, melt temperature and volume, i.e. the state variables, of the moulding compound during the mould formation process. If this applies for one specific point of a moulding, i.e. at the point where the measurement is taken, then experience has shown that the same interrelationships between the three state variables will apply at all times for other points of the moulding as well. It is for this reason that production monitoring of injection moulded parts using these sensors (primarily pressure probes in the mould cavity, which are not without their problems in the rough operating conditions) has gained ground almost everywhere. It has also been confirmed in practice that this type of monitoring is equivalent to 100% control and is recognised as such by customers (9).

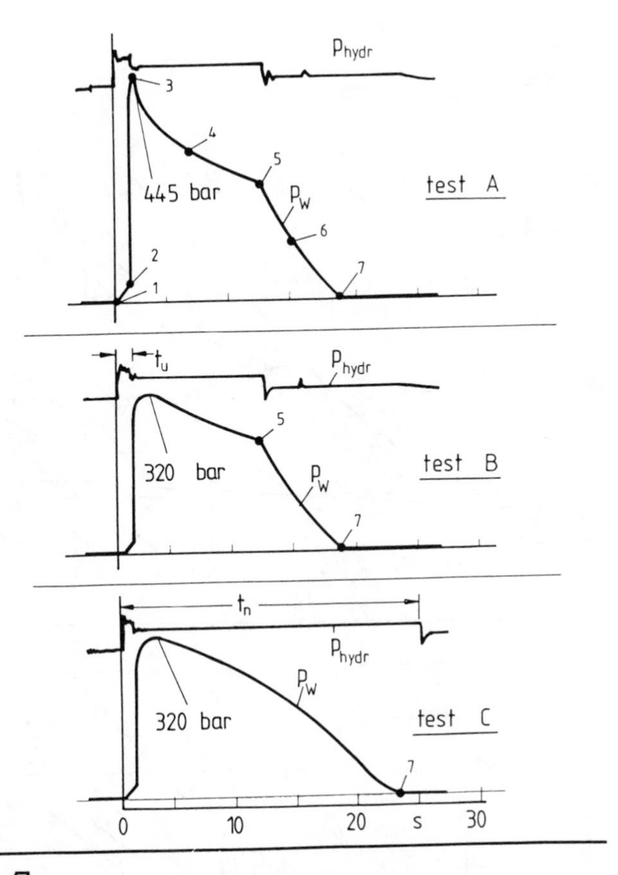

Cavity pressure in injection moulding

Fig. 4: Pressure profile in the mould and conversion into the PVT diagram (Source: IKV)

Transfer of cavity datas in the diagram of state

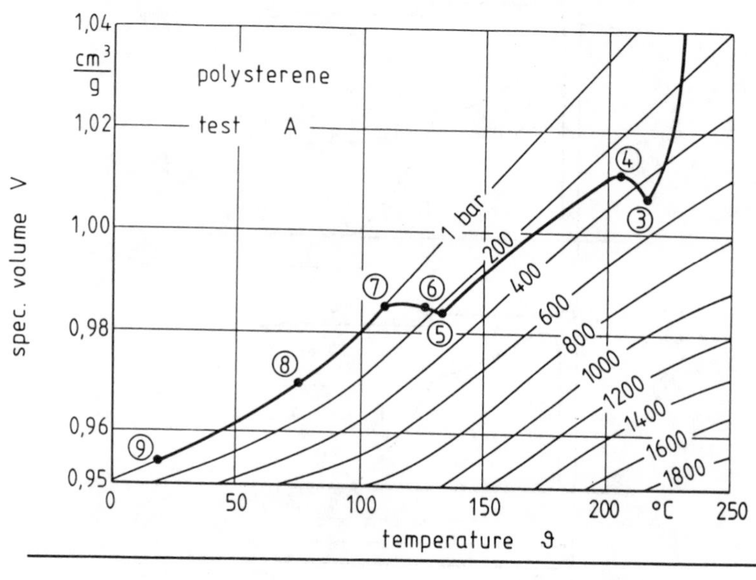

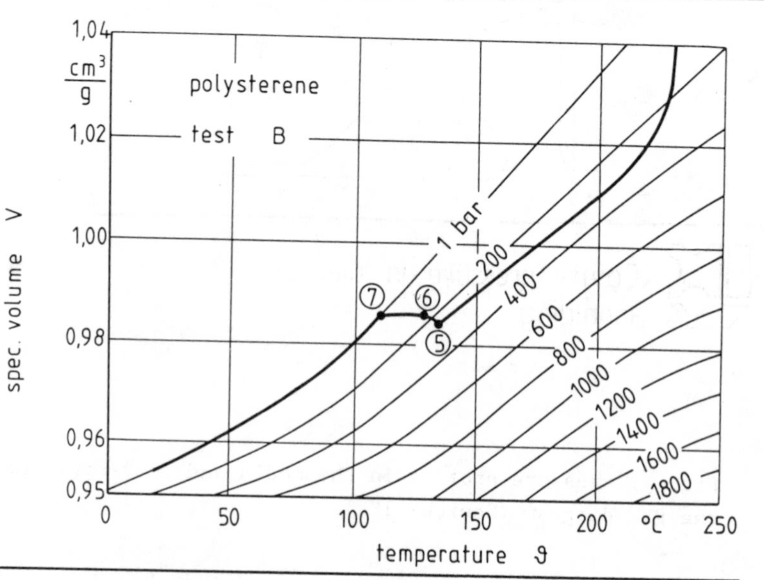

If the machine is computer controlled it is possible to have an optimisation strategy on this basis implemented in the computer - a so-called PVT-control (10,11). This smooths out temperature fluctuations in the moulding compound automatically through spontaneous pressure intervention and thus leads to moulded parts with very highly constant dimensions.

Thermosetting plastics with glass fibre and other fillers and also elastomers are now being processed on this type of computer controlled plant as well, although the plants need to be fitted with specially adapted plasticising cylinders and screws to take these moulding compounds. Process monitoring is performed in the same way whilst the process controls and their underlying strategies differ considerably. This is not to be examined here, however (for information consult 12, 13, 14).

Precisely the same procedure is adopted for foamed and non-foamed RIM parts, even though the production units differ considerably and no process optimisation has been drawn up on account of the lack of process models.

Fibre composites

The chief technique applied to manufacture moulded parts with long fibre reinforcement is compression moulding using SMC (sheet moulding compound). The preforms, which contain the ready-prepared, though not yet cured moulding compound and the fibres, are placed in the mould after being cut to size and baled. The action of the mould closing causes the moulded part to be formed into shape and it is then cured through the heat of the heated mould. Highly heat-resistant, large-format, shell-type mouldings can be formed in this way, such as are suitable for bodywork parts. In this case, however, it is essential for a process sequence which has been seen to give the correct results to be observed with exceptional precision. This is only possible in fully automatic plants, as is shown in a schematic diagram in Fig. 5. An industrial plant based on this proposal (15), which has already been producing parts for a year, shows that it is indeed possible to manufacture identical parts simply by strictly observing the process sequence (16). The pressure probes installed in the mould cavity, or a sensor that monitors the closing displacement and closing velocity of the compression mould over time (20), also supply information on the uniformity of the moulded parts. When applying fibres to fibre composite parts, which are made up of layers of endless fibres and resin and where the fibres are applied by winding or by tape-laying, it is absolutely essential to have the same configuration and quantity of fibres at identical points on each and every moulded part. This is not guaranteed by the method generally used to build up such configurations today, i.e. the hand lay-up method which is first practised on test patterns. In addition, non-destructive tests are only possible to a limited extent. For this reason work of this type is being increasingly assigned to robots now that robots have been successfully applied for such tasks in research for a number of years (17). Robots can be easily programmed by the teach-in method and once a path has been fed in the robot will repeat it faithfully. The moulds for the robot do have to be manufactured individually, however, since there is no industrial supply as yet. The schematic diagram in Fig. 6 shows the tasks of this type that a robot can perform. It is also possible for the robot to carry out a wide range of procedures in succession with an automatic tool change. The robot can also serve a wide range of moulds in succession with a recognition system helping it to select the corresponding process sequence program required in each case (Fig. 7). This means that the robot can be located in a closed cell, thereby allowing physiologically harmful resins and other materials to be processed. The robots are simply fitted out with relatively cheap and robust tactile sensors.

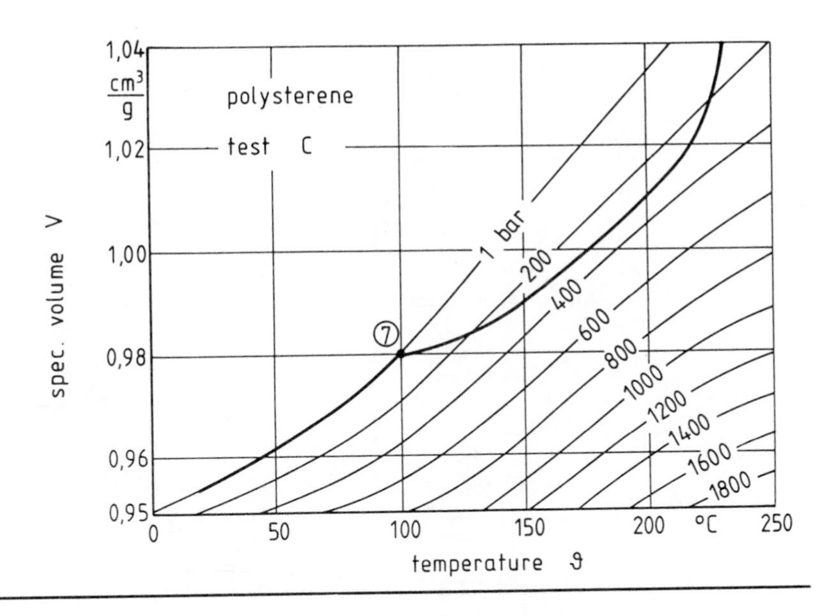

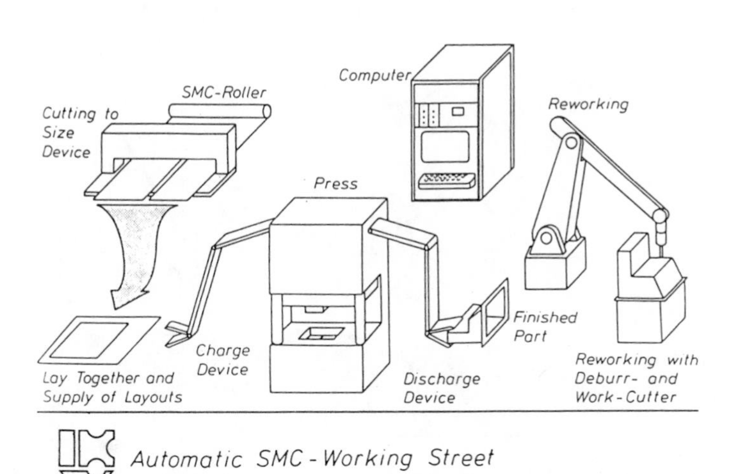

Fig. 5: Interlinked production of SMC compression moulded parts (Source: IKV)

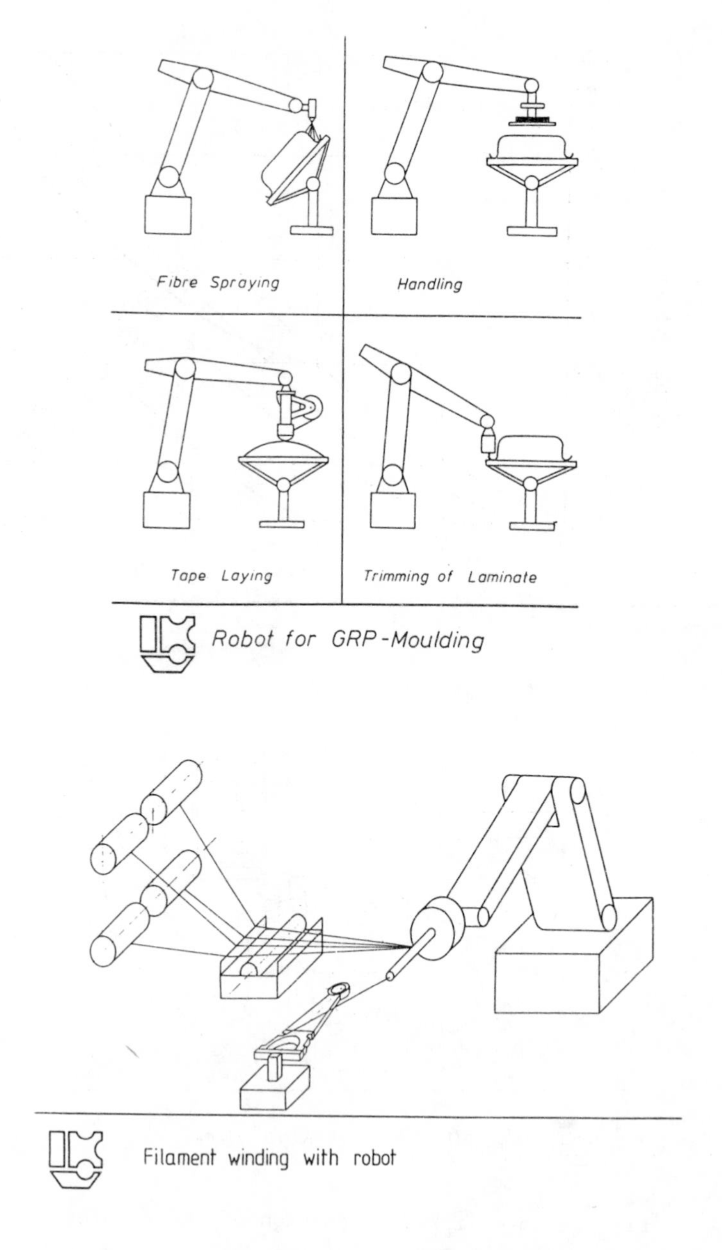

Fig. 6: Robots for the production of fibre composite parts

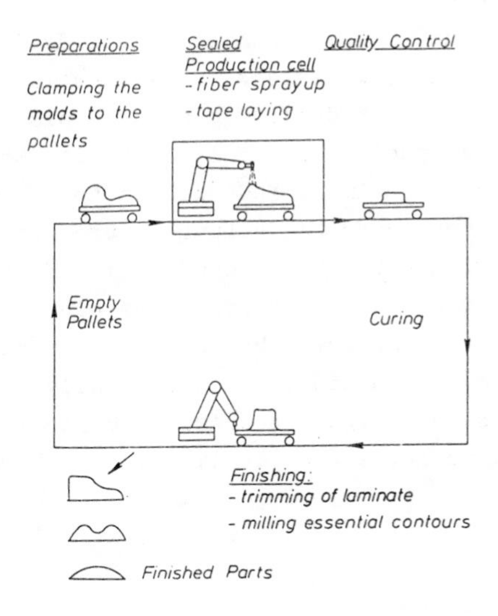

Fig. 7: "Hand lay-up" with the aid of a robot (Source: IKV)

Extrusion

Endless profile and film manufacture no longer manages without process monitoring as a rule. Production units for blown or flat film, pipes, window profiles and the like are highly automated now. In highly refined processes, such as the elaborate manufacture of fine film for high-grade packaging of X-ray films and recording tapes (particularly where this involves biaxial in-line stretching), it is essential for the process sequence to be monitored with exceeding care and be kept constant. Although units of this type are controlled by master computers, process control cannot as yet be realised for the process as a whole. Instead, there are large numbers of subordinate feedback control circuits for highly specific functions which have to keep particular process parameters constant.

The same applies for calendering lines, used to produce film. The requirements placed on even cheap packaging film are generally very high, since the slightest deviation in thickness, for instance, will lead to malfunctions in subsequent processing. Production plant of this type generally runs at a specific point of operation which has been established by experiment and checked to see whether it will allow the highest possible output in production. Once this point of operation has been found it is adhered to and the dimension produced for as long as possible. Sensors are employed in order to make sure that no faulty products are manufactured during production, which generally only has low-level supervision. Figure 8 shows a monitoring configuration of this type for a profile line in a schematic diagram.

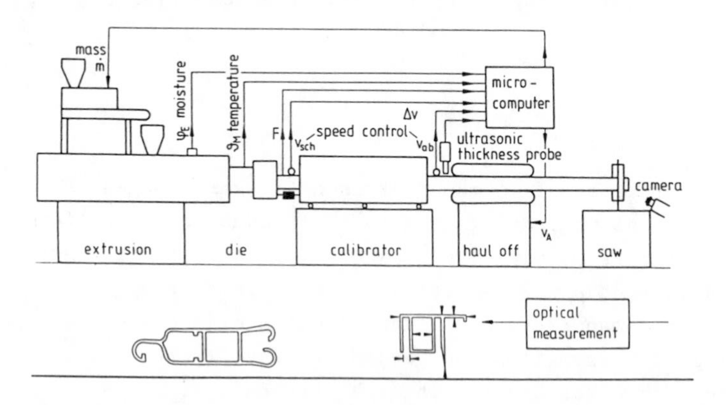

Fig. 8: Sensor-monitored profile production with a computer controlled extrusion line (Source: IKV)

3. Incorporation of scientific findings into planning and production.

Plastics companies have largely worked on an empirical basis to date. Scientific findings are only taken up into practice very hesitatingly since staff have had virtually no background of suitable training in the past. Only now has the recognised job of "Plastics Moulder" been introduced which involves an apprenticeship. Once again, however, the plastics moulder is not in a position to convert scientific findings into practice. Only medium-sized and big firms have staff with scientific capabilities in research departments, and these again are only in small numbers. Such staff have thus generally been too busy to engage in the systematic further training for themselves which the rapid developments in this new field really require, or to train the staff beneath them. In view of the sparse treatment still given to the fundamentals of the production and material science of plastics in present-day engineering courses, it has to be considered whether there are not other means that could be employed. One potential approach would be expert programs for highly specific tasks, such as the design of moulded parts or machine setting.

Machine setting has so far been conducted solely on the basis of experience. The machine setter works at the machine thereby costing machine time. The quality of the setting is consequently dependent upon the qualification of the setter. The amount of time involved is considerable and neither quality nor profitability is generally optimal. We are thus thinking of applying the correlations between production and properties currently being worked out at our Institute (15) here as well. The data and functions already established have been found to be machine-independent and hence universally applicable. It is possible to take the setting data on the machine and, with the aid of process models, convert it into the process variables of melt pressure, melt temperature, shear or expansion rate and temperature gradient (cf Fig. 9). The internal properties of crystallisation and spherulite size, and orientation and inherent stress are then expressed in terms of this data. This data, in turn, is known to influence the final properties of the manufactured products.

The findings to date can be summarised as follows: Plasticising units have no notable influence on final properties if the units are correctly set up and optimally

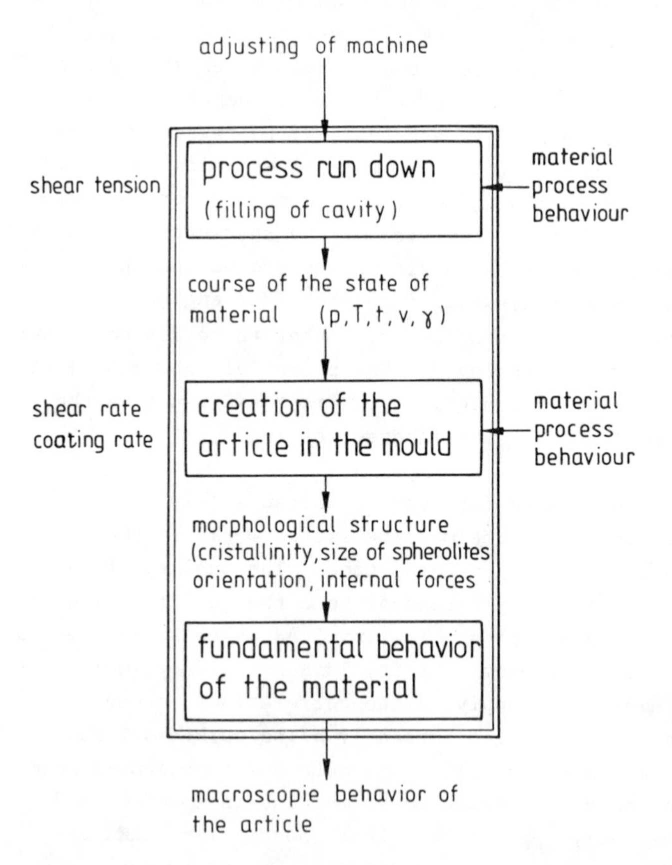

Fig. 9: Schematic diagram of the correlation between machine setting data, process parameters, internal properties and final product properties (Source: IKV)

adjusted. This applies first and foremost to the screws, although the choice of screw has a considerable influence in the case of sensitive thermoplastics, such as polypropylene. This is due to the fact that the molecular weight can be rapidly reduced through shear and also through temperature (cf Fig. 10)(16). Excessive temperatures are far more critical than shear. If, however, too much heat of

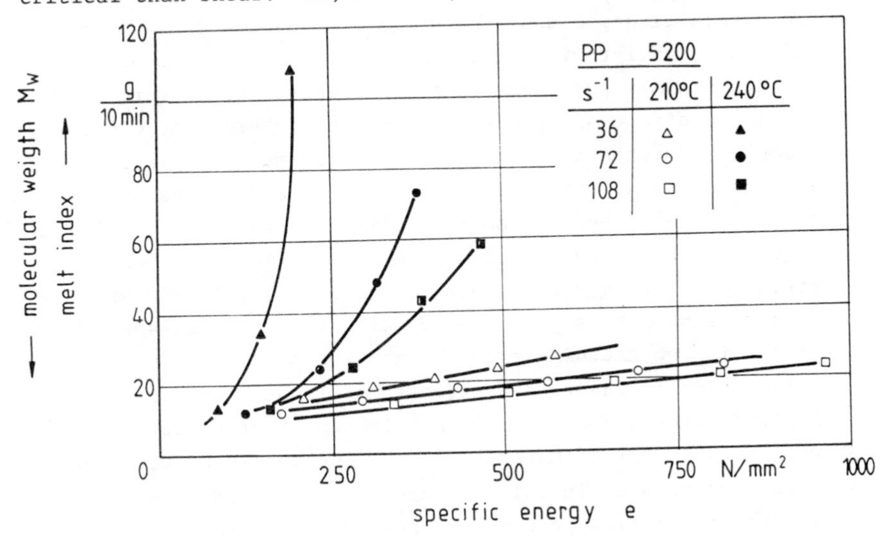

Fig. 10: Reduction of molecular weight through extruder processing (Source: IKV)

dissipation is produced, such as through the use of the wrong screw, then the MFI (a potential practical measure) can alter by a full order of magnitude. This also leads to very much poorer product properties - injection moulding compounds, in particular, can take on very brittle behaviour. The same applies, as is well known, to thermoplastics produced by polycondensation if these moulding compounds are heated up in a moist state, such as in the extruder screw. If one follows the raw material manufacturer's recommendations regarding setting, equipping the plant and the state of the pellets, however, then it can generally be assumed that the influence of plasticisation on final properties can be neglected on screw machines.

All the moulding steps, by contrast, have a very lasting influence. These include the flow and cooling processes that take place in an injection mould, or the deformation experienced by a profile shortly before and after output from the die, and cooling.

As expected, it is seen that with semi-crystalline plastics the crystallisation and spherulite size have a dominating influence, whilst with amorphous thermoplastics the orientation has a marked effect on final properties. Figure 11 shows the effect of orientation in a polystyrene by way of example. Orientation forms in injection moulded parts on account of shear stress. This shear stress develops as a result of melt shearing during filling of the mould cavity. Figure 12 shows the correlation between orientation and strength. Strength and shear stress are thus linked via orientation. It is possible to calculate the mean shear stress in the moulding from the machine setting. This calculation is based on a process model which describes the filling process and which works with the representative viscosity by way of material data. The representative shear stress calculated in this way can be integrated along the length of the flow path and expressed in the form of a mean shear stress.

To date the crystalline structure component has been regarded as the sole factor influencing the final properties of semi-crystalline thermoplastics, such as modulus and hardness. We have seen, however, that it is probably not so much the crystalline component in itself but rather the fine spherulitic structure component that has a decisive influence on mechanical end properties. This fine spherulitic structure develops under the surface of injection moulded parts as a result of the high cooling rates generally applied and would seem to behave in the same way as the crystalline grain size in metals. A correlation also exists between crystallisation content and spherulite size, which we have not analysed as yet. Figure 13 shows the correla-

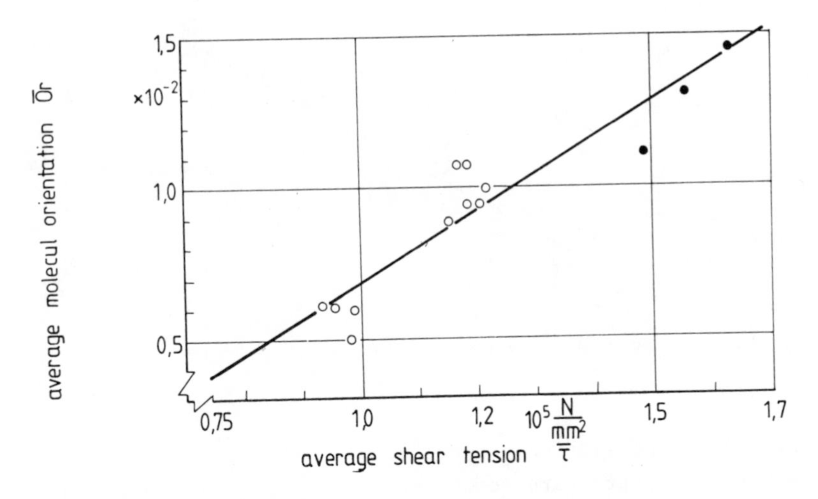

Fig. 11: Development of orientation in injection moulded parts in amorphous thermoplastics as a result of shear stress in the moulding compound created during the mould filling process (Source: IKV)

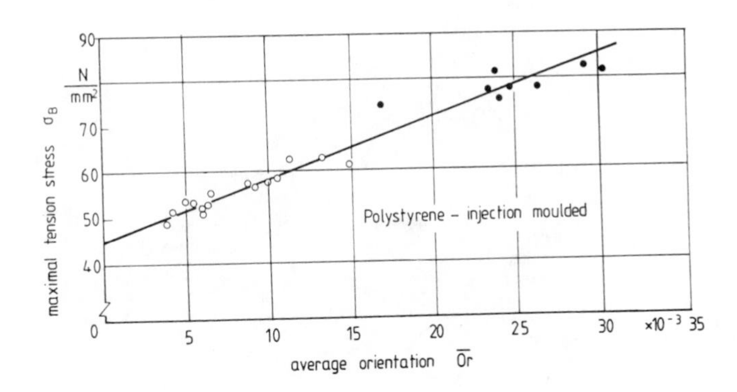

Fig. 12: Tensile strength of injection moulded parts in polystyrene as a result of the orientation created by shear stress (Source: IKV).

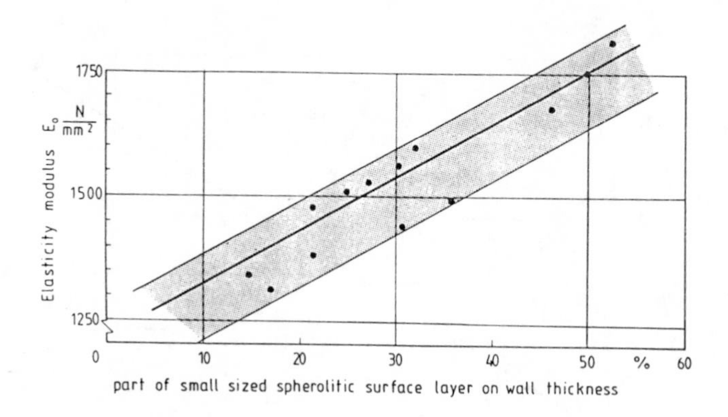

Fig. 13a: Correlation between modulus a) and the fine spherulitic structure component in the edge zones of injection moulded parts (Source: IKV)

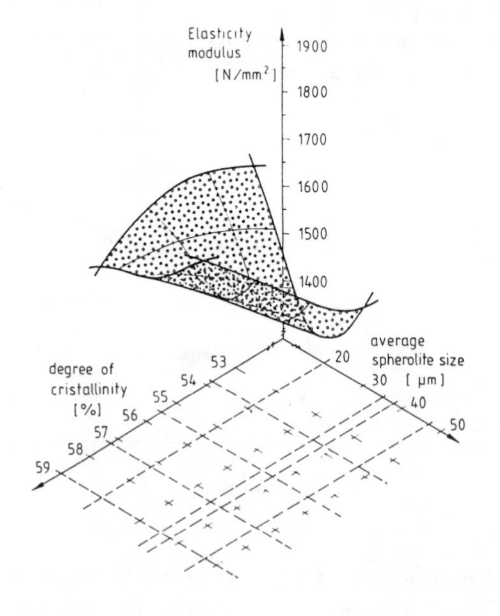

Fig. 13b: Correlation between modulus b) and spherulite size and crystallisation content

tion between the mechanical properties of modulus and hardness and the fine spherulitic structure.

We have also seen that shrinkage is influenced by this fine crystalline structure component (Fig 14) (17).

It is our intention to establish these correlations for the different groups of materials and to set them down in a correlation model. It should thus be possible in a number of years to predict properties right at the design stage. This will be built into the expert programs so that when the designer is designing a part he can modify his part or gate position etc. directly on the CAD unit screen through simulation calculations with the correlation model until such time as he has achieved a part with optimum properties. He then designs the moulds for these and has the processing data and machine setting calculated.
The same procedure will be adopted in extrusion where a process model is already available to run on medium-sized data recording units. This is based on finite elements and is designed to establish the course of flows through dies and the cooling in calibration dies (19, 20). The model for the die flow uses the KZM model (21), which describes both viscous flow and elastic behaviour. The heating due to

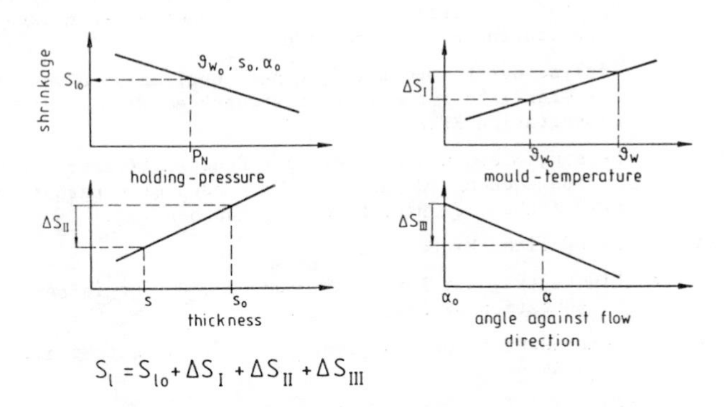

Fig. 14: Influence on shrinkage in moulded parts made of semi-crystalline thermoplastics (Source: IKV)

dissipation and the resultant change in melt properties is also taken in over a number of runs. Since the orientation or crystallisation and inherent stress due to cooling have already been established with the two programs, only the influence on final properties still needs to be added from the correlation model in order to allow quality predictions to be made for profiles, films and pipes.

Results are so far available from the special research programme for one amorphous and one semi-crystalline, unfilled homopolymer. The next step is a corresponding investigation of copolymers and blends with the same base polymers being taken again.

Literatur

1. W. Haack — Der Rechnereinsatz bei der Konstruktion von Spritzgießformteilen Dissertation RWTH Aachen 1983
2. J. Schmitz — Anleitung zum methodischen Konstruieren von Spritzgießteilen Dissertation RWTH Aachen 1984
3. F.M. Jäger — Synthetic Nr. 3 /1984
4. G. Langecker — Kunststoffe 73 (1983), H. 10, S. 559-563
5. FMI in Japan — Japan Plastics Jan. 1985
6. W. Jürgens — Untersuchungen zur Verbesserung der Formteilqualität beim Spritzgießen teilkristalliner und amorpher Kunststoffe Dissertation RWTH Aachen 1969
7. S. Stitz — Analyse der Formteilbildung beim Spritzgießen von Plastomeren als Grundlage für die Prozeßsteuerung Dissertation RWTH Aachen 1973
8. H. Recker — Messen, Steuern, Regeln in der Kunststoffverarbeitung Herausgegeben vom Institut für Kunststoffverarbeitung an der RWTH Aachen, C. Hanser Verlag, München
9. A. Nicolay — Persönliche Mitteilung
10. H.O. Hellmeyer — Ein Beitrag zur Automatisierung des Spritzgießbetriebes Dissertation RWTH Aachen 1977
11. A. Matzke — Gemeinschaftsforschungsvorhaben des IKV mit Mitgliedsfirmen 1983/84
12. F. Buschhaus — Automatisieren beim Spritzgießen von Duroplasten und Elastomeren, Dissertation RWTH Aachen 1982
13. W. Janke — Rechnergeführtes Spritzgießen von Elastomeren Dissertation RWTH Aachen 1985

14. A. Matzke — Proze rechnereinsatz beim Spritzgießen - ein Beitrag zur Erhöhung der Flexibilität in der Fertigung
Dissertation RWTH Aachen 1985

15. H. Derek — Zur Technologie der Verarbeitung von Harzmatten
Dissertation RWTH Aachen 1982

16. Dipl. - Ing. Haldenwanger — Private Mitteilung

17. A. Mayer — Extruderbaureihen: Ein Beitrag zur Auslegung und Optimierung von Einschneckenextrudern
Dissertation RWTH Aachen 1984

18. W.B. Hoven-Nievelstein — Die Verarbeitungsschwindung thermoplastischer Formmassen
Dissertation RWTH Aachen 1984

19. B. Gesenhues — Rechnerunterstützte Auslegung von Fließkanälen für Polymerschmelzen, Dissertation RWTH Aachen 1984

20. J. Wortberg, G. Menges, J. Schmidt — Berechnung des Wärmeübergangs an extrudierten Profilen unter realen Wärmeübergangsverhältnissen, Abschlußbericht zum DFG-Forschungsvorhaben Wo 302/1-1

21. B. Bernstein, E.A. Kearsley, L.J. Zapas — Trans. Soc. Rheol., 7 (1983), S. 391 ff

Part VIII

STRUCTURE AND MORPHOLOGY

SOME FACETS OF ORDER IN CRYSTALLINE POLYMERS AS REVEALED BY POLYETHYLENE

A. Keller

H.H. Wills Physics Laboratory,
University of Bristol, Bristol U.K.

INTRODUCTION

A CASE FOR THE STUDY OF CHAIN FOLDING IN POLYETHYLENE

This article will relate to the organisation of long chain molecules in the crystalline state. The subject has become so extensive that it is not possible to write about it comprehensively in a comparatively brief survey without merely recapitulating what has been said many times elsewhere, and this in a rather superficial manner. I shall therefore choose some selected aspects of recent topicality, such as being pursued in our laboratory and not yet presented together if at all, and confine myself to examples relating to chain folded crystallisation in the case of one chosen substance, polyethylene. Nevertheless, I do not intend this article to be a specialist's progress report alone, and shall endeavour to retain somer broader perspectives. The present Introduction is specifically devoted to this latter purpose.

My first point relates to the potential significance of the topic, the prerequisite of any research activity with claim for attention. I feel that in the present instance, this is particularly opportune, because I seem to sense that the topic in question is about ceasing to attract the wider interest which in my opinion it deserves for its intrinsic scientific interest and importance for application. This is due partly to the much drawn out contraversies without tangible conclusions on the one hand, and seemingly endless belabouring of details on the other, both of which have tended to obscure the intrinsic fascination of the phenomenon of chain folded crystallisation. Chain folding is, and remains, an extraordinary manifestation of molecular behaviour on any standard, transcending polyethylene, and in fact the closer confines of traditional polymer science. Another reason for the appearently waning interest in the topic is the decline of scientific respectability of polyethylene and similar large volume low priced commodity thermoplastics, with the new chemically and thus electronically more complex speciality polymers with more sophisticated behaviour patterns taking the limelight. In what follows, I shall first present a defense for studying these, by some views outmoded, issues on an allegedly outmoded material.

At the risk of restating the commonplace I wish to recall the importance of long chain molecules themselves. They are part of our daily life in the form of household goods, packaging materials, garments, or to quote more high tech. applications, they serve as aircraft and rocket components, video tapes and are essential ingredients of micro-circuitry. All of the above are products of technology, and have affected our living and thus, are having an impact on society. But unlike many other forms of technologically useful classes of material, long chain molecules have an even more profound significance, in as far as they are the main constituent of living matter, which nobody, not even the most antagonistic critique of our scientific age can query. Anybody puzzled by the ultimate enigma of his own existence and, to bring it down to practicality, concerned with his own physical well being, must recognise the central part which long chain molecules play in at all.

The ultimate reason for the ubiquitousness and indispensability of long chains lies, according to my reasoning, in two interconnected facts. The first of these is that the molecular chains are very long and the chain members are linked by chemical valence forces. This for self evident reasons renders them to be structural materials par excellence, be it for technologically fabricated objects or for sustaining, encapsulating or linking portion of living organisms. The second reason for the overriding role of long chain molecules lies in their intrinsic variability. The simpler of these is the physical, that is, conformational variability of the chain which in macroscopic terms imparts the characteristic flexible (including rubbery) properties, coupled with toughness and strength, to technological plastics or living tissues. Even more profound is the limitless possibility of constitutional (chemical) variability which can be realised along a long chain through even a relatively few differing constituents. Thanks to this variability, technological plastics can be tailored to highly specific requirements, and more importantly, living matter acquires its potential for storing and transmitting information and thus to reproduce itself, and for regulating life processes with utmost subtlety and specificity.

Having sketched this broader overall scenario it may seem rather prosaic to return to such an apparently ordinary substance as polyethylene, the source of my examples to quote from. This I am doing not only because I consider polyethylene all that important per se, at least in the global perspectives raised above, (not doubting that it is a technologically very significant material), but because it serves as one of the simplest models of how an idealised long chain would behave, providing a model also for all other kinds of long chain in the multitudinous variety of functions and behaviour outlined above. Polyethylene is possibly the nearest real world approximation to the abstraction of an ideal chain molecule. It consists of one kind of repeat unit only, which is chemically and physically as unspecific as it can possibly be in a real substance. Its internal homogenity can be made nearly perfect and its end groups (which are by necessity always different from internal chain members in any polymer) are as little dissimilar, hence alien, to the rest of the chain as they can possibly be. Further, it is highly flexible and thus provides full scope to the intrinsic conformational variability of long chains, which for my purposes will be manifest by its ability to become organised in the form of a crystal. Also, it is available in a very large range of chain lengths from oligomers (paraffins) through to polymers of extreme lengths.

As just stated, my main theme will be how polyethylene chains can organise themselves when forming a solid, which in practice amounts to crystallisation. I shall assert that in this respect the behaviour of polyethylene is as intrinsic to that of an idealised chain as it can be realised, and that any pattern which emerges from its study will also hold for other, more complex chain-molecules.

I am realising of course that when taking a polymer of uniform composition, such as polyethylene, as my model I have already discarded constitutional variability which, as stated above, endows macromolecules with the ultimate potential to govern our entire existence. As well known, extensive efforts are vested in the latter subject. Nevertheless, as long as such a basic organisational feature as crystallisation by chain folding remains only partially understood, if understood at all, even in the simplest of all carbon based long chains, regarding the structure of the fold, the reason for and mechanism of the folding process, the full appreciation of the vastly more complex organisational behaviour of chains with multiple and even sequentially variable constituents will always be lacking at its very base.

Neither is the understanding of chain folding an ultimate aim in itself. The eventual performance of the polymer, whether for technological use or biological purpose, is largely determined by it in a way which is second only to the chemical constitution of the chain itself. As seen in a broader perspective, the elucidation of chain folding is to serve this more general purpose for which polyethylene is offering an auspicious starting point.

CRYSTAL STRUCTURE AND CHAIN FOLDING

My main topic will be crystallisation through chain folding. I shall first provide a brief overview on polymer crystallisation in general, to indicate where and how my specific topic fits into the broader picture.

A crystal involves a lattice, i.e. a repeating structure in three dimensions. Here we can divide the macromolecules in two extreme classes: periodic and aperiodic molecules. In periodic molecules the periodicity is already provided along the chain direction thanks to the repeating chemical constitution, and if the chains are in an extended form so that the repeating units are in a linear progression a one dimensional lattice results automatically. (Fig. 1a). Extension of the lattice in two and three dimensions then arises through appropriate packing of different chains. It is to be remarked at this place that chain folding in the sense to be discussed here represents a 'short-cut' in the development of lateral order where this is achieved, at least in two dimensions, by the same chain folding back on itself in the sense of (Fig. 1b). The so called crystal structure is then represented by the atomic arrangement within the unit cell itself, the repeating, to a large extent intramolecular motive creating the lattice.

If the molecules are aperiodic, as for example in the all important globular proteins, we may still have crystallinity, in fact macroscopic single crystals. This will arise if the chains themselves are all strictly identical in terms of both composition and length. In this case the repeat unit is the whole molecule itself which is the repeating element giving rise to the lattice (Fig. 2). Here the molecule is in a particular convoluted form, determined by the chemically specific units along a given chain. The details of this convoluted chain path are referred to as 'protein folding' which thus appears to be totally different from the folding of our concern such as in (Fig. 1b), a distinction, while self evident, not usually drawn in the texts on polymer crystals. Our type of chain folding (Fig. 1b) is a way to facilitate the establishment of a lattice, while the protein folding, so important in biology,

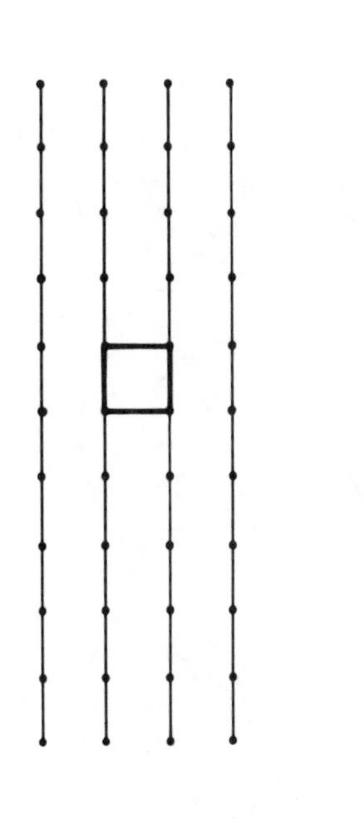

Fig. 1 a) Concept of a lattice from parallel periodic chains, dots representing geometric repeat units, the square in bold lines the unit cell. b) Chain folding in aid of forming the lattice. In both cases the representation is two-dimensional.

is a conformational feature of an otherwise aperiodic molecule (for further comments see ref. 1). The situation in fig. 1b and 2 represent two extremes in a spectrum of possibilities which include situations where the monomer repeat is broadly but not strictly periodic (for example random co-polymers), or where folding such as in fig. 1b may arise even if the molecular repeat is largely aperiodic, but nevertheless some pattern of repeating units exists with specific groups at which the chains prefer to fold. For the present our concern will be confined to the idealised situation of fig. 1b, that is to periodic folding in strictly periodic molecules of which polyethylene is the simplest example, or to cases where there are specific, periodically occuring interactions between groups (say hydrogen bonds) which are best satisfied intramolecularly by periodic chain folding, of the type in fig. 1a.

Fig. 2 Schematic representation of a lattice from a globular (protein) molecule. Here each molecule, although aperiodic is strictly identical. The convoluted chain path is the protein folding.

MODES OF CRYSTALLISATION

In a usual fuseable and/or soluble periodic chain, crystallisation can arise through three sources.

1) Concurrent with polymerisation, the so called nascent crystallisation.

2) Orientation induced crystallisation.

3) Supercooling induced crystallisation from the random state.

1) Arises unavoidably during polymerisation of any crystallisable polymer. In exceptional cases it can give rise even to macroscopic single crystals.

2) Arises under the influence of externally induced orientation. Amongst others it leads to the well known 'shish-kebabs' (Fig. 3) and is a prominent feature of fibre formation along some specific routes (see refs. 2 and 3).

3) Will be the concern of the rest of the article.

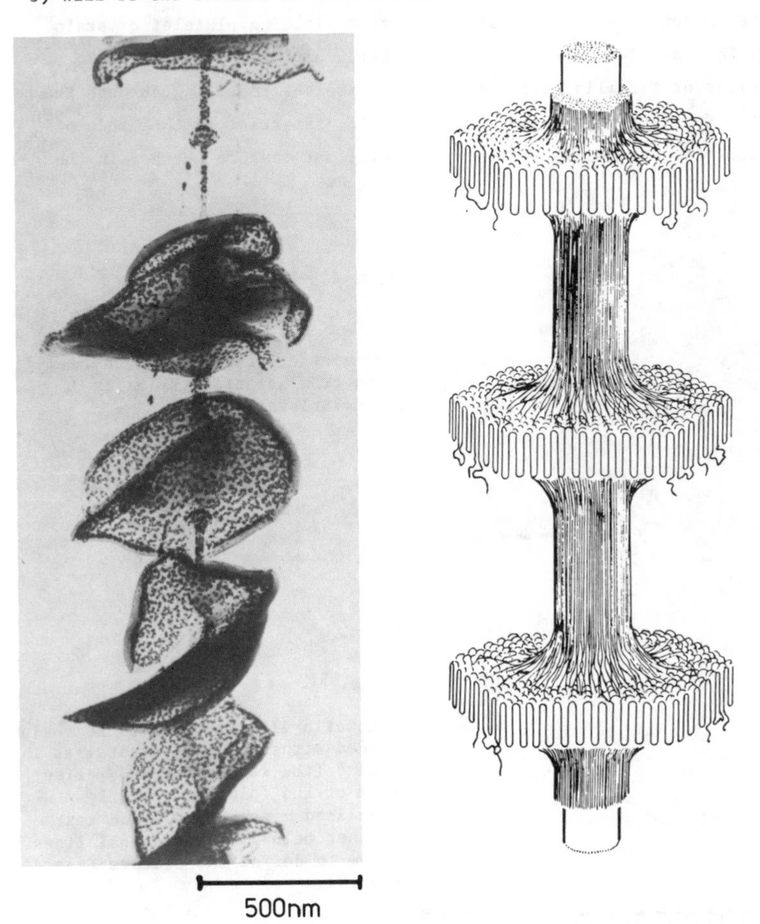

Fig. 3 a) Electron micrograph of a shish-kebab type crystal of polyethylene grown by oriented crystallization from solution(34).

b) Diagram of a shish-kebab showing fibre-platelet structure with molecular connections between chain-folded platelets and central core (after Pennings(35)).

It is by now an indisputable fact that random chains crystallise by forming lamellar crystals when supercooled either from solutions or from the melt (see eq. refs. 1, 4, 5). Figs. 4 and 6 provide two illustrations for each case. In addition, it needs stating that from very highly supercooled solutions (achievable with poorly or slowly crystallisable polymers, and only quite exceptionally with polyethylene), gel forming crystallisation results, as opposed to the conventional particulate suspension yielding platelet crystals as in Fig. 4. The junctions of these gels are being envisaged as in Fig. 7, that is, micellar or fibrillar type crystals, themselves resulting through the confluence of a few, and only a few, chains. Again, important as this mode of crystallisation seems to become presently, it will not feature further in the present article.

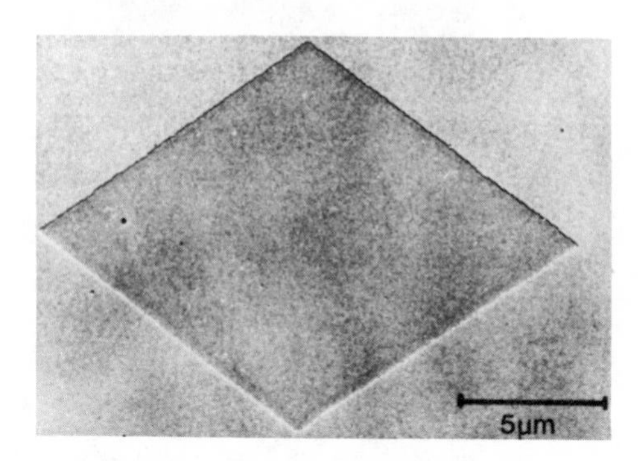

Fig. 4.

Monolayer crystal of polyethylene grown from solution (by Sally J. Organ, Bristol)

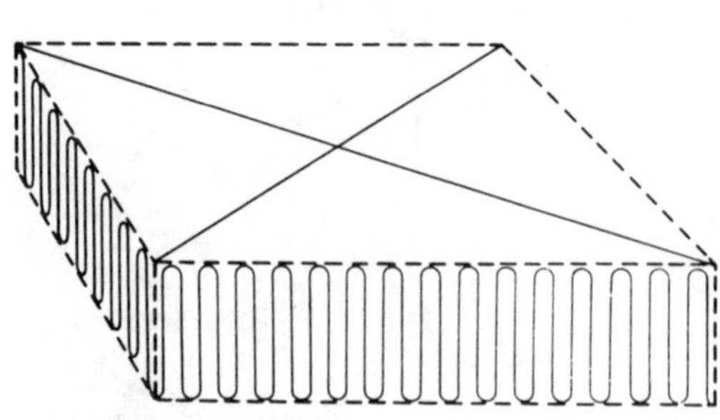

Fig. 5.

Schematic illustration of the chain folded structure of a crystal as in Fig. 4 (the regularity and perfection of the folding is highly idealised - see Fig. 13 and text further below). The diagonal lines serve to define the sectors (see Fig. 8).

Fig. 6. Example of morphologically distinct crystal lamellae as grown from the melt. In the present example the crystals consist largely of linear polyethylene growing in a melt of a blend constituted by linear polyethylene (20 %) and by a slightly branched ('linear-flow') polyethylene (27, per 1000 C atoms ethyl branches), where the latter crystallizes more slowly and with a finer scale texture. The larger crystals are therefore the results of segregation of two chemical identical polymer species of slightly different isomeric constitution. Electron micrograph, using the permanganic etching technique by Bassett et al[36]. (From ongoing researches in Bristol[33,34].

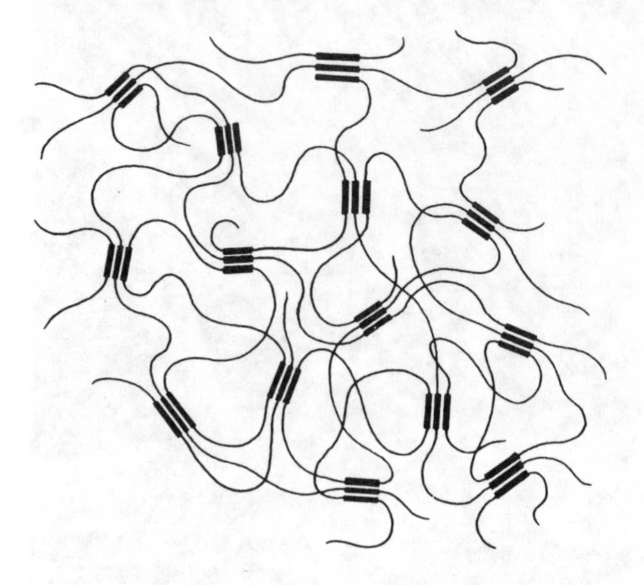

Fig. 7. Model of fringed-micellar crystal texture, as revived recently due to works of gel forming crystallization[4].

CHAIN FOLDED CRYSTALLISATION; SOME BASIC FEATURES

The argument leading to the inevitability of chain folds in crystals such as Fig. 4 (in the sense of Fig. 5) is well known and needs no recapitulation. Only a few statements will be made to serve as reminder and for sake of emphasis. Chain folding, in the sense of Fig. 5 is, as already stated, one of the most remarkable phenomena amongst polymers and possibly amongst materials in general. It was totally unforeseen from a priori knowledge and its discovery is due to direct observation and straightforward deductions therefrom, all a posteriori rationalisations notwithstanding. It is totally general and has strikingly verifiable consequences such as, for example sectorization (Fig. 8).

The quantity characterizing it is the fold length ℓ which corresponds, or is closely related to the layer thickness. It is directly measurable electronmicroscopically or through small angle X-ray scattering (SAX) or low frequency Raman spectroscopy (LAM), the three entirely different methods yielding results in close agreement. (I am ignoring arguments regarding possible differences between SAX and LAM for the present paper). ℓ is a variable and is determined by the supercooling in a surprisingly reproducible

Fig. 8. Morphological manifestation of sectorisation, in accordance with predictions of the scheme in Fig. 5, in a monolayer solution grown crystal of polyethylene(38).

manner. ℓ can change: it increases after crystallisation by heat annealing, or in many instances, during the crystallisation process itself (isothermal thickening) beyond the initial value arising during the primary crystallisation (ℓ_g*). In both instances the increase in ℓ corresponds to an increase in lamellar thickness, hence to chain refolding, a remarkable phenomenon itself.

In the case of polyethylene, used as a model substance, ℓ is very much larger for crystallisation from the melt than for crystallisation from solution even when referred to identical supercoolings (ΔT). It is usually, even if not unquestionably, assumed that ℓ from solution crystallisation corresponds to ℓ_g* (that is to the primary fold length) while that from the melt is affected by isothermal thickening (8-11). While such isothermal thickening is known to occur in the latter case, the origin of this difference, not to speak of the correct quantitative value for ℓ_g*, so important for basic theories, has remained unsubstantiated. Some of the most recent results to be included in this article will be addressed to this issue.

Another issue of some uncertainty and very considerable controversy relates to the nature of the fold and fold surface. In Fig. 5 the folds have been drawn sharp with adjacently re-entrant stems. Undoubtedly this is an idealisation. Even so, there is considerable opposition to this view claiming appreciable randomness along the fold surface involving both irregularity in the folds and non-adjacency in stem re-entry (Fig. 9). My second specific topic will be addressed to this latter issue.

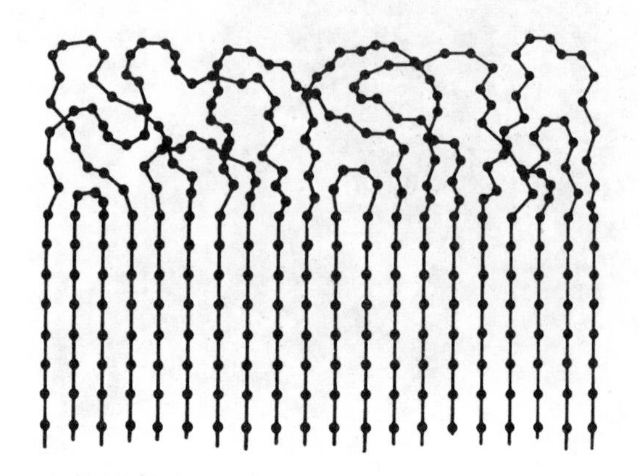

Fig. 9. Sketch of a disordered fold surface: the switchboard model in a two-dimensional representation(39).

The origin of the fold surface controversy is rather deeply rooted. The ultimate reasons for the objection to appreciable fold regularity, at least as I see it, seems to lie in the widespread incredulity that long chain molecules should be able to organise themselves into neatly folded arrays from the random entangled state in which they are being generally envisaged whilst amorphous. The third part of the new material to be reported here will provide evidence for the unexpected mobility of the chains, together with the surprising 'purposefullness' by which chains 'want' to fold, at least in the case of long paraffins where also, and chiefly, the transition between extended and folded structures will be demonstrated on increasing the chain length.

To sum up, I shall be concerned with the following issues in the light of recent results.

1) The primary fold length, including the issue of isothermal thickening;

2) The regularity and adjacency of the fold structure;

3) The transition between extended chain crystals in paraffins and chain folding in polyethylene.

These three topics will then be followed by some new material on morphologies and structure hierarchies.

Coupled with the above some theoretical issues will also arise to which I shall first devote some introductory remarks.

ON CRYSTAL GROWTH THEORIES

A few remarks may be opportune, particularly in the light of recent developments of how theories of polymer crystal growth fit into the wider subject of crystal growth in general.

By current views crystals can grow by three mechanisms.
a) By secondary nucleation. b) By screw dislocation mechanism. c) Via equilibrium surface roughening.
a) was historically the first (e.g. Stransky (12)) and required the formation of a stable two-dimensional patch, the so called secondary nucleus on an otherwise perfectly smooth crystal face which then can grow by lateral accretion covering the face with a new layer of atoms or molecules. Further advance of the phase will then require repeated secondary nucleations on successively completed faces. Experimentally it was found, however, that most crystals can actually grow at much smaller supercoolings (or supersaturation) than required by the formation of a secondary nucleus, hence in general the latter cannot be the rate determining step. Developments which followed circumvented formation of a secondary nucleus as the rate determining step in crystal growth. One of these (b) due to Frank (13) invoked the ledge on an otherwise smooth crystal surface, such as would arise from a screw dislocation emerging at a given crystal face, as providing the site for continually proceeding crystal growth. As well known, such a ledge will wind up into a spiral terrace structure centred on the screw dislocation, a well documented observation on most growing crystals (Fig. 10). The success of mechanism b) temporarily overshadowed the third mechanism, c), orginating from about the same period (Burton, Cabrera and Frank (14)). Accordingly, above absolute zero temperature a crystal surface will never be quite smooth, but will possess a certain degree of equilibrium roughness. If sufficiently prominent, as will be at high enough temperature (about what later became termed roughening transition), this roughness can be the source of crystal growth bypassing the need for secondary nucleation. This mechanism of growth came into the forefront during the last decade. Its most conspicuous manifestation is the loss of straight facets and the appearance of rounded crystal forms (e.g. Jackson (15)).

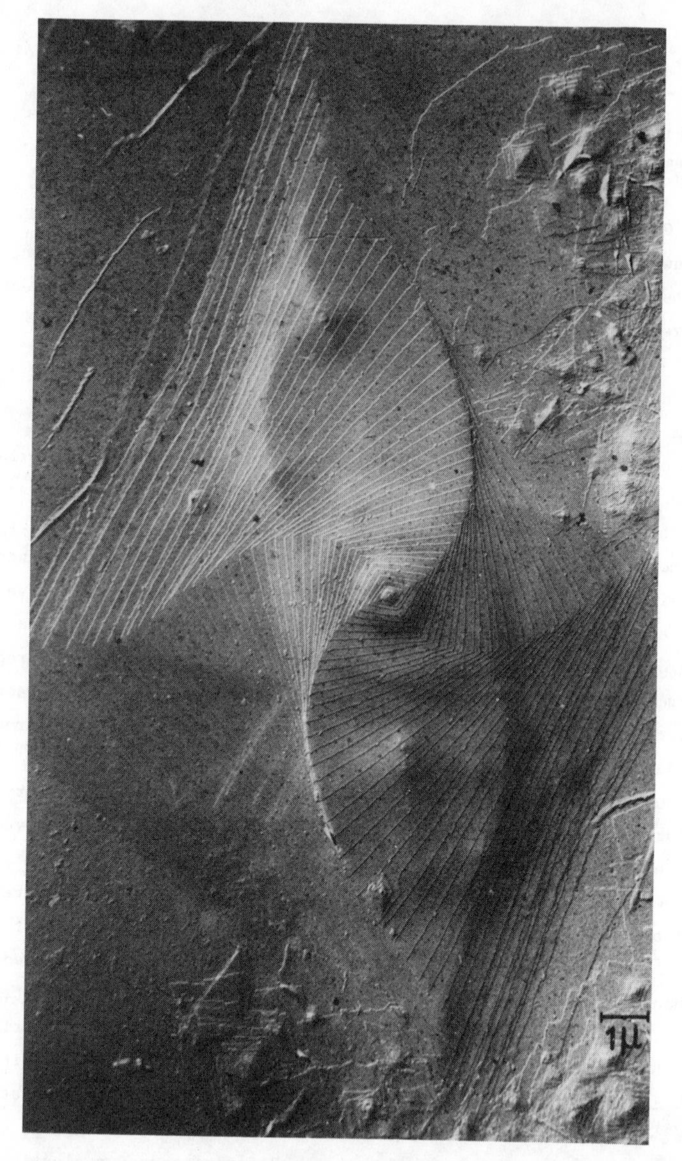

Fig. 10. Multilayer polyethylene crystal displaying spiral terrace growth initiated by a (central) screw dislocation. This particular crystal is special in as far as the overlaying terraces themselves rotate consecutively in a regular manner thus displaying spiralling also on the next highest level of the morphological hierarchy. Electron micrograph, replica(37).

Within these wider perspectives the familiar theories for polymer crystal growth, due to Lauritzen and Hoffman (see e.g. 16), fall in class a), that is, are based on secondary nucleation. It is true that polyethylene crystals display spiral terraces as required by mechanism b) (Fig. 10), but what matters for the present purpose is not the advance of the basal face which is controlled by the screw dislocation (where in this case the Buergers vector is the layer thickness), but that of the ledge, that is, the side surface of the layer which is the site of chain folding. It is the advance of this layer edge, and with it chain folding itself, which by prevailing theories is secondary nucleation controlled (e.g. Hoffman et al (16). However, observations on crystals grown at elevated temperatures (on an absolute temperature scale, as opposed to supercooling) have increasingly brought in prominence crystals with curving lateral faces such as in Fig. 11. Curved faces, that is non faceted growth would, as stated above, be expected from crystal growth mechanism based on equilibrium surface roughness. This recognition has triggered off a new approach to polymer crystal growth (Sadler (18)). While in my opinion it is unlikely that it is to replace existing theories based on the secondary nucleation principle, (there are only too many observed features in favour of the latter for this to be realistic) but may explain the growth of polymer crystals, together with its chain folded nature, in the highest temperature (absolute) ranges where the observed curvature of the lateral faces would make the appropriateness of nucleation induced crystal growth, which in its basic forms relies on formation and spreading of the secondary nucleus along crystallographically defined low index facets, questionable. Also the observed, broadly inverse dependance between ℓ and ΔT can be accounted for by the theory based on surface roughening. This means at least that much, that this correlation between ℓ and ΔT need not be the prerogative of only one theory, and the same applies to the form of the temperature dependence of the linear grwoth rates (G) of the crystals ($G \sim \exp(-K/T\Delta T)$; here K is taken as a constant not to be discussed further at this place) featuring prominently in the developments of past theories. Whatever the case, the introduction of the surface roughening concept to polymer crystal growth has given a new momentum to the theoretical side of polymer crystal research and provides a welcome unification of polymer crystal growth with the wider field of crystal growth studies in general. In brief, in traditional substances crystal growth has been concerned essentially with habit features, while in polymers so far with the issue of chain folding (and associated layer thickness), where the nature of the lateral habit has played no major part. Clearly, the lateral habit development must

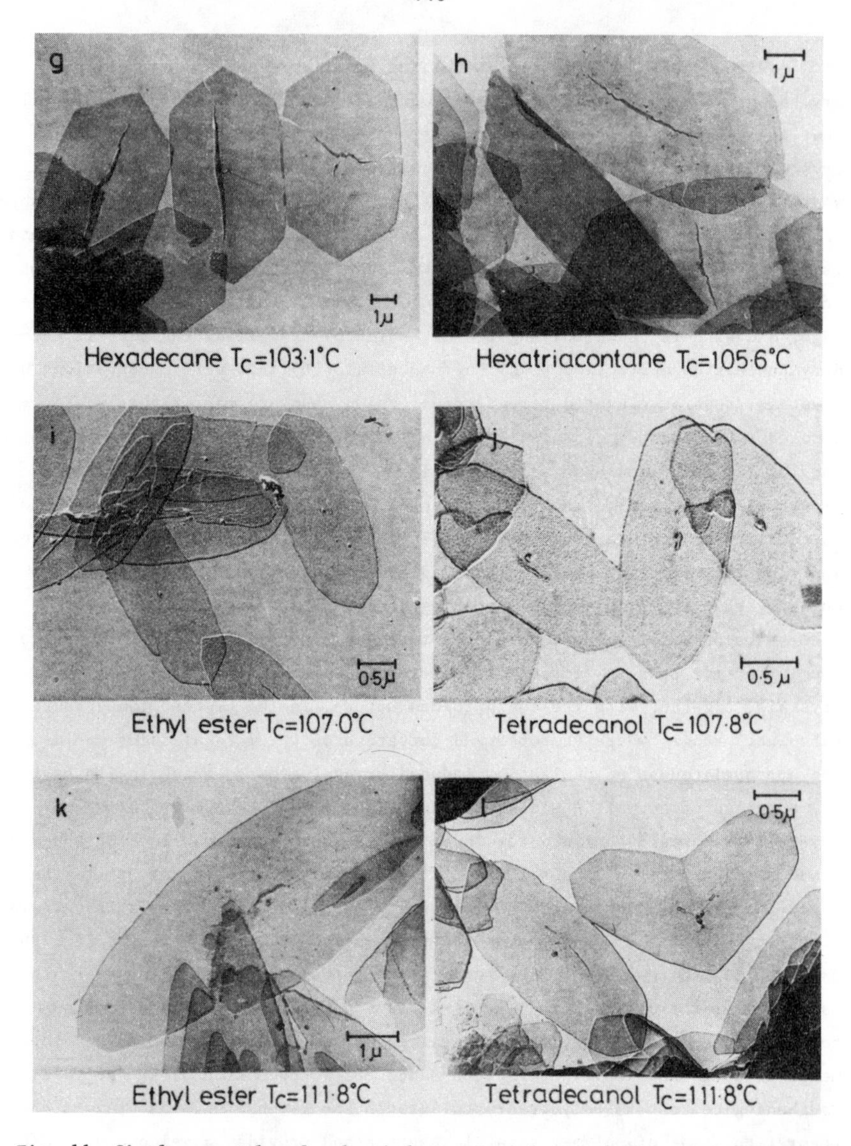

Fig. 11. Single crystals of polyethylene grown from solution at increasingly higher temperatures, T_c, in the sequence g-l, through appropriate choice of solvents, where through decreasing solvent power the realizable T_c values (absolute) can be raised consecutively. The crystals display increasing curvature in their lateral faces on increasing T_c. Electron micrographs[17].

also be part of the picture, and it is only appropriate that it is now being involved, irrespective of what the final outcome of the present theoretical developments will be.

PRIMARY FOLD LENGTH - ISOTHERMAL THICKENING

The importance of the primary fold length (ℓ_g*) has already been emphasised, together with the differences which exist between solution and melt crystallisation regarding the value of ℓ_g*, the latter yielding much higher values for equivalent supercooling contrary to any of the theoretical predictions. Further, the role of isothermal thickening has long been recognised as capable of increasing the ℓ values beyond ℓ_g*, still in the course of crystallisation. The latter is of particular significance as it may put any ℓ_g* value as measured on a given crystal preparation in doubt, thus jeopardising the validity of crystallisation theories which are all based on alleged ℓ_g* values and on their recorded broadly inverse variation with supercooling (ΔT) (more precisly $\ell_g^* \sim Q/\Delta T + \delta\ell$ where Q we take as a constant, not to be discussed specifically, and $\delta\ell$ an additive term small in magnitude but significant in princple). On the other hand it may offer an explanation for the differences between melt and solution crystallisation regarding ℓ values, raising the hope that once isothermal thickening becomes accountable in quantitative terms equivalance between solution and melt crystallisation may be achievable. Finally, isothermal thickening is an important phenomenon in its own right and any furhter information on it is highly desirable. It is with this threefold aim (namely safeguarding the reality of the $\ell_g^* \sim Q/\Delta T + \delta\ell$ relationship, the closing of the gap between melt and solution crystallisation, and acquisition of further knowledge on isothermal thickening), that an extensive programme has been initiated in our laboratory several years ago, which is currently leading, to what appears, conclusive results. As the work cannot be developed analytically at this place, only the end results will be quoted. These are as follows:

1) There is no sizeble isothermal thickening in crystallisation from solutions, thus the ℓ_g* values as obtained from such preparations are safeguarded, reaffirming the existing theoretical framework against suggestions to the contrary (see ref. 17).

2) There is substantial isothermal thickening in melt crystallisation, attributable to the higher absolute temperature of the crystallisation from the melt as compared with that from solution even for equivalent supercooling, presumably due to the higher chain mobility within the crystal. Once the effect of

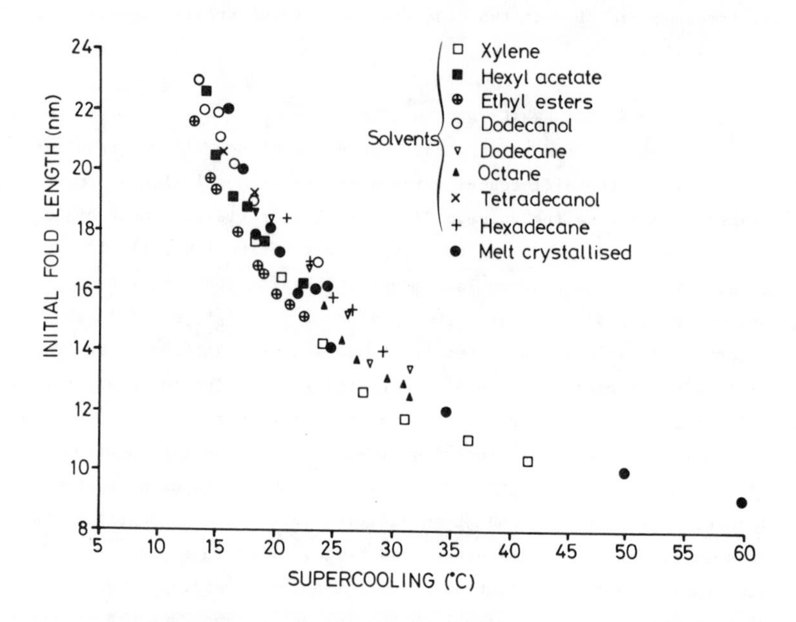

Fig. 12. Initial fold length (ℓ_g*) against supercooling for both melt and solution crystallization (the latter involving a variety of solvent). The plot reveals numerical identity regarding ℓ_g*, in addition to the identical functional dependence regarding ℓ_g* v. ΔT between the two classes of crystallization, once the effect of isothermal thickening in the case of melt crystallization has been recognized and accounted for[19].

this thickening has been assessed the ℓ v. ΔT curves become all coincident (Fig. 12), thus achieving at last the long outstanding unification of crystallisations from melts and solutions (19).

3) Isothermal thickening itself turned out to be a two stage phenomenon. The first stage consists of a stepwise increase of the thickness of the primary lamella by a significant factor (broadly by 2). This is then followed by a gradual, in fact logarithmic, increase with time. This two stage sequence is particularly noteworthy as lamellar thickening has been traditionally considered as a continuous process proceeding logarithmicaly with time.

Nevertheless, occasionally discontinuous increases, often by a factor of 2, have been reported (e.g. ref. 20). Both these effects, the stepwise and the slow gradual increase have important implications for the mechanism of chain refolding within the crystal (21). We now see that reports on the two types of thickening processes do not represent conflicting claims but that both can exist, and in fact are two consecutive stages of the same overall phenomenon.

ARE THE FOLDS REGULAR OR RANDOM

The importance of this issue has already been indicated. In ultimate generality the question which hinges on it is as follows: Is chain folding merely a consequence of the statistical deposition of a random coil along a growing crystal face, or is there a special inducement for the chains to deposit stem after stem in a sequentially regular manner with the necessary intervening of short, hence sharp and possibly regular folds? If it were the latter, is a given chain capable of performing this act configurationally and kinetically? Clearly from the point of view of the random chain statistics in the amorphous state only the first alternative may seem feasible. If the second alternative pertained this could only be through the intervention of the additional driving force provided by crystallisation. Historically, this question has been put in a manner which has given the impression that the two alternatives are mutually exclusive, which has led to a rather unfortunate polarisation of views. For the pertinent argumentation I need to refer to the relevant Faraday Discussion (ref. 22). Here I shall make a few additions from latest works, both in the present and following chapter. The present chapter will contain conclusions from a purely analytical approach on the structure of specific polyethylene samples as such, in continuation of previous works, while the chapter to follow will take up a new and altogether different line.

It will be recalled that the principal approach to the structure problem during the last decade has relied on the introduction of isotopically doped guest molecules in the host matrix assuming that they will behave just as the rest of the material regarding crystallisation behaviour, and investigate these doped species by methods such as can record them as distinct entities within the isotopically different matrix. In the present case the isotopically distinct species is deuterated, as opposed to the usual protonated polyethylene, or vice versa, and the methods of investigation are neutron scattering and infra red spectroscopy. Neutron scattering at small angles provides information on the molecular trajectory, and infrared spectroscopy on stem-stem interactions, hence on the corresponding stem environment. In both cases the isotopically distinct guest molecule is added in sufficiently low concentratic for it to be representative of an isolated molecule of this particular isotopi species.

Before listing the conclusions (this is not the place to develop the arguments analytically) it will be stated that to our knowledge our own studie in question are the only ones where both techniques, namely neutron scattering and infra-red spectroscopy, have been used in combination and this on the same specimen types. The consistency of the results obtained by the two provides confidence in the validity of the conclusions arising. The explicit results quoted will refer to solution grown crystals, some comments on melt crystallisation will be made at the end.

Neutron scattering

The basic principles need no recapitulation. The method relies on coherent elastic scattering where the deuterated species provides the required contrast within the protonated environment and vice versa. Further, the larger the scattering angle (still within the low angle range) the smaller will be th molecular detail on which information can be obtained; or conversely the smallest angles provide information of the more global characteristics of the chain. Results on solution grown crystals can be summed up as follows:

a) The global confirmation (radius of gyration, R_g, from the smallest angles). The most salient feature for the present argument is the near invariance of R_g on increasing the molecular weight.

b) Local chain trajectory (from intermediate angles). The overall shape of the chain within the crystals conforms most closely to that of a sheet. The thickness of the sheet increases with molecular weight, but the sheet itself cannot be constituted by a single molecule (that is a given isotopic species) but needs to be diluted by about 50 % through other chains (a requirement fro the measured absolute scattering intensity).

It follows from a) and b): the chain deposits in the form of a chain-folded sheet along the crystal face, a given chain occupying only one half of the total sheet area (or volume). Once the sheet length exceeds a given value, that corresponding to a molecular weight of about 20000, it folds back upon itself forming double and multiple sheets as the molecular weight increases further. This effect we term 'superfolding' and leads to a multiply stacked sheet structure (see e.g. ref. 23).

c) At widest angles (still within the small angle range) we can obtain some information about the mutual stem arrangement within the sheet structure.

It emerges that there must be a clustering of stems belonging to a given molecule, consistent with the overall 50 % dilution. The stems are adjacent within the clusters, the total degree of adjacency amounting to 75 %, with the clusters of adjacent stems being separated by larger distances. The final structure resulting from a combination of a), b) and c) is shown by Fig. 13 (24).

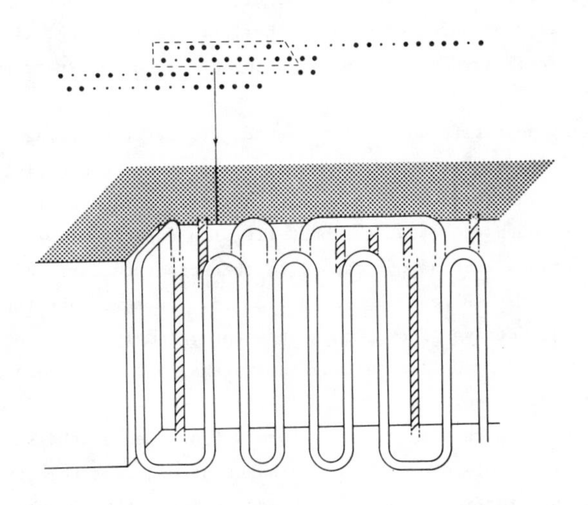

Fig. 13. Perspective diagram of the arrangement of the part of one molecule (unhatched chain) and its neighbours (hatched chain) within a solution grown single crystal (top basal face shaded). The corresponding stem arrangement is shown above, in projection, with heavy dots corresponding to the unhatched chains in the perspectieve view. The arrangement in question corresponds to 50 % stem dilution combined with 75 % adjacency in the stem reentry arrived at by the combined neutron scattering and infra-red spectroscopic study on samples containing isotopically doped marker molecules[24,26].

Infra-red spectroscopy

Adjacent, symmetrywise non-equivalent stems produce splitting in appropriate adsorbtion bands. In a mixture of deuterated and protonated species such splitting will only occur in the case of adjacent deposition of stems of one and the same isotope. Thus the existence or non-existence of band splitting, and beyond that, the amount of splitting and detailed band shape can give information on stem environment, hence on the mutual arrangement of stems belonging to a given molecule. In the original works by Krimm (and coworkers (24) only doublets featured as split bands. However, technical improvements (low temperature recording, use of higher resolution and subtraction capability of Fourier Transform recording, and application of deconvolution procedures) revealed a multiplicity of band splitting dependent on molecular weight and conditions of crystallisation. While this, no doubt, complicates the interpretation, it is also a source of much further information. Fig. 14 shows the kind of spectral detail involved, also comparing the recorded spectrum with calculations.

The calculations just referred to involve the effect of grouping of stems of the same isotopic constitution, hence in the present case deriving from the same molecule. The details of the splittings are very sensitive to such groupings. It was found (26) that the best fit with experiment was obtained for stem groupings such as correspond to the structure derived from neutron scattering, that is, as represented in Fig. 13. The agreement represented in Fig. 14 in fact derives from such a structure, namely multiple sheets with statistically arranged groups containing in average 75 % fold adjacency within an overall 50 % dilution.

In final analysis, we thus have a structure which embodies the sheet like deposition of a given molecule, the predominance of adjacent re-entry, with the sharp folds this implies, coupled with larger separation between such groups of adjacent stems with concomittant large, possibly loose fold loops. We also note that two entirely different techniques yield the same results. It needs remarking that the ever increasing power of the infra-red technique, at least in probing the nearest neighbour stem environment, is becoming ever more apparent; in fact currently it is this latter technique which is gradually taking up the leading role in continuing studies. These latter studies include such issues as to what happens during chain refolding in molecular terms, which is the phenomenon which, featured in the preceeding chapter.

The above resumê referred to one kind of solution grown crystal. The results on melt crystallisation, confined to fast cooled specimens to avoid isotopic segregation, are somewhat different, in as far as the global confirmation which the chain had possessed in the melt is largely retained (R_g does not change on crystallisation) but variable amounts of stem clustering (referred to as 'subgroups' or 'variable clusters') are still formed within the apparently unaltered overall molecular envelope (27). Here clusters again signify stem adjacency, but in reduced amounts compared to the solution grown

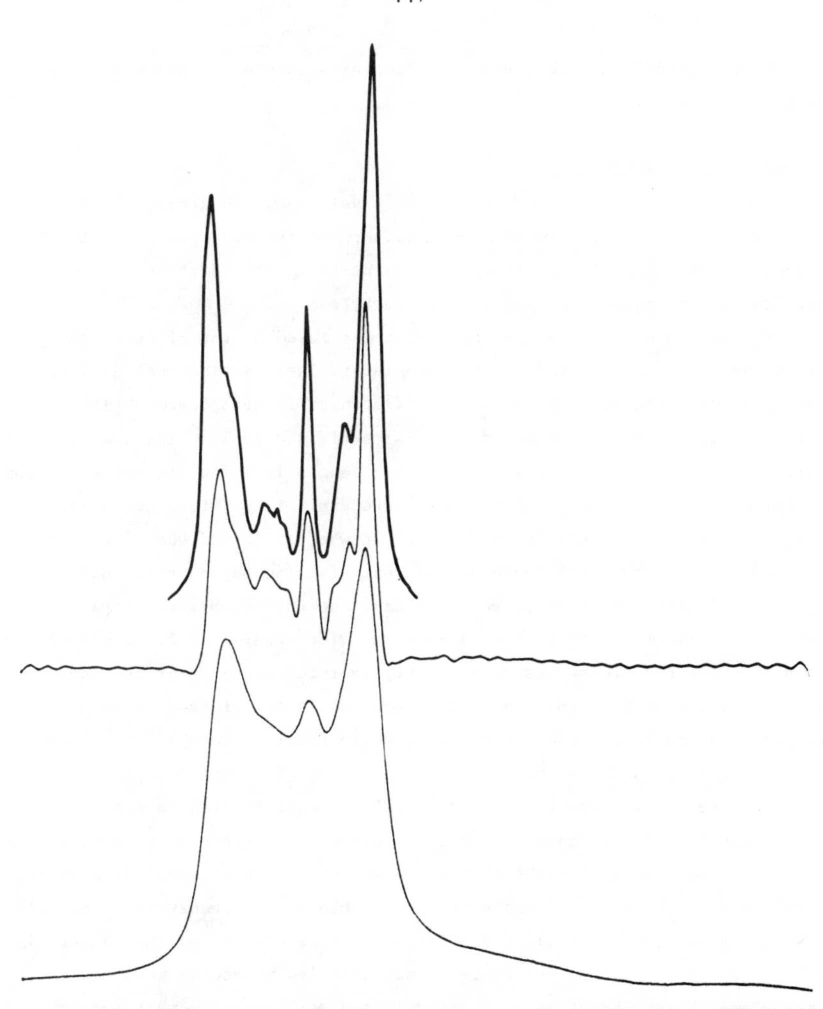

Fig. 14. An example of infra-red evidence for structure of the type as in Fig. 12 obtained on solution crystallized polyethylene doped with 3 % deuterated polyethylene of M_w = 386000.

Bottom: Experimental IR spectrum as obtained
Middle: After deconvolution
Top: Calculated spectrum based on a model as in Fig. 13 but with 11 sheets.

crystals, the actual numerical values differing according to the source of investigation or calculation.

FROM PARAFFINS TO POLYETHYLENE

The crystal structure of paraffins is well known in atomic detail. Here the chains are extended, forming layers with the methyl end groups at the layer surface, the methylene groups along the chains being all in the same crystallographic register as those in polyethylene.

The layers of extended chains are then stacked on top of each other but in this case in exact crystallographic register. There are several different stacking modes corresponding to slight differences in methyl group packing which give rise to different polymorphs in paraffin crystals. The question which clearly arises is: what happens as the chain length is being increased and that of polymeric polyethylene is approached? More explicitly, will the chains stay straight or will they fold? If the latter, at what stage and how? This subject has a long history by which the onset of chain folding beyond some chain length is definitely indicated. Nevertheless, the unavailability of paraffins of strictly uniform length prevented these earlier studies to be as definitive as could be desired. There was more and highly interesting progress with the analogous system of polyethylene oxide (28), not to be enlarged on in detail at this place, where nevertheless the chains while forming sharp fractions, were still not strictly uniform and contained rather specific end groups.

The present section will refer to quite recent progress along the above route of enquiry. It was made possible through the availability of paraffins of strictly uniform length (by the stringent criteria of traditional organic chemistry) up to much greater lengths than available so far thanks to a new method of synthesis by Whiting et al. (28,29). At the time of writing the longest is $C_{360}H_{722}$, which is well in the range we may call low molecular weight polyethylene. These paraffins were crystallised both from the melt and from solution and the 'layer' thickness determined by X-ray diffraction and low frequency Raman spectroscopy. In this case we know the chain length exactly to the precision of a single carbon atom, hence we should be able to tell when a chain folds and how.

The results which have emerged are definitive and clear cut even at this preliminary stage and deserve quoting (30). By the criterion of measuring the straight chain portion by the same methods as in polyethylene, the paraffin of $C_{150}H_{302}$ is already capable of folding, and such folding occurs with increasing propensity for longer paraffins for crystallization both from melts and solution. The fold length which has emerged is always such as to correspond close to an integer fraction of the total chain length. Thus, if we define the ratio n_c/n_{LAM} as that formed by the number of carbon atoms in the chain (n_c) to that contained by the straight chain portion (n_{LAM} as assessed by the Raman LAM method) we obtain values such as: for $C_{150}H_{302}$ (solution) n_c/n_{LAM} can be 0.99 or 1.93; for $C_{198}H_{398}$ (solution) it can be 2.0 or 3.01; for $C_{246}H_{494}$ it can be 0.99 or 1.98, for $C_{390}H_{782}$ 5.09 or 1.93 for solution and slow cooled melt respectively, to quote only a few figures from the much larger number available[30,31]).

Of the many consequences embodied by results, such as just quoted, one is the fact that even quite short chains will fold. Further, that folding will be such as to complete a full fold, which implies that the end groups will 'want' to be at the fold surface, even when it is only a methyl group, which, hardly differs from the methylene chain member itself. Additional results, not to be itemised here, show that higher crystallisation temperatures lead to longer fold lengths, and further, that crystals with a given fold length can increase their fold length through annealing, in both instances the closely integral relation between fold length and total chain length being retained.

But perhaps the most significant feature is the closely integer value of n_c/n_{LAM} itself. As seen from the above examples they only exceed the exact integer by 2-5 %. This means that there can only be a very few carbon atoms possibly 2-4 within a given fold. It follows that the folds must be sharp, and consequently the stem reentry adjacent. We can therefore confidently assert that sharply and adjacently re-entrant folding does exist and that paraffins of uniform length crystallise by folding in this manner. The latter applies at least up to four folds, that is five stems as the example for $C_{360}H_{722}$ shows, and there is no reson to assume that this is the upper limit.

SOME UNIFYING FEATURES

The direct attack on the polyethylene structure, at least in its form of solution grown crystals, has led to the picture of chain folded sheets where a given, sufficiently long molecule forms runs of adjacent stems, such adjacently re-entrant fold sequences being separated by wider gaps along the sheets. Such sheets still from the same molecules can then be stacked, one next to the other (consistent with superfolding). There is no direct information on the fold itself, except that those connecting adjacently re-entrant stems are likely to be sharp (even if for no other reason than space requirement) while those linking more remotely spaced clusters of stems are necessarily long and thus most probably possessing considerable amount of looseness.

The new results on long and uniform paraffins lead to the inescapable picture of the chains forming folded clusters. The sheet nature is not yet assessable from evidence available, but that much is practically certain that the fold stems must be adjacently re-entrant and the folds themselves sharp.

The two classes of evidence seem now readily combinable in a unified picture for the 'real life' polyethylene. The adjacently re-entrant sharply folded clusters of the paraffins would correspond to the limited runs of adjacent stems in Fig. 13, an arrangement the molecule endeavours to take up in the crystal but if the chain is sufficiently long it will be prevented from doing so through the intervention of other molecules and/or multiple nucleation of the same molecule along the given crystal face. This picture will thus combine both stem adjacency and sharp folds with remote stem re-entry and long, loose folds with amorphous characteristics such as required by a considerable body of experimental evidence. The former, as emerging from the paraffin results would be the intrinsic trend in crystallisation, while the latter the consequence of the mutual interference of chains, possibly together with the competing effects of nucleation and propagation of stem deposition in a given chain, according to circumstances. There is probably no fixed limit to the length of adjacently folded runst which is likely to be determined by crystallisation conditions, molecular weight and concentration. In addition to the above scheme, there is room for other features, such as non-uniformity of the fold length and the emergence of silica for both of which evidence exists; both of these would contribute to the roughness of the fold surface, usually inferred from studies on polyethylene.

ON CRYSTAL REGULARITY AND STRUCTURE HIERARCHGY

Fig. 6 is an example from melt crystallisation, and so are Figs. 15, 16, 17. All three have been obtained under rather special circumstances (32,33), nevertheless serve to illuminate some general points. The samples underlying these figures are really blends but of identical materials as far as chemical composition is concerned. The material of both components is polyethylene, yet one component, in this case present in much smaller quantities, is a linear polymer, while the other, the majority component, is a branched polymer, more precisely containing 1.4 % ethyl branches (on molar basis) being of the so called 'linear low' type. The blend components are fully miscible in the melt, but the linear component phase segregates by crystallisation when the melt is

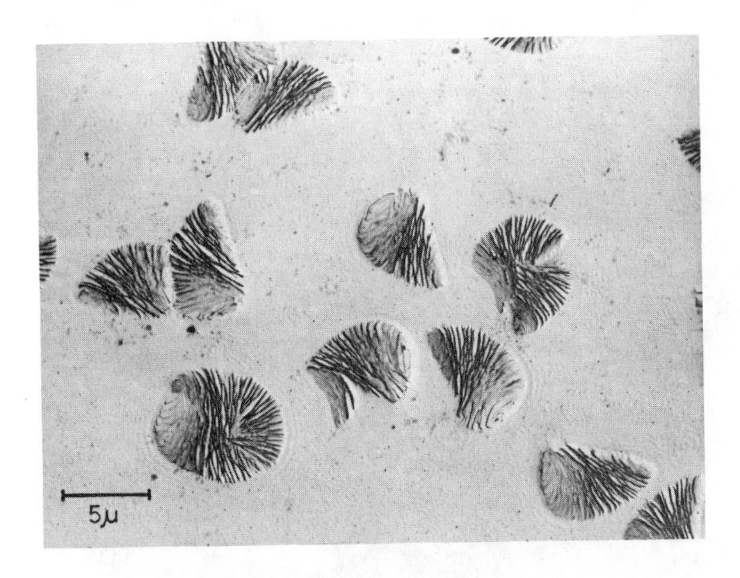

Fig. 15. Morphologically distinct sheaf development in a 2 % : 98 % blend of linear and slightly branched (14 ethyl branches per 1000 C atoms) 'linear low' polyethylene. Further particulars as in caption of Fig. 6. Overall view.

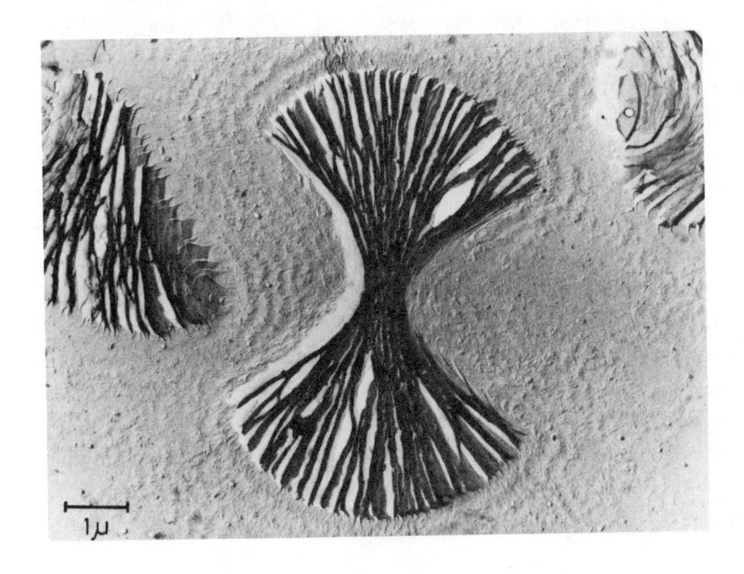

Fig. 16. As in Fig. 15 showing the early development of a sheaf at higher magnification.

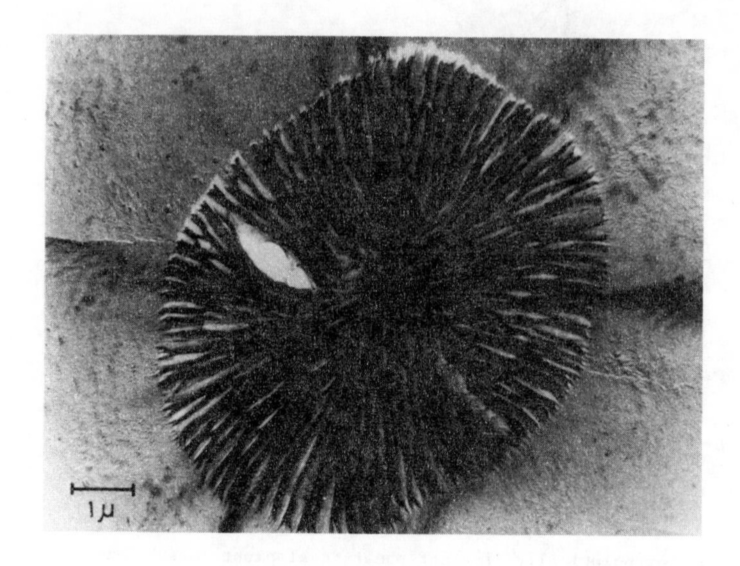

Fig. 17. As in Fig. 16 with a more highly developed sheaf.

stored at appropriately high temperatures for sufficient long time (32,33). Th whole phenomenon can be regarded as crystallisation of the linear polymer within a polymeric solvent of its own chemical constitution. When cooling to room temperature also the branched component will crystallise, however, yielding a much finer texture so that the two species remain identificable electron microscopically through using appropriate etching techniques which distinguish between the two, preferentially attacking the less well developed crystals of the chemically more vulnerable branched component.

So far the system, while no doubt important in its own right as a form of blend, may appear rather specialised. Nevertheless the results permit some generalisations. The 'solvent' itself being a high molecular weight polymer th system may be considered as a good approximation for crystallisation from the molten state, at least from the point of view of mobility, disentanglement and transport of long chains, even if not in terms of competition in deposition along the crystal face (in which latter respect the different chain types are clearly not equivalent).

We see that, in the first place, highly distinct, individual lamellae can form under the above circumstances with most of the regular features which characterise the usual chain folded single crystal when formed in isolation from non-polymeric solvent. Hence the crystallising linear polymer chain must have the ability to disentangle from its polymeric environment to produce these distinct morphological entities.

Secondly, we note that even if the concentration of the linear material is low (2 per cent in cases of Figs. 15, 16, 17) the different molecules seem to be able to find each other, so to speak, and convene into lamellar clusters many microns apart. While we have no absolute guarantee that all the linear chains have precipitated in this way, and that there are none left in portions which are most remote from the crystals formed by the linear material, that much seems certain that at least a substantial amount, if not all, of the linear material is contained by the spacially well separated sheaves. This implies that the corresponding long chain molecules must have migrated over distances of many microns to have become drawn into the crystals. This would indicate a very high degree of chain mobility and long range diffusion, most likely by the reptation mechanism. Such chain migration would be aided by the crystallisation force exerted by the growing crystals through entanglements which transmit such forces but do not significantly interfere with the crystallisation process itself. The importance of the last argument is obviously far reaching. Admittedly, in its present form it is more intuitive than rigorous.

Finally there are the lamellar superstructures themselves. As apparent from Figs. 15, 16, 17, they give rise to sheaves and eventually to spherulites (not illustrated here) leading to the superlamellar architecture characterising a typical crystalline polymer. It is the sum total of these hierarchical features which constitute the solid technological material and determine its macroscopic properties. Beside providing a glimpse into this world of the higher organisational hierarchies through Figs. 15, 16 and 17 the subject will not be pursued here further beyond drawing attention to the many faces of order which a long chain molecule is capable of displaying as illustrated here for the case of polyethylene.

REFERENCES

1. Keller, A. in 'Polymers, Liquid Crystals and Low-Dimensional Solids'. Edts. March, M. and Tosi, M. p. 3-142, Plenum Press, 1984.

2. Barham, P.J. and Keller, A., J. Materials Sci. in the press.

3. Pennings, A.J. and Meihuizen, K.E. in 'Ultra-High Modulus Polymers' Edts. Cifferi, A. and Ward, I.M., Appl. Sci. Publ. 1979.

4. Keller, A. in 'Structure-Property Relationships of Polymeric Solids' Edt. Hiltner, A. p. 25, Plenum Press, 1983.

5. Wunderlich, B., Macromolecular Physics, vol. 1, Academic Press, 1973.

6. Keller, A., Reports on Progress in Phys. part 2, vol. 31, 623, 1968.

7. Bassett, D.C., Principles of Polymer Morphology, Cambridge University Press, 1981.

8. Hoffman, J.D. and Weeks, J.J. 'J. Natn. Bur. Stand.' A, vol. 66, 11, 1965.

9. Dlugosz, J., Fraser, G.V., Grubb, D., Keller, A., Odell, J.A., and Goggin, P.L., 'Polymer' vol. 17, 471 (1976).

10. Chivers, R.A., Barham, P.J., Martinez-Salazar, J., and Keller, A., 'J. Polymer Sci. Phys. Edtn.' vol. 20, 1717 (1982).

11. Martinez-Salazar, J., Barham, P.J., Chivers, R.A. and Keller, A., 'J. Materials Sci.' vol. 20, 1616, 1985.

12. Stranski, I.N., 'Z. Physik. Chem.' vol. 136, 259, 1928.

13. Frank, F.C., 'Discussion Faraday Soc.' vo. 5, 48, 1949.

14. Burton, W.K., Cabrera, N. and Frank, F.C., 'Phil. trans. Roy. Soc. A, vol. 243, 299, 1951.

15. Jackson, K.A. in 'Treatise on Solid State Chemistry' Edt. N.B. Hannay, vol. 5, 233 (1975).

16. Hoffman, J.D., Davis, G.T. and Lauritzen, J.I., in 'Treatise on Solid State Chemistry', Edt. Hannay, 13, vol. 3, Chapter 7, Plenum Press, 1976.

17. Organ, S.J. and Keller, A., 'J. Materials Sci.' vol. 20, 1571, 1985.

18. Sadler, D.M., 'Polymer' vol. 24, 1401, 1983.

19. Barham, P.J., Chivers, R.A., Martinez-Salazar, J., Keller, A. and Organ, S.J., 'J. Materials Sci.' vol. 20, 1625, 1985.

20. Keller, A. and Priest, D.J., 'J. Polymer Sci.' B, vol. 8, 13, 1970.

21. Dreyfuss, P. and Keller, A., 'J. Polymer Sci.' B, vol. 8, 253, 1970.

22. Faraday Discussion No. 68 'Organization of Macromolecules in the Condensed Phase' 1979.

23. Sadler, D.M. and Keller, A., 'Science' vol. 203, 263, 1979.

24. Spells, S.J. and Sadler, D.M., 'Polymer' vol. 25, 739, 1984.

25. Bank, M.I. and Krimm, S.J., 'J. Polymer Sci., A-2, vol. 7, 1785, 1969.

26. Spells, S.J., 'Polymer Communications' Vol. 25, 162, 1984.

27. Sadler, D.M. and Harris, R., 'J. Polymer Sci. Phys. Ed.' vol. 20, 562, 1982.

28. Paynter, O.I., Simmonds, D.J. and Whiting, M.C., 'J. Chem. Soc. Chem. Commun.' p. 1166, 1982.

29. Bidd, I., and Whiting, M.C., J. Chem. Soc., 'Chemical Communications' in the press.

30. Ungar, G., Stejny, J., Keller, A., Bidd, I. and Whiting, M.C., 'Science' in the press.

31. Ungar, G. unpublished.

32. Norton, D. and Keller, A., 'J. Materials Sci.' vol. 19, 447, 1984.

33. Norton, D., Ware, I., Rosney, C. and Keller A., unpublished.

34. Hill, M.J., Barham, P.J. and Keller, A., 'Colloid Polymer Sci.' vol. 258, 1023, 1980.

35. Pennings, A.J., 'J. Polymer Sci., Polymer Symp.' vol. 59, 55, 1981.

36. Bassett, D.C., and Hodge, R.M., 'Proc. Roy. Soc. Lond.' A, vol. 359, 121 (1978).

37. Keller, A., Kolloid-Z.Z., Polymere, vol. 219, 118, 1967.

38. Bassett, D.C., Frank, F.C. and Keller, A., 'Phil. Mag.', vol. 8, 1753, 1963.

39. Fischer, E.W., Kolloid Z.Z. Polymere, v. 231, 458, 1969.

INVESTIGATION OF THE CRYSTALLIZATION PROCESS OF POLYMERS BY MEANS OF NEUTRON SCATTERING

E.W.FISCHER

Max-Planck-Institut für Polymerforschung, Mainz

Synopsis

The morphology of semicrystalline polymers can be described to a good approximation as consisting of lamellar crystallites separated by amorphous regions. The way in which a single macromolecule traverses the crystalline and amorphous phases can be evaluated from neutron scattering studies of mixtures of deuterated and protonated samples of the same polymers.

In the small angle range $q<0.03\ \text{Å}^{-1}$ only the radius of gyration R_g of the molecules can be measured, whereas in the intermediate angle range $0.03<q/\ \text{Å}^{-1}<0.3$ two important integral numbers can be obtained about the spatial correlation of crystalline stems. The evaluation of the neutron scattering data in this range can be performed without introducing detailed structural models. The only assumption made is that the molecular structure can be described as consisting of "clusters" of crystalline stems which belong to the same molecule. It is shown that this cluster model can be verified experimentally for the cases of polyethylene oxide, polypropylene and polyethylene.

The spatial correlation of the crystalline stems within a cluster can be evaluated from the scattering data by introducing a direct correlation function c(x). The method was checked by Monte-Carlo calculations and applied to the neutron scattering data of PEO in the wide angle range. The fraction of stems occupying adjacent sites in the crystal lattice depends strongly on crystallization conditons.

1 INTRODUCTION

Semicrystalline polymers with a high degree of crystallinity generally consist of stacks of lamellar crystals separated one from another by amorphous layers. This structural model has been well established by numerous electron microscope and small angle X-ray scattering studies. The conformation of the single molecule in the semicrystalline state is still a matter of controversy, however. Various models have been proposed [1] and Fig.1 shows the schematic diagrams of two extreme examples [2],[3]. The models differ especially with regard to the

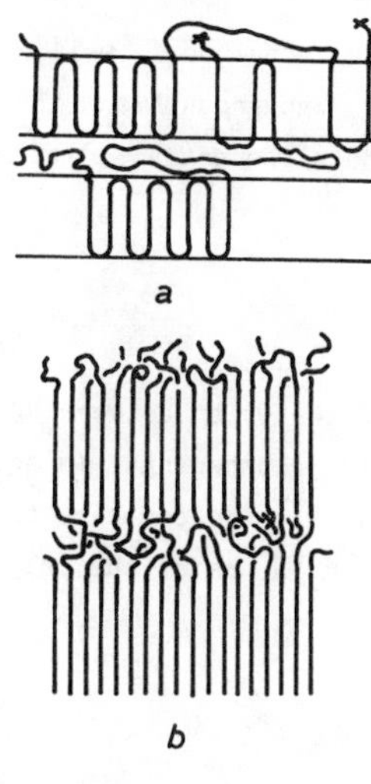

Fig.1: Two examples of proposed models for the structure of the amorphous regions in semicrystalline polymers a) Central core model with 8 adjacent re-entry folds in the cores [2],
b) "Switchboard" model [3].

assumptions made about the nature of the so-called fold surface and about the number of "re-entries" of a chain molecule into the same lamella. The structure of the amorphous regions has important implications, especially with regard to the mechanical properties including the deformation behaviour.

On the other hand information about the molecular conformation are most important for developing a suitable theory of crystallization, which is able to explain the dependence of morphological structure, degree of crystallinity and crystallization rate on various experimental parameters like crystallization temperature T_c, orientation of the molecules, molecular mass and chemical structure of the polymers.

Direct information about the chain conformation can be obtained from neutron scattering studies of mixtures consisting of deuterated polymers and co-crystallized with protonated chains of the same polymer. The technique allows the evaluation of the single chain structure factor P(q), where $q=(4\pi/\lambda)\sin\theta/2$. The main aim of the following paper is to show what kind of information can be obtained from the various ranges of the scattering vector q without detailed assumptions about the structure. So we ask ourselves how far a straight-forward evaluation of the scattering data will yield relevant structure parameters. The proposed methods are applied to polyethylene oxide, polyethylene and polypropylene, and the results are discussed which wêre obtained so far.

2 THEORY

The principle of the neutron scattering technique is well known [4]. After a correct subtraction of the incoherent "background" and the contributions due to density fluctuations, the differential scattering cross section per unit volume $d\sigma/d\Omega$ is given by

$$\frac{d\sigma}{d\Omega}(q) = c_D(1-c_D)K\, n_w\, P(q) \tag{1}$$

provided that there exists no thermodynamic interaction between H/H and D/D molecules (no segregation or intermolecular clustering). In eq.(1) c_D is the concentration of deuterated molecules, K is the contrast factor, n_w the degree of polymerization, and P(q) the form factor of the polymer molecule:

$$P(q) = \frac{1}{n_w^2} < \sum_{i,j} \exp\left[iq(\underline{R}_i-\underline{R}_j)\right] > \tag{2}$$

where $\underline{R}_i$ is the vector pointing to the i-th H atom of the chain.

For the evaluation of the data it is useful to introduce the reduced scattering intensity

$$J(q) = \frac{d\sigma/d\Omega}{c_D(1-c_D)K} = n_w\, P(q) \tag{3}$$

or a "scattering function"

$$F_n(q) = n_w\, P(q)\, q^2 \tag{4}$$

This function is very often used for comparison of measured and calulated data.

In principle, the form factor P(q) can be calculated by eq.(2) for any given model and compared with the measured data. It turned out, however, that the experimental results in the q-range, where most measurements have been performed, can be explained with quite different models by adjusting one or more suitable parameters [1]. Therefore we are concerned with the question which information can be obtained from the scattering data in a straight-forward manner without the introduction of detailed structural models. It is assumed only that the crystalline stems are incorporated in crystalline lamellae, see fig.2 and that the distances between stem centers within a "cluster" of stems located in one lamella and belonging to one molecule are much smaller than the distances between stem centers belonging to different lamellae. Details of the chain re-entry or chain tilting are not taken into account at the moment. We assume further that the sample has a high degree of crystallinity w_c and that a well pronounced long spacing reflection in observed by small angle X-ray scattering.

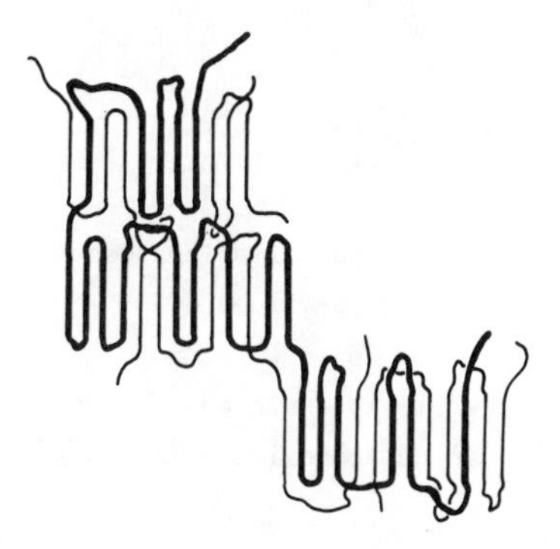

Fig.2: Schematic representation [5] of a single macromolecule traversing different lamellar crystals. Clusters of crystalline stems situated in different lamellae are connected by tie molecules. Note that the drawing is not in correct scale:The length of the crystalline stems is about 50 times greater than their lateral distances.

With the approximation of complete crystallization the coherent differential scattering cross section per unit volume can be written as

$$\frac{d\sigma}{d\Omega}(q) = c_D(1-c_D)\,\frac{K}{n_w} < \sum_{i,j}^{n_{st}} \exp(i\underline{q}\cdot\underline{r}_{ij}) \sum_{n,m}^{N_s} \exp(i\underline{q}\cdot\underline{x}_{nm}) > \qquad (5)$$

where $\underline{r}_{ij}=\underline{r}_i-\underline{r}_j$ are the difference vectors between the hydrogen sites of a single stem and $\underline{x}_{nm}=\underline{x}_n-\underline{x}_m$ are the difference vectors between the centers of the N_s stems of length n_{st} belonging to one molecule. The angular brackets refer to ensemble averaging and therefore also to orientational averaging over all positions of stacks of lamellae.

According to the model shown in fig.2, we divide the N_s stems among ν clusters, where the αth cluster contains N_c^{α} stems. Within one cluster the crystalline stems are connected by "amorphous" loops or chain folds; the various clusters are linked by tie molecules. These intramolecular clusters should not be confused with intermolecular clusters caused by segregation effects [6].

The reduced intensity for a conformation such as that in fig.2 can be written as [5]

$$J(q) = \frac{d\sigma/d\Omega}{c_D(1-c_D)K} = \frac{1}{n_w}\left(\nu\langle F_c^2\rangle + \sum_{\alpha\neq\beta}^{\nu}\langle F_c^{\alpha}(F_c^{\beta})^{*}\exp(i\underline{q}\cdot\underline{R}_{\alpha\beta})\rangle\right) \qquad (6)$$

where the $\underline{R}_{\alpha\beta}$ are the difference vectors between the centers of clusters α and β. Here the structure factor $\langle F_c^2\rangle$ of the clusters is defined as the average over the single cluster structure factors

$$\langle(F_c^{\alpha})^2\rangle = \langle\sum_{i,j}^{n_{st}} \exp\left[i\underline{q}\,\underline{r}_{ij}\right] \sum_{n,m}^{N_c^{\alpha}} \exp\left[i\underline{q}\,\underline{x}_{nm}^{\alpha\alpha}\right]\rangle \qquad (7)$$

The eq.(6) can be discussed with various approximations depending on the q-range under consideration. At very small q-values

$q < R_g$ and $q < \pi/L$

where R_g is the radius of gyration of the whole molecule and L is the long spacing, the form factor of the molecule can be approximated as usually by

$$1/P(q) \sim 1 + \frac{1}{3}q^2 R_g^2 \qquad (8)$$

In contrast for $q>\pi/L$ the second term in eq.(6) can be neglected [5] and in this q-range the scattering by the seemingly uncorrelated clusters is observed as described by eq.(7).

In order to simplify the treatment we consider a molecule consisting of ν clusters each of N_c stems. A distribution of the cluster sizes N_c has been treated in a recent paper [5]. It is further convenient to factorize the orientational averaging in eq.(7) and to rewrite the reduced intensity J(q) in the form [6),7),8]

$$J(q) = n_{st}P_s(q)\frac{1}{N_c}\sum_{m,n}^{N_c} J_o(qx_{mn}) \tag{9}$$

where J_o is the Bessel function of order zero which appears as a consequence of the fact that all stems in one cluster are parallel. The form factor $P_s(q)$ of the stems is given by

$$P_s(q) = \frac{1}{n_{st}^2}\sum_{ij}^{n_{st}} \frac{\sin(qr_{ij})}{qr_{ij}} \tag{10}$$

The question of spatial averaging will be discussed in more detail in a subsequent paper. Model calculations [9] have been shown that eq.(9) is a good approximation if $q>\pi/L$ and if L is larger than the lateral extensions of the single clusters.

The scattering of the uncorrelated clusters depends only on the number N_c of stems per cluster and on the mutual arrangement of the stems. We therefore introduce a stem correlation function

$$H(q) = \frac{1}{N_c}\sum_{m\neq n}^{N_c} \langle J_o(qx_{mn})\rangle \tag{11}$$

and consequently

$$J(q) = n_{st}P_s(q)[1+H(q)] \tag{12}$$

The stem correlation function H(q) is most important for the decision between different models as pictured in fig.1. It can be evaluated straight forward from the experimental data by means of eq.(12) if the form factor $P_s(q)$ of the stems is known. This can be calculated exactly from the crystal structure data by eq.(10).

It has been argued against the consideration above that they are based on the assumption of complete crystallization, that means

$$n_w = \nu\, N_c n_{st} \tag{13}$$

There are good reasons to believe, however, that eq.(12) is still right for semicrystalline polymers if the non-crystalline sequences are taken into account by modifying the stem for factor $P_s(q)$. Instead of calculating $P_s(q)$ according to eq.(10) for a completely stretched chain one has to perform the summation in eq.(10) also over the "loose" ends of the stems, which represent the "amorphous" sequences. The effect of this modification on the stem correlation function H(q) will be rather small, since in the intermediate q-range H(q) is completely governed by the average distance of the stems as we will show below.

Without any special assumption about H(q) one obtains from eq.(11)

$$H(q) = \frac{1}{N_c} \sum_{m \neq n}^{N_c} \left(1 - \frac{(qx_{mn})^2}{4} + \cdots\right)$$

or

$$H(q) \approx (N_c - 1) - \tfrac{1}{2} N_c R_{cc}^2 q^2 \tag{14}$$

with the radius of gyration of the stem centers in a cluster

$$R_{cc}^2 = \frac{1}{N_c} \sum_{1}^{N_c} x_{n,0}^2 \tag{15}$$

where $x_{n,o}$ is the vector from the center of gravity of the cluster to the nth center of a stem.

Using this approximation one can easily determine the number N_c of stems per cluster and the dimensions of the cluster by plotting H^{-1} vs.q^2 as shown in fig.3.

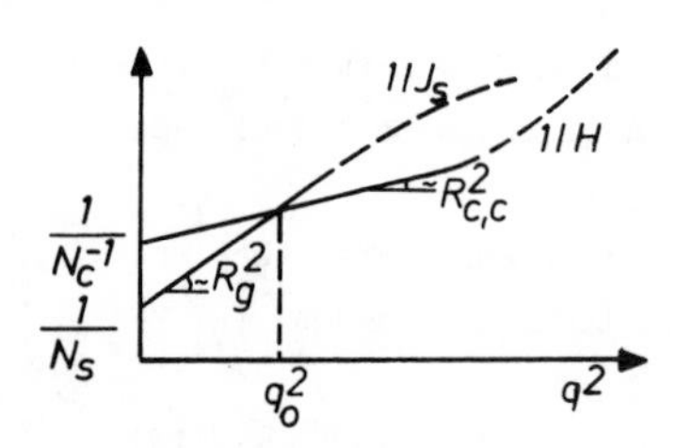

Fig.3: Schematic plot of the normalized intensity $J_s(q)$ for the entire molecule and the correlation function H(q) of the stem centers within a single cluster, both in the Zimm representation.

For a distribution of cluster sizes the averaging has been described in detail in a previous paper [5].

3 RESULTS

3.1 Small angle range

The evaluation of R_g according to eq.(8) may be distorted by phase separation between H- and D-molecules which frequently occurs during crystallization. In so far as reliable information could be obtained, it turned out that the gyration radius in the crystalline state often does not differ markedly from that in the melt or in dilute solution [10]. In the case of polyethylene oxide quickly crystallized by quenching to T_c=40°C only slight deviations of the radius of gyration R_g (cryst) from the value R_g (melt) have been found [11] as shown in fig.4. For small undercooling $\Delta T=T_m-T_c$ during crystallization, that means for large long spacings L, large deviations of R_g (cryst) from R_g (melt) are observed [11].

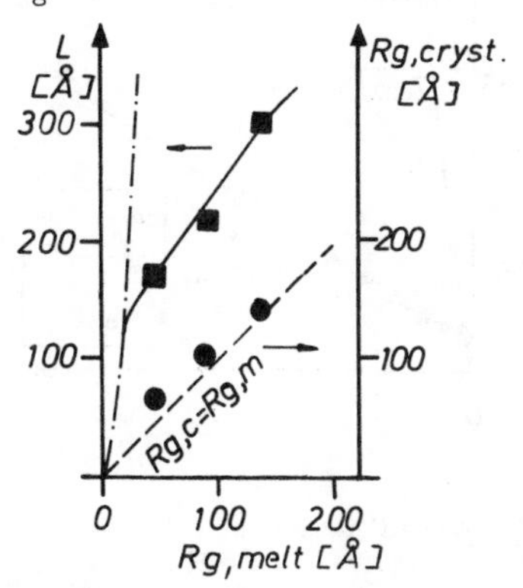

Fig.4: The long spacing L and the radius of gyration $R_{g,cryst}$ of polyethylene oxide quickly crystallized by quenching to T_c=40°C in dependence on the radius of gyration $R_{g,melt}$ in the melt. R_g values are weight-averages. The broken line gives approximate values L for extended chain crystals.

If we restrict our considerations to crystallization conditons where R_g remains approximately invariant during crystallization, the most simple explanation may be based on a model which we call "Erstarrungsmodell" (solidification model) [7],[12]. In this rather naive picture it is assumed that the crystallization occurs only by straightening of suitably oriented sequences of the coil which are incorporated into the crystalline lamella. This should happen in such a way that the "volume" occupied by that part of the chain which is incorporated into one lamella is not changed appreciably during crystallization. The model also implies that the number of entanglements which are present in the melt has not changed, they are just shifted to the amorphous regions.

3.2 The intermediate q-range

From the intermediate q-range $0.03<q/\text{Å}^{-1}<0.3$ one may expect to obtain more detailed information about the spatial correlations of the crystalline stems belonging to the same molecule. The application of the method described by fig.3 to experimental neutron scattering data was quite successful in all cases so far studied [5], that means for polyethylene oxide, polyethylene and isotactic polypropylene. For example fig.5 shows the experimental data for PEO (M_w=125 000) quickly crystallized by quenching to T_c=40°C. In fig.6 the same procedure was applied to polyethylene crystallized by quenching. In both figures the inserts show the normalised intensity J(q) for the small angle range and for comparison the lines $J(q)^{-1}$ versus q^2 are also drawn in the plots. As one can notice a clear distinction is possible between the (uncorrelated) cluster scattering in the intermediate q-range and the scattering by the correlated clusters in the SANS range.

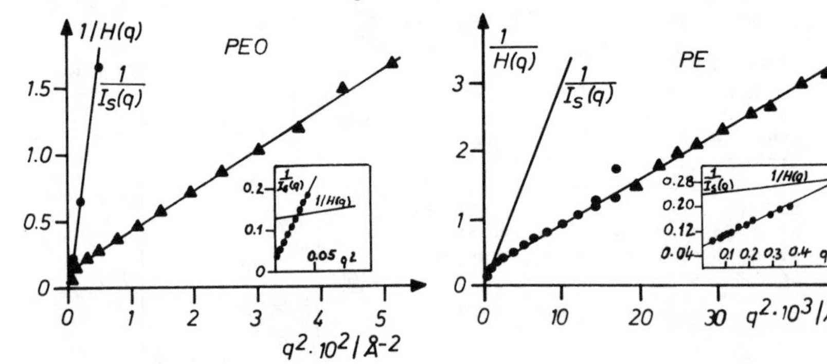

Fig.5: Evaluation of $\langle N_c \rangle$ and $\langle R_{cc}^2 \rangle$ from $H(q)^{-1}$ vs. q^2 for polyethylene oxide quickly crystallized by quenching to T_c=40°C. Long spacing L=300 Å. The insert shows the experimental data for J(q) vs. q^2.

Fig.6: As in fig.9 for polyethylene crystallized by quenching from the melt.

● Schelten et al. [13],

▲ Stamm et al. [12]

3.3 The large q-range

The radius of gyration R_{cc}^2 of the stem centers depends on the packing densities of the stems within the cluster, that means on the average distance $\langle a \rangle$ between the stems. Assuming a linear arrangement of stems along a lattice plane one obtains approximately

$$\langle a \rangle^2 = 12\, R_{cc}^2/(N_c-1)^2 \tag{16}$$

For the two cases of fig.5 and 6 one obtains $\langle a \rangle$=9.1 Å for PEO and 14.3 Å for PE. These values clearly indicate that a structure model based on the assumption of pure adjacent re-entry must be excluded.

In order to evaluate details of the stem correlation function h(x) the wide angle scattering range has to be considered. The evaluation of the scattering data in a straight forward manner by Fourier-Bessel-transform is complicated by the formation of the "clusters" which we discussed above. The problem can be solved by use of the so-called "direct correlation function" c(x) which was introduced by Ornstein-Zernike in the theory of critical opalescence. Preliminary results obtained by this method are reported elsewhere [10].

REFERENCES

1. See, for example, Organization of Macromolecules in the Condensed Phase, Faraday Discuss. Chem. Soc. 68 (1979)

2. Hoffman,J.D., Guttman,C.M. and DiMarzio,E.A.,Farad.Disc.68,177 (1979)

3. Yoon,D.Y. and Flory, P.J., Farad.Discuss. Chem.Soc., 68, 288 (1979)

4. Ballard, D.G.H. and Schelten,J., in Developments in Polymer Characterization, Vol.2, J.V.Dawkins, Ed., Appl.Science, London, 1980

5. Fischer, E.W., Hahn, K., Kugler, J., Struth, U., Born, R. and Stamm, M., J.Polym.Sci., Phys.Ed. 22, 1491 (1984)

6. Fischer, E.W., in IUPAC Macromolecules, H.Benoit and P.Rempp, Eds., Pergamon, London, 1982

7. Fischer, E.W., Pure Appl.Chem,. 50, 1319 (1978)

8. Dettenmaier, M., Fischer, E.W. and Stamm, M., Colloid Polym. Sci., 258, 343 (1980)

9. Brereton, M.G., unpublished results

10. Fischer, E.W., Polymer J., in press

11. Kugler, J., Struth, U., Born, R., Fischer, E.W., Hahn, K., in prep.

12. Stamm, M., Fischer, E.W., Dettenmaier, M. and Convert, P., Farad. Disc. Chem. Soc, 68, 263 (1979)

13. Schelten, J., Ballard, D.G.H., Wignall, G.D., Longmann, G. and Schmatz, W., Polymer 17, 751 (1976)

LAMELLAR ORGANIZATION IN POLYMER SPHERULITES

D.C. BASSETT

J.J.Thomson Physical Laboratory, University of Reading, Reading, U.K.

SYNOPSIS

The technique of permanganic etching allows the study of representative lamellar organization within melt-crystallized polymers. Detailed study of several polymers reveals common patterns of ordering. Spherulites are commonly constructed from a framework of individual dominant lamellae. These branch and splay apart, while between them are subsidiary lamellae formed from later-crystallizing molecules. Many properties will depend on this ordering and its associated microsegregation.

1 INTRODUCTION

Many of the commercially important properties of crystalline polymers are controlled not just by molecular structure but also by molecular organization into the morphology of samples. This is the principle reason why the same polymer can reveal substantially different modulus, tensile strength, fracture behaviour, dielectric strength or thermal conductivity for example, depending upon processing conditions. A major component of the morphology is lamellar crystals, typically 10nm and more thick. For these to be adequately resolved and studied, transmission electron microscopy is required. Unfortunately, direct study of a polymer specimen is rarely successful because of radiation damage. However, indirect study of surfaces is possible provided they contain sufficient detail. Permanganic etching[1,2] is a solution to the problem of providing such detail in a variety of polyolefines, (PE, PP, PS) polyaromatics (PEEK) and vinyl polymers. All the polymers studied by us have been profusely lamellar, with their lamellae arranged in similar ways. In a number of cases the organization has been shown to involve micro-segregation of shorter and/or more branched molecular species. Detailed textural characterization has been made: in polyethylene as a function of most crystallization conditions[3-6] and in isotactic polystyrene[7,8] as a function of growth rate and molecular weight. Measured distances for lamellar organization disagree with predictions for cellulation theory.

2. RESULTS AND DISCUSSION

The approach we have adopted with all the polymers studied is to begin to examine growth at low supercoolings. This offers the advantages of comparatively simple objects and ready distinction between lamellae grown isothermally and thinner ones produced on quenching. The greater complexities of morphologies grown at more usual, lower temperatures are then more easily appreciated. It has nevertheless, been found that there is textural continuity across these growth conditions.

A basic feature, observed in all the homopolymers is that it is individual lamellae which grow out initially to create the skeleton of objects, be they axialites, immature or mature spherulites. In polyethylene these initial, or dominant, lamellae are mostly non-planar, sometimes ridged but typically S-shaped, so that they tend to confine the uncrystallized melt into columns before it crystallizes as subsidiary lamellae. There is thus the appearance of cellular crystallization[3], although the width of a dominant lamella is not a good measure of the dimensions of 'cells': the separation between adjacent dominant lamellae is better[6]. The size of 'cells' is not commensurate with the length δ = Diffusion Coefficient/Growth Rate[6] which controls cell-size in constitutionally supercooled metal alloys. Moreover, the later crystallization of the 'cell interior' is the inverse of cellular morphology in metals.

The planar dominant lamellae of isotactic polystyrene[7] do not suggest closed growth cells, nor have we found evidence of lamellar fibrillation in association with the change from axialites to mature spherulites[7]. Instead, one notes the branching and splaying apart of dominant lamellae which generates the structures observed in the three polyolefines. The splaying, it is suggested[7], is due to pressure from cilia i.e. uncrystallized portions of molecules between branching lamellae.

Subsidiary lamellae have been shown, especially in polyethylenes grown slowly, to consist of shorter[3,5] and/or more branched molecules[8]. In fig. 1 there is correspondingly greater etching of such subsidiary lamellae, grown at a lower temperature. Work is in progress to quantify the nature and location of preferentially etched polymer.

In highly doped melts, objects form which optically appear to contain discrete fibres. Using electron microscopy one often finds that these are several lamellae viewed from the side, but there is still continuity to single dominant lamellae and (ridged) subsidiary lamellae (fig. 2). The separation λ of such 'fibres' (fig. 3) in isotactic polystyrenes doped with various atactic species has been measured at 190°C as functions of growth rate, molecular weight and isotactic content[9] (fig. 3). There are major differences between the observed and the predicted δ, especially in the variation with molecular weight. Such measurements, down to 180°C, the growth rate maximum in polystyrene, have also failed to provide evidence for the involvement of cellulation in spherulitic growth. Cellulation may, however, occur in addition to the dominant/subsidiary mechanism at sufficiently low crystallization temperatures, e.g. 130°C in polypropylene[10]. Observations which may indicate this are under investigation.

ACKNOWLEDGEMENTS

I am indebted to Dr. A.M. Hodge for fig. 2 and to Dr. A.S.Vaughan for figures 3 and 4.

REFERENCES

1. Olley R.H., Hodge A.M. and Bassett D.C. 'A Permanganic Etchant for Polyolefines'. J.Polm.Sci. (Phys. ed.) 17, 627-643 1979.
2. Olley R.H. and Bassett D.C. 'An Improved Permanganic Etchant for Polyolefines'. Polymer, 23, 1707-1710, 1982.
3. Bassett D.C. and Hodge A.M. 'On Lamellar Organization in Certain Polyethylene Spherulites'. Proc.Roy.Soc. A359, 121-132 1978.
4. Bassett D.C.and Hodge A.M. 'On the Morphology of Melt-Crystallized Polyethylene. I. Lamellar Profiles' Proc. Roy.Soc. A377 25-37 1981.
5. Bassett D.C. Hodge A.M. and Olley R.H. 'On the Morphology of Melt-Crystallized Polyethylene. II.Lamellae and their Crystallization Conditions', Proc.Roy.Soc. A377, 39-60 1981.
6. Bassett D.C. and Hodge A.M. 'On the Morphology of Melt-Crystallized Polyethylene. III. Spherulitic Organization'. Proc.Roy.Soc. A377 61-67 1981.
7. Bassett D.C. and Vaughan A.S. 'On the Lamellar Morphology of Melt-Crystallized Isotactic Polystyrene'. Polymer (in press) 1985.

8. Bassett D.C., Khalifa B.A. and Olley R.H. 'Crystallization and Annealing of Polyethylene Copolymers. I. Atmospheric Behaviour'. to be submitted to Polymer.

9. Vaughan A.S., Ph.D. Thesis, University of Reading, 1984.

10. Bassett D.C. 'Electron Microscopy and Spherulitic Organization in Polymers'. CRC Critical Reviews 12, 97-163, 1984.

Figure 1

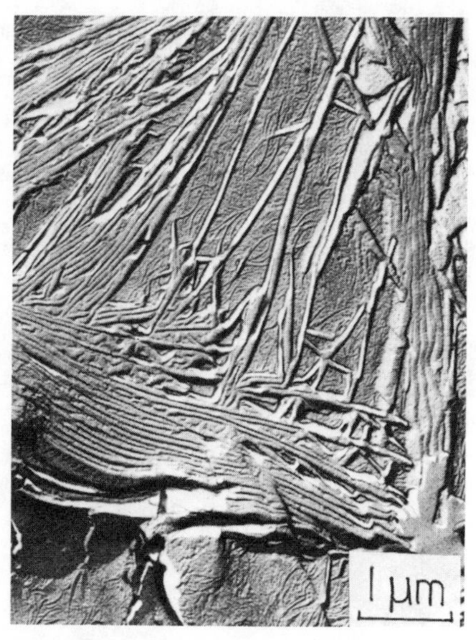

Figure 2

Fig. 1 Differential permanganic etching of subsidiary lamellae in linear polyethylene crystallized partly at 128°C and partly on quenching.

Fig. 2 Field, including individual dominant and regions of subsidiary lamellae, in linear polyethylene at 30,000 molecular weight crystallized at 130°C from a 1:1 melt with low density polyethylene.

Fig. 3 Growth tips of a polystyrene spherulite crystallized at 190°C.

Fig. 4 Growth rate data for isotactic polystyrenes crystallized at 190°C from melts doped with atactic polymer relating λ, the distance between adjacent dominant lamellae to the fraction of isotactic polymer and molec.wt. In fig. 4c the upper line has $M_w = 3.6 \times 10^5$ for the atactic component, the lower line has $M_w = 6.0 \times 10^5$ for the isotactic polymer.

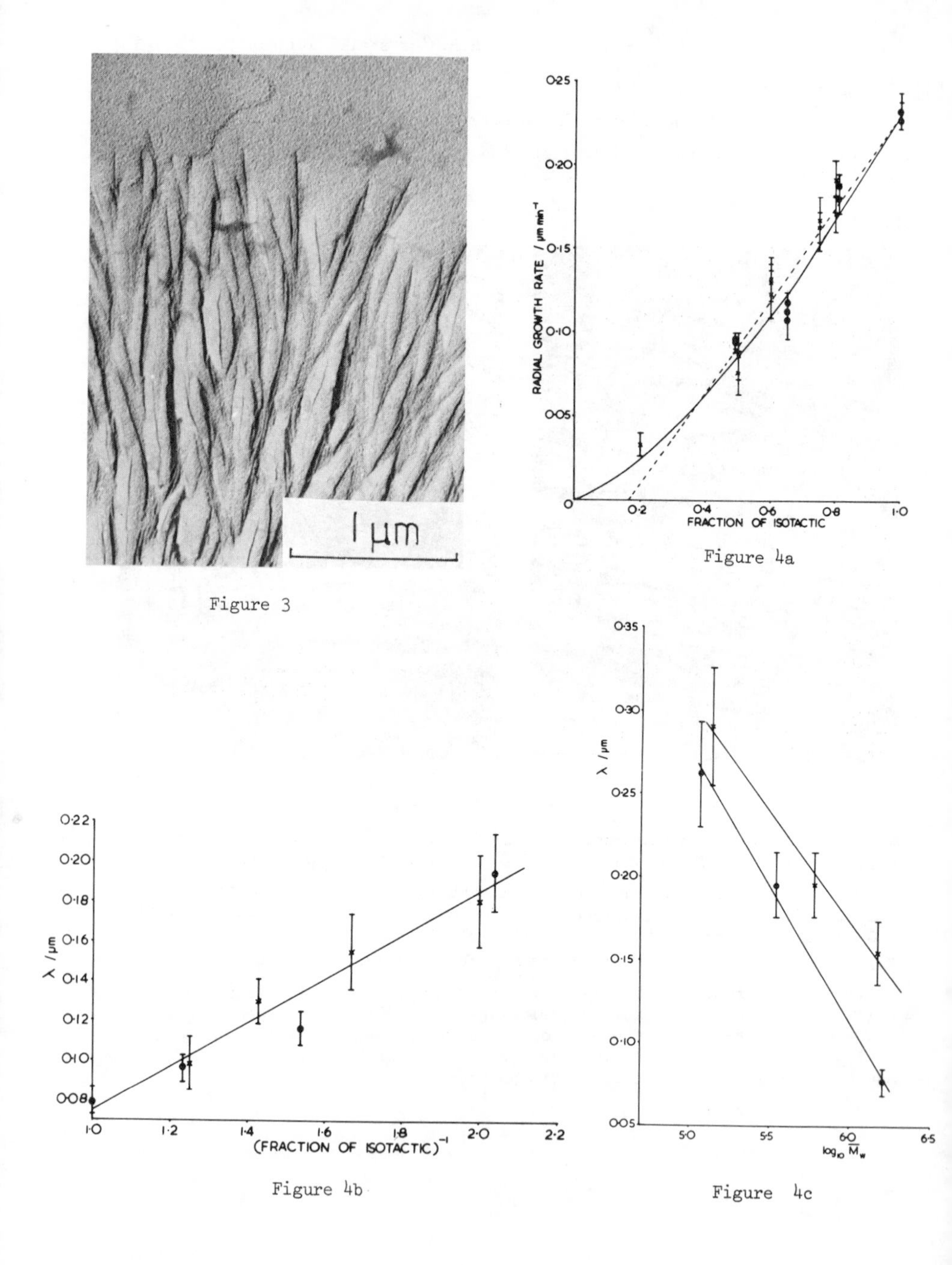

Figure 3

Figure 4a

Figure 4b

Figure 4c

CONSIDERATIONS ON THE CRYSTALLIZATION WITH CHAIN FOLDING IN POLYMERS

C.G. Vonk

DSM Central Laboratory, P.O. Box 18, 6160 MD Geleen, Netherlands

Synopsis

In 1962 Flory (J.A.C.S. 84, 2857) showed that the fraction of stems in a lamellar polymer crystal, that is connected by folds in one of the lamellar end planes, is about ½. It is here emphasized that the derivation of this ratio is rigorous, and does not allow deviations such as discussed in literature of late. In extending the arguments to solution-grown crystals it is argued, that also here folding must be considered as a compromise, resulting from the inability of the chains to make large movements in lateral directions in the solution. Chain folding is furthermore shown to be promoted by the local flow pattern resulting from the overcrowding in the amorphous or the solvent phase. The fraction of the stems connected to tie molecules at one of the crystal faces in a melt grown crystal is found to be close to 1/4. It is shown that crystallization can proceed without anisotropy being induced in the amorphous phase.

1 INTRODUCTION

The overcrowding effect to be discussed here, was probably recognized already soon after the discovery of the folded chain morphology by Keller[1] in 1957, but was first treated in a quantitative way by Flory[2] in 1962. The effect is visualized in fig. 1, which shows a polymer crystal in the case that no chain folding takes place. As the average cross section of the chains in the plane B, because of the random orientation of the chains, is larger than in the plane A, the chains would have to fan out in the way shown in the figure. In a lamellar crystal of large lateral dimensions, such as occuring in most semi-crystalline polymers, this obviously is not allowed; this implies, that between the planes A and B the chains have to fold back. Here, the fraction of the chains penetrating from plane A to plane B will be indicated as the penetration factor f. Flory derived that the value of this factor is about ½; because of the importance of this derivation in the present context, it will be repeated here in a somewhat extended form.

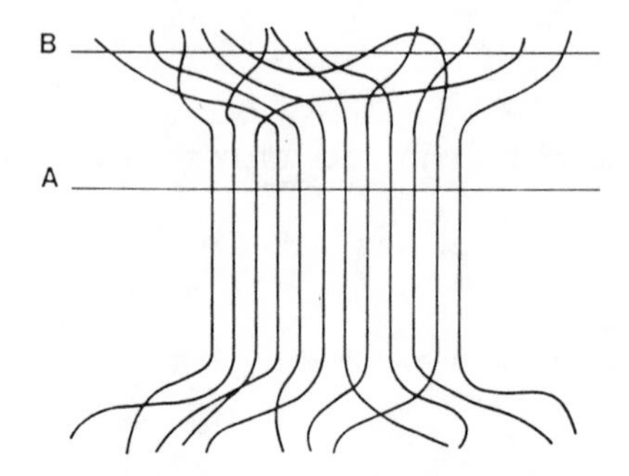

Fig. 1. Crystallization of a polymer without chain folding.

2 CALCULATION OF THE PENETRATION FACTOR

Let the chains be represented by straight line segments of length ℓ, connected to each other end-to-end at pivot points P. Choosing in an arbitrary way one end of each chain as its starting point, one may consider each pivot point as the starting point of one segment, running in the chain direction. Let F be a plane in the sample and x the coordinate perpendicular to it. A segment making an angle θ with the x-direction will cut through F if the distance of its pivot point to F is between 0 and $\ell\cos\theta$ (fig. 2). In an isotropic sample only a fraction $\frac{1}{2}\sin\theta d\theta$ of all pivot points in this region carries segments for which the

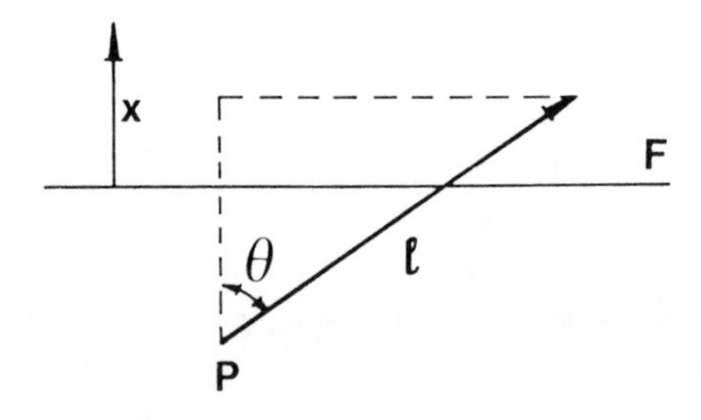

Fig. 2. Position of a chain segment ℓ intersecting a plane F.

angle θ is between θ and $\theta + d\theta$. If A_a is the effective chain cross section in the amorphous phase, the number of pivot points per unit volume is $1/(\ell A_a)$. The number of segments per unit area intersecting F from below will then be given by

$$\frac{1}{\ell A_a} \int_0^{\pi/2} \ell\cos\theta \cdot \tfrac{1}{2}\sin\theta \, d\theta = \frac{1}{4A_a}$$

The same number of segments intersects F from above, so that the total number N_a of segments crossing F per unit area is

$$N_a = \frac{1}{2A_a} \tag{1}$$

To find the penetration factor we notice that the number N_c of chains per unit area intersecting the plane A in fig. 1 is given by

$$N_c = \cos\alpha/A_c,$$

where α is the tilting angle between the stems and the normal to the lamellar surface (zero in fig. 1), and A_c is the effective cross section of the chains in the crystalline phase. With the relation $\rho_a/\rho_c = A_c/A_a$, where ρ_c and ρ_a are the densities of the crystalline and amorphous phases respectively, we find from the condition $f = N_a/N_c$ the relation

$$f = \rho_a/(2\rho_c \cos\alpha) \tag{2}$$

Ignoring the difference between ρ_c and ρ_a, and assuming $\alpha = 0$, one finds $f = \frac{1}{2}$, the result generally quoted in literature.

In the derivation of this equation no restrictions are imposed on the choice of the segments, the only condition being an isotropic distribution of the orientations of the chain segments. More specifically, the derivation is not invalidated by the occurence of orientation correlations between successive segments in the chain. Flory assumed that such correlations were not allowed; on this assumption, ℓ should be chosen equal to the statistical chain element, which in polyethylene is of the order of 18 CH_2 groups[3]. Modelling an actual chain by segments of such length would obviously involve an approximation, which might affect the result as represented by eqn. (2) in a way which is difficult to predict. However, as the correlation between successive chain segments does not enter into the considerations, we may choose for ℓ the shortest line segment allowing a description of the chain trajectory. As such, in polyethylene the lines connecting the midpoints of successive C-C bonds seems most suitable. The validity of eqn. (2) than is only limited by the condition of isotropy in the amorphous phase.

The above restriction made by Flory probably has induced other investigations to consider eqn.'s (1) and (2) as approximations, which might be improved on the basis of more sophisticated considerations. Thus Frank[4] discussed some causes for deviations from the value of f = ½; also, other investigators[5,6,7] among whom also Flory[5], concluded to a penetration factor of 1/3. This however can be shown to be a direct consequence of the cubic lattice model, which these investigators used to simulate the chain trajectory in the amorphous phase. As a chain is supposed to pass through each lattice point, 2 of the 6 edges starting at each lattice point will carry a chain segment. Thus, in an isotropic amorphous phase only one third of the edges in all three coordinate directions will be occupied. As in the crystalline phase all edges in the stem direction are supposed to be occupied by chain segments, only 1/3 instead of ½ will be able to penetrate into the amorphous regions. We therefore feel that, though the lattice approximation may be quite usefull in predicting the relative abundance of loops of various lengths, it apparently fails in predicting the correct value of the penetration factor.

3 CRYSTALLIZATION FROM SOLUTION

It is tempting to extend the present argumentation to crystallization from solution, as was also done by Flory. Here, the number of chains per unit area passing through an arbitrary plane apparently is reduced by the factor c, which is the volume fraction of the polymer in solution. The penetration factor thus is to be represented by

$$f = \frac{c\rho_s}{2\rho_c \cos\alpha} \tag{3}$$

where ρ_s is the density of the solution. In this way, the tendency to crystallize from solution in large lamellae with a very high fold fraction (1-f) is based on a quantitative argument. This argument must however be considered with caution, and the underlying assumption has to be examined. This assumption is, that the chains, emanating from the crystal surfaces, have the same concentration in the near vicinity of the crystal as in the surrounding liquid. An alternative mode of crystallization would be one in which the chains emanating from the crystal crowd together to form a phase in which the concentration of the polymer as a function of the distance to the crystal face drops gradually to the value pertaining in the solution. In this situation, the penetration factor would equal the value of about ½, occuring in melt-grown polymers, and no additional folding would be required. Flory[2], in considering the thermodynamic consequences of the crystallization from solution, came to the conclusion that at a certain stem length a maximum degree of chain folding would always lead to a more stable situation than any process involving only partial crystallization.

This point of view would however hold equally well for crystallization from the melt, in which under normal circumstances always an amorphous fraction occurs. As we know from later investigations[8], also in solution crystallized polyethylene of normal molecular weight the crystallization never is complete.

That neither complete crystallization occurs, nor the crowding together discussed above, may be explained by assuming that the chains are much more easily displaced in the chain direction by reptation, than in directions perpendicular to the chain direction. The latter movements would be impeded not only by the friction in the solvent, but much more effectively by the other chains in the solution, which limit the movements of a particular chain by entanglements. Crystallization may than be assumed to proceed through the reeling in of the dangling chain ends, which are displaced by reptation. Those chain ends, which are not yet crystallized when the crystallization has proceeded sufficiently far to relieve the overcrowding, will, after removal of the solvent, give rise to amorphous regions. Thus, the degree of chain folding would, at least to a first approximation, be determined by eqn. (3); for the proces of crystal growth with chain folding the well-known theories as described by Wunderlich[9] might still be applicable.

In this view, chain folding is not a natural tendency of the polymer chains, but should rather be considered as a compromise between the energy gain on crystallization on the one hand, and kinetic restrictions on the other. A possible factor, inducing the actual proces of chain folding, will be discussed in the following section.

4 INDUCTION OF CHAIN FOLDING DURING CRYSTALLIZATION

In taking a closer look at the crystallization from the melt, we suppose that in the melt there exists a degree of local order involving the occurence of regions, in which the chains are approximately parallel. One may assume the crystallization to be preceded by the reorientation of such bundles, by which the chain segments are brought in the correct direction for crystallization. Once this is achieved, one may expect the crystallization to proceed by the generally accepted nucleation-and-growth mechanism[9]. Already during the first phase of orientation, a contraction of the material in the plane in which a lamella is developing, must take place, as in the reoriented material the number of intersections of the chains with this plane is reduced by a factor of about 2.

In fig. 3, in which the crystallization is thought to proceed from left to right, this contraction is indicated by arrow A. It may be thought to be compen-

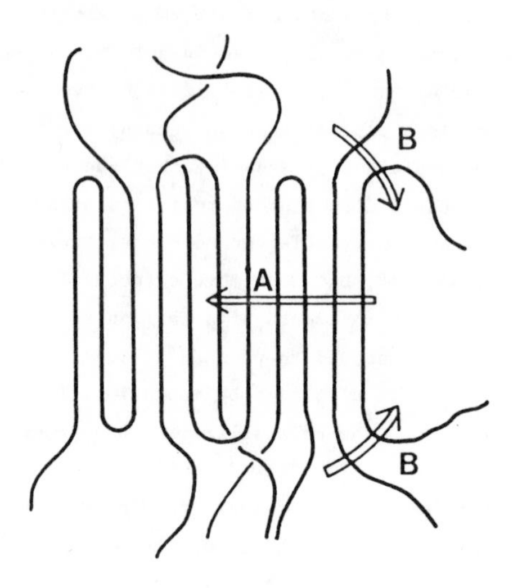

Fig. 3. Local flow pattern during crystallization.

sated by local flow around the edges of the growing crystal, indicated by the arrows B. These flows will have just the right direction for promoting the folding of the next stems which are to crystallize. Thus one may presume that crystallization will proceed if this flow is just sufficient to maintain the degree of chain folding necessary to relieve the overcrowding effect.

5 TIE MOLECULES

An important aspect in relation to product properties is the number of tie molecules between the lamellae. Let a fraction g of the stems in the crystalline regions be connected to a chain end in the adjoining amorphous region at one side, and let a fraction h of the chains penetrating into this amorphous region, but not terminating in a chain end, be connected to a stem in the next lamella. The ratio t of these tie molecules to the number of stems will than be given by

$$t = (f - g)\,h \qquad (4)$$

In considering the factor h, we first find the average length of the chains in the amorphous regions. We may assume that each chain traverses

crystalline and amorphous regios in turn, and thus on the average shows a crystallinity equal to the overall value.
Ignoring for the moment the difference between the crystalline and amorphous densities, and assuming the tilting angle α to be zero, we find from eqn. (2) that the average length of the crystalline chain portions is 2C, and that therefore the average length of the amorphous chain segments must be 2A. Here A is the (number-average) thickness of the amorphous layers.

Next, we may compare this average length of the amorphous chain segments, which according to a recent study in our laboratory[10] for a linear polyethylene sample of normal molecular weight (M_n = 20,000) is of the order of 100 Å, with the length of the statistical chain element. This has been estimated to be about 18 CH_2 groups[3], corresponding to a length of 23 Å. We may thus safely assume that the chains, penetrating into the amorphous regions and not terminating in chain ends will make a random walk, which will bring half of them to the next lamella, and the other half back to the lamellae from which they emanated. Thus, we may estimate h to be ½.

To find the factor g, we assume a random distribution of the chain ends in the whole polymer. Than, the number of chain ends per unit surface area in an amorphous layer of thickness A will be $2A/(LA_a)$, where L is the (number-average) chain length, and A_a again is the effective cross section of the chains in the amorphous regions. As the number of stems per unit surface area in the crystalline region for $\alpha \cong 0$ equals $1/A_c$, and as only half of the chain ends under consideration is connected to these stems, we find $g = (2A/LA_a)/(2/A_c) = (A/L)(A_c/A_a)$. For $\rho_c = \rho_a$ this reduces to $g = A/L$, which for the polymer discussed above ($A \cong 50$ Å, $L \cong 2000$ Å) is of the order of 0.025. Considering eqn. (4), and observing that generally $f \cong$ ½, one may conclude that in normal polymers the molecular weight has but little effect on the number of tie molecules, the ratio t being about 1/4. This conclusion is not much affected by assuming the chain ends to be completely concentrated in the amorphous regions, which may increase g by a factor of 2-4.

6 ORIENTATION OF THE AMORPHOUS PHASE

In the previous considerations the amorphous phase was assumed to be isotropic; in the case of a prefered orientation in this phase the penetration factor would deviate from the value predicted by eqn. (2). One might suppose that here a prefered orientation of the chains in the directions parallel to the lamellar surfaces would take place: as the chains are straightened in the direction of the normal to the lamellae in the crystalline regions, the chains in the amorphous regions might be expected to orient themselves parallel to the lamellae, in order to maintain the overall shape and conformation they had before crystallization. That this shape is however not completely preserved follows from the occurence of birefringence in the semi-crystalline material. In general, the crystalline regions would be too thin to show individual effects in the optical microscope; thus, the birefringence must be the net effect of the average orientations in the crystalline plus amorphous phases. The fact that this is not zero, but has a value which is clearly related to the orientation in the crystalline phase, implies, that the crystalline orientation is at least not fully compensated by an opposite amorphous orientation. In the following it will be shown that this situation is indeed possible and can be understood on the basis of the limited degree of chain folding, following from the overcrowding principle.

In an imaginary crystallization proces one may think of the lamellae to be thin to such a degree, that the crystallization only affects the direct environment of the new lamellae, without disturbing the chain conformations at larger distances, and thus without giving rise to orientation of the amorphous regions. Because of the overcrowding principle, the folding ratio $1 - f$ has to be about $\frac{1}{2}$ also in these thin lamellae; here the folding would involve the reeling in over small distances of part of the chains. Next, one may imagine the crystal to thicken; in this proces the folds already present would have to be displaced to the new top and bottom planes of the lamellae. However, as long as the effect of chain ends can be ignored, the number of the folds will remain the same, and the penetration factor will keep the value pertaining to an adjacent amorphous phase of random orientation.

Reversing the argument, we may conclude that, because the penetration factor keeps this value, also the amorphous regions remain isotropic. It would of course be of no consequence whether the crystals actually grow in the way described here, or are born directly in the situation prevailing after the crystal thickening. Apparently crystals of any thickness can develop without introducing prefered orientation in the remaining amorphous material. The net orientation in a lamellar stack, which gives rise to the birefringence, must be ascribed to the chain segments which are reeled in during the crystallization; in this proces, the previously isotropic distribution of these segments is replaced by a sharp orientation in the stem direction of the crystalline lamellae.

The circumstance that the amorphous layers remain isotropic does not imply that crystal thickening can continue unlimited. Here, a factor obstructing further growth is the competition between neighbouring lamellae in incorporating chain segments of the tie molecules. In this respect it is of interest to notice that thickening of the crystals actually does take place on annealing of linear polyethylene; however, according to our observations[11], this proces is accompanied by the melting of a part of the lamellae. One might interpret this as the result of the competing forces exerted through the tie molecules on neighbouring lamellae, because of which the least stable ones (which generally will be the thinnest ones) dissolve at the expense of the more stable species.

REFERENCES

1. Keller, A., Phil. Mag. 2, 1171 (1957).

2. Flory, P.J., J. Am. Chem. Soc. 84, 2857 (1962).

3. Treloar, L.R.G., 'The Physics of Rubber Elasticity', 3rd ed., Clarendon Press, Oxford (1975).

4. Frank, F.C., Far. Disc. Chem. Soc. 68, 7 (1979).

5. Flory, P.J., Yoon, D.J. and Dill, K.A., Macromol. 17, 862 (1984).

6. Guttmann, C.M., DiMarzio, E.A. and Hoffman, J.D., Polymer 22, 1466 (1981).

7. Mansfield, M.L., Macromol. 16, 914 (1983).

8. Domszy, R.C., Glotin, M. and Mandelkern, L., J. Pol. Sci., Pol. Symp. 71, 151 (1984).

9. Wunderlich, B., 'Macromolecular Physics', II, Acad. Press, New York, (1976).

10. Vonk, C.G. and Pijpers, A.P., J. Pol. Sci., Pol. Phys. Ed., To be published.

11. Vonk, C.G. and Koga, Y., J. Pol. Sci., Pol. Phys. Ed., To be published.

CHAIN MOBILITY IN PHASE TRANSFORMATIONS OF INORGANIC POLYMERS

J.H. Magill*

Max Planck Institut für Polymerforschung
University of Mainz, 6500 Mainz, FRG
West Germany

SYNOPSIS

Besides crystallization temperature, chain length and mobility are paramount in polymer crystallization. It has been well illustrated using several methods for following spherulitic and bulk crystallization that the kinetics of crystallization strongly depend upon molecular weight until the chain length exceeds the mean entanglement molecular weight, at comparable degrees of undercooling. For longer chains there is a gradual 'transition' in kinetics which is almost independent of molecular mass. These extensive data for poly(tetra-methyl-p-silphenylenesiloxane) demonstrate that spherulite crystallization involves chain features much shorter than the entire chain. Furthermore, it is suggested that the growth at the solid-melt interface, whatever the undercooling (above or below the maximum in the crystallization curve) involves localized ('segmental') chain features. These effects are clearly manifested in the morphology-molecular weight dependence of poly TMPS where it is found that the crystallinity decreases with chain length particularly in bulk crystallization at equivalent undercoolings, the latter being where a proper comparison of properties should be made.

In the case of some polyphosphazenes which exhibit thermotropic behaviour, chain mobilization is manifested during the transformation, T(1), from the crystalline to the thermotropic state where chain extension seems to occur without the application of pressure to the system. Such mesophase behaviour has been characterized and recently reported for several polyphosphazenes. The trans-

* Permanent address: School of Engineering, University of Pittsburgh, Pittsburgh, Pa. 15261, USA

formation occurs from an orthorhombic phase of relatively well-defined crystal thickness upon heating through, T(1), into the mesophase region $T(1) < T < T_M$. Upon cooling, a significant change takes place to a more highly crystalline orthorhombic state comprised of much larger crystallites --- presumably of a chain-extended nature, and of an enhanced enthalpy and higher transformation temperature, T(1). This demonstration of chain mobility involves a volume change of some 6 % for poly(bistrifluoroethoxyphosphazene) which has been investigated in detail at this time. Other systems are being studied too.

INTRODUCTION

Macromolecules comprised of inorganic elements in the chain backbone feature amongst the most interesting materials. Historically, their development has been slow[1,2], but now with the growing realization of their chemical versatility, technological value and scientific interest, a new era in macromolecular science and chemistry is dawning.
Organosilicon[3] and phosphonitrogen[4] polymers, amongst others[5], comprised of elements other than carbon and hydrogen from the periodic table, are attacting much attention as relatively strong and stable, flexible chain polymers with liquid crystal[6] forming ability amongst some of their special attributes*, Hybrid[7] siloxane and phosphazene polymers as well as copolymers of these two types are the focus of much polymerization chemistry aimed at developing new synthetic routes[8] and properties, despite the fact that the fundamental inorganic chemistry of small molecules is not all that highly developed. Nevertheless, appreciable progress has been made even producing superconducting polymers[9] such as poly(sulfurnitride), $(SN)_n$, and 'hightech' ceramic components which obtain from polysiloxane degradation[10]. Many notable advances in the creation and development of 'inorganic' macromolecules have evolved since the commercialization of the first polysiloxane.
However, as new materials are made and investigated, progress is assured as world-wide interest spreads in this category of inorgano-backbone polymers, creating new incentives and opportunities. Overviews are found in references 1 to 5 and 14, amongst others.

---

* Thermal stability, low toxicity and flamability, specific solvent resistance or solubility, electrical and photoresist potential, prosthetic applications and drug transporting ability, as well as uses as O-ring, gaskets, fuel lines, hydraulic and high temperature fluids, insulating materials and so on.

In this paper, discussion will be confined to a few polysiloxanes and polyphosphazenes which relate to investigations made mostly at the University of Pittsburgh. As the title suggests it will concentrate on manifestations of chain mobility and its implications in the condensed state where chain flexibility is important during crystallization and in thermotropic phase transitions. The article will be divided into two parts. One will serve to highlight some aspects of the level of understanding that has developed for semi-crystalline polysiloxane homo- and copolymers. Since morphology and crystallization kinetics are strongly related with polymer fabrication conditions, the effect of crystallization temperature, T_c, time t, molecular weight MW, polymer chemistry and composition are revealed though physical properties. For this reason, any 'model system' selected for extensive study must possess some innate qualities. It must be thermally stable, exhibit controllable nucleation and growth rates over a wide temperature span and be uniquely representative of polymeric behaviour. In the writers opinion the polysiloxanes fulfil this role.

The polyphosphazenes serve to illustrate another aspect of polymer crystallization namely that of liquid-crystal forming ability. Although specimens currently used have broad molecular weight distributions, their inherent behaviour with thermal treatment (temperature and time) provides guidelines for further study as well-characterized specimens become available. In any case, the role of 'inorganic' materials in the polymer hierarchy which stems from polyethylene to diamond (perhaps), present an interesting challenge to all of us in the exploration and understanding of these contemporary things. The differences which lie within the framework seems to be one of degree rather than kind.

MATERIALS

Poly(tetramethyl-p-silphenylenesiloxane) poly(TMPS) fractions[11,12] ranging from the monomer to 1.4×10^6 daltons. The material is illustrated by the chamical formula: -

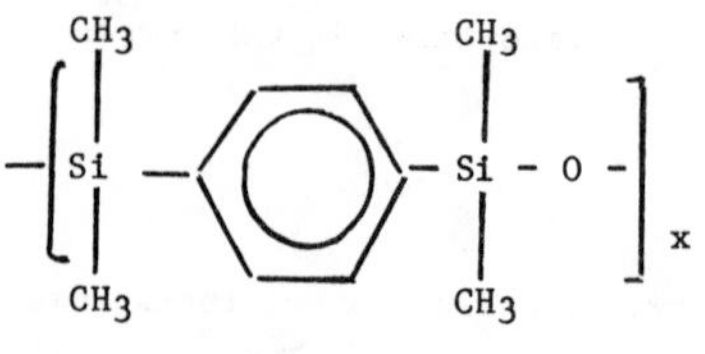

$$-\left[\underset{CH_3}{\overset{CH_3}{|}}\!\!\!\!Si - C_6H_4 - \underset{CH_3}{\overset{CH_3}{|}}\!\!\!\!Si - O - \right]_x$$

The polysiloxane copolymers are poly(tetramethyl-p-silphenylenesiloxane) (dimethylsiloxane) i.e. poly (TMPS) (DMS). The crystalline blocks are variable in

length and the other DMS component is typically about 18 monomers.
The formula is:

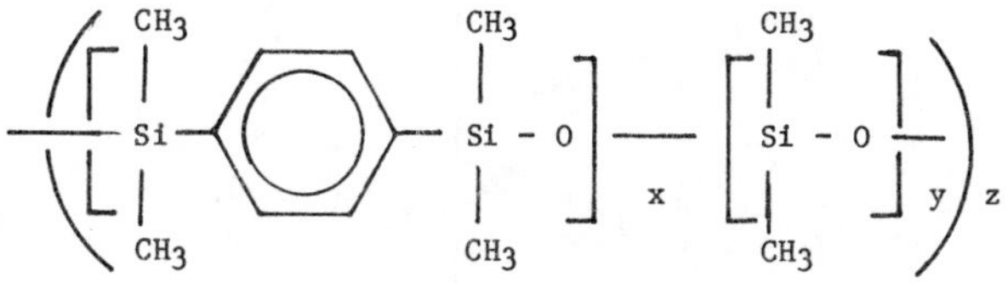

$$-\left(\left[\begin{matrix}CH_3\\|\\Si\\|\\CH_3\end{matrix}-\bigcirc-\begin{matrix}CH_3\\|\\Si\\|\\CH_3\end{matrix}-O\right]_x-\left[\begin{matrix}CH_3\\|\\Si\\|\\CH_3\end{matrix}-O\right]_y\right)_z$$

The wt. fraction of x to y varies from 100/0 to 30/70 and the molecular mass is typically about 10^5 daltons.

Polyphosphazenes[14] are represented by the formula:

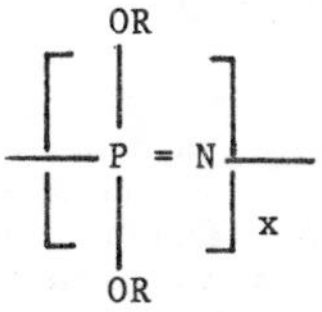

$$-\left[\begin{matrix}OR\\|\\P\\|\\OR\end{matrix}=N\right]_x-$$

where R used here is - CH_2CF_3 for the alkoxy polymer, PBFP and R is - C_6H_5 for the acryloxy specimen PBPP respectively.
They may have polydispersities between 2 and 10 fold depending on preparative methods.

CRYSTALLIZATION KINETICS

The first guidelines in the understanding of the relationship between crystallization kinetics, T_c or undercooling ΔT_c, and t evolved from studies on a complex but slowly crystallizing material - rubber[15]. Basic principles were established in this and later work[16].

In the case of the polysiloxanes --- poly(tetramethyl-p-silphenylene-siloxane) --- poly (TMPS) -- has proved to be a very manageable material in regards to its crystallization behaviour.
Figure 1 illustrates spherulitic growth rates[11,12] as a function of molecular weight and temperature. The range over which polyethylene crystallizes isothermally is indicated for comparison on the same graph. Intricate details[17,18] of polyethylene crystallization have been modelled using such a limited range of data! Although the influence of molecular mass on spherulitic growth is apparent from a surface energie analyses* of these data (see Figure 2), which were

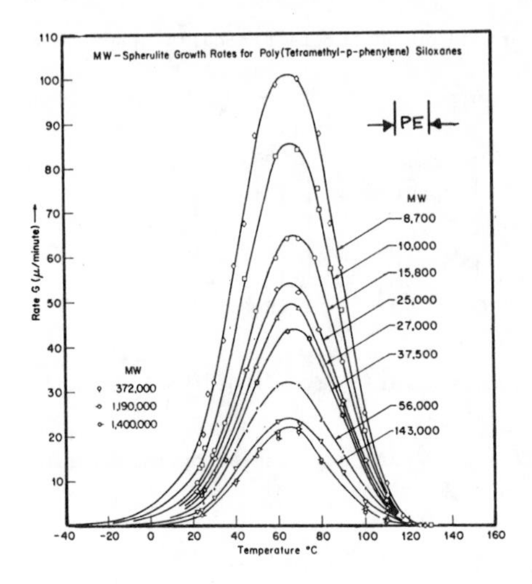

Fig.1: Spherulitic growth rates (μm/min) as a function of crystallization temperature T_c, and molar mass for poly-TMPS fractions. The isothermal crystallization range for PE is indicated between the arrows in the graph.

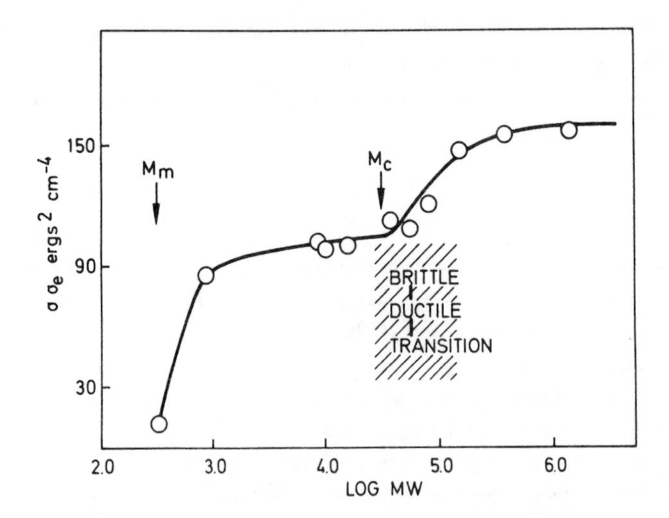

Fig.2: Product of lamellar surface energies, $\sigma\sigma_e$ (erg^2/cm^4, σ lateral and σ_e end-surface, as a function of log(molecular weight) for poly(TMPS) fractions. The monomer weight M_m, single stem, M_c (from zero shear viscosity measurements). A ductile-brittle fracture region is broadly indicated above M_c.

subsequently extended to lower molecular masses, including the monomer itself. A more interesting cross-plot (Figure 3) illustrates the effect of chain length on potential crystallizability. Only a few selected curves are shown here as isothermal growth rate plots.

Results at constant undercooling yield similar trends suggesting that above the initial molecular mass between entanglements, Mc, the rate is severely curtailed. It seems that Mc is rate limiting so far as spherulitic growth is concerned in poly (TMPS) in this 'plateau ' region.

This factual information has largely been overlooked by theoreticians who model systems where very limited information is available. Crystallization theories have ignored the direct role of entanglements per se even though they dominate crystallization behaviour and control crystallinity in many commercial plastics.

The transformation from a brittle to ductile transition (low to high molecular mass range) is undoubtedly matched by the onset of trapped tie molecules or segments within the crystallizing lamellae comprising the spherulites (see Figure 4). The trend in many physical properties is in line with the observation that all sample fractions of poly (TMPS) exhibit lamellar but attenuated lamellar textures are found on going from the free volume controlled crystallization regime to that dominated by chain entanglements were lamellar dimensions are curtailed. There is no doubt about this kind of transition which is now claimed to be apparent in other systems including polyethylene[20] and nylon-6[21].

Clearly, entrapped entanglements, concentrated mainly at the lamellar surfaces (crystals) and in interfaces (spherulites), manifest their presence in several ways.

ENTANGLEMENT-PROPERTY CORRELATIONS

Aside from the change in fold surface energy, σ_e, with molecular weight it has been established that crystallinity[22,23] falls off sharply with

---

* This analysis shows that the lamellar surface (or interfacial) energy doubles approximately on passing through the entanglement region. On either side of it, there is a smooth increase in this parameter which tends to level off above 10^6 molecular mass (see Figure 13/also reference 19). It must be emphasized that the observed increase in interfacial energy (between lameller) has nothing to do with the Regime I → Regime II transition[18]. Note that this latter occurs with respect to temperature not molecular weight.

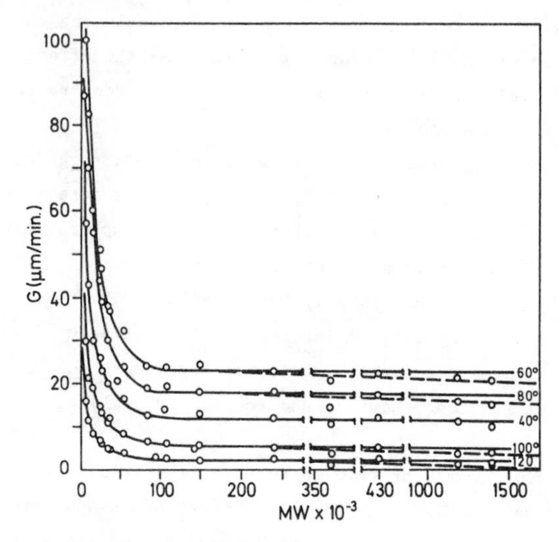

Fig.3: Selected isothermal cross-plots for poly(TMPS) fractions: G(μm/min) vs. MW (ref. 12)

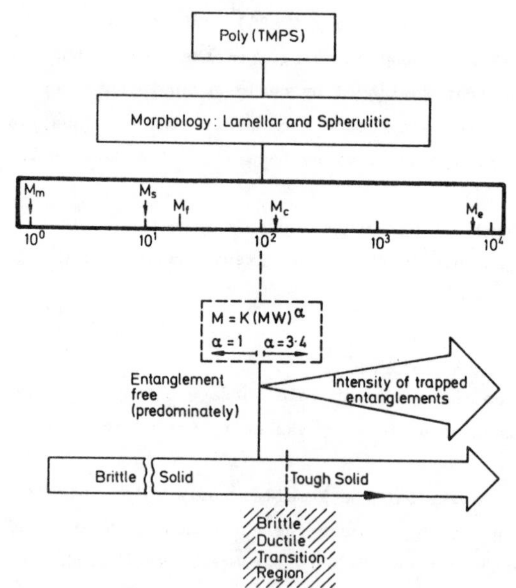

Fig.4: Schematic of poly(TMPS) system depicting important transitions in chain length or properties. The range of material masses, expressed as D.P. , is indicated by M_m, M_s, M_f, M_c and M_e which denote monomer, single fold, several folds, molecular weight between entanglements, final molecular weight tested respectively.

increasing molecular mass (Figure 5) as noted for polyethylene and polyoxides[23]. Correspondingly for poly (TMPS) homopolymer these is a substantial increase in lamellar thickness for fractions crystallized above and below the entanglement molecular weight, even after data are corrected for undercooling using a well defined temperature - molecular weight calibration such as that given by Hoffman-Weeks plots[26] to obtain equilibrium melting temperatures for each fraction investigated. The lamellar interface is clearly molecular weight dependent.

Early work[24] suggested this trend in behaviour but its validity is confirmed[25] now that (in situ) SAXS measurements have been made at temperatures at the National Center for Small Angle X-ray Scattering, Oak Ridge, Tennessee. The enhancement of lamellar thickness paralleled by a decrease in crystallinity measured by DSC and X-ray, is a clear manifestation of the presence of entrapped disordered entanglements above, M_c. The much higher crystallinity in poly (TMPS) below M_c, and the sharp dependence of spherulitic growth rate on chain length in the free volume dominated regime (where brittleness is paramount downward to the monomer itself as illustrated in Figure 4), undoubtedly demonstrates that few if any entanglements function as ties between lamellae in low molecular weight fractions. Although our fractions of poly (TMPS), $< 5 \cdot 10^3$ molecular weight, are not pure enough to conduct discrete chain folding experiments, the work of Kovacs and coworkers[27] on polyethyleneoxides, and more recently by Keller and associates[28] using paraffins up to 390 °C atoms, clearly demonstrates that folding in these systems is distinct and is controlled by the crystallization temperature*. Although the crystallinity has not been measured for these oligomeric species it must be less than 100 %, since any fold surface however sharp, by its very nature must functions as a crystal defect created kinetically during the formation of well defined lamellae disposed perpendicular (or tilted) on the basal plane. It is anticipated that chain folds or chain ends (unevenly stacked) perturb the chain core packing even if only superficially. Specimen density may not change.

---

* A new communication on this topic has just been published for well defined polyethylenes (high molecular weight paraffins). Discrete folding is obtained but chain ends may lie inside or outside the crystalline core, depending upon molecular mass and conditions. See K.S. Lee and G. Wegner, Die Makromol. Chemie (Rapid Comm.), 6, 203 (1985).

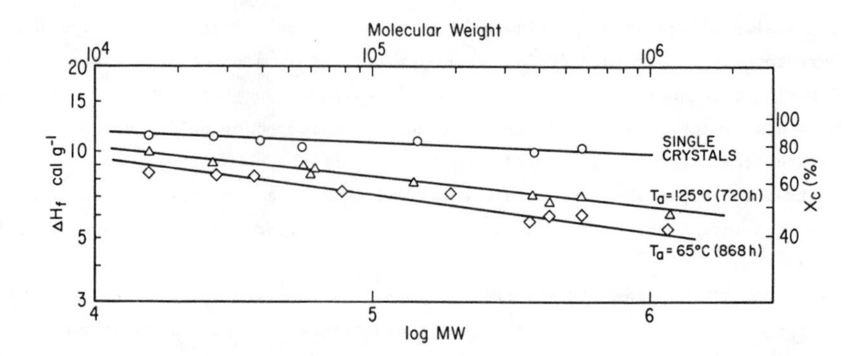

Fig.5: Crystallinity($\%X_c$) as a function of log (Mw) for isothermally crystallized fractions from (i) solution at 30°C and (ii) melt crystallized, at roomtemperature followed by annealing at 125°C(720h) and at 65°C(868h) respectively.

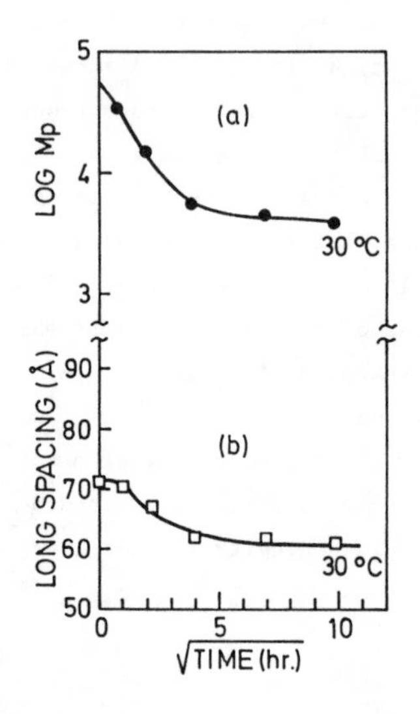

Fig.6: (a) Molecular weight change with (etching time)$^{\frac{1}{2}}$.
(b) Change in crystal thickness with (etching time)$^{\frac{1}{2}}$ for poly-(TMPS) using the same HF solution (48%) as a etching agent.
(after ref. 33)

CHEMICAL ETCHING AND SURFACE MORPHOLOGY

The role of selective surface etching has been exploited for several polymer systems to gain insight into the extent of surface order in polyethylene[29], polypropylene[30], nylon-66[31] and poly (TMPS)[32,33]. Only the latter example will be emphasised here since the chain scission of 'single' crystal surfaces and spherulitc interfaces have been characterized using many complementary experimental methods. Solution[33] and gaseous[34] HF etching have been used with supportive results, the gaseous treatment in the presence of He diluent being much faster than the solution techniques. Besides, this is a dynamic process where the etchant is continually replaced, so that stagnant layers of reagent do not build-up vicinal to the polymer surface. The more vulnerable surface (folded or looped interface) is preferentially attacked, leaving behind a core of crystaline rods with a small-angle periodicity 8-10 Å shorter than the starting crystals of 72 Å (see Figure 6(a)). This dimensional change corresponds (on average) to two TMPS repeat units for samples of molecular weight of 50,000. This average value has been well documented using several techniques* which have probed the core remanents of the crystals during, as well as at the end of the reaction, to provide clear-cut evidence[34] from surface spectroscopic chemical analysis (ESCA).
Figure 7(a) demonstrates the kinetics of etching. Since this ESCA technique is sensitive to $\leqslant$ 50 Å probing depth (the crystal thickness initially is 72 Å) less than a single crystal stem is examined in this highly selective procedure. Noteworthy is the fact that the crystallinity increases from just over 80 % to around 96 % crystallinity during this etching process.. These measurements confirm earlier investigations[33] which demonstrated that a parallel change was obtained in 48 % HF solution etching carried out at 30 °C where a plateau in the rate of surface removal is sustained after 100 to 200 hours reaction time (Figure 6(b)) with a concomintant molecular weight change. The core residue in this instance (by GPC analysis) consisted of single items, about 8 monomers in length, but there are some others about 50 % longer (presumably arising from buried folds). Undoubtedly the surface attack comprises selectively cutting Si-O bonds and removal of detached fragments into the gas phase*.

In the case of bulk crystallized poly (TMPS) the kinetic curve[34], Figure 7(b), depicting the Si/F intensity ratio with etching time is almost

---

* If there is a GC-MS system available to study the distribution of fragments in the effluent HF gas stream, the writer would like to hear about it!

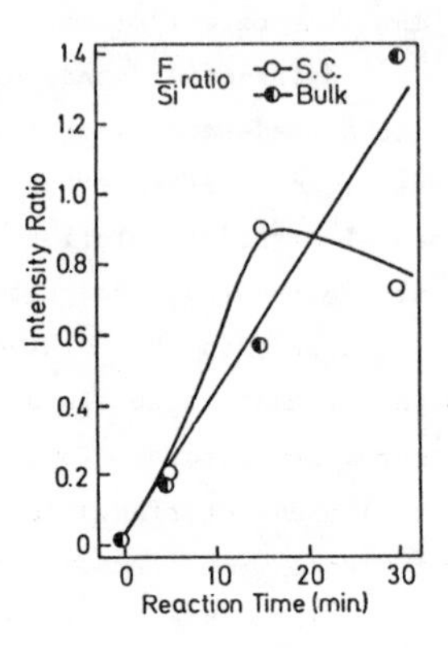

Fig.7: Variation of ESCA (F/Si) bond intensity ratio versus etching time (min) using HF/He (90/10 v/v) on poly(TMPS) (o)solution grown crystals (◐)bulk crystalized specimen of the same molecular weight (see ref. 34).

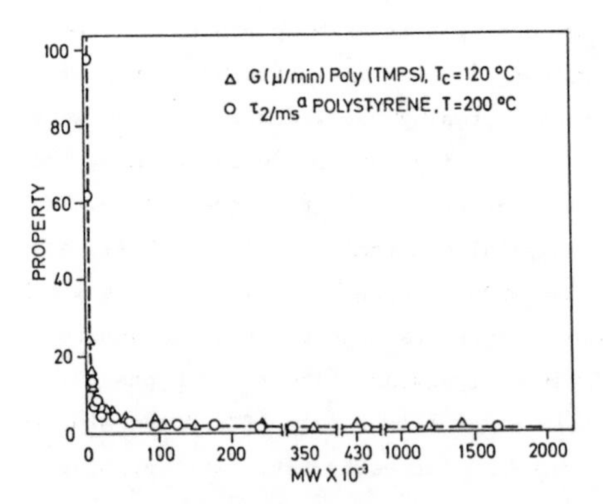

Fig.8: Plots of G (μm/min) for poly(TMPS) (11,12,19) and τ_2/ms (X10) for NMR measurements for atactic polystyrene(35) fractions, both as a function of MW. (Adopted from ref. 12 and 35)

linear, contrasting sharply with the plateau or arrest noted for single crystals in Figure 7(d). The difference in response is distinct and depicts the more complex nature of the interfacial regions found between lamellae where disorders are more abundant, complicated and some less accessible than free crystal surfaces. No doubt Si/F ratio in Figure 7(b) will plateau out, given sufficient reaction time in the HF/He gas stream. We will revert to a further discourse on this topic for polysiloxane copolymers which also presents another case of surface segregation with the more mobile DMS phase.

CHAIN MOBILITY (OR IMMOBILITY)

Evidence for the control of spherulitic growth at MW's > Mc by entanglements is evident even though crystallization theories ignore the role of entanglements in transport, which term is introduced in an *ad hoc* manner anyhow. The relatively free chain which pervades the polymer melt below M_c indicates that growth depends strongly on its length which correlates with the free volume of chain ends, see ref. 19, Figure 134. An analogous situation is observed in some properties of amorphous polymers. Recently, it has been noted that the relaxation time, τ_2, of atactic polystyrene[35], over the same wide molecular weight interval, closely parallels the spherulitic growth rate pattern present here in Figure 8. This strong correlation (irrespective of the plot type) serves to illustrate that participation of polymer segments in crystallization must mimic (to a large extent) the trends depicted for amorphous chain motions, irrespective of the conceptualization of the processes theoretically in the melt. Chain segments* (must be less than a substrate length (~ 10 units) in poly (TMPS) and presumably of similar mean dimensions in polystyrene[36] which has a large number of repeat units (300) between entanglements, i.e. in excess of De Gennes[37] prediction of 100. Even so, a general formalism may be reinvoked now[11] with confidence to describe chain mobility as:

$$\log (\eta_T/\eta_R) = \log (\xi_T/\xi_R) = - \log (J_T/T_R) = - \log (G_T/G_R) = \log (\tau_T/\tau_R)$$

where the η's, ξ's, J's, G's and τ's denote viscosity, friction, jump frequency,

*i.e. More than an order of magnitude shorter than the length between entanglements (see Figure 4), and comparable with a Kuhn segment site.

** It must be emphasized that bulk crystallization is too complicated to treat in this manner since the nucleation rate (which is a variable depending upon sample purity) is not unique to for a specified molecular mass and chemistry.

growth rate and relation times respectively. The subscripts refer to the parameter measured at some temperature T with respect to another reference temperature, R. The proper normalization of these quantities should lead to a phenomenological description of crystallization of commercial polymers which have $Mw > Mc$ on average. However it must be emphasized that the bulk crystallization kinetics can hardly be treated** in this manner, unless the nucleation density of primary nuclei (with temperature and time) are known. Even so, it is obvious that the strong dependence and striking analogy between these two types of measurements only serves to emphasize that the motion of small groups of monomer units are actively blocked by entanglements which influence behaviour and properties. Only some are trapped during solidification since a number* can 'wriggle-free' or reptate.
Naturally there are many well above the M_c chain length (purely for statistical reasons) since multiple surface nucleation along the crystalline substrate (lamellar) length must be more predominant. As a consequence surface roughness**[38,39] should be a function of molecular mass as well as crystallization temperature. Hitherto, both factors have not been considered in theoretical treatments of crystal growth.

POLY (TMPS)/(DMS) COPOLYMERS

It is appropriate to switch now to copolymers where it has been demonstrated that the more mobile DMS*** components segregate to the crystal surface during crystallization from solution.

CRYSTALLIZATION KINETICS

The temperature and composition dependence of the crystallinity (Figure

* This must be much less than that encountered in flow and cannot involve complete chains, i.e. macroscopic transport to a large degree.

** This signifies interlamellar 'roughness' in addition to that normally considered on the growth front. It is of course necessary to distinguish between the roles of equilibrium and non-equilibrium roughness in both cases.

*** It should be emphasized that the short DMS segments can only crystallize at cryogenic temperature and probably have glass transitions[40] comparable with their oligomeric counter parts (i.e. $\leqslant -140$ °C).

9) points to the fact that both of these parameters play a significant role in the phase transformation.

Not only does the 'mobile' non-crystallizing impurity (DMS) depress the melting temperature, but the spherulitic growth rate* is surpressed signifying that the DMS segments are excluded in this process. A clear indication of the changing morphology with composition is obtained because of the parallelism which exists between (interfacial) surface energy and composition (see Figure 143 ref. 19). The molecular weight of each weight fraction being approximately constant[32,33] and results are unaffected by contributions from chain ends. Of course changes in the distribution of amorphous crystalline material in the system, due to compositional variables is another matter, at the heart of this presentation. It is this distribution which is responsible for the physical properties.

The spherulitic texture of these copolymers is striking.

In fact they develop considerable extinction regularity (become banded) as the composition of DMS is increased (see Figure 10). Of course as the crystallinity drops to low levels (< 20 %), this optical regularity tends to diminish but does not disappear.

At anyrate, it does seem that the numerous theories to explain banding[41] in spherulitic materials still require some modification where block copolymers are concerned.

SOLUTION GROWN CRYSTALS

An even more striking example of DMS mobility stems from the fact that this component is surface segregated during crystallization from dilute solution. This is illustrated in Figure 7 of reference (42) where the discrete tetragonal habit of the poly (TMPS) homopolymer becomes niched and its outline distorted as the DMS content of the copolymer system increases. In concert with this change, the thickness increases considerably and screw dislocation growths disappear completely about the 50/50 weight percent level. Although gold decoration[43] and other metal shadowing techniques reveal a changing surface morphology, the 'mobility' of the segregated DMS component prevents definite conclusions to be made. However selective surface etching with HF has proved to

* Since segregated impurities are expected to depress physical parameters as the composition of non-crystallizable (impurity rejected) species in a polystyrene blend increases for example[45].

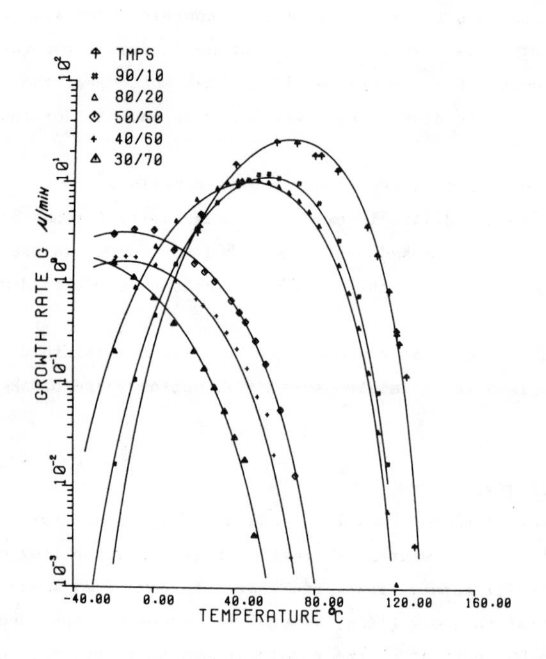

Fig.9: Sperulitic growth rate of poly(TMPS)(DMS) copolymers as a function of composition (♦) TMPS; (✱) 90/10; (△) 80/20; (◇) 50/50; (+) 40/60; (▲) 30/70; respecitively. The homopolymer is including for comparison (ref. 46).

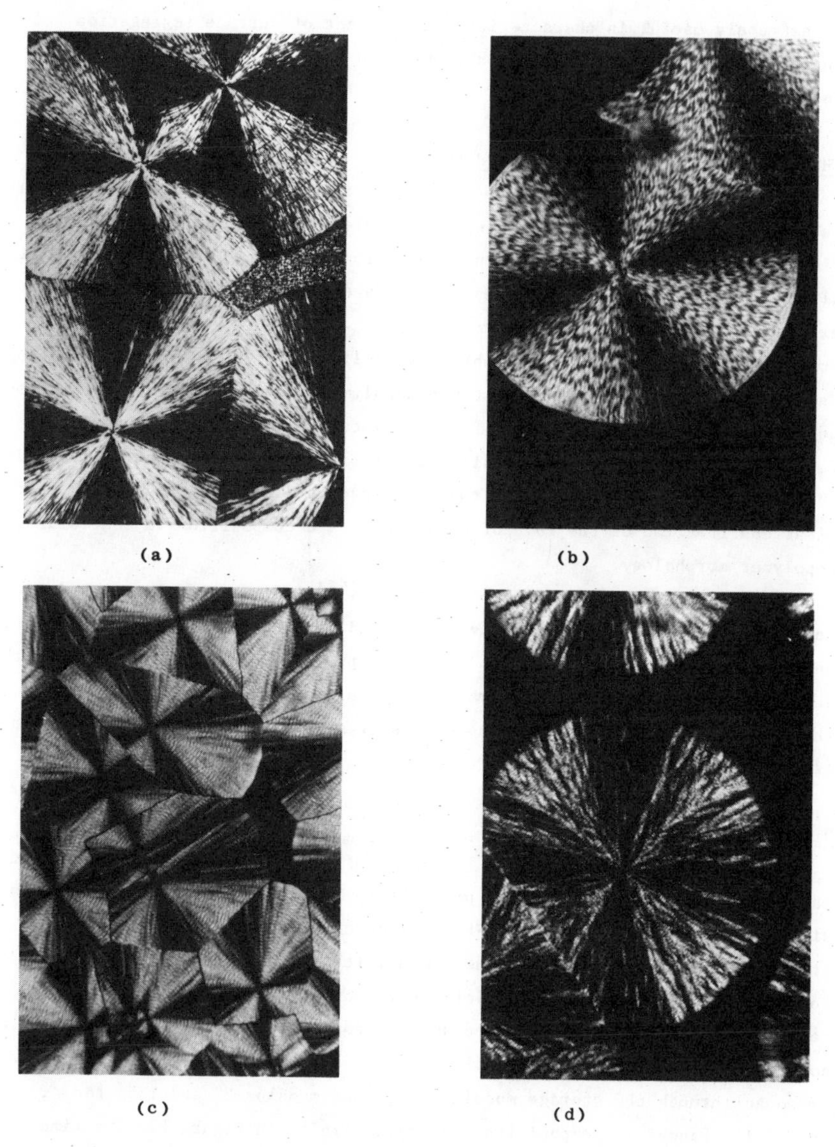

Fig.10: Variations in spherulite morphology of poly(TMPS)/(DMS) copolymers with composition.
(a) 100/0 (60°C) X150 magnification
(b) 90/10 (60°C) X150 magnification
(c) 50/50 (60°C) X400 magnification
(d) 30/70 (50°C) X375 magnification

be extremely useful in characterizing the extent of surface segregation. Small-angle X-ray scattering has also substantiated this trend in crystal mats (32-34) and bulk preparations (24-44).

CHEMICAL ETCHING AND SURFACE MORPHOLOGY

In summary here, it can be stated and illustrated that the non-crystallizable component in poly TMPS/DMS copolymer clearly concentrates in surface. Figure 11(a) illustrates the dependence of long period on composition and Figure 11(b) shows the change in thickness which take place with reaction time as the exposed crystal surfaces are removed.
Strikingly, there is an identifiable crystalline core which exhibits a tetragonal structure[42] reminiscent of the homopolymer itself. Electron diffraction mesurement made on the initially formed crystals before etching attest to this fact. Noticeable changes in the diffraction patterns parallel the increase in DMS component in accordance with the etching experiment. In other words there is clear confirmation of the role of copolymer chain composition in controlling copolymer morphology.

Again, surface spectroscopic evidence[46] in the form of ESCA and FTIR techniques, coupled with supportive small and wide angle X-ray methods, thermal analysis and GPC characterization that have provided definitive answers to the surface segregation of 'mobile' material in a complex crystallizable system where the DMS component is well above its crystallization range and far from its glass transition region.

OTHER PROPERTIES

Glass temperatures are an indicator of forms of disorder in polymeric materials. The magnitude of this property in DSC and other techniques is sensitive to molecular weight too. From a mechanical properties viewpoint the stress-strain behaviour, including the ability of the poly (TMPS)/DMS copolymers to neck and draw, distinctly depicts their morphological state (see Figure 3 ref. 47). Changes in dynamical mechanical properties with composition (wt.% of noncrystallizable DMS) also establish that a strong correlation in mechanical response through the storage modulus, E', loss modulus E" and loss tan δ, parallels changes in composition and crystallinity in Figure 12. The time scale and magnitude of the mechanical response too mimics the chain mobility in the surface of interfacially segregated amorphous regions of these two phase systems. From a morphological point of view the drawability of these and related copolymers distinctly shows that considerable amounts of amorphous contents can

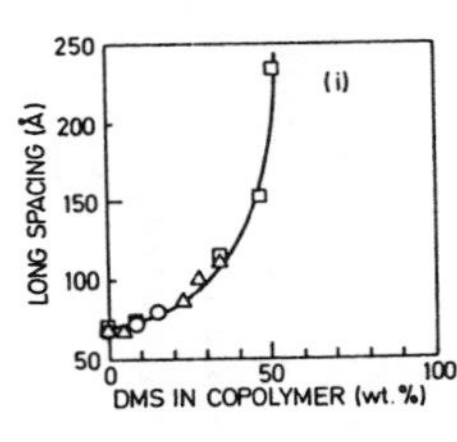

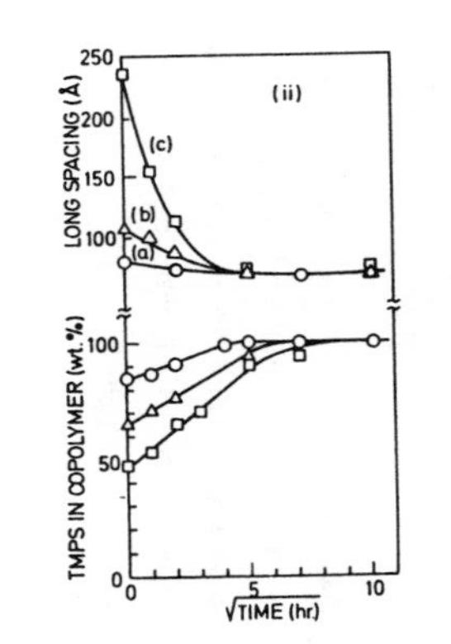

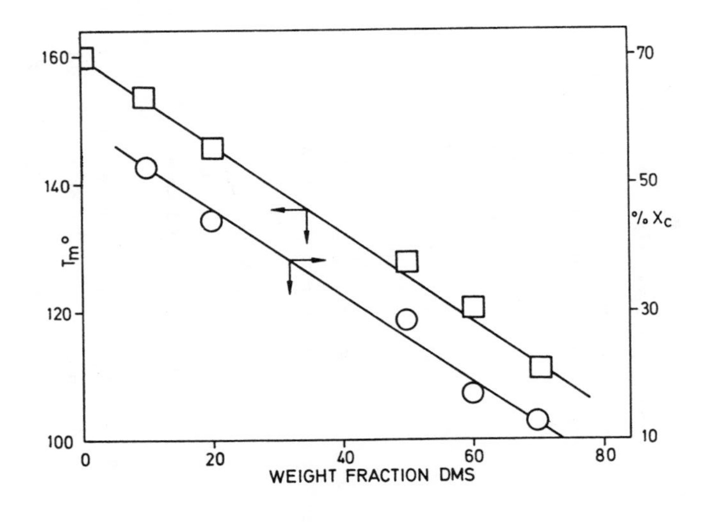

Fig.12: Dependence of crystallinity(X_c), —O—, and melting temperature(T_m^o), —□—, on copolymer composition.

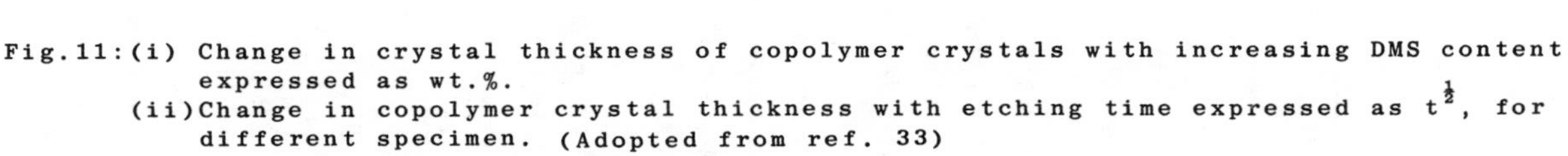

Fig.11: (i) Change in crystal thickness of copolymer crystals with increasing DMS content expressed as wt.%.
(ii) Change in copolymer crystal thickness with etching time expressed as $t^{\frac{1}{2}}$, for different specimen. (Adopted from ref. 33)

coexist within an optically ordered phase (spherulite) giving rise to well-defined structures (Figure 10) even when the amorphicity exceeds 80 %. Note that this condition holds tenaciously in these siloxane copolymers which obeys the composition rule[47] for two phase systems, where the sample T_g is expressed

as:

$$\frac{1}{T_g} = \frac{w_1}{T_{g,1}} + \frac{w_2}{T_{g,2}}$$

In this simple equation w_1 and w_2 represents the weight fractions of the poly(TMPS) and (DMS) components with glass temperatures $T_{g,1}$ and $T_{g,2}$ respectively.

CHAIN MOBILITY AND PHASE TRANSITIONS

Polyphosphazenes[14] have recently brought a new dimenson to the study of macromolecules without carbon in the backbone. Their chain flexibility[48] compares very favorably with the polysiloxanes[49] which also have been synthesised with significantly different side group chemistry and dimensions. In this respect, some polymers have liquid crystal forming ability like the aryloxy- and phenoxy-phosphazenes.
To illustrate the behaviour of these materials poly(bis(trifluoroethoxy) phosphazene) PBFB and poly(bis(phenoxy)phosphazene) PBPP will be used.

SINGLE CRYSTAL MORPHOLOGY

Like the polysiloxane homopolymers, the polyphosphazenes can be prepared in the crystalline platelets from dilute solution[51,52]. Their behaviour is more complicated than the siloxanes in that they form several polymorphs and also undergo a thermotropic transition upon heating which results in erratic crystal thickening. This is a striking example of chain mobility in the condensed phase in PBFP well below their final melting of 245 °C approximately (see for example Figure 13).
PBFP plate-like crystals exhibits a well defined morphology but three polymorphic forms and one mesoform are found depending upon the crystallization conditions and molecular weight. PBPP solution grown crystals also exibit one mesoform and two polymorphic forms when prepared from dilute solution. A typical morphology is illustrated in Figure 14(a).

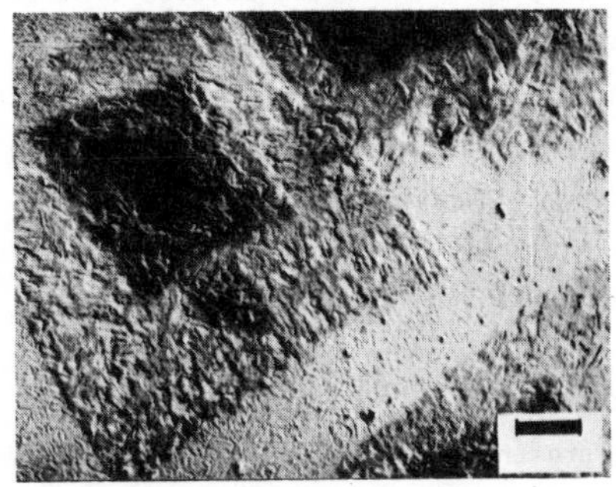

(a) (b)

Fig.13: Solution grown monoclime crystal preparation of PBFP from: (a) THF and (b) after 200°C (30 min) displaying uneven thickning on annealing. (from ref. 52)

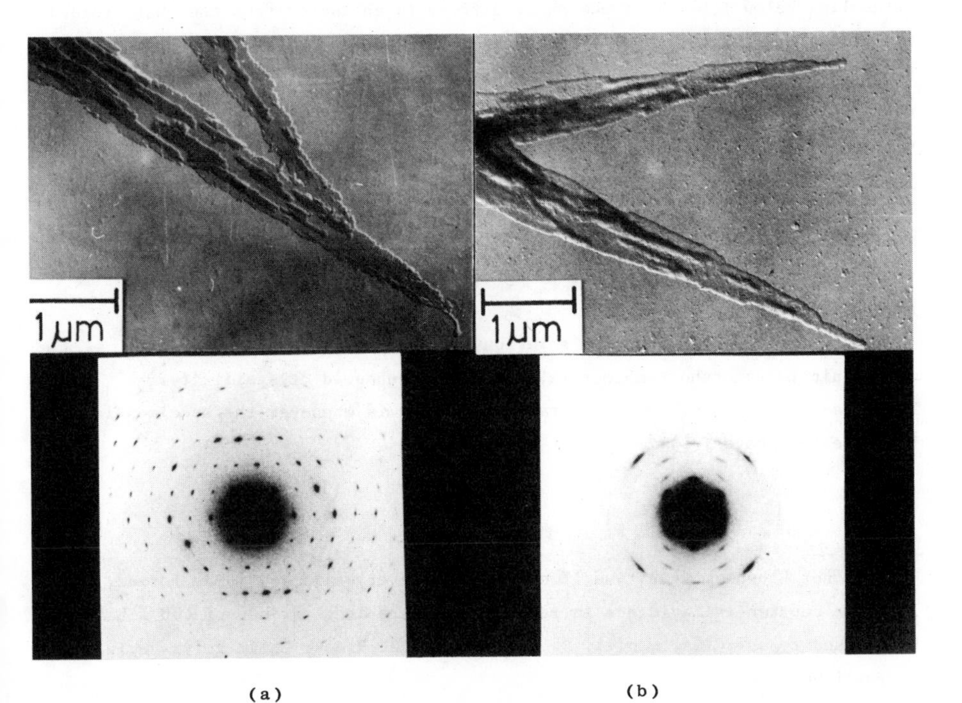

(a) (b)

Fig.14: PBPP solution grown crystals from xylene (a) as cast with electron diffraction insert and (b) after heating above T(1) ~ 150°C with electron diffraction insert of the orthorhombic phase transformation from the monoclinic form in (a).

Presumably this uneveness derives from defects in different parts of the crystal. Naturally, entanglements where present will have an impeding influence on the chain thickening process. The kinetics of this process of mesophase transformation is now being investigated by small-angle X-ray and synchroton radiation, in order to gain new insights into this fascinating changes which are still not well understood. Electron diffraction and dark field microscopy has already revealed that dislocations and/or stacking faults exist within polyphosphazene crystals. Chain migration may be 'nucleate' at some of these defects.

These prominent changes in crystal thickness occur around 70 °C, the initial T(1) or thermotropic transition of the polymer in this chain folded form. This change is accompanied by a substantial increase in volume[53] as the crystals pass into the 2D thermotropic state where the molecules appear to chain extended* and then reverts to a well ordered 3D highly crystalline orthorombic phase on cooling below T(1). This new phase differs in character from the chain-folded (starting) crystals of 150 Å thickness. In like manner, heating the PBFP polymer above its melting temperature rapidly destroys all semblance of order, some of which it quickly acquires again upon cooling (i.e. there is limited undercooling with transformation into the 2D mesophase again. Further cooling produces a chain extended* crystal structures similar to that obtained on cycling through T(1).

However the morphology which obtains under these conditions bears a striking resemblance to the high pressure hexagonal L.C. phase reported[54] in the crystallization of polyethylene.

In the case of PBPP, the solution grown crystals transform from a monoclinic to an orthorhombic form, again with enhanced crystallinity. The phase change is nicely illustrated in Figure 14 whenever the specimen is heat treated above T(1).

---

* It is not known at this time if extended chain crystals are fully formed, but there is substantial evidence in favor of it where long periods of 630 Å have been found by us. (J.A. Magill, J. Peterman and U. Rieck, Coll. Polym. Sci., to be published.

**It is well known that stiff molecules exhibit mesophase behaviour in solution and even in the melt but polyphosphazenes with a Mark-Houwink Sakarada exponent around 0.6 in solution do likewise, which is surprising.

The similarities between PBPP and PBFP are striking since the T(1) of PBPP (~ 150 °C) is still far below its melting point. Besides, the phenoxy side groups on the phosphorous, are more closely held to the flexible** (-P = N-) backbone linkages which must take on a 'rod-like' appearance whenever the side groups undergo considerable torsional gyrations, gaining much configurational entropy[55] on passing into the mesoform.
The transitional enthalpy change at T(1) is an order of magnitude larger than at T_m (with a corresponding increase in entropy). After specimens have passed into the mesophase and are cooled they become embrittled (but not degraded) and are more highly crystalline[56] than in their initial solution-cast state.
The ductile-brittle change accompanying this heat treatment of polyphosphazenes can only reflect a drastic change in the internal morphology as amorphous material is converted into a more highly crystalline state which transformation seems to be accompanied by the elimination of crystalline defects.

SPHERULITIC MORPHOLOGY OF POLYPHOSPHAZENES

Like other macromolecules polyphosphazenes form spherulites, Figure 15(a) from solution. However they crystallize as aggregates from the melt on cooling, nucleating rapidly a relatively small undercoolings (see Figure 15(b)). This in a consequence of their passing through the thermotropic state where chain extension features predominately and considerably stability is inferred through this process.
X-ray diffraction (unpublished results) indicates that the aggregates are more highly crystalline than the spherulites prepared from solution, and they are comprised of relatively large crystals as ascertained from fracture surface morphology[57] using the SEM, X-ray diffraction and electron diffraction from thin film-specimens. A similar morphological and property trend is observed in PBFP and PBPP polymers, namely that a change from the as-cast state into a thermotropic phase is followed by a transformation into an orthorhombic form of higher crystallinity (greater than 80 %). These morphological changes manifest themselves via numerous physical parameters detected through dilatometry, birefringence, dynamic mechanical measurements and DSC measurements to mention only a few.

OTHER MEASUREMENTS

Once PBFP (for example) has passed into the thermotropic state it can be recycled with structural improvement taking place. The volume change[53] at first order transition T(1) and T_m respectively, is 6 % approximately,

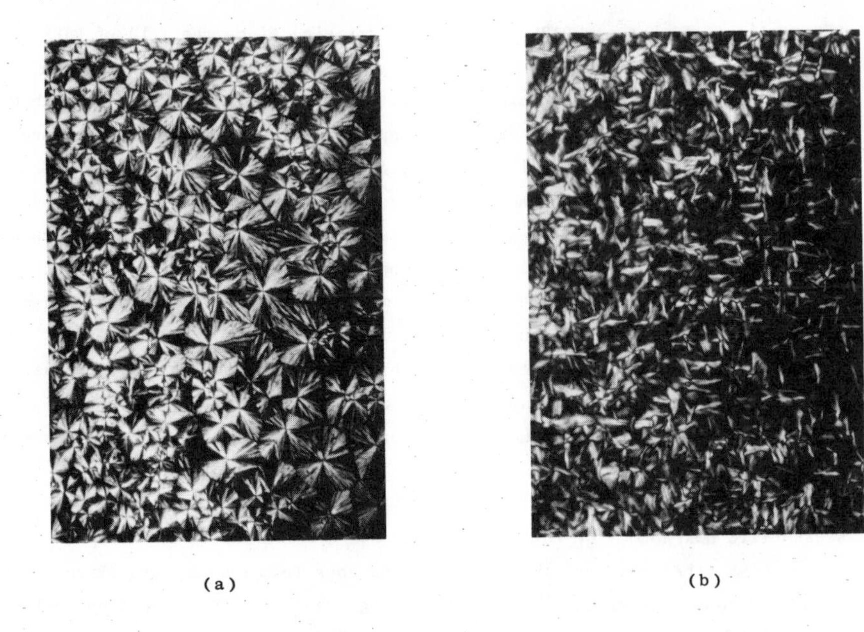

Fig.15: Spherulite of PBPP (a) solution preparation from (b) Needle-like aggregates formed after fusion above 245°C and cooling to roomtemperature.

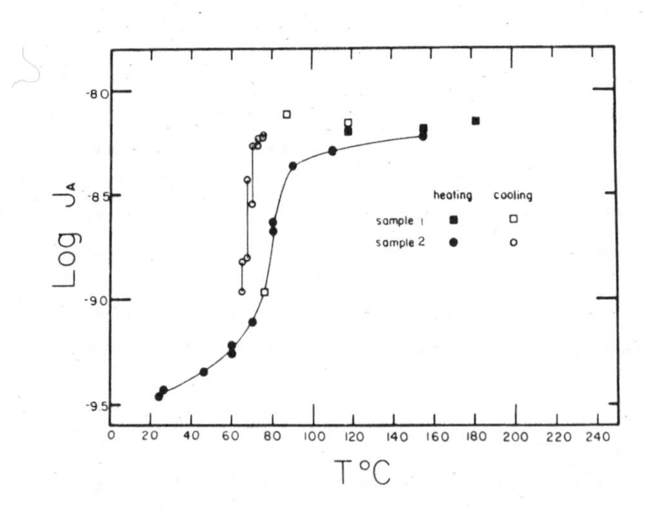

Fig.16: Log (creep compliance), J_A (cm^2/dy), vs. temperature for heating- and cooling PBFP. (ref. 53)

signifying considerable expansion and/or contraction in the vicinity of these transitions. The 'softening' of the sample on passing through T(1) is indicated from the substantial increase in the creep compliance J(t) (typically a factor of seven) portrayed in Figure 16. A correspondingly large drop in dynamic modulus has also been found[58] in several polyphosphazenes which exhibit a thermotropic transition. Figure 17 also illustrates this behaviour for PBPP where a drop in modulus occurs at a higher temperature in PBPP because of its 'stiffer' yet 'flexible' backbone. Note however that no T(1) transformation has been observed by us for randomly substituted side group phosphazenes. These merely show the typical fall in E* expected of an amorphous polymer in the plateau region. However some 'blockiness' associated with the polymerization process may raise the plateau level above the truly amorphous level, even though crystallinity (3D) is be required to produce the phase change in crystallographic terms. However well-defined molecular weight fractions are essential to a definite study of polyphosphazenes. This project is underway.

CONCLUSIONS

Morphology depends on polymer history and inherent chemical and molecular weight characteristics of any system. The spectrum of molecular weights possible allows different behaviour across the chain length scale. Oligomeric and shorter chain molecules can crystallize in well defined chain folded crystals of thickness depending strongly upon the crystallization temperature. Longer molecules may crystallize in lamellar structures, the extent of surface or interlamallar disorder depending strongly upon the density of entanglements pinned or trapped during crystallization. There is a striking analogy between chain mobility near a crystal substrate during spherulite growth and the relaxation process taking place in the polymer melt. The role of free volume below the critical molecular weight M_c, and that of entanglements about it, strongly influence polymer properties (crystallinity and morphology) in an indisputable manner. Under comparable crystallization conditions, the crystallinity (influenced by trapped entanglements) strongly parallels changes in the interfacial characteristics (texture) of samples of homopolymer or copolymers of polysiloxanes. All textures in our polysiloxanes are spherulitic and lamellar even under conditions where the crystallinity is well below 20 %. In poly (TMPS)/(DMS) copolymers which form solution grown lamellae for all wt. percent compositions in the range 100/0 to 30/70, a striking change in 'crystal' thickness and small-angle long period parallels the change in surface morphology largely due to the surface segregated DMS component. This surface layer or corresponding interfacial

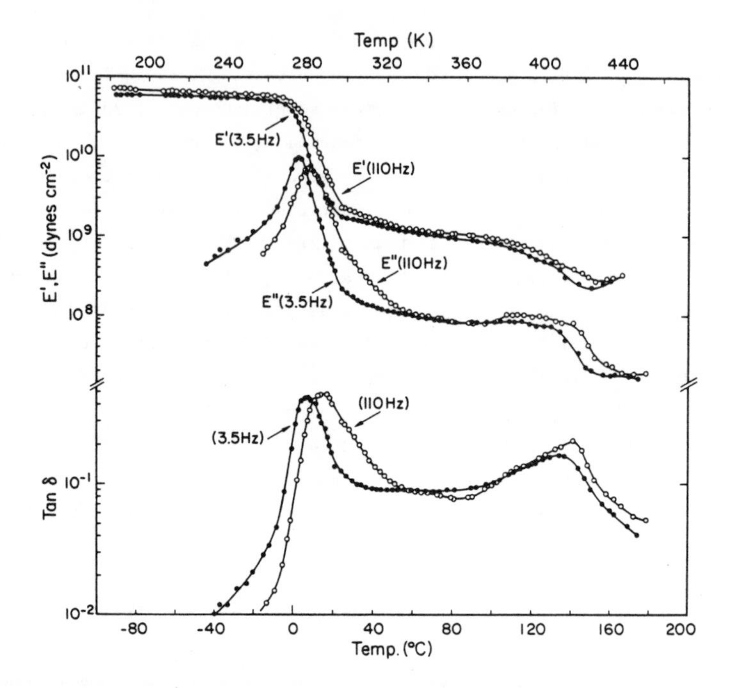

Fig.17: Thermomechanical transitions in PBPP (oriented) at 3.5 and 110 Hg respectively (ref. 58).

region in melt crystallized specimens can be selectively etched away using HF (solution or gaseous reagent) revealing a crystalline core of striking uniformity. Manifestations of the fact that amorphous interfaces exist in copolymers and high molecular weight homopolymers have been well documented for polysiloxane systems using many experimental techniques. The stability of the polymers, their ability to crystallize isothermally over an extensive range of temperatures for a wide range of molecular weights place them in an unique position --- maybe as a paradigm in this class of materials. Clearly, it seems that the morphology of a polymer system is a widely varying entity ranging from a chain folded texture at one of the molecular mass spectrum to a predominately amorphous material at the other end, controlled by processing.

Likewise, these is a need to emphasize that in polyphosphazenes --- a new class of 'inorganic' polymer --- exhibit unique mobility in the condensed state as they undergo crystal thickening on passing into the thermotropic state --- a transition region which depends greatly upon the kind of substituents attached to the phosphorous backbone atom and to a lesser degree upon the casting medium. Apparently chain extension occurs on passing into the liquid crystalline state where the side-groups become mobilized promoting a volume change of some 6 % in the process for PBFP for example, thereby creating sufficient free volume for chain extension to a hexagonal thermotropic phase which converts into a chain extended 3D lattice upon cooling from this phase or from the molten state ($> T_m$, final clearing point of the system). Analogously chain extension occurs in polyethylene under higher pressure with a remarkable similarity to the morphology of the L.C. phases.

ACKNOWLEDGEMENTS

The author is honored to make this presentation on the occasion of the retirement of two internationally recognized polymer scientists, Drs. Ron Koningsveld and Chris Vonk. He also thanks the National Science Foundation (Polymer Program) and Office of Naval Research (Chemistry Program), for support of this work. In addition he expresses indebtedness to the Alexander von Humboldt Foundation for support during a sabbatical leave from the University of Pittsburgh, 1984-85 when this paper was written.

REFERENCES

1. H.R. Allcock, Phosphorous-Nitrogen Compounds, Acad. Press. N. York (1972).
2. N.H. Ray, 'Inorganic Polymers', Acad. Press. N. York (1978).
3. A. Noshay and J.E. Mc Grath 'Block Copolymers', Acad. Press. N. York (1977).
4. S.V. Vinogradova, D.R. Tur and I.I. Minosyants, Russ. Chem. Revs. (Eng. Trans.) 53, 87 (1984).
5. C.K. Ober, J-I Jin and R.W. Lenz, Ad. in Polymer Sci, 59, 103 (1984).
6. N.S. Schneider, C.R. Desper, J.J. Beres, 'Liquid Crystalline Order in Polymers', Chap. 9 (A. Blumstein Ed.) Acad. Press. N. York (1978).
7. C.E. barrahar, Jr., J. Chem. Education, 58, 921 (1981).
8. D.P. Tate, T.A. Antkowiak, Kirk-Othmer Encyclopedia Chemical Technology (3rd Ed. chap. 10), Wiley Interscience 936 (1980).
9. P.M. Grant, R.L. Greene, W.D. Gill, W.E. Ridge, G.B. Street, Mol. Cryst. Liq. Cryst. 32 171 (1976).

10. T. Nishikawa, Shin Nisso Kako Fine Chemicals publication, August 31, (1983).
11. J.H. Magill, J. Applied. Phys., 35, 3249 (1964).
12. J.H. Magill, J. Polymer Sci., A2 7, 1187 (1969).
13. R.L. Merker, M.J. Scott., G.G. Haberland, J. Polymer Sci., A2, 2, 31 (1964).
14. H.R. Allcock, Chem. and Eng. News, pp 22-36, March (1985).
15. L.A. Wood and N. Bekkedahl in High Polymer Physics Symp. (H.A. Robinson, ed.,) p 258, Remen Press, Chem. Pub. Co., N. York (1948).
16. J.M. Schulz, 'Polymer Materials Science', Prentice Hall, Englewood bliffs, N. Jersey (1974).
17. See for example, Dis. Faraday Soc., Symp. 68 (1979).
18. J.D. Hoffman, G.T. Davis, J.I. Lauritzen, Jr., 'Treatise on Solid Plate Chemistry' (N.B. Hannay Ed.), vol. 3, Chap. 7, Plenum N.Y. (1976).
19. J.H. Magill in 'Treatise on Materials Science and Technology' (Ed., J.M. Schulz), vol 10c, Chap. 1 pp 181-295 Acad. Press. N.Y. (1977).
20. J. Rault in 'Plastic Deformation of Amorphous and Semi-Crystalline Materials' (B. Escaig and C.G. Sell. Ed.) Les Ulis, Fr. (1982).
21. J.C. Hser, and S.H. Carr, Poly. Sci. Engng., 19, 436 (1979).
22. J.H. Magill, IUPAC Meeting Proceedings p 580, U. Mass. July 12-16 (1982).
23. E. Ergoz, J.G. Fatou and L. Mandelkern, Macromolecules, 5, 147 (1972); I.G. Voigt-Martin and L. Mandelkern, J. Polymer. Sci. (Polymer Phys. Ed.) 22, 1901 (1984).
24. S.S. Pollack and J.H. Magill, J. Polymer Sci. A2, 7, 551, (1969).
25. J.H. Magill, J.M. Schultz and J.S. Lin. Bull. Amer. Phy. Soc. (DHPP), Abstract AV8, March 25-28, Baltimore, MD; J. Colloid and Polymer Sci., (submitted 1985).
26. J.D. Hoffman and J.J. Weeks, J. Chem. Phys. 42 4301 (1965).
27. A.J. Kovacs, C. Straupe, A. Gönthier, J. Polym. Sci. (Polym. Symp., 59, 31 (1977), see also ref. 17.
28. G. Ungar, J. Stejny, A. Keller, I. Bidd and M.C. Whiting, Polymer, in press (1985).
29. T. Williams, D.J. Blundell, A. Keller and I.M. Ward J. Polym. Sci., A-2, 6, 1613 (1968).
30. C.W. Hock, J. Polymer Sci. A2, 4, 227 (1966).
31. Y.K. bhoun and J.P. Bell Bull. Amer. Phys. Soc. (DHPP), Abs. AS2 and AS3, p 240, March 26-30, Detroit, Michigan (1984).
32. N. Okui and J.H. Magill, Polymer 17, 1086 (1976).
33. N. Okui, J.H. Magill and K.H. Gardner, J. Applied Phys. 48, 4116 (1977).

34. J. Gardella, Jr., J. Chen., J.H. Magill, D.M. Hercules, J. Amer. Chem. Soc., 105, 4536 (1983).
35. R. Kimmich, Polymer, 25, 187 (1984).
36. J.D. Ferry 'Viscoelastic Properties of Polymers', (2nd Ed.). J. Wiley, N. York (1970).
37. P.G. De Gennes 'Scaling concepts in Polymer Physics', chapt. 8, Cornell, Ithaca N.Y. (1979).
38. D. Sadler and G.H. Gilmer, Polymer, 25, 1446 (1984).
39. D. Sadler, Polym. Comm., 25, 196 (1984).
40. J.M.G. Cowie and I.J. Mc. Ewen, Polymer, 14, 423 (1973).
41. H.D. Keith and F.J. Padden, Bull. Amer. Phys. Doc. (DHPP), Abs. AR5, p271 March 21-24 (1983).
42. M. Kojima and J.H. Magill, Macromolecular Science (Physics) 10, 419 (1974).
43. M. Kojima, J.H. Magill and R.L., Merker, J. Polymer Sci., A2, 12, 317 (1974).
44. J.H. Magill, Die Makromol. Chemie (submitted 1985).
45. H.D. Keith and F.J. Padden, Ir., J. Applied Phys., 35, 1286 (1964).
46. J. Gardella Ir., J.A. Schmidt, R.L. Chim and J.H. Magill, paper presented at 35th Pittsburgh Conference on Analytical Chemistry and Applied Spectroscopy, Abs. # 921, March 5-9, Atlantic City, N. Jersey (1984).
47. H.M. Li and J.H. Magill, Polymer, 19, 829 (1978).
48. H.R. Allcock, R.W. Allen and J.J. Meister, Macromolecules 9, 950 (1976).
49. K.H. Gardiner, J.H. Magill and E.D.T. Atkins Polymer, 19, 370 (1978).
50. H. Finkelmann and G. Rehage, Ad. Polymer Sci., 60/61, 100 (1984).
51. M. Kojima and J.H. Magill, Polymer Comm. 24, 329 (1983).
52. M. Kojima, W. Kluge and J.H. Magill, Macromolecules, 17, 1421 (1984).
53. T. Masuko, R.L. Simeone, J.H. Magill and D.J. Plazek Macromolecules, 17, 1421 (1984).
54. D.C. Bassett and B. Turner, Phil. Mag. 29 925 (1974).
55. B. Wunderlich and J. Grebowicz, Advances in Polymer Sci., 60/61 1 (1984).
56. J.H. Magill, M. Kojima and G. McManus (unpublished work).
57. M. Kojima and J.H. Magill, J. Materials Science (to be published 1985).
58. I.C. Choy and J.H. Magill J. Polymer Sci. (Chemistry Ed.) 19, 2495 (1981).

Ultra-drawing of High-Molecular-Weight Polyethylene Cast from Solution IV. Effect of Annealing/Re-Crystallization

C.W.M. Bastiaansen, P. Froehling, A.J. Pijpers and P.J. Lemstra
DSM, Central Research, P.O.Box 18, 6160 MD Geleen, Netherlands

Synopsis

The ultra-drawability of spun/cast UHMW (ultra-high-molecular-weight)-Polyethylene is lost gradually upon annealing, but almost instantaneously upon melting/re-crystallization. This phenomenon poses severe limitations on processing of so-called "dis-entangled" polyethylene, as will be discussed below.

Introduction

In previous papers (1,2,3) it has been shown that spinning/casting from semi-dilute solutions of UHMW-PE drastically improves the effective drawability of UHMW-PE, even after complete removal of solvent from the as-spun/cast structures.

This phenomenon resulted in a technological process, now often referred to as gelspinning, for the production of high-strength/high-modulus PE fibres (4,5), a process originating from DSM Central Research.

From the influence of the initial polymer concentration in solution on the maximum drawability of the spun/cast UHMW-PE structures, it was concluded (2) that the enhanced drawability was related to a reduction in the concentration of "trapped entanglements" in the generated gel or solid structure. In this model it is assumed that trapped entanglements act as cross-links that are semi-permanent on the time scale of the drawing experiment. In the case of melt-crystallized UHMW-PE, the high entanglement density per chain is prohibitive for ultra-drawing, and spinning/casting from semi-dilute solutions provides an optimum with respect to morphology and entanglement density.

In this paper the effect of annealing and re-crystallization on the ultra-drawability will be discussed.

Experimental

Hostalen Gur-412, an UHMW-PE grade from Hoechst/Ruhrchemie, M_w appr. 1.5×10^3 kg/mole, was used in our studies. Cast films were made by dissolving the polymer in xylene or decalin, concentration 1.5% w/v, and after complete dissolution the hot solutions were poured into stainless steel trays. The solvent was evaporated at room temperature and residual traces of solvent were removed by extracting the films with dichloromethane.

Annealing was performed by wrapping the films in aluminum foil and immersing in silicon oil or heating in a press at controlled temperatures and times.

Stress-strain measurements were recorded using an Instron tensile tester fitted with a temperature regulated oven. Either dumbbell-shaped or rectangular-shaped pieces of film were used. The draw ratio was determined from the displacement of ink-marks or by using a strain-gauge. The samples were drawn at constant strain rates of 0.05 respectively 0.1 sec^{-1}.

The melting behaviour was determined using a Perkin-Elmer DSC-2, adopting a standard heating rate of 5°C/min.

Transmission-Electron-Microscopy was performed using a Philips EM 420T operating at 100kV. Trimmed samples were treated with chlorosulphonic acid according to Kanig (6) at 56°C for 18 hours. Thin sections were obtained by cryo-ultramicrotomy (Reichert Ultracut E/FC4) at -60°C, and subsequently staining with uranyl acetate.

Results and Discussion

Figure 1 shows a typical stress-strain curve of solution-cast vs. melt-crystallized UHMW-PE.The constraints which limit the drawability of melt-crystallized UHMW-PE have been removed to a large extent by casting or spinning from dilute solution as visualized in figure 2 in the proposed model of trapped entanglements, acting on the one hand as semi-permanent crosslinks on the time scale of the drawing experiment but providing connectivity between individual single crystals on the other hand.If the polymer concentration is too low, below the overlap concentration ϕ^* , indivudual single crystals will precipitate and in-line spinning/drawing is impossible.This model is simple and combines well-known facts from polymer rheology (the entanglement density will decrease in solution) and polymer morphology (single crystals of PE are ductile,Statton (7)).

Spinning from semi-dilute solutions provides an optimum with respect to morphology and entanglement density; the drawability can be adjusted with the initial polymer concentration in solution and hence the fibre properties can be varied at will with the experimental conditions.

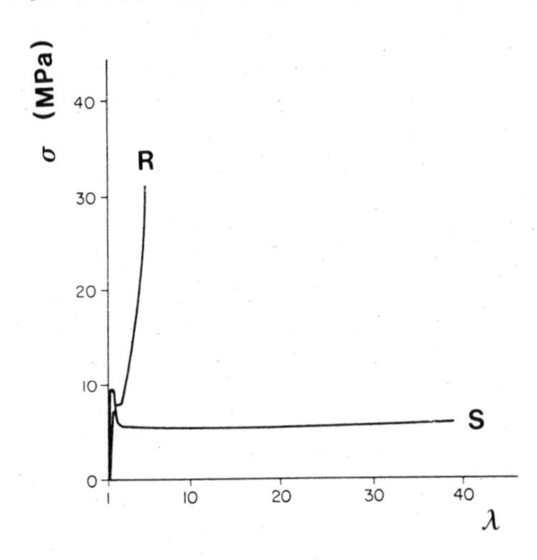

Figure-1: Stress-strain behaviour of melt-crystallized (R) vs. Solution cast UHMW-PE (S), $T_{draw} = 90^oC$, initial polymer conc. 1.5%

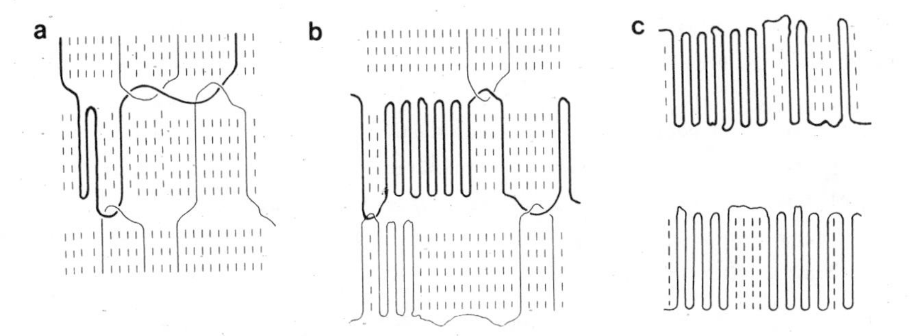

Figure-2: Chain topology, dependence on crystallization conditions:
a) from the melt ; b) from solution $\phi^* > \phi^*$; from solution $\phi^* < \phi^*$

Figure 3 shows the TEM-micrographs of respectively solution-cast (a) and melt-crystallized (b) UHMW-PE films.The solution-cast sample, cast from a 1.5% solution in xylene, shows a stacked lamellar structure in contrast to randomly oriented lamellar crystals in the melt-crystallized sample.

The effect of annealing on the drawability of solution-cast films (1.5 % initial polymer concentration) is shown in figure 4 whereas in figure 5 some corresponding stress-strain curves are depicted.Figure 6 shows the melting behaviour of various annealed samples.As can be inferred from figures 4, 5 and 6 , annealing reduces the drawability in the solid state, measured here at 110°C, especially when partial melting occurs during the annealing process.

A slow decay in drawability as a function of annealing time is understandable in terms of re-entangling during partial melting and hence the destruction of a favourable intermolecular chain topology.Figure 7 shows the corresponding changes in the morphology upon annealing.Figure 7a still shows the stacked lamellar organization (annealing at 132°C for 24 hours, compare also with figures 4 and 5) whereas figure 7c (annealing at 134°C for about 1 hour) is almost indistinguishable from a melt-crystallized sample (compare with figures 3b and 4).

Surprising however is the almost instantaneous decay in drawability if the sample is heated above the melting-point and re-crystallized as shown in figure 8.The drawing behaviour (fig.8) and morphology is indistinguishable from a standard melt-crystallized sample obtained through compression-moulding of UHMW-PE powder.It is difficult to envisage this decay in drawability as a consequence of the reformation of an entanglement network in the melt at this short period of time (less than 1 minute is sufficient to lose completely the ultra-drawability (see figure 8)) and for these high-molecular-weight samples (see also ref.9).Local reorganization of long-chain macromolecules is probably sufficient to completely destroy the favourable drawing characteristics of spun/cast UHMW-PE.

An important consequence of this phenomenon is that carefully prepared samples possessing ultra-drawing characteristics in the solid state, for instance single crystals and/or disentangled reactor powder (10) can lose this favourable property almost immediately upon heating for short times above T_m.Hence, processing should be limited to temperatures below the melting point (solid state extrusion etc.).

References

1)P.Smith and P.J.Lemstra , Coll. and Polym.Science 258/7 , 891 (1980)

2)P.Smith, P.J.Lemstra and H.C.Booy , J.of Polym.Sci., Polym. Phys.Ed., 19 , 877 (1981)

3)P.Smith , P.J.Lemstra, J.P.L.Pijpers and A.M.Kiel , Coll. and Polym.Science, 259/11 , 1070 (1981)

4)DSM/Stamicarbon USP 4,344,908/4,422,933/4,430,383 and 4,436,689

5)P.J.Lemstra,R.Kirschbaum,T.Ohta and H.Yasuda , Symposium Kyoto Aug. 1985

6)G.Kanig , Prog. Coll. Polymer Science , 57 , 176 (1975)

7)W.O.Statton , J.of Appl. Physics , 38 , 4149 (1967)

8)T.Kanamoto,A.Tsuruta,A.Tanaka and R.S.Porter , Polymer J. ,15 , 327 (1983)

9)P.J.Lemstra and R.Kirschbaum , Polymer "Special Issue-Speciality Polymers", 26 , 1372 (1985)

10)P.Smith ,Symposium Morphology of Polymers , Prague , July 1985

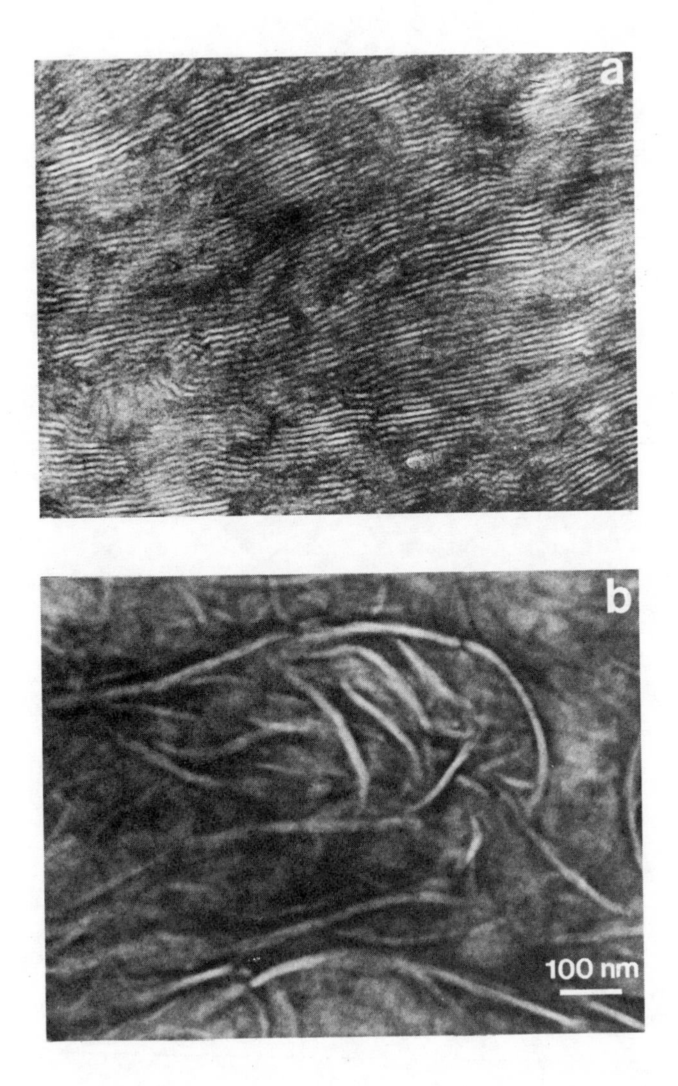

Figure-3: TEM-micrographs, Solution-cast (a) and Melt-crystallized (b)

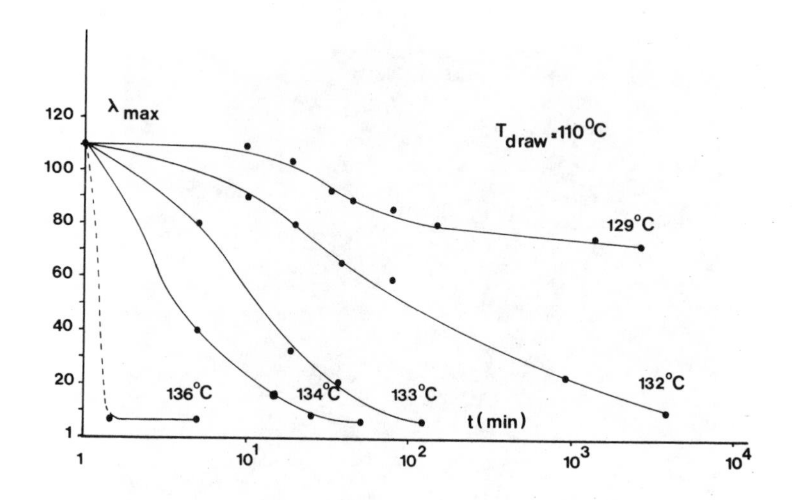

Figure-4. Effect of annealing on the maximum drawability (λ_{max}) as a function of time and temperature

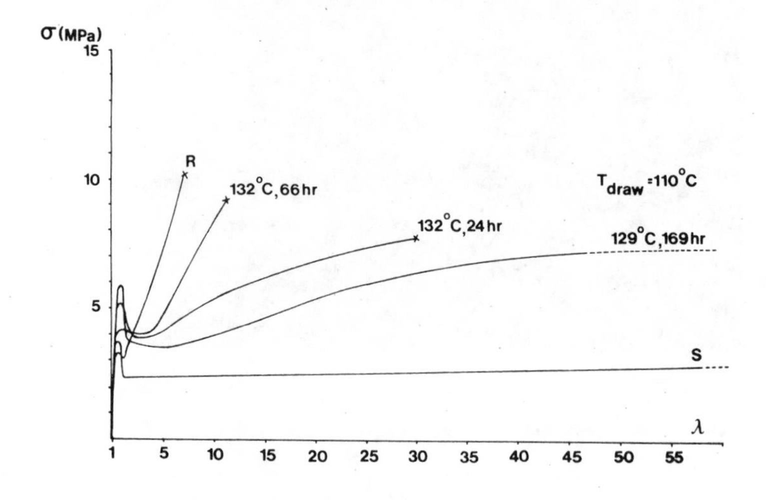

Figure-5. Stress-strain behaviour of annealed samples , S =solution-cast and R=melt-crystallized

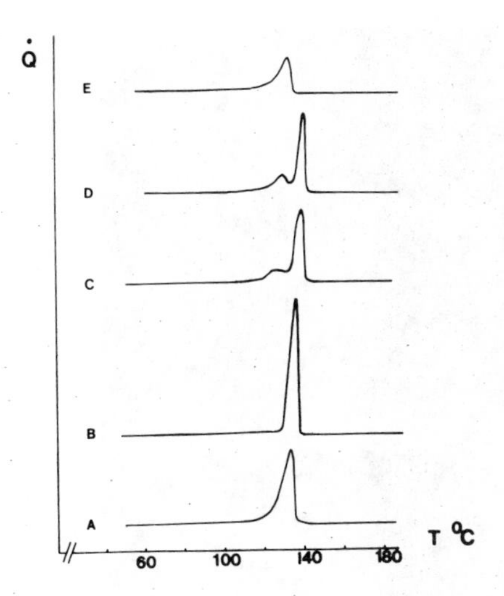

Figure-6. Melting behaviour of solution-cast (A) and annealed samples; annealing temperature/time: B) 129°C/120hrs ,C) 132°C/24hrs D) 132°C/66hrs and E) 134°C/1hr.

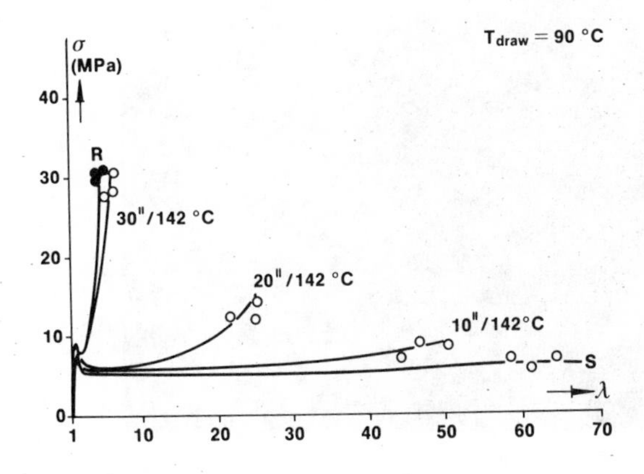

Figure-8. Effect of melting/re-crystallization on drawability. S=solution-cast, R=melt-crystallized, drawing temperature 90°C. The heating time at 142°C is indicated in the graph.

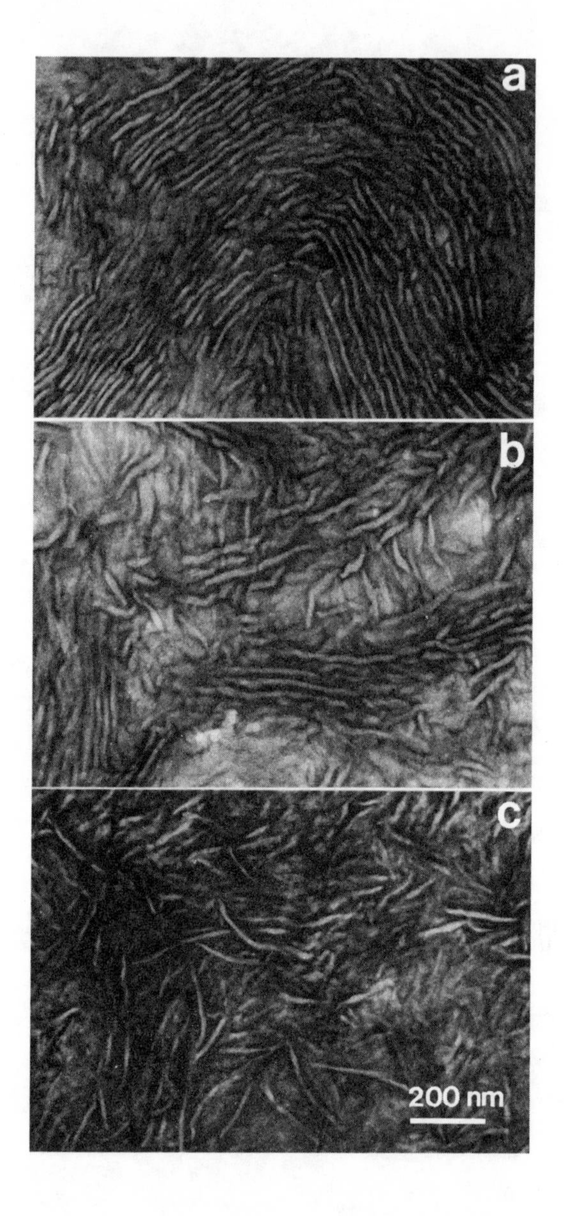

Figure-7: TEM micrographs of annealed samples

a)Annealing at 132°C/24hrs; b)Annealing at 132°C/66hrs and c) Annealing at 134°C/1hr.

MICROHARDNESS OF SEMICRYSTALLINE POLYMERS

F.J. BALTA CALLEJA* and H.G. KILIAN†

*Instituto de Estructura de la Materia, CSIC,
Serrano 119. 28006 Madrid. Spain.
†Abteilung Experimentelle Physik,
Universität Ulm, D-7900 Ulm. FRG.

SYNOPSIS

A new approach to justify the dependence of microhardness of semicrystalline polymers, specially polyethylene, upon average crystal thickness is given. A basic model of deformation under local compression is proposed in terms of the heat dissipated by the plastically deformed crystals and the volume of crystals destroyed under the indenter. The results indicate that the mechanism of deformation is dictated by the initial mosaic-block structure controlling the generation of a final system of shear planes. Though the analysis is approximate, acceptable parameters give satisfactory agreement with observation.

1. INTRODUCTION

In recent years there has been increasing recognition of the importance of the surface structure of polymers as means of developing materials with improved physical properties. Within this context the study of identation microhardness (MH) has emerged as a potential useful technique which can yield information regarding the microstructure of semicrystalline polymers at various morphological levels[1,2]. This is possible because microindentations respond to specific changes in microstructure. Specifically, microhardness-critical stress needed to plastically deform the surface of semicrystalline polymers-has been shown to depend on structural parameters such as intermolecular distances within the crystals, lattice perfection and crystal dimensions[3]. In semicrystalline polymers microhardness measurements may also throw light on the changes in the overall chain orientation[4-7]. There are two important aspects concerning the hardness study in polymers. First, there is the question of a quantitative justification for the observed dependence of MH on lamellar thickness and the information which can be obtained, in principle, by the microhardness technique regarding the volume fraction of crystalline material. This leads directly to the problem of relating MH to microdeformational mechanisms as this may permit a comparatively limited information to be put to optimun use. Secondly, there is the relationship between microscopic (microhardness) and macroscopic (modulus, yield stress) mechanical properties, where microstructural information has been valuable in correlation both types of properties[8]. A previous review of this subject has been given recently[9].

The main concern of this paper is the development of a description of microhardness of semicrystalline polymers, derived thermodynamically, in terms of the average crystal thickness. The relevance of the latter structural parameter is highlighted as describing accurately the surface plastic behaviour of polymers under local compression.

2. CONCEPT OF MICROINDENTATION HARDNESS: BACKGROUND

The method most commonly used in determining the microhardness of a polymer surface is the static indentation test[10]. It involves the formation of a permanent impression after loading the surface of the material with a sharp indenter of known geometry. The microhardness is calculated as the ratio of peak contact load, p, to the projected area of impression, A:

$$H = k^* \, p/A = k \, p \, \delta^{-2} \qquad (1)$$

where $k^* = 9.272 \times 10^5$, and δ is the penetration depth of the indenter within the surface (for a Vickers pyramid $\delta = \sqrt{2A/7}$ ($|H| = \mathrm{MNm}^{-2}$; $|p| = \mathrm{N}$; $|\delta| = \mu\mathrm{m}$). The hardness, so defined, is considered to be indicative of the irreversible deformation processes characterizing the polymer material. These include phase transformations and twin formation, at low strains, and chain tilt and slip, crack formation and chain unfolding at larger strains[3]. The indentation stresses, although highly concentrated in the plastic region surrounding the contact, may extend into the more distant elastic matrix. Thus the material under the indenter consists of a spherical zone of plastic deformation (~4-5 times the penetration distance, δ, of the indenter below the specimen surface) surrounded by a larger zone of elastic deformation. For conventional instruments using forces p~0.1N, the penetration distance (residual impression) left by a sharp indenter is for most commercial polymers, of the order of a few μm. However with the new instruments operating down to 10μN, the penetration distance can be reduced to a few hundred Å [11]. Impressions, in this case, should be observed with help of an electron-microscope. To differentiate between reversible and irreversible contributions contact deformation measurement of the elastic recovery after unloading is required[10]. However, pyramid indenters provide a contact pressure which is nearly independent of indent-size and are less affected by elastic release than other indenters. The time-dependent contribution to plastic deformation during loading can be described by a law of the form $H=Bt^{-k}$, where B is a coefficient depending on temperature and loading-stress and k furnishes a measure for the rate of creep of the material. In previous investigations we have mostly adopted a MH value at 0.1 min because it approaches Tabor's relation[3,5,9]. After load release a long delayed recovery can, often, be detected[9]. The relaxation of the residual impression depth increases with the volume fraction of "amorphous" material.

3. HARDNESS RELATING TO MICROSTRUCTURE: A THERMODYNAMICAL APPROACH

Semicrystalline polymeric solids show a distinct morphology of stacks of crystalline lamellae intercalated by "amorphous" disordered regions. The lamellae themselves posses a mosaic-block superstructure with liquid-like lattice distortions (paracrystallites) and a high concentration of lattice defects at the grain boundaries reducing the lateral cohesion of stacked lamellae in the solid. From a mechanical viewpoint the material may be regarded as a composite consisting of alternative hard (crystalline) and soft-compliant (amorphous) layers. The hardness of such a system can be approximated to:

$$H = w_c H_c + (1-w_c)\, H_a \qquad (2)$$

where H_c and H_a are the intrinsic hardness values of the lamellar crystals and amorphous layers respectively. The additivity law expressed by eq. 2 has been verified for mixtures of two polyethylenes with well differentiated values[12].

Since $H_c \gg H_a$ ($H_a \sim 0.5 MN/m^2$)

the heat dissipated by the plastically deformed crystals can be written as:

$$\Delta\phi_{cd} = \Delta\phi / V_{cd} \qquad (3)$$

where $\Delta\phi$ is the total enthalpy and V_{cd} is the volume of crystals destroyed under the stress field of the indenter. If we assume that the volume of destroyed crystals is proportional to the volume of material physically displaced under the indenter:

$$V_{cd} \simeq \alpha A \delta \qquad (4)$$

The mechanical work performed during indentation must be equal to:

$$W = p\ \delta = \Delta\phi \qquad (5)$$

By combination of eqs. 1-5 one is led to the simple relationship:

$$H_c = H_o\ \Delta\phi_{cd}\ ;\ H_o = (k^*/w_c)\alpha \quad (p=const.) \qquad (6)$$

Since the indentation of the polymer surface involves a yielding process one expects a substantial destruction and/or severe modification of a volume fraction of lamellae localized at the surface. For $p{\sim}0.1N$ values of $V_{cd}{\sim}10^3\mu m^3$ have been derived for melt crystallized polyethylene[3]. It is interesting to express V_{cd} in terms of the modified microstructural units left after deformation. Let us assume that the destruction of lamellae is heterogenous and involves a generation of a more or less dense system of shear-planes, wherein the energy of deformation is grossly dissipated (fig. 1). Assuming the original crystal blocks to be cubic in shape the relative fraction of "shear planes" will be, in a first approximation:

$$surface/volume \approx \ell_c^{-1} \qquad (7)$$

Hence the "dissipation volume" of the crystalline material can be approximated to:

$$V_{cd} = V_o\ (1 + b_1\ell_c^{-1}) \qquad (8)$$

One can then express eq. 6

$$H_c = H_o \frac{\Delta\phi}{V_{cd}} = \frac{H_o\Delta\phi^*}{(1+b_1\ell_c^{-1})}\ ; \quad (\Delta\phi^* = \frac{\Delta\phi}{V_o}) \qquad (9)$$

This expression is of great interest because it correlates the "crystal hardness" of a polymeric material to the thickness ℓ_c of its building mosaic-block units. It is noteworthy that for $\ell_c\to\infty$, H_c approaches to $H_o\Delta\phi^*$, which is the maximum possible value of dissipated energy through plastic deformation. For polyethylene[9] $H_o\Delta\phi^* \simeq 170MNm^{-2}$. Eq. 9 offers a quantitative measure for intrinsic plastic deformation of semicrystalline polymers, predicting a clear dependence of H_c with the size of the constituent units.

4. COMPARISON OF CALCULATED AND EXPERIMENTAL DATA

Fig. 2 illustrates the plot of H_c as a function of ℓ_c^{-1} for different lamellar structures crystallized from the melt. The experimental data[3,8] show a good agreement with the calculated (solid) curve according to eq. 9 using b_1=20nm.

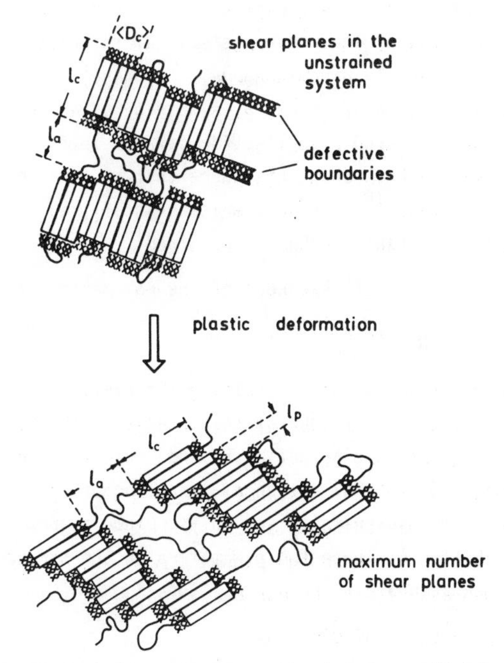

Fig. 1. Model of lamellae deformation beneath the stress-field of the indenter

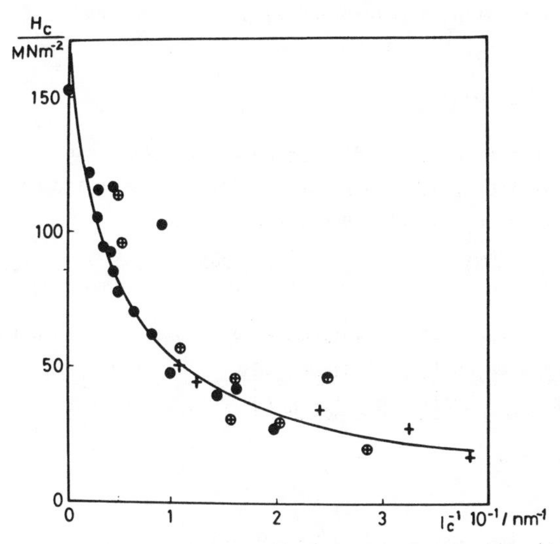

Fig. 2. H_c against reciprocal crystal thickness including defective boundaries (eq. 9)

Eq. 9 offers, in fact, a direct description of the hardness of a large stack of crystal lamellae in terms of their average crystal thickness, including the defective surface boundary (non-homogeneous microphase)[13,14]. Paraffins can, likewise, be described by means of eq. 9, simply by using the chain length parameter. The value of b_1=200 Å applies for samples melt crystallized at atmospheric pressure. We have shown elsewhere that b_1 is a decreasing function of crystallization pressure[15]. For PE samples crystallized, for instance, at $p=150 MN/m^2$ a fitting constant b_1=10nm is required[15].

Since $H_a \ll H_c$ the overall hardness of the polymeric material will be:

$$H = H_c w_c = H_o \Delta\phi^* \, w_c/(1+b_1 \ell_c^{-1}) \tag{10}$$

where w_c is the molar fraction of crystalline material. Eq. 10 offers a convenient model relating the plastic behaviour of the material with its microstructural parameters. Structural changes affecting ℓ_c and w_c can thus, be adequately interpreted by using eq. 10. For instance, annealing of PE in range II produces a substantial lamellar thickening ℓ_c and only a minor increase[16] in w_c. The independent contribution of these two parameters in the hardness of annealed PE can thus be followed accurately through eq. 10.

5. MECHANISM OF PLASTIC DEFORMATION

Let us finally discuss the local micromechanism of plastic deformation from the initial existing mosaic-block structure into the final modified one. The dissipated work related to the volume of the crystals can be defined as:

$$\Delta\phi/V_{cd} = p\,\delta/V_{cd} \tag{11}$$

We further assume that plastic deformation of crystals mainly proceeds by a multitude of shearing planes. Therefore, on the basis of a cubic symmetry for the cross-section of a crystal block we are led to

$$\Delta\phi/V_{cd} = \frac{\Delta h}{V_o}\left(\frac{4a_o}{\ell_p}\right) \qquad \frac{\Delta h}{V_o} : \text{defect enthalpy density related to the unit volume } V_o \tag{12}$$

where a_o is the average lateral intermolecular distance within the crystals and ℓ_p describes the average lateral dimension of the final "crystal block", with invariant thickness ℓ_c, after plastic deformation. From eqs. 11 and 12:

$$\ell_p = \frac{4\Delta h}{V_o} \frac{V_{cd}}{p\,\delta} a_o \tag{13}$$

Furthermore, by using eqs. 8 and 13 we obtain:

$$\ell_p = K(1+b_1\ell_c^{-1})\, a_o/\delta \quad ; \text{ where } \quad K = \frac{4\Delta h}{p} \tag{14}$$

This equation relates the average lateral dimension of the newly created blocks, ℓ_p, to the original mosaic-block dimensions before plastic deformation (fig. 1). Table I collectes the ℓ_p and ℓ_c values for a series of polyethylene samples with varying chain-defect concentration X_{nc}, together with the average value of the lateral coherent dimensions of the system D_c prior to deformation[17]. The ratio $n_a=D_c/\ell_p$ is proportional to the number of surface defects (boundaries) produced after plastic deformation.

These data show first, that the final dimension ℓ_p is practically independent from the initial mosaic-block length; secondly, that ℓ_p corresponds to 4-5 intermolecular distances yielding elementary "crystal rods" constituted by 16 to 25 stems. The ratio n_a decreases substantially with decreasing size of the original crystal block dimensions. Fig. 3 illustrates the correlation found between n_a and the molar degree of crystallinity. These results support the concept that the maximum number of shearing planes increases progressively with rising degree of crystallinity. Thus for a fully crystalline lamellar system with a thickness ℓ_c, the number of shearing planes below the indenter tends to become infinite despite of having a constant final limiting size ℓ_p. On the other hand, for low crystallinities the original mosaic blocks are so small that they are nearly unmodified after plastic deformation. The present results characterize the mechanism of local plastic deformation of semicrystalline systems as exclusively governed by the primary mosaic-block structure regulating the intrinsic "solid state" deformation mechanism.

Fig. 4 illustrates the large increase of dissipated energy with rising molar mass-fraction of crystalline material. The solid line was computed according to (Table 1):

$$W_{diss}C^* = 4\ell_c\ell_p W_p \tag{15}$$

The contribution of the dissipated energy from the smallest crystals to the overall deformation energy of the total system is negligible. The solid state deformation process occurs predominantly for very large crystal sizes entailing large crystallinity values.

In summary, the discussed results justify the obtained correlation between crystal hardness and lamellar thickness in semicrystalline polymers (polyethylene). Microhardness, thus, depends mainly on the plastic deformation of the lamellar crystals which are viewed as non-homogeneous solid microphases.

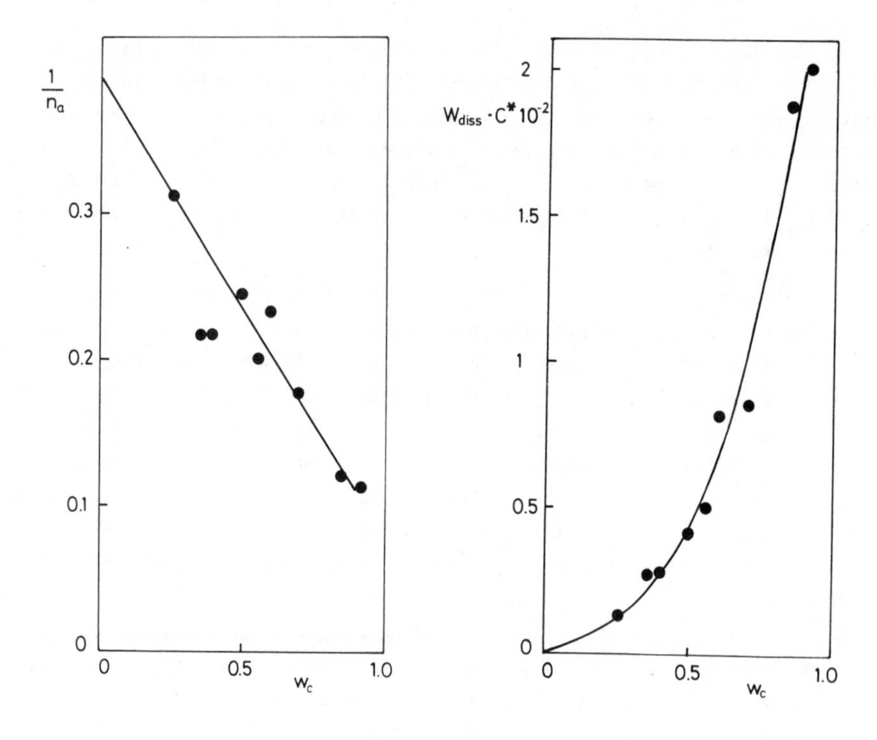

Fig. 3. Plot of reciprocal number of shearing planes after crystal destruction vs. molar degree of crystallinity.

Fig. 4. Increase of dissipated energy during crystal destruction as a function of molar fraction of crystalline material

Table 1:

Average crystal dimensions before (ℓ_c, D_c) and after (ℓ_p) plastic deformation and number of shearing planes (n_a) and dissipated energy (W_{diss}) after crystal destruction for different polyethylene samples with varying number of chain defects (X_{nc}) and molar degree of crystallinity (W_c). The quantity w_p is the molar crystallinity including the crystal surface defective parts. Parameters are defined in ref. 14.

x_{nc}	ℓ_c/nm	$\langle D_c \rangle$/nm	w_p/mol%	w_c/mol%	ℓ_p/nm	n_a	W_{diss}C*
0.0019	26.0	17.1	0.997	0.922	1.95	8.8	202
0.007	27.0	14.85	0.979	0.849	1.78	8.3	188
0.0176	12.0	10.9	0.908	0.708	1.95	5.6	85
0.0263	10.8	9.85	0.829	0.605	2.30	4.3	82
0.0304	10.0	8.05	0.789	0.56	1.60	5.0	50
0.0361	9.0	6.45	0.731	0.502	1.58	4.1	41
0.048	7.0	7.55	0.612	0.396	1.63	4.6	28
0.0534	5.8	9.8	0.561	0.355	2.12	4.6	27
0.069	5.0	5.0	0.425	0.254	1.57	3.2	13

ACKNOWLEDGMENTS

Grateful acknowledgment is due to CAICYT, Spain, for the support of this investigation.

REFERENCES

1. Baltá Calleja, F.J. Colloid & Polymer Sci., 254, 258-266, 1976.
2. Bowmann, J. and Bevis, M. Colloid & Polymer Sci., 255, 954-966, 1977.
3. Baltá Calleja, F.J., Martinez-Salazar, J., Cacković, H. and Loboda-Cacković, J. J. Materials Sci., 16, 739-751, 1981.
4. Baltá Calleja, F.J. and Bassett, D.C. J. Polymer Sci., 58C, 157-167, 1977.
5. Baltá Calleja, F.J., Rueda, D.R., Porter, R.S. and Mead, W.T. J. Materials Sci., 15, 765-772, 1980.
6. Rueda, D.R., Baltá Calleja, F.J. and van Hutten, P.F. J. Materials Sci. Lett., 1, 496-498, 1982.
7. Rueda, D.R., Baltá Calleja, F.J., Garcia, J., Ward, I.M. and Richardson, A. 19, 2615-2621, 1984.
8. Martinez-Salazar, J. and Baltá Calleja, F.J. J. Materials Sci., 18, 1077-1082, 1983.
9. Baltá Calleja, F.J. "Advances in Polymer Science", 66, 117-148, 1985
10. Lawn, B.R. and Howes, V.R. J. Materials Sci., 16, 2745-2752, 1981.
11. Bangert, H., Wagendrizted, A. and Aschinger, H. Philips Electron Optics Bull., 119, 17- , 1983.
12. Martinez-Salazar, J. and Baltá Calleja, F.J. J. Materials Sci. Lett. (in press)
13. Holl, B., Heise, B. and Kilian, H.G., Colloid & Polymer Sci., 261, 978-992, 1983.
14. Kilian, H.G., Colloid & Polymer Sci., 261, 374-380, 1984.
15. Baltá Calleja, F.J., Rueda, D.R., Garcia, J., Wolf, F.P., Karl, V.H. J. Materials Sci. (in press)
16. Rueda, D.R., Martinez-Salazar, J. and Baltá Calleja, F.J. J. Materials Sci. (in press)
17. Martinez-Salazar, J. and Baltá Calleja, F.J., J. Crystal Growth, 48, 283-294, 1980.

MODEL CALCULATIONS FOR WAXS PROFILES FROM THE POLYMER CRYSTALLINE PARTICLE SIZE DISTRIBUTION

G.BODOR[x] AND D.HOFFMANN[xx]

[x] Polymer Research Institute, Budapest H-1950
[xx] Institute of Polymer Chemistry, GDR Academy of Sciences
Teltow DDR-15

SYNOPSIS

The paper calculates the WAXS profiles from the supposed crystalline particle size distributions.

It is shown, how the second integrals of the crystalline particle size distributions are depending on the distribution functions. These second integrals are the Fourier coefficients of the particle size line broadening functions.

The usual line profile analysis is reversed in this experiment for a better understanding of the whole line broadening process.

1. INTRODUCTION

The crystalline particle size distribution determination from WAXS profiles calculates the second derivatives of the Fourier coefficients of the line profiles.

The shape of these Fourier coefficient functions is very interesting: their second derivative must be a positive value or zero, as negative particle size does not exist.

Sometimes a hook effect appears in the Fourier coefficient functions, sometimes this function does not approaches to zero. The reasons for all kind of such phenomena could be detected by model experiments, where we can arbitrarily choose the crystalline particle size distribution parameters.

2. CRYSTALLINE PARTICLE SIZE DISTRIBUTION ASSUMPTIONS

The range of the calculation was set to 0...60 nm. For the distribution function, we used 60 or 16 points.

The following functions were calculated:

2.1. Gauss distribution function,

$$I_G(X) = I_G(0) \exp(-4*0.683*Z1^2) \quad \ldots(1)$$

where

$$Z1 = (X-X0) / H1 \quad \ldots(2)$$

X0 is the position of the maximum of the distribution on the X (nm) scale,

I0 the maximum value at X0

H1 is the full width at half of the maximum intensity.

2.2. Cauchy distribution function,

$$I_C(X) = \frac{I0}{1+4\ Z2^2} \qquad \ldots(3)$$

where

$$Z2 = (X-X0)\ /H2 \qquad \ldots(4)$$

H2 is the full width at half of the maximum intensity, for the Cauchy function.

2.3. Voigt approximation function,

$$I_W(X) = \frac{I0}{1 + 4Z1^2 + (1-A)\ 16\ Z1^4} \qquad \ldots(5)$$

where A is the lineform parameter, it's value is between 0.743 and 1, the first value gives a Gaussian, the last a Cauchy type profile as marginal cases (1,2).

We used A = 0.85.

2.4. Pearson VII function,

$$I_P(X) = \frac{I0}{\left\{1 + 4Z1^2 * (2^{1/M} - 1)\right\}^M} \qquad \ldots(6)$$

where M is a shape parameter (3), in our case M was 3.

2.5. Voigt function,

the convolution of the Gaussian and of the Cauchy functions:

$$I_V(X) = \int_{-\infty}^{\infty} I_G(U) \quad I_C(X-U)\, dU \qquad \ldots(7)$$

where U is a help-parameter in the same range as X.

2.6. Bimodale distribution functions

As we can suppose bi- or multimodale-distributions, we investigated the influence of bimodale crystalline particle size distribution for the calculations.

Some typical distribution functions are presented in fig. 1 a, b, c.

3. CALCULATION OF THE FOURIER COEFFICIENTS

The crystalline particle size distribution is the second derivative of the A_t^S curve versus R /4/. A_t^S is the cosine part of the particle size broadening of the Fourier coefficient F/t/ of the f /x/ WAXS profile, R is the distance function /5/.

The second integral of the crystalline particle size distribution was calculated. The result, A_t^S gives very interesting results: a plot against R gives on the abscissa $\bar{D}$, the average crystalline particle size (given by our assumptions), there is no hook effect and in all cases the initial A_t^S values are producing the same crystalline particle size value. Examples are shown in fig 2 a, b, c.

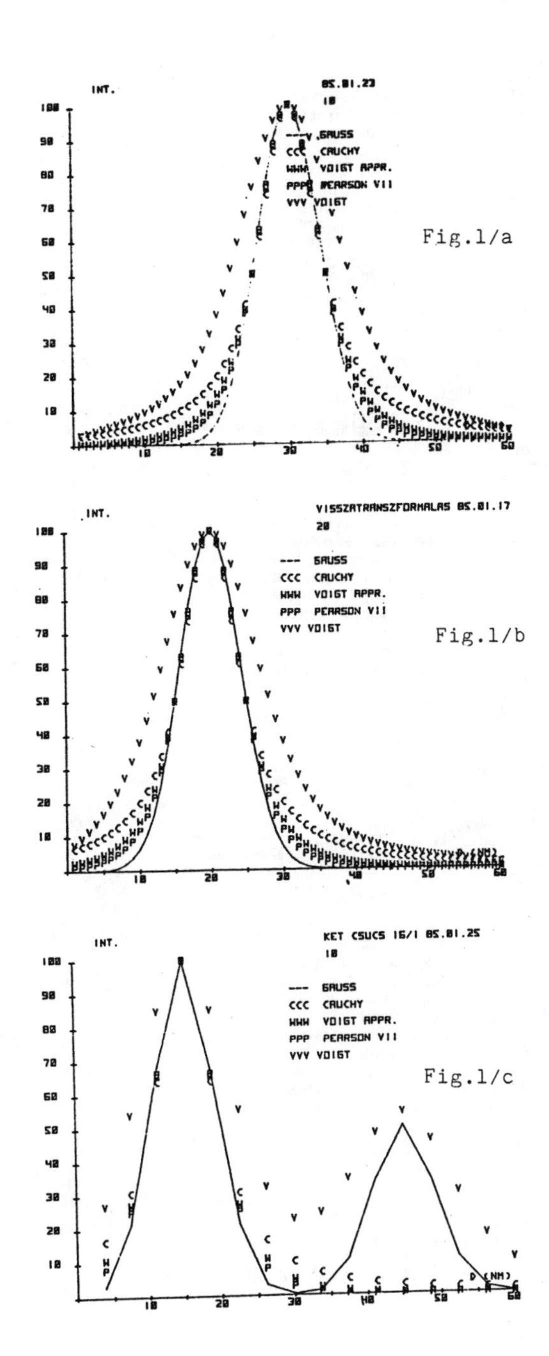

Fig.1. Assumed crystalline particle size distributions /CPSD/. Five function types. a./ 60 points, $\overline{D}$ = 30 nm, b./ 60 points, $\overline{D}$ = 20 nm c./ 16 points. For the Gaussian bimodal with $\overline{D}_1$ = 15 nm and $\overline{D}_2$ = 45 nm, 2:1 ratio. For other types monomodal, $\overline{D}$ = 15 nm.

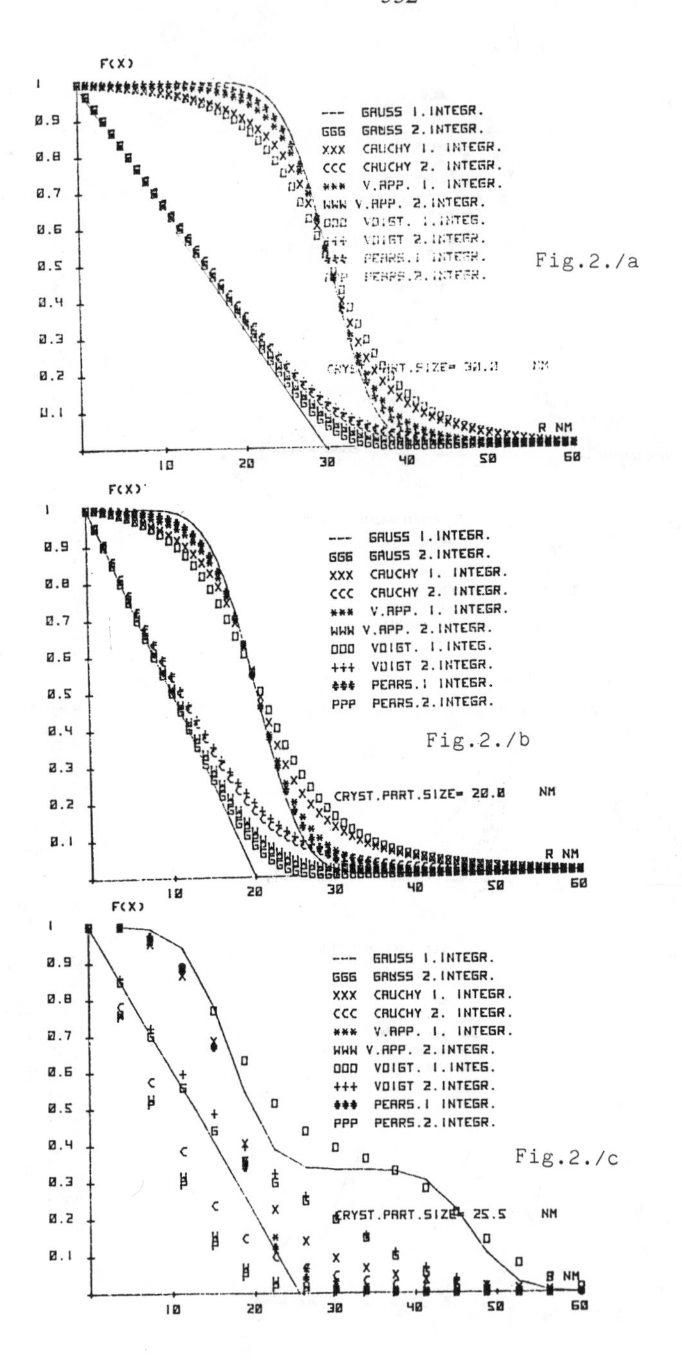

Fig.2. The first and second integrals of the CPSD of the fig.1. The $\overline{D}$ is extrapolated from the Gaussian function.

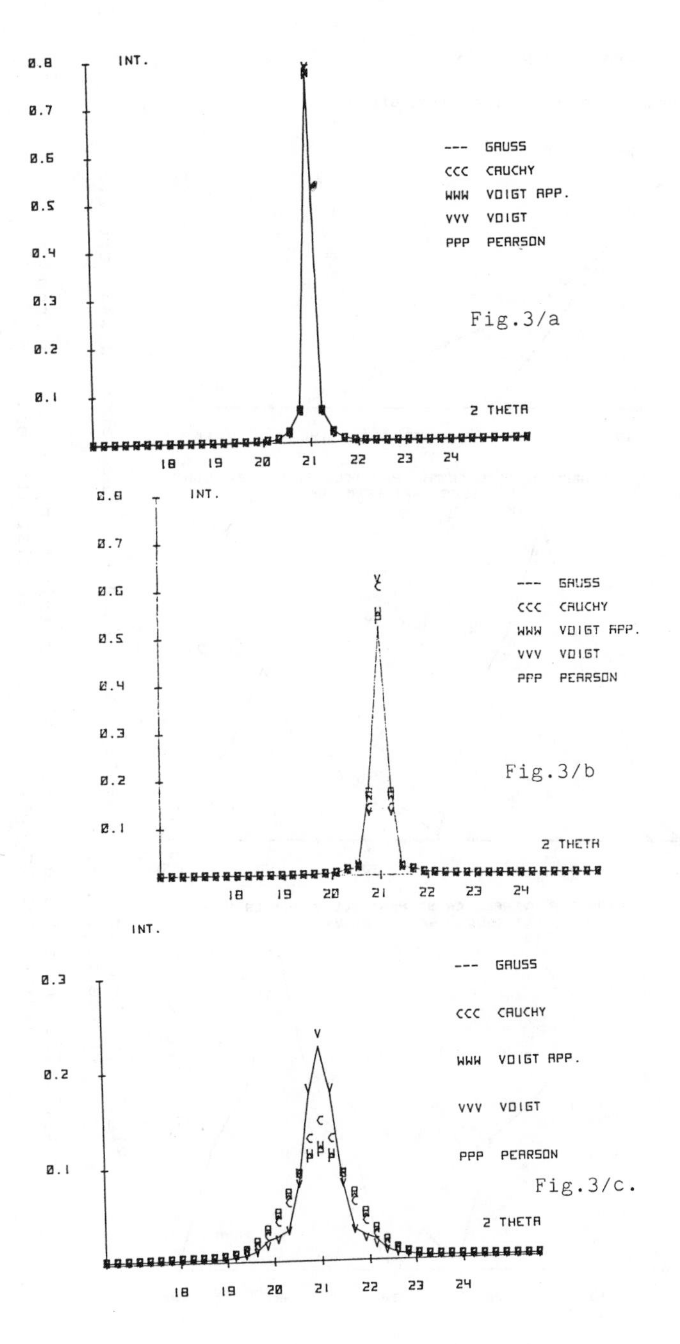

Fig.3. The WAXS profiles from the second integrals.

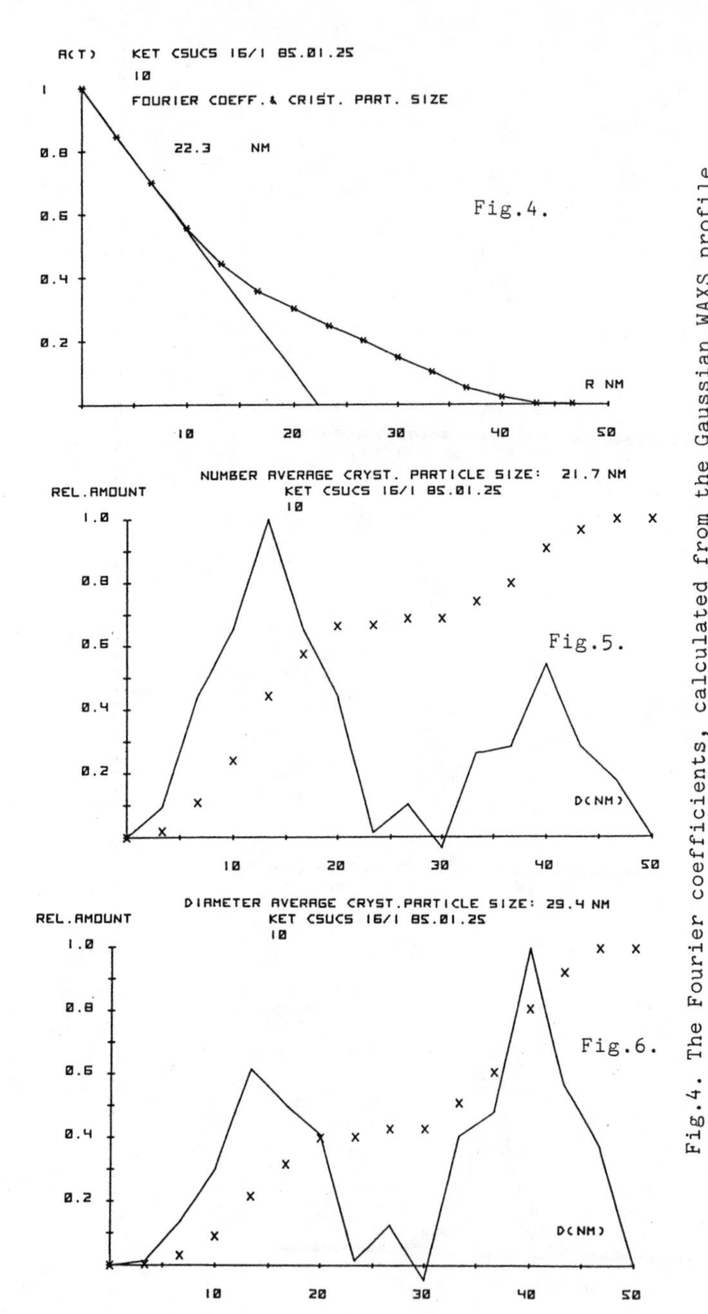

Fig.4. The Fourier coefficients, calculated from the Gaussian WAXS profile of the fig. 3/c.

Fig.5. The crystalline particle size distribution, calculated from the Fourier coefficients, fig.4. Number distribution.

Fig.6. The same as in fig.6., size distribution.

4. THE WAXS PROFILE

The theoretical WAXS profile is calculated from the A values. A plot of the intensity vs. 2θ is given for each experiment. From such profiles the integral breadth was used again for the determination of the average crystalline particle size data. The profiles are given in fig. 3. The Fourier coefficients (from fig. 3.) are on fig. 4., the crystalline particle size number distribution on fig. 5., size distribution on fig 6. The results are in a very good aggreement with the input data.

REFERENCES

1. Shirane, G., D.E.: Phys. Rev., 125, 1158 (1962)

2. Meisel, N.: Exp. Techn. der Physik. IXX, 23 (1971)

3. Hiusman, R. and Heuvel, H.M.: J.Polym. Sci. Polym. Phys.Ed. 14, 941-954 (1976)

4. Bertaut, E.F.: Comp. Rend. 228, 187 (1949)

5. Bodor G.: X-ray Line Shape Analysis. A Means for the Characterization of Crystalline Polymers. Advances in Polymer Science 67. Springer-Verlag Berlin, Heidelberg 1985.

INFRARED SPECTROSCOPY ON PET YARNS

H.M. Heuvel and R. Huisman*

*Akzo Research Laboratories, Arnhem, Fibre Research Department, The Netherlands

SYNOPSIS

An infrared technique is presented which provides quantitative information on the physical structure of PET yarns. Especially a more detailed picture of the molecular arrangement in the amorphous regions can thus be obtained.

1 WHY INFRARED ON PET YARNS ?

The physical structure of yarns of semi-crystalline polymers can be described adequately by a two-phase model consisting of alternating amorphous and crystalline regions as illustrated in fig. 1. The crystalline regions can be studied in detail using x-ray diffraction techniques. Combinations of overall quantities such as density and sonic modulus with these crystalline data provides information on the amorphous regions. This information, however, is confined to a kind of average parameters such as the average orientation of the molecules in the amorphous regions, but does not give any details. To understand the mechanical properties, a detailed insight into the arrangement of the amorphous regions is needed. Figure 2 shows two features which are thought to be of major importance with respect to mechanical properties. Re-entry of the molecules at the boundaries of the crystals causes loss of coherence between the molecules and is therefore unfavourable for strength and modulus. The other feature is the chain length distribution of the tie-molecules. If this distribution is broad, the molecules will not co-operate in bearing a load exerted on the yarn. Successive breakage or slippage of one molecule after the other will take place, resulting in low strength and low modulus.

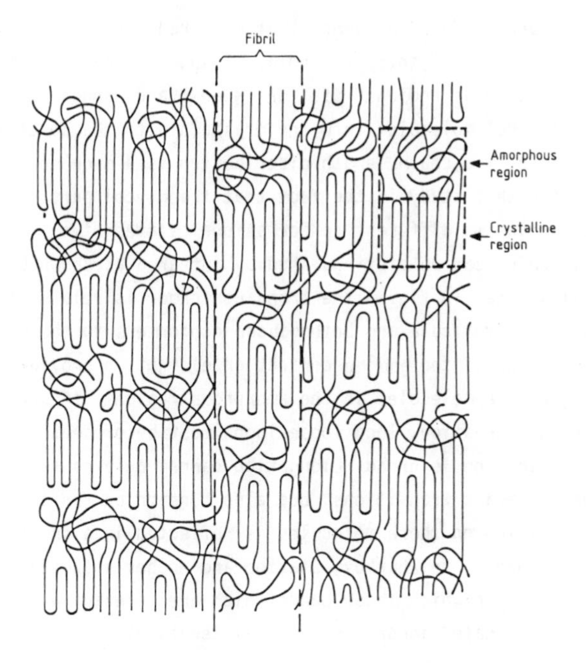

Fig. 1. Two-phase model of a semi-crystalline yarn.

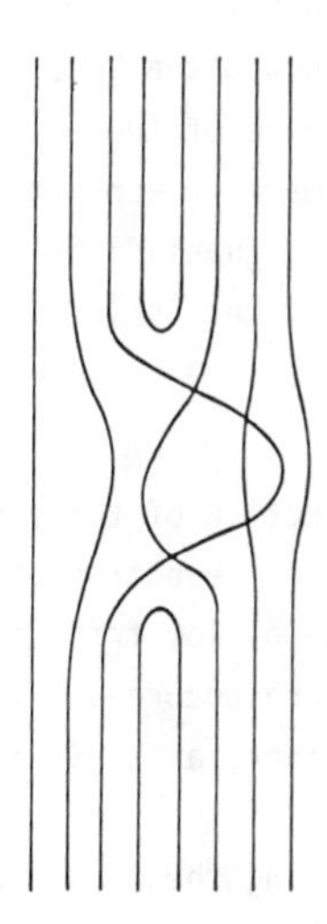

Fig. 2. Details of the amorphous regions.

The literature produced evidence that infrared spectroscopy is able to provide information on re-entry (or folding) and tie-chain length distribution in the amorphous regions. Statton, Koenig and Hannon (1) associated the re-entry of the PET molecules into the crystals with the so-called fold band at 989 cm^{-1}. The tie-chain length distribution is a much more complicated aspect. Helpful for this study is the fact that the glycol part of the PET molecule can occur in two different conformational states, i.e. trans and gauche (or cis), which can be distinguished by infrared as shown by Ward et al (2-6). As illustrated in fig. 3 the trans conformer is related to the straight parts of the molecule and both gauche conformers to the bended parts. Therefore the reaction of the gauche content on elongation provides information on the uncoiling of the molecules in the amorphous regions. If uncoiling hardly takes place during elongation, long loose loops must be present and the tie-chain length distribution is broad. Elongation cannot only cause uncoiling, indicated by gauche-trans transitions, but also tension on the taut tie-molecules. The increase of tension on these molecules is reflected in a shift of the 972 cm^{-1} trans band according to Zhurkov, Statton and Mocherla (7-12). In fig. 4 a PET spectrum is given in which those bands which are relevant with respect to the foregoing are indicated according to the literature.

2 THE APPROACH FOR QUANTITATIVE INFORMATION ON PET YARNS

As the literature clearly showed that the infrared technique is capable of providing a better detailed picture of the molecular arrangement in the amorphous regions, an attempt was made to apply this technique for obtaining a quantitative analysis of yarns. Both quantification and working with yarn samples required a special approach. Here only some remarks on these two points are made.

2.1 Yarns

Due to the circular cross-section of the yarns much scattering of the infrared radiation occurs. Therefore a spectrophotometer is required with excellent sensitivity in the region of low transmittance. To obtain also information on orientation, spectra are recorded with the infrared beam polarized parallel and perpendicular to the fibre axis. So for each yarn two spectra are scanned.

Samples are prepared by winding the yarns very carefully in such a way that on both sides of a small metal frame a perfectly smooth layer with the

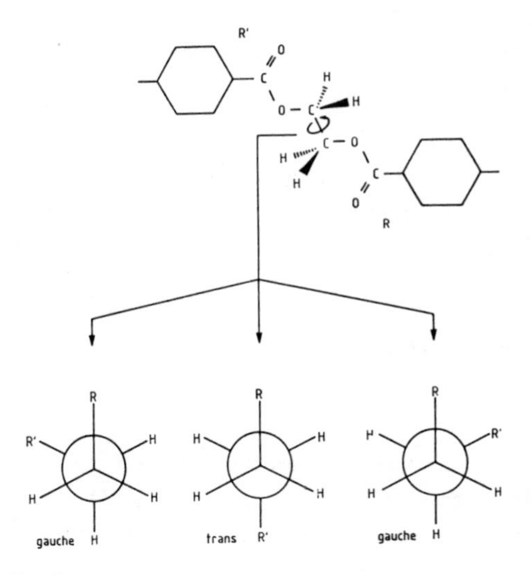

Fig. 3. Trans and gauche conformers of the PET molecule.

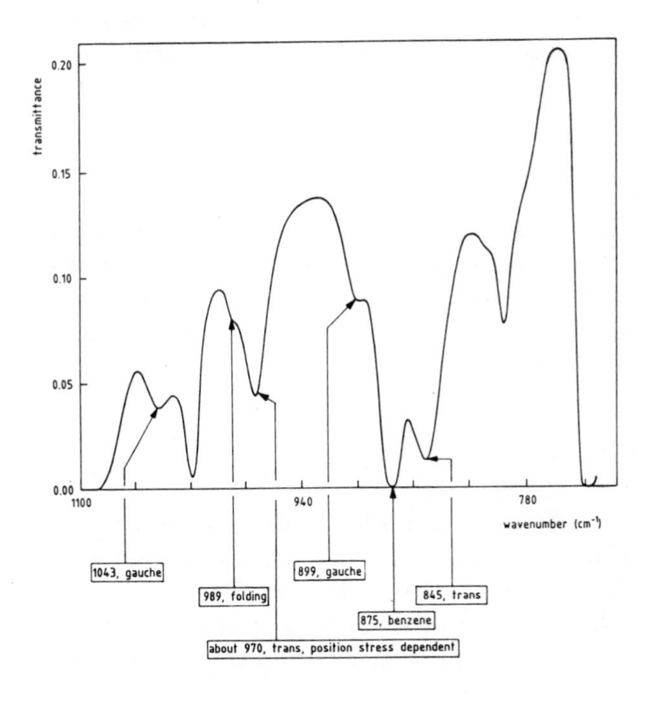

Fig. 4. Some band assignments according to the literature.

thickness of one monofilament is formed. Yarns can be wound at constant length or elongated to the extent required.

2.2 Quantification

The parallel and the perpendicular spectra are fitted simultaneously with a computer model applicable to yarns with quite different structures. The transmittance T is calculated as

$$T = 10^{-A} \tag{1}$$

where the absorbance A is described as

$$A = p + q\sigma + \sum_{i=1}^{n} P_i\ (\sigma) \tag{2}$$

where σ is the wavenumber, $p + q\sigma$ the equation used for the base line, $P\ (\sigma)$ the symmetrical bell-shaped Pearson VII function (13, 14) and n the number of bands needed for fitting the spectra. In fig. 5 an example of a fit for a PET tyre yarn is presented.

To acquire useful information on the relative amounts of the various molecular arrangements involved, the spectra have to be normalized on an internal reference banc which takes into account the mass of sample in the infrared beam. To this end use is made of the 875 cm^{-1} benzene band. The procedure developed turned out to be fully reliable.

As it is the aim of this paper to show how infrared can be used to study physical molecular arrangements in the amorphous regions of yarns and to demonstrate the relation of this structure with process conditions and mechanical properties, no further details about the development of this infrared technique are given. A detailed description will shortly be published elsewhere (15). Here only the main results essential for a better understanding of the physical yarn structures will be discussed.

3 MAIN FEATURES OF THE MODEL DEVELOPED

For all bands the isotropic areas, being proportional to the number of molecules active in the vibration transition involved, are calculated with the computer program mentioned before. Also all dichroisms, measures of the orientation of those parts of the molecules which are involved in the vibrations concerned, are calculated. The bands which are of particular interest for the physical interpretation of the results are indicated in fig. 6. For a few bands, however, some further comments are needed.

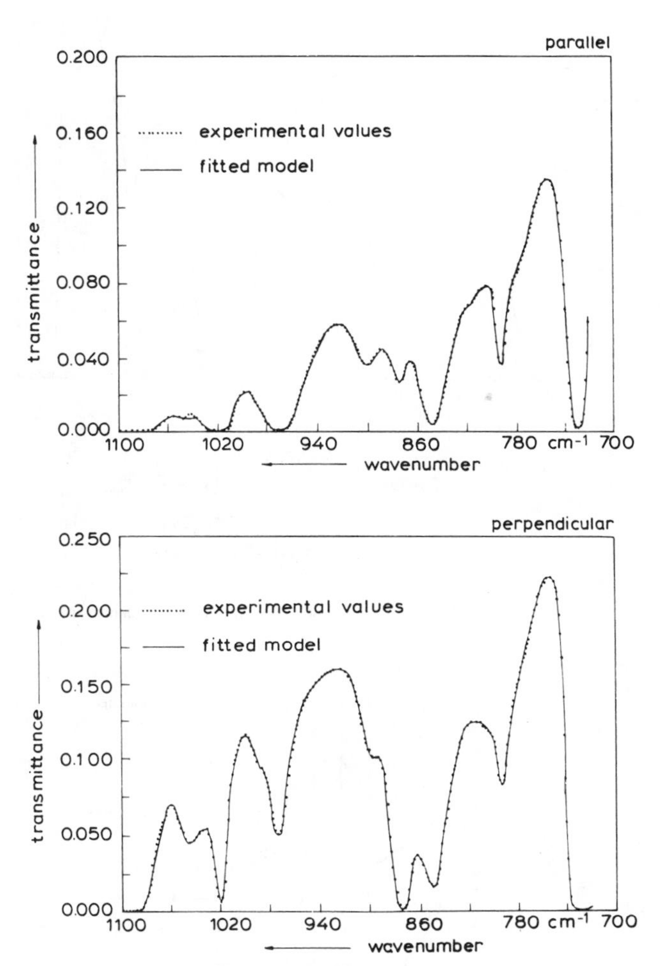

Fig. 5. Fitting results of a parallel and a perpendicular spectrum of a PET yarn.

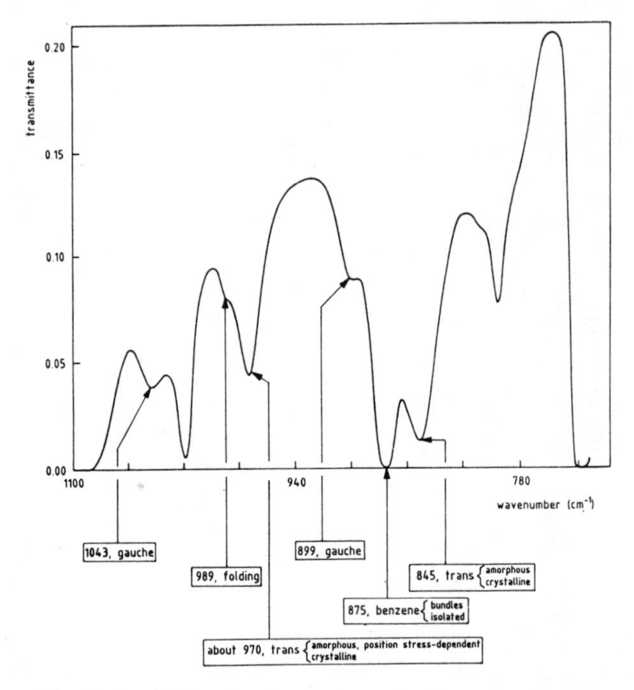

Fig. 6. Some band assignments supplemented with our own results.

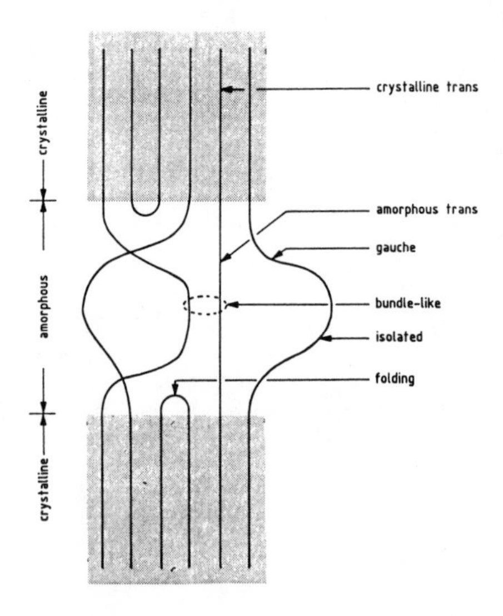

Fig. 7. Various molecular arrangements in the two-phase model.

These comments are presented below together with an illustration of some molecular arrangements shown in fig. 7.

3.1 Trans, 972 cm^{-1}

This band proved to be composed of two components, one relating to the amorphous parts of the molecules, the other to the crystalline parts. Both peak positions shift with increasing tension to lower wavenumbers but the shift of the amorphous band is larger.

3.2 Benzene, 875 cm^{-1}

This band turned also out to be composed of two contributions. In this case one is related to bundle-like aggregates (crystals as well as aggregates too small to be detected by x-ray), the other to more isolated molecules.

The orientation of both individual components provides interesting information. These orientation factors can be added, using a special weighing factor, to give a good overall orientation factor.

Indications were found that the small bundles, detected by infrared in x-ray-amorphous yarns can serve as nuclei for later crystallization upon heating.

3.3 Trans, 845 cm^{-1}

Here, too, there are two components, a crystalline and an amorphous one. This band is most suitable for obtaining crystallinities.

4 SUMMARY OF THE INFORMATION WHICH CAN BE OBTAINED

4.1 Single measurement

The performance of one single measurement, parallel and perpendicular, on a yarn wound at constant length gives the following information:

4.1.1 Crystallinity: this information is not unique, as it can also be determined by a combination of x-ray diffraction and density.

4.1.2 Folding or Re-entry: unique information.

4.1.3 Coiling: through the amount of gauche, unique information.

4.1.4 Association: one of the 875 benzene bands, unique information.

4.1.5 Orientation: on the 875 benzene bands. Unique information for the separate contributions of associated bundles and isolated molecules. An overall measure of the orientation can be obtained as the weighed sum of both components.

4.2 Multiple measurement

By carrying out measurements on various samples, prepared by winding a yarn at a variety of elongations, the following unique additional information is obtained:

4.2.1 Tie-chain length distribution: by means of uncoiling (amount of gauche) of the molecules upon straining.

4.2.2 Stress on taut tie-molecules: by means of the shift of the 972 cm^{-1} trans peak position.

5 APPLICATIONS

Industry has shown much interest in the effect of the winding speed on the physical structure of PET yarns. Usually yarns are made in a two-step process. The first step is the spinning process in which the molten polymer is pressed through spinneret holes and the threads are wound, after cooling in a blowbox, on a spool some metres below the spinneret. The winding speed is the running parameter of the examples to be presented. In the second step the undrawn yarns have to be drawn to achieve acceptable mechanical properties. In two applications the effect of the winding speed on the physical structure of both undrawn and drawn yarns is shown.

5.1 Undrawn yarns

In fig. 8a the amount of crystalline material, measured as the crystalline component of the 845 cm^{-1} trans band, is presented as a function of the spinning speed. This result coincides with that obtained by the combination of x-ray diffraction and density measurements. Figure 8b shows that the re-entry of the molecules starts as soon as crystallization has commenced.

Figure 9 presents the decrease of the gauche conformer content as obtained from the isotropic absorbance at 899 cm^{-1} with spinning speed. Apparently the stress in the spinning line is high enough to cause partial uncoiling of molecules to an extent which increases with winding speed. It can be seen that at about 3000 m/min, the speed at which crystallization starts, already an appreciable straightening of the molecules has taken place. This straightening favours the association of molecules into bundles, either in crystals or in smaller bundles, as shown in fig. 10, where the fraction of associated molecules is presented as a function of winding speed.

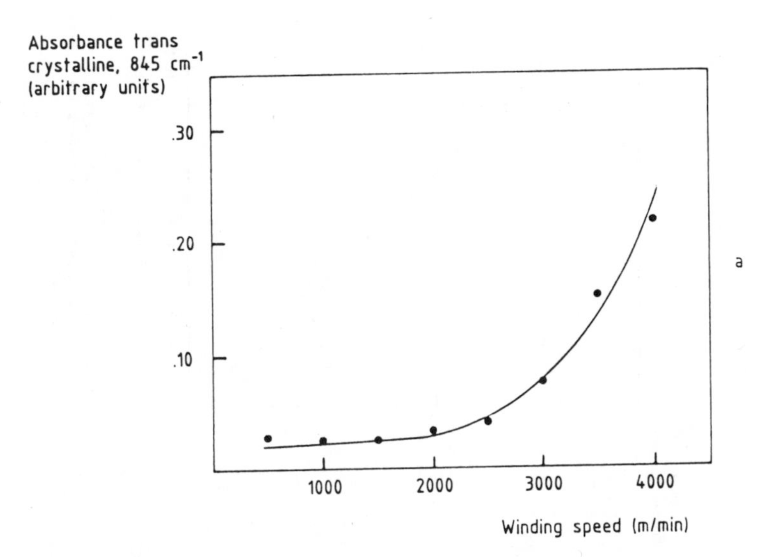

Fig. 8a. Area of the crystalline trans band as a function of winding speed for undrawn PET yarns.

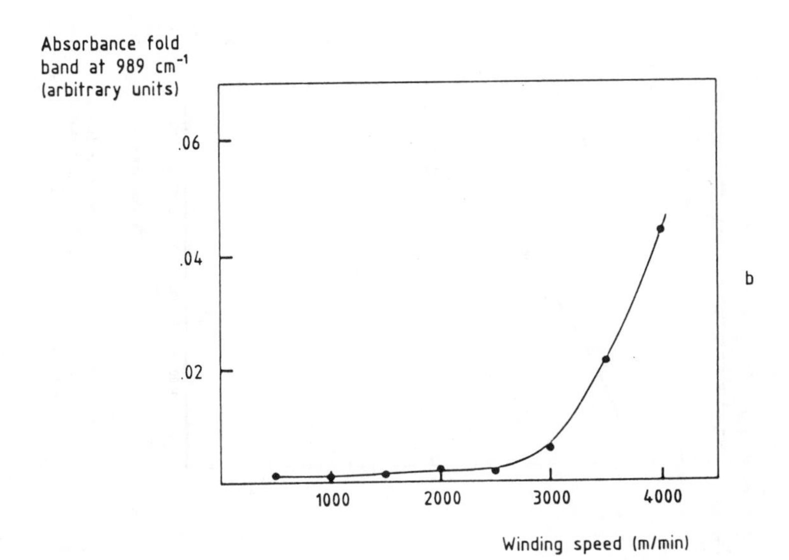

Fig. 8b. Area of the fold band as a function of winding speed for undrawn PET yarns.

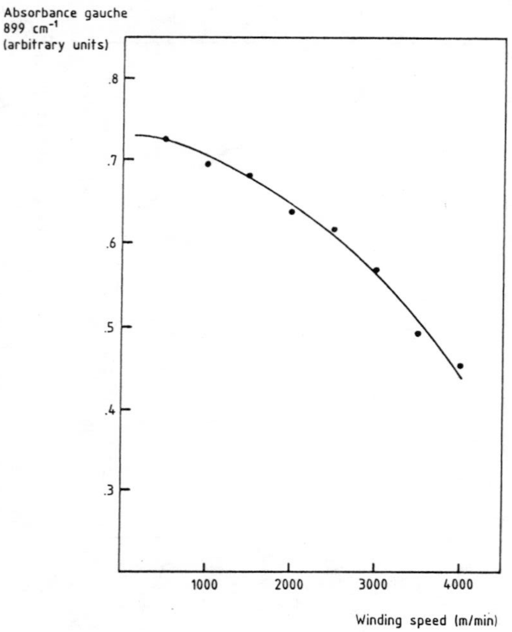

Fig. 9. Area of the gauche band as a function of the winding speed for undrawn PET yarns.

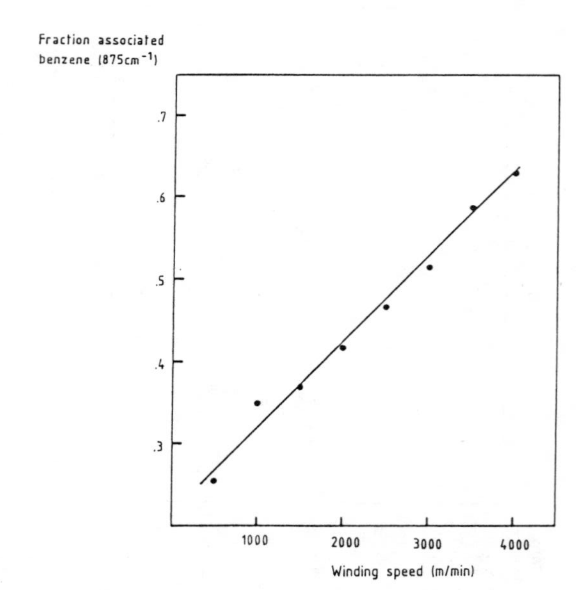

Fig. 10. Fraction associated benzene as a function of winding speed for undrawn PET yarns.

In fig. 11 the orientation of the associated parts of the molecules is given as a function of the winding speed together with the orientation of the isolated parts. Apparently at low speeds the associated as well as the isolated parts of the molecules are nearly randomly oriented. At an increase of the spinning speed the molecules are more and more associated in bundles which become better and better aligned with respect to the fibre axis, whereas the remaining isolated parts, the loose loops, are found to stay unoriented or to orient in the direction perpendicular to this axis.

From these results it follows that in the process studied first uncoiling, orientation and association of molecules increase with winding speed followed by crystallization at speeds higher than 3000 m/min.

5.2 Drawn yarns

Three yarns, spun at 500, 2000 and 4000 m/min have been drawn to the same elongation at break and compared with respect to their physical molecular structure as well as to their mechanical properties. Fig. 12a reveals that single measurements indicate that the yarn spun at the highest speed is more crystalline, whereas fig. 12b shows that this yarn has also more re-entry of molecules at the boundaries of the crystals.

Figs. 13 and 14 present results of multiple infrared measurements. The decrease of the gauche content in fig. 13 shows that uncoiling takes place to a higher extent in the yarn spun at 500 m/min, indicating that this yarn has a relatively narrow tie-chain length distribution. From fig. 14 it will be seen that the shift of the peak position of the 972 cm^{-1} trans amorphous band per per cent of elongation is about the same for all three yarns. However, the level of the high speed spun yarn is lower, indicating that the taut tie molecules in that yarn have already been stressed in the non-elongated situation.

Combining these results a picture of the physical structures of two drawn yarns, originally wound at different speeds, can be made as given in fig. 15. This result can account for some significant mechanical differences: Yarns spun at higher speeds have:

a lower strength (much folding, broad tie-chain length distribution);
a higher initial modulus (frozen-in stress on taut tie molecules)

and a flag at the end of the stress-strain curve (broad tie-chain length distribution.

So it has been demonstrated how this new analysis can contribute to the interpretation of differences in mechanical yarn properties by providing a better insight into the amorphous regions.

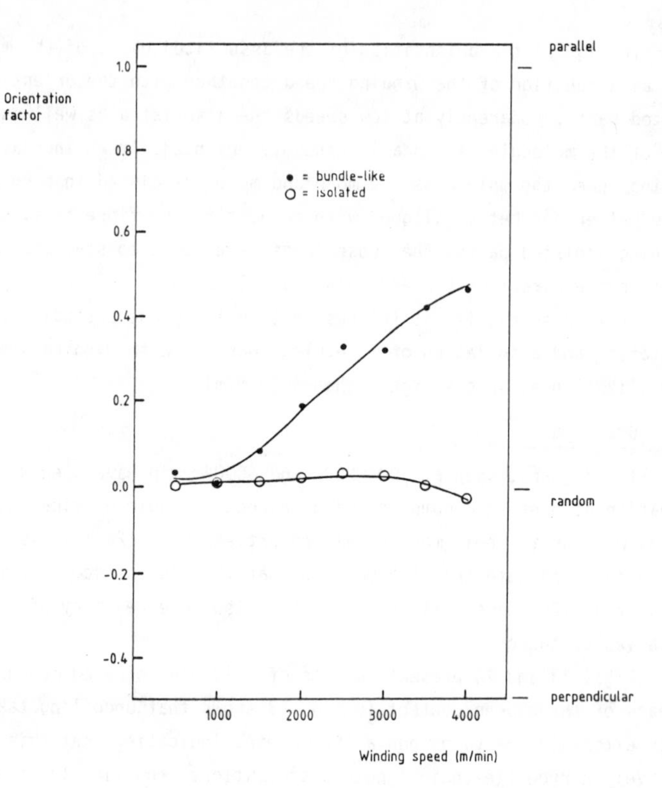

Fig. 11. Orientation with respect to the fibre axis versus winding speed for undrawn PET yarns.

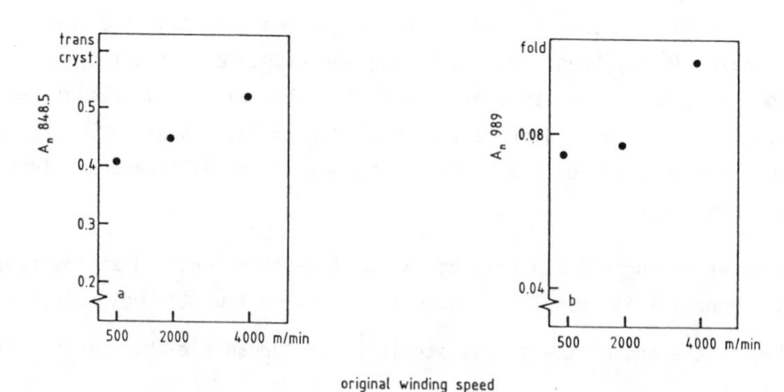

Fig. 12a. Area of the crystalline trans band for drawn PET yarns as a function of the original winding speed.

Fig. 12b. Area of the fold band for drawn PET yarns as a function of the original winding speed.

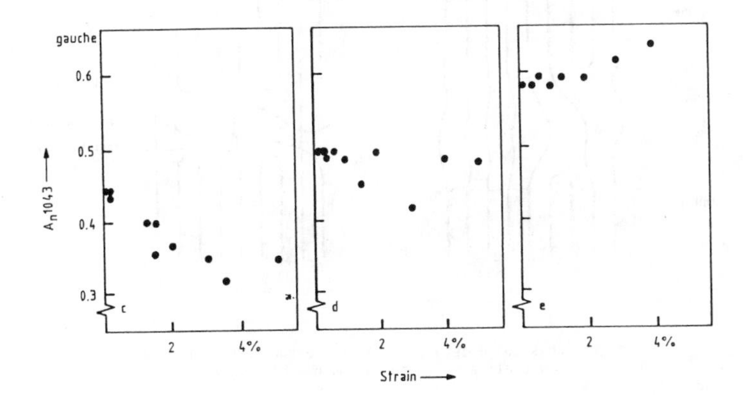

Fig. 13. Area of the gauche band as a function of strain of the sample. Drawn yarns, originally wound at: c = 500 m/min, d = 2000 m/min and e = 4000 m/min.

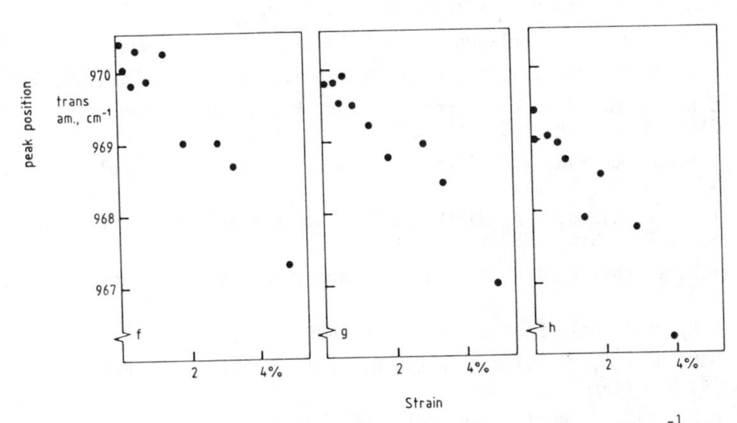

Fig. 14. Peak position of the trans amorphous band near 970 cm^{-1} as a function of sample strain. Drawn yarns, originally wound at: f = 500 m/min, g = 2000 m/min and h = 4000 m/min.

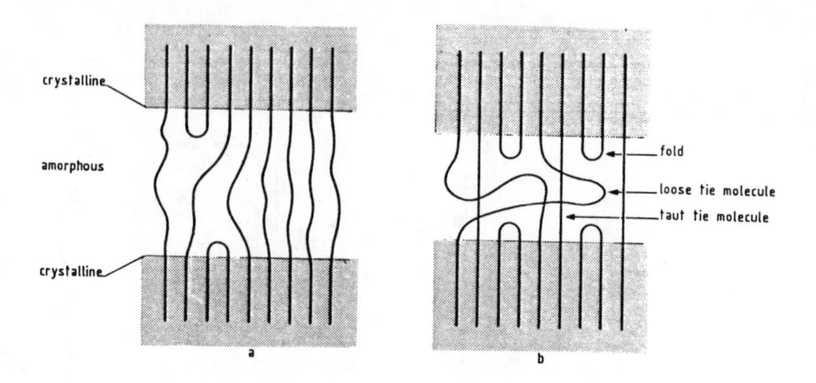

Fig. 15. Physical molecular model for drawn yarns, originally wound at: a = low speed and b = high speed.

REFERENCES

1. W.O. Statton, J.L. Koenig, M. Hannon, J. Appl. Phys., 4290 (1970).
2. Ward, I.M., Chem. Ind. 1956, p. 905.
3. Ward, I.M., Chem. Ind. 1957, p. 1102.
4. Grime, D and Ward, I.M., Trans. Faraday Soc. 54, 959 (1958).
5. W.W. Daniels and R.E. Kitson, J. Pol. Sci., 33, 161 (1958).
6. A. Garton, D.J. Carlsson and D.M. Wiles, Text. Res. J., 51, 28 (1981).
7. V.I. Vettegren, I.I. Novak, J. Pol. Sci., Pol. Phys. Ed., 11, 2135 (1973).
8. S.N. Zhurkov, V.I. Vettegren, I.I. Novak and K.N. Kashincheva, Dokl. Akad. Nauk USSR, 176, 623 (1967).
9. K.K.R. Mocherla, W.O. Statton, J. Appl. Polym. Sci., Appl. Pol. Symp. 31, 183 (1977).
10. K.J. Friedland, V.A. Marikhin, L.P. Myasnikova and V.I. Vettegren, J. Pol. Sci., Pol. Symp. 58, 158 (1977).
11. V.M. Voroboyev, I.V. Razumovskaya and V.I. Vettegren, Polymer, 19, 1267 (1978).
12. R.P. Wool, Pol. Eng. and Sci., 20, 805 (1980).
13. H.M. Heuvel, R. Huisman, K.C.J.B. Lind, J. Pol. Sci., Polym. Phys. Ed., 14, 921 (1976).
14. M.M. Hall, J. Appl. Cryst., 10, 66 (1977).
15. H.M. Heuvel, R. Huisman, to be published in J. Appl. Polym. Sci.

INTERACTION BETWEEN CRYSTALLIZATION AND ORIENTATION

H.G. ZACHMANN and R. GEHRKE

Institut für Technische und Makromolekulare Chemie, Universität Hamburg

SYNOPSIS

Investigations on polyethylene terephthalate have shown that, at the same temperature, the amorphous oriented material crystallizes much faster than the unoriented one. However, the densities obtained at the end of crystallization are for both kinds of samples the same. The degree of crystallinity of the oriented material was determined by extending Ruland's method. By comparison of the X-ray results with the densities, conclusions are drawn on the density of the non-crystalline regions.

Using synchrotron radiation, the rate of partial melting and recrystallization has been studied by wide angle and small angle X-ray scattering. The oriented polyethylene terephthalate melts more slowly and crystallizes more rapidly than the unoriented one. It is concluded that crystal thickening proceeds by melting rather than by chain diffusion.

If polyamide-6 is oriented before crystallization, a γ-modification is formed which is thermally more stable and exhibits a 002 crystal reflection of much higher intensity than a material first crystallized and afterwards oriented. Orientation before crystallization is obtained not only by high speed spinning but also by normal spinning followed by quenching and drawing.

1 INTRODUCTION

Crystallization and orientation interact with each other in many different respects. Such interactions are not only of scientific interest but also of technological importance. For example, as it is well-known, by special kinds of deformation high strength materials are obtained.

When we discuss these interactions we have to mention first the stress induced (or orientation induced) crystallization known from some rubber-like materials[1,2]. Such materials crystallize when they are extended by applying stress; after the stress is removed, they melt. This effect is due to an increase of the melting point caused by a decrease of the entropy of melting. In the unoriented state, these materials are not able to crystallize.

With materials which do crystallize in the unoriented state, another effect can be observed. Such materials can be oriented prior to crystallization either by drawing in the glassy state or by elongation during extrusion ("fast spinning") or by a shear gradient in the melt. It has been found that the orientation may influence the rate of the subsequent crystallization[3,4,5] as well as the final degree of crystallinity[5], the morphology[6,7], and the crystal modification[8] obtained. Also in this case, the crystallization after orientation sometimes is referred to as "orientation induced" crystallization. However, this term is misleading because, in contrast to the case mentioned first, stable crystals may be formed also without orientation of the chains; the orientation only influences the course of crystallization.

A third kind of interaction between crystallization and orientation is observed if the material is oriented after it has been crystallized. The maximum orientation one can reach as well as the mechanical properties and the morphology obtained depend strongly on the conditions of orientation[9,10]. In addition, the treatment prior to deformation is important. By reducing the number of entanglements by gel spinning[11] or by a special melt extrusion process[12] the maximum drawing ratio can be increased markedly.

In the following some results are discussed concerning the influence of orientation on the kinetics of phase transitions, on the final degree of crystallinity and on the crystal modification obtained. In addition, the influence of crystalline order on the relation between draw ratio and orientation is treated.

2. INFLUENCE OF ORIENTATION ON THE KINETICS OF PHASE TRANSITIONS

The influence of orientation on the kinetics of crystallization was studied by Zachmann and Althen[5]. Amorphous polyethylene terephthalate was drawn at 92°C so that different values of birefringence were obtained. During the drawing process the samples remained almost amorphous. Afterwards, they were crystallized isothermally. Fig. 1 shows the half-time of crystallization τ as a function of the crystallization temperatures T_c. The parameter written at each curve is the birefringence before crystallization Δn_o. As one can see, the half-time of crystallization decreases markedly with increasing birefringence at any given temperature T_c. Similar results have been obtained by Smith and Steward[3] at higher temperatures of crystallization and by Alfonso et al.[4] for fibres oriented during extrusion by using high take up velocities. Under all conditions orientation of the chains increases the rate of crystallization.

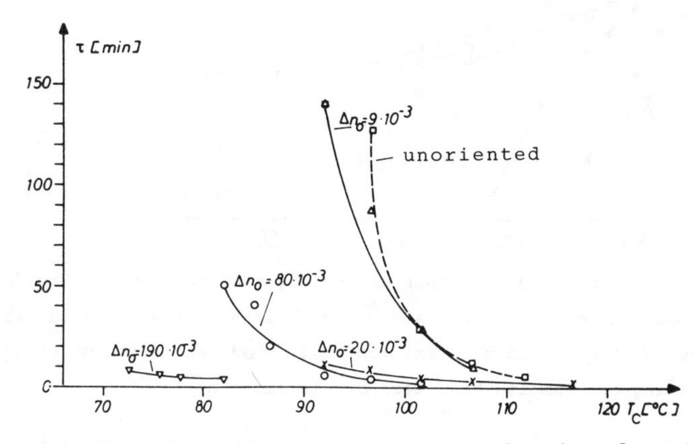

Fig. 1: Half time of crystallization τ as a function of crystallization temperature T_c for polyethylene terephthalate showing different values of birefringence Δn_o[5].

The opposite is true for melting. Gehrke et al.[13,14] have studied the melting process under isothermal conditions by placing a sample in the beam of synchrotron radiation, heating it within a few seconds to a temperature slightly above the melting point and following the change of small angle scattering with time. Fig. 2 shows scattering power Q which, according to equ. 1, is a measure for the degree of crystallinity as a function of melting time for oriented and unoriented samples. Compared at the same temperature, for the melting of the oriented sample a longer time is needed than for that of the unoriented one. The delay in melting of the oriented sample is due to a free enthalpy of activation which has to be overcome. This effect has been predicted earlier[15,16] as a consequence of entropy effects.

If a sample of polyethylene terephthalate is crystallized at a comparatively low temperature (130°C) and is annealed afterwards at higher temperatures, melting and recrystallization occur[17]. Gehrke and Zachmann[18] have studied this processes by following the small angle scattering using synchrotron radiation. Fig. 3 shows the change of small angle scattering after heating a sample, previously crystallized at a lower temperature, to 250°C. A scattering

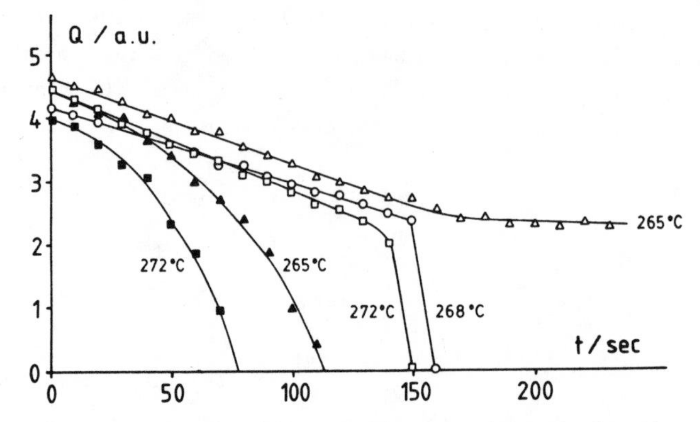

Fig. 2: Scattering power Q, in arbitrary units,as a function of time during melting of polyethylene terephthalate[14]. The melting temperature is written at each curve. □○△ oriented material (Δn_o=0,19), ■▲ unoriented material. Prior to melting the samples have been crystallized at 255°C .

curve was taken every 20 seconds. The large peak at the right side of each pattern is the primary beam. At the beginning of the experiment a small angle peak with comparatively low intensity is observed. During the annealing the scattering curve changes. We have evaluated two quantities: The long period L, obtained from the position of the maximum, and the scattering power Q, that is the total integrated intensity. As it is well-known, the scattering power is determined by the volume of the scattering material V, the degree of crystallinity within the scattering regions (spherulites) w_{cs} and the density difference between crystalline and non-crystalline regions $\rho_c-\rho_a$

$$Q = \text{const} \cdot V \cdot w_{cs} \, (1-w_{cs})(\rho_c-\rho_a)^2 \qquad (1)$$

At the bottom of the figure we see the change of the temperature and that of the scattering power Q with time. After raising the temperature, Q first increases due to a change in the density difference $\rho_c-\rho_a$ which is caused by the different thermal expansion of the crystalline and the amorphous regions. This increase is followed by a decrease of Q caused by partial melting. Next an increase of Q which is due to recrystallization is observed.

Fig. 4 shows the change of the degree of crystallinity during annealing of different samples previously crystallized at 120°C. The values were obtained from change of Q after correction for the thermal expansion. The temperature

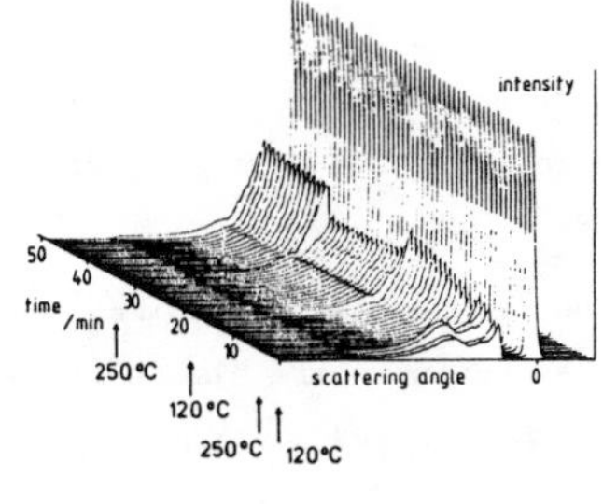

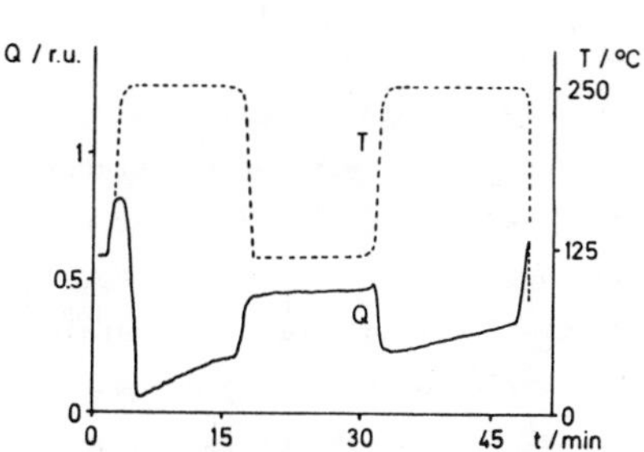

Fig. 3: Change of small angle X-ray scattering of polyethylene terephthalate after stepwise heating and cooling[19]. The sample was crystallized previously at 125°C. Above: scattering diagrams. Below: time dependence of temperature T and scattering power Q.

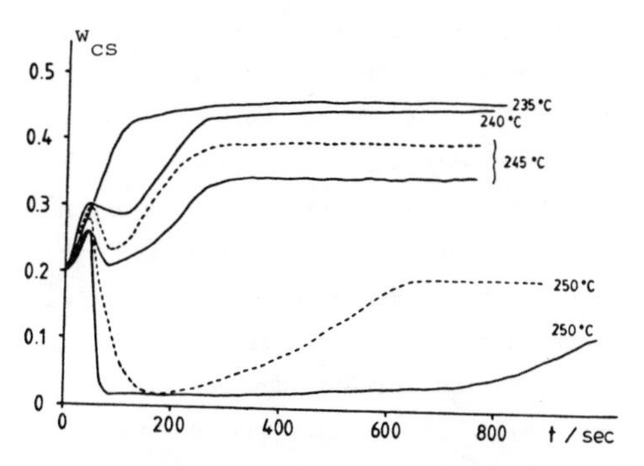

Fig. 4: Change of the degree of crystallinity w_{cs} with time during melting and recrystallization of polyethylene terephthalate[14] at the temperature written at each curve. The samples were previously crystallized at 120°C. ——— unoriented samples, ----- oriented samples.

written at each curve is the annealing temperature. The dotted curves refer to oriented, the full curves to unoriented samples. As one can see, with oriented samples, the partial melting proceeds more slowly than with unoriented ones, exactly as it is observed with total melting above the melting point. Moreover, in the oriented samples the crystallization proceeds more rapidly than in the unoriented ones, as in the case of crystallization from the completely amorphous state.

Obviously, orientation influences the rate of partial melting and recrystallization. From this it is concluded that, after stepwise increase of temperature, recrystallization takes place in such a way that crystal lamellae melt completely and crystallize again. This conclusion is also supported by the observed influence of the molecular weight on the rate of recrystallization as well as by studies of the changes in the shape of the scattering curve[14,20].

3. INFLUENCE OF ORIENTATION ON THE DEGREE OF CRYSTALLINITY

As has been shown above, the orientation prior to crystallization increases the rate of crystallization. Does it also increase the degree of crystallinity obtained at the end of crystallization?

Fig. 5 shows the densities of the polyethylene terephthalate samples from fig. 2 crystallized to the end. One sees that these densities are almost independent of the birefringence before crystallization Δn_o. However, from this it does not follow that the degree of crystallinity is independent of Δn_o. Biangardi[23] has found that the density of amorphous polyethylene terephthalate increases with increasing birefringence Δn_o. The results are shown in fig. 6. The upper curve refers to samples which were oriented by drawing at temperatures slightly above the glass transition temperature, the lower curve to fibers extruded with different take up velocities. The larger values for the samples oriented by drawing may arise partly from some small crystals or paracrystals formed during drawing. However, in any case the density of the oriented amorphous material is larger than that of the unoriented one. To what extent is the density of the noncrystalline regions in the oriented semicrystalline material increased?

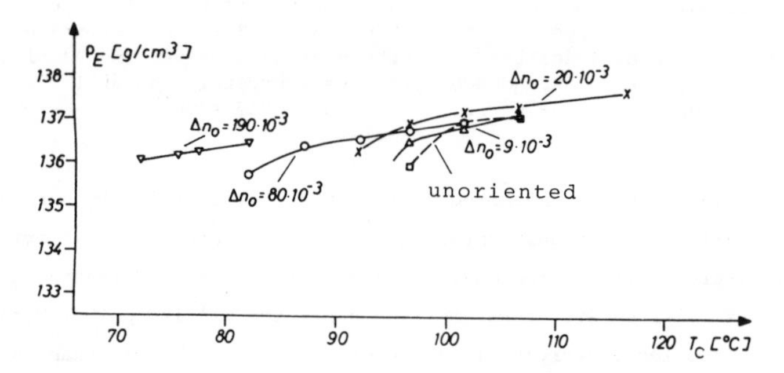

Fig. 5: Density of polyethylene terephthalate obtained at the end of isothermal crystallization as a function of the crystallization temperature T_c[5]. Δn_o is the birefringence of the sample before crystallization.

Correct results on the degree of crystallinity of oriented polymers are obtained by X-ray wide angle scattering measurements. Gehrke and Zachmann[22] have performed such measurements by using a device that permits to obtain the scattering at different azimuthal angles and to average over all angles. Fig. 7 shows the wide angle scattering diagram of an unoriented sample and that of an oriented

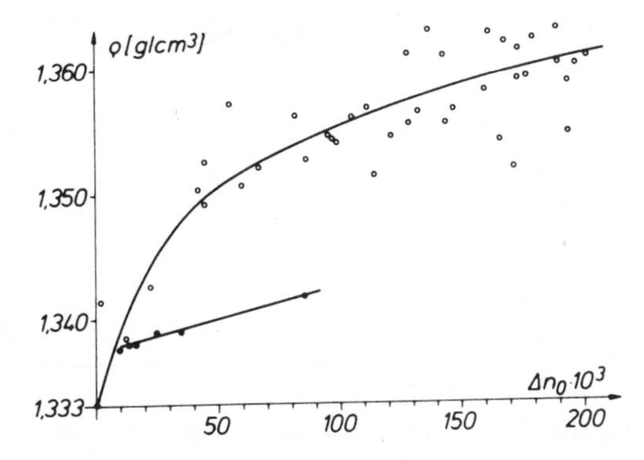

Fig. 6: Density of oriented amorphous polyethylene terephthalate as a function of the birefringence Δn_o[23]. -o- samples oriented by drawing above the glass transition temperature, -•-samples oriented by fast spinning.

sample crystallized at the same temperature and averaged over all azimuthal angles. The degrees of crystallinity and the densities of the samples are given in the figure. From the results it can be concluded that the density of the non-crystalline regions in the oriented semicrystalline sample is given by the lower curve in fig. 6 rather than by the upper one. In addition, it follows that the orientation of the material prior to crystallization does not increase the degree of crystallinity obtained at the end of the crystallization process.

4. INFLUENCE OF ORIENTATION ON THE CRYSTAL MODIFICATION

Polyamide-6 crystallizes in a so-called α-modification and in a γ-modification. The γ-modification is obtained mainly at low crystallization temperatures and in fibers which have been spun with high take up velocity[24,25]. Recently, Tidick and Zachmann[8] have shown that the formation of the γ-phase is generally enhanced by an orientation of the material prior to crystallization. The fiber was extruded with a comparatively small take up velocity so that no orientation ocurred. After extrusion the fiber was quenched and remained almost completely amorphous. Then it was drawn at room temperature to a draw ratio of $\lambda = 4.6$. The birefringence obtained was $\Delta n_o = 30 \cdot 10^{-3}$. At last the fiber was annealed at 150°C for 10 min, cooled down to room temperature, annealed again at 170°C, cooled down to room temperature and so on. Fig. 8 shows the scattering patterns obtained at room temperature after each step of annealing. Before annealing (upper row in

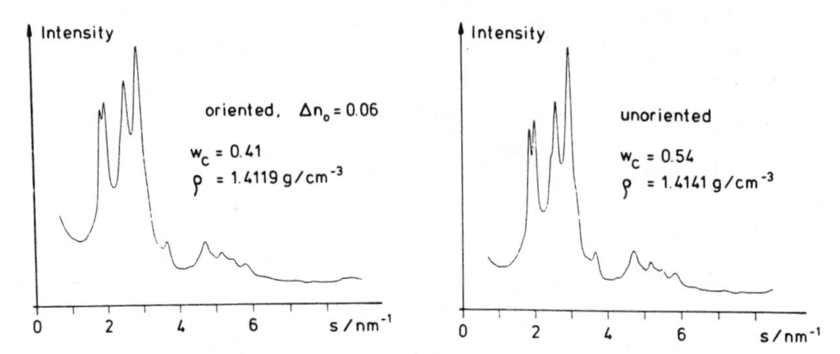

Fig. 7: X-ray wide angle scattering diagram of oriented and unoriented polyethylene terephthalate crystallized at 240°C for 3 hours[22]. ρ is the density, w_c the degree of crystallinity obtained from the scattering diagram.

fig. 8a) one sees mainly an amorphous halo with a small 002-reflection of the γ-phase. After annealing at 150°C (second row) the intensity of the 002-reflection has increased markedly. The intensity increases further at higher annealing temperature up to 210°C. From this temperature on the intensity decreases and the γ- 002 -reflection disappears completely at 225°C.

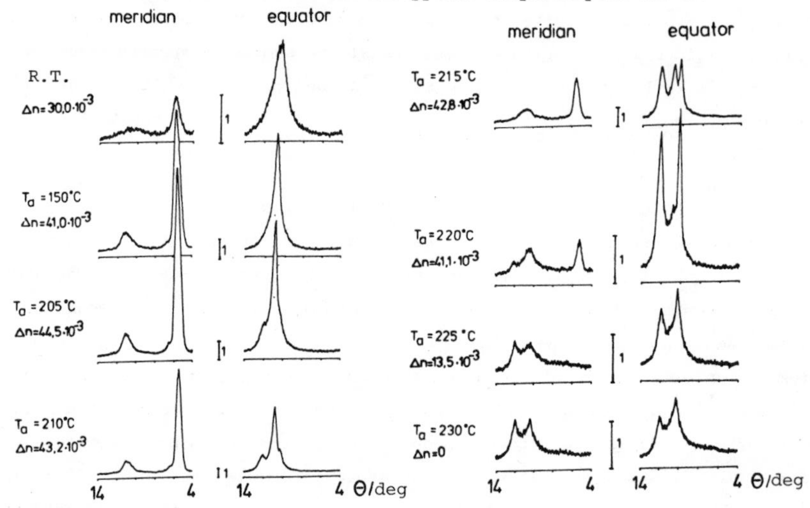

Fig. 8: Wide angle X-ray scattering of Polyamid-6 at the meridian and at the equator measured at room temperature after subsequent annealing at the temperatures T_a[8]. Thermal pretreatment: the samples were extruded, quenched and drawn ($\Delta n_o = 30\cdot 10^{-3}$).

It is important to mention that the γ-modification formed in unoriented materials disappears at 215°C while that formed in the material oriented by fast spinning disappears at 220°C. This means that an orientation of the material in the amorphous phase prior to crystallization leads not only to the highest amount of γ-modification but also to the thermally most stable form of this modification.

The results obtained show also that it makes a great difference if the material is first crystallized and then oriented or if it is first oriented and then crystallized. If polyamide is first crystallized and then oriented one obtains the usual mixture of α- and γ-modifications as in the unoriented material.

5. INFLUENCE OF CRYSTALLINE ORDER ON THE RELATION BETWEEN DRAW RATIO λ AND BIREFRINGENCE Δn

If amorphous polyethylene terephthalate is drawn without necking at elevated temperatures, the birefringence obtained depends not only on the draw ratio λ but also on the temperature and rate of drawing (see fig. 9). The smaller the rate of drawing and the higher the temperature, the smaller the orientation obtained[26]. This is due to the fact that the network points which fix certain points of the chains and induce orientation upon drawing are formed by chain entanglements. With increasing temperature and decreasing drawing rate these entanglements become less and less effective in fixing the chains. The birefringence obtained is also dependent on the molecular weight[27]. The lower the molecular weight, the smaller the birefringence. The largest birefringence is obtained if one chooses the lowest temperature and the highest drawing rate at which no necking occurs. The corresponding results are represented by the curve which is designated by max in fig. 9.

The situation is completely different for crystalline polyethylene terephthalate[28]. The results obtained are shown in fig. 10. The temperature and the rate of drawing, the molecular weight and the catalyst content have been varied considerably. The points representing the results are lying in a quite good approximation on a single line. Moreover, this line is identical with the curve designated by max in fig. 9 for the amorphous samples. We think that this result can be explained by assuming that the amount of chain entanglements does not change during crystallization and that the chain entanglements are also the net work points during deformation of the crystalline material. The crystals which are disrupted and reorganized during the deformation process are not able to act as net work points; they only prevent the slipping of the chains. Therefore, no

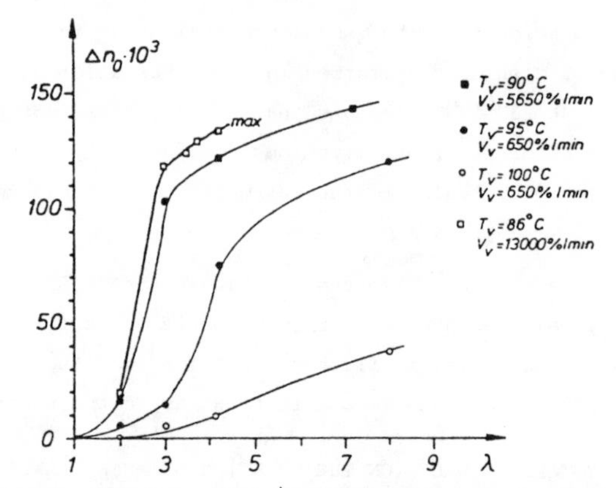

Fig. 9 : Birefringence Δn_o of amorphous polyethylene terephthalate[23] drawn at the temperature T_v with the rate V_v as a function of the draw ratio λ.

influence of the drawing conditions on the result is obtained.

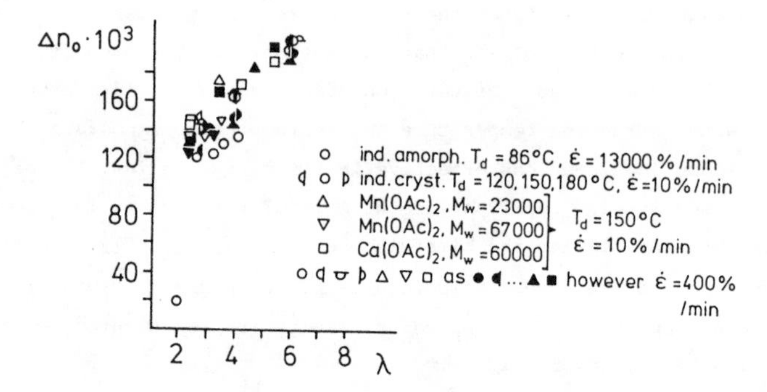

Fig. 10: Birefringence Δn_o of amorphous and crystalline polyethylene terephthalate as a function of the draw ratio λ for samples with different molar masses, different drawing temperatures and different drawing rates[28].

With drawing of crystalline polyethylene, Capaccio et al.[29] have found that the Youngmodulus only depends on the draw ratio and is not influenced by the other conditions of drawing and the molecular weight. As the Young modulus and the birefringence are, under usual conditions, related to each other, this result is in good agreement with the result of fig. 10. Obviously, for all crystalline polymers, within a wide range of drawing conditions, the draw ratio alone determines the orientation obtained, while for amorphous materials other parameters have to be considered.

5. DETERMINATION OF ENTANGLEMENTS

Entanglements play an important role in the orientation of polymers and therefore they influence the processibility and the mechanical properties. Unfortunately, there is no straightforward method to measure the amount of entanglements. Indirect methods are the plateau-modulus and nuclear magnetic resonance. According to Smith and Lemstra[11] it is also possible to get information about the amount of entanglements by measuring the maximum draw ratio which can be obtained. The results presented in figs. 9 and 10 show that there might exist another possible method: the measurement of the dependence of birefringence on draw ratio under conditions where the entanglements cannot be opened during drawing. This condition is fulfilled in amorphous material if drawing takes place at low temperatures and high rates. It is generally fulfilled in crystalline materials far enough below the melting point. The higher the birefringence at a given draw ratio, the larger the amount of entanglements.

REFERENCES

1. Gent, A.N.; J.Polym.Sci.Part A 3 1387 (1965)
2. Koch, M.H.J., J. Bordas, E. Schöla and H.Chr. Broecker; Polymer Bulletin 1, 709 (1979)
3. Smith, F.S. and R.D. Steward; Polymer 15, 283 (1974)
4. Alfonso, G.C., M.P. Verdona and A. Wasiak; Polymer 19 711 (1978)
5. Althen, G. and H.G. Zachmann; Makromol.Chem. 180, 2723 (1979)
6. Pennings, A.J., J.M.A.A. van der Mark and H.C. Booij; Kolloid Z. u. Z. Polymere 236, 99
7. Mackley, M.R. and A. Keller; Phil.Trans.Royal Soc. (London) 278, 29 (1975)
8. Tidick, P. and H.G. Zachmann, to be published, see also P. Tidick, Dissertation University of Hamburg 1985
9. Capaccio, G., A.G. Gibson and I.M. Ward in "Ultra-High Modulus Polymers", edited by A. Ciferri and I.M. Ward, Applied Science Publishers, 1979, p. 1
10. Zachariades, A.E., W.T. Mead and R.S. Porter in "Ultra-High Modulus Polymers", edited by A.Ciferri and I.M. Ward, Applied Science Publishers 1979, p. 77
11. Smith,P.,P.J. Lemstra and H.C. Booij; J.Polym.Sci., Polym.Phys.Ed. 19 , 877 (1981)
12. Bayer, R., to be published
13. W. Prieske, C. Riekel, M.H.J. Koch, H.G. Zachmann; Nucl.Instr. & Methods, 208, 435 (1983)
14. Gehrke, R., Dissertation, University of Hamburg 1985
15. Zachmann, H.G.; Kolloid-Z. u. Z.f. Polymere 206, 25-29 (1965)
16. Wunderlich, B.; "Macromolecular Physics", Vol.3, Academic Press 1980

17. Zachmann, H.G. and H.A. stuart; Makromol.Chem. 49, 148 (1960)

18. Gehrke, R. and H.G. Zachmann, to be published, see also ref. 14 and 19

19. Elsner, G., C. Riekel and H.G. Zachmann; Adv.Polymer Sci. 67 1 (1985)

20. Zachmann, H.G., D. Wiswe, R. Gehrke and C. Riekel; Makromolekulare Chem. Suppl. 12 (1985)

21. Zachmann, H.G., G. Elsner and H.J. Biangardi in "Rheology" Vol. 3, Edited by G. Astarita, G. Marrucci and L. Nicolais, Plenum Publishing Corporation 1980, p. 275

22. Gehrke, R. and H.G. Zachmann, to be published

23. Biangardi, H.J.; Habilitationsschrift, Berlin 1980

24. Reichle, A. and A. Prietzschk; Angew.Chem. 74, 562 (1962)

25. Miyasaka, K. and K. Ishikawa; J.Polymer Sci. Part A-2, Vol.6, 1317 (1968)

26. Biangardi, H.J. and H.G. Zachmann; Colloid and Polymer Sci., 62 71 (1977)

27. Günther, B. and H.G. Zachmann; Polymer 24 1008 (1983)

28. Zachmann, H.G. and B. Günther; Rheol.Acta 21, 427 (1982)

29. Capaccio, G., T. A. Crompton and I.M. Ward; J.Polym.Sci.Polym.Phys.Ed. 14, 1641 (1976)

Neutron Scattering of Poly(ethylene terephthalate)

J.W. Gilmer, D. Wiswe, and H.G. Zachmann, Institute for Technical and Macromolecular Chemistry, University of Hamburg

J. Kugler and E.W. Fischer, Institute for Physical Chemistry, University of Mainz

Through the use of deuterium labeling, small angle neutron scattering (SANS) has been found to be an extremely effective tool for elucidating the structure of polymer molecules in the condensed matter state. In this study, the molecular morphology of poly(ethylene terephthalate) (PET) has been determined under four different conditions: (1) in the glassy state, (2) in the semicrystalline state, (3) after cold-drawing, and (4) after undergoing partial transesterification.

Experimental

Poly(ethylene terephthalate) (PET) was synthesized in both its deuterated and nondeuterated form from ethylene glycol and dimethyl terephthalate. Deuterated and hydrogenated PET were then blended by codissolving them in hexafluoroisopropanol followed by reprecipitation in a methanol nonsolvent. The precipitate was then dried in a vacuum oven, melt-pressed under vacuum at either 280° C or 250° C and then quenched in ice water to prevent crystallization.

The samples prepared at 280° C were melt-pressed for different periods of time to investigate the transestiferication process which occurs at this temperature. As a result of this process, the H and D polymer molecules are transformed into H-D blockcopolymers. The block molecular weight and its radius of gyration are then obtainable from the neutron scattering profile.

Three sets of samples were prepared at 250° C where the transesterification process is greatly minimized. One of the three sets of samples prepared at this temperature received no further treatment. A second set of samples was then crystallized by annealing at 180° C for 5 minutes. A third set of samples was cold-drawn at 45° C, subsequent to pressing, to observe the effect of the necking process on PET.

Neutron scattering experiments were then performed on all samples with the D-11 and the D-17 at the ILL in Grenoble. The single chain scattering function P(q) and the weight average degree of polymerization Z were derived from the following expression for the absolute scattering intensity

$$R(q) = x\ (1-x)\ (a_H - a_D)^2\ N\ Z^2\ P(q)$$

where x is fraction of deuterated species present, a is the neutron scattering length per monomer repeat unit of either the deuterated (D) or the nondeuterated (H) species, N is the number of molecules per unit volume, and q is the scattering vector, proportional to the sine of the scattering angle. For samples which have undergone transesterification, the single chain scattering function obtained is that for a D or H block instead of that for the entire molecule.

The z average radius of gyration Rg is then obtained from the small q dependence of the single chain scattering function given by

$$P(q)^{-1} = 1 + (R_g^2\, q^2 / 3) + \ldots$$

The factor of 3 in the above equation drops out when only a component of the radius of gyration in a given direction is considered.

Results and Discussion

SANS results were obtained from samples melt-pressed for different lengths of time at 280° C. The amount of transesterification that had taken place during pressing was then determined by calculating the molecular weight from the extrapolated value of the neutron scattering intensity at zero q. The SANS determined molecular weight as a function of the annealing time at 280° C is presented in Fig. 1.

The noncrystalline and the semi-crystalline samples prepared at 250° C were investigated in both the small and intermediate angle regions. The scattering in the Guinier region from both sets of samples (Fig. 2 and 3) exhibited a linear Zimm plot with a slight increase in the radius of gyration upon crystallization from 6.5 nm. to 7.2 nm. The molecular weight values of just over 27,000 were obtained for both samples represent a 40% reduction in the block length due to the occurrence of transesterication during melt-pressing.

In the intermediate angle neutron scattering profiles from the crystallized homopolymer blanks, the presence of a long period of about 13 nm. was evident (Fig. 4). For both the noncrystalline and semi-crystalline samples, a Kratky plot presentation of the intermediate angle scattering intensity (Fig. 5) exhibited a plateau indicative of a coillike structure. (Although a polymer chain may be folded in a lamella, it is still possible that it exhibit Gaussian like behavior in certain ranges of q.) For the noncrystalline sample, a polydispersity value (M_w/M_n) of 1.96 was calculated, which is in good agreement with the anticipated value for PET of 2.0. For the semi-crystalline sample, the plateau region observed extended over a smaller range of q than that of the noncrystalline sample; the high q departure from the plateau behavior probably resulted from the rodlike character of any chain portions contained in a lamella. In addition, the height of the plateau was slightly lower for the crystallized sample, in agreement with the small angle data where a slightly expanded radius of gyration was observed.

The SANS was observed from PET samples (Fig. 6) which were cold-drawn at 45° C at 2 mm/min. A draw ratio of 4.9 was obtained for PET under these conditions. In Fig. 7, a comparison has been made of the SANS determined radius of gyration values with both that obtained for the unoriented sample and those predicted in the case where the polymer chain deforms in proportion to the macroscopic extension of the sample (the affine case). The failure of the PET molecules to deform affinely could easily result from factors such as chain slippage during draw.

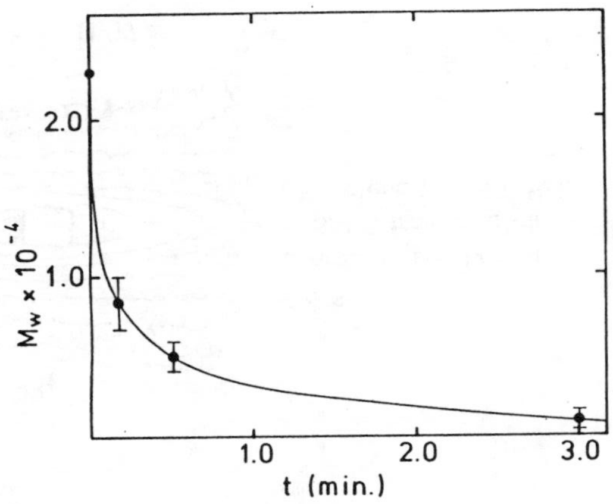

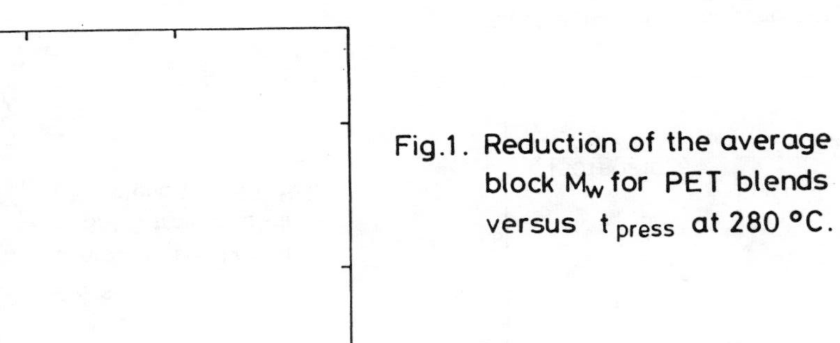

Fig.1. Reduction of the average block M_w for PET blends versus t_{press} at 280 °C.

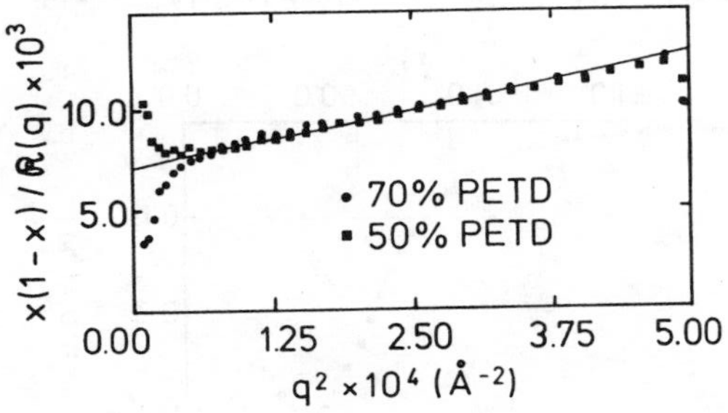

Fig. 2. Zimm plot of noncrystalline PET blends.

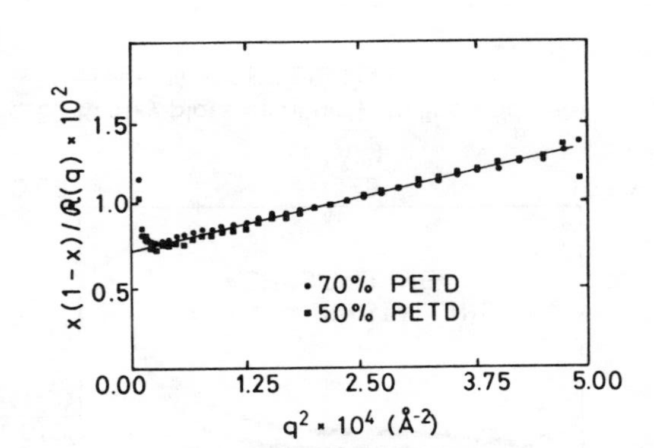

Fig. 3. Zimm plot of PET blends after crystallization (300 s, 180 °C).

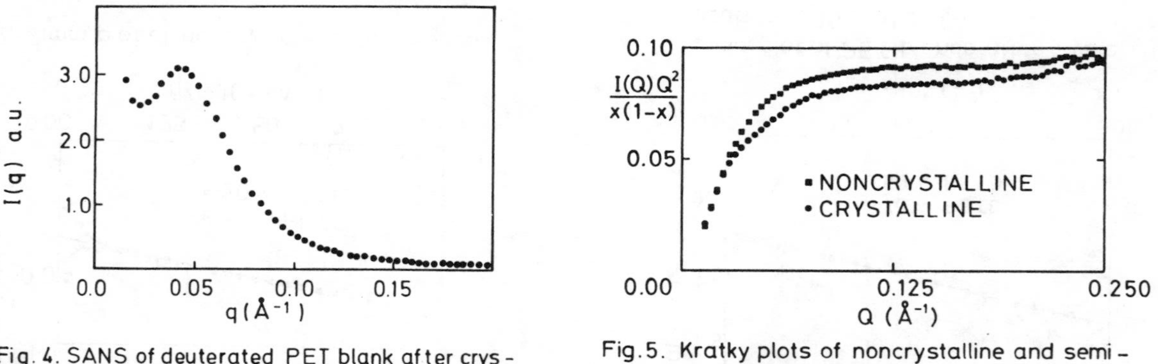

Fig. 4. SANS of deuterated PET blank after crystallization (300 s, 180 °C).

Fig. 5. Kratky plots of noncrystalline and semicrystalline PET blends.

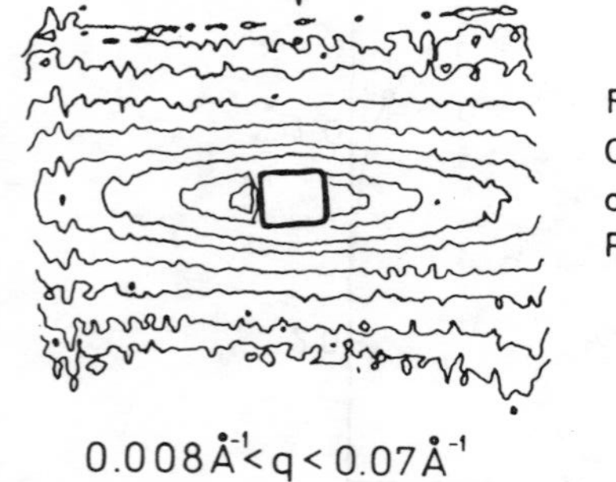

Fig. 6.
Contour intensity plot of SANS from cold-drawn PET blend (50:50, H:D).

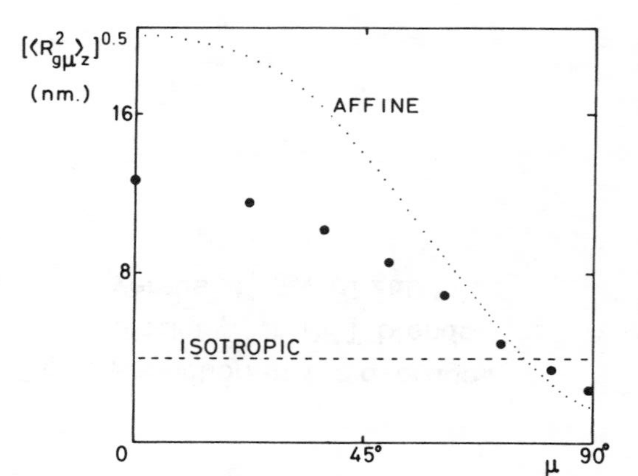

Fig. 7. SANS determined component of R_g versus mu for cold-drawn PET blends.

THE SIMILARITY BETWEEN CELLULOSE AND ARAMID FIBRES

M.G. NORTHOLT

Akzo Research Laboratories, Corporate Research Department, P.O. Box 60,
6800 AB Arnhem, The Netherlands.

SYNOPSIS

The common aspects of well-oriented cellulose and aramid fibres with regards to structure and mechanical properties are discussed. It is shown that the mechanism of axial extension is similar for both kinds of fibres. The elastic extension of the fibres is determined by the chain modulus, the modulus for shear between the chains and by the orientation distribution of the chains.

1. INTRODUCTION

The structure and tensile deformation of regenerated and native cellulose fibres have been reexamined, see refs. 1 and 2. It has been found that the axial properties of these fibres can be interpreted to a large extent by a series model of crystallites arranged end to end. Although this model has primarily been developed for the understanding of the mechanical properties of aramid fibres built up of rigid chains, it also provides a description of the course of the birefringence and dynamical modulus as a function of the extension of well-oriented cellulose fibres. A careful examination of the various structural aspects of both kinds of fibres furnishes an explanation for this result. A full exposition of the series model is given in refs. 3 and 4.

2. STRUCTURAL ASPECTS

Well-oriented cellulose fibres and aramid fibres show nearly the same kind of small-angle X-ray diffraction pattern, which appears on the film as a equatorial streak of continuous decreasing intensity, while the so called two-point meridional diffraction pattern so characteristic of semi-crystalline fibres, is absent. The equatorial scattering is caused by elongated voids having the largest dimension parallel to the fibre axis. The lateral dimension of

these voids is about 5 nm in cellulose fibres and 3 nm in the aramid fibre poly-(p-phenylene terephthalamide), abbreviated here as PpPTA . A Hosemann analysis of the meridional reflections in the wide-angle X-ray diffraction patterns of both kinds of fibres shows the broadening of the reflection profiles to be of the Cauchy type, which is typical of a paracrystalline lattice.
We find for the highly oriented cellulose fibres Cordenka EHM® and Fortisan® a crystallite size of nearly 20 nm and for the distortion parameter about 0.7%. For PpPTA fibres these values are 54 nm and 1.9%.

These results for well-oriented cellulose fibres have been confirmed by other investigators (ref.5) and indicate that there is no need for the introduction in these fibres of a distinct second phase of amorphous nature having different elastic properties. More likely is the existence of defect domains characterized by a higher concentration of lattice distortions. In the case of PpPTA fibres the rigidity of the chain excludes the existence of an amorphous phase. In a way this argument also holds for cellulose fibres. It has generally been accepted now that the natural polymer cellulose can be regarded as a more or less semi-rigid chain, or as a chain with restricted flexibility. Calculations by Sarko (ref.6) of the most probable conformation of an isolated chain show that the cellulose molecule is not likely to exist in anything but a highly extended conformation due to intramolecular hydrogen bonds between the successive anhydroglucose units.

Next the cellulose and PpPTA molecules have the capability of further hydrogen bonding into a sheet-like arrangement of chains. On a larger structural scale this leads to other features common in both kinds of fibres. Ballou, Dobb et al. and Warner (ref. 7,8,9) have reported that a lateral texture may occur in aramid fibres, viz. the hydrogen-bonded planes may have a radial, random or tangential orientation in the filament. This phenomenon has also been observed in certain kinds of cellulose fibres, see refs. 10 and 11.

To summarize, we have encountered the following molecular and structural aspects in cellulose as well as in aramid fibres: a) non-flexible chains, b) hydrogen bonding between chains thereby forming a sheet-like structure, c) lateral texture in the filament, d) similar SAXS-patterns indicating fibrils seperated by elongated voids and the absence of a two-phase structure, e) paracrystallinity and f) strongly anisometric dimensions of the crystallites. Most of these phenomena may be ascribed to the wet-spinning process according to which the cellulose and aramid fibres are made.

3. TENSILE PROPERTIES

As a result there are also common aspects in the tensile curves of these fibres, which are depicted in figs. 1a and b. At a certain strain level, i.e. at about 1% for the cellulose fibres and about 0.5% for PpPTA fibres, a sort of yield phenomenon is observed, which is presumably caused by the disruption of hydrogen bonds between the fibrils. Beyond this yield point the tensile curve adopts a concave shape which is attended with an increase of the dynamic modulus.

According to the theory developed for the elastic extension of one-phase crystalline fibres, see refs. 3 and 4, the dynamic modulus E is given by

$$E^{-1} = e_k^{-1} + \langle\sin^2\phi\rangle(2g)^{-1} \qquad (1)$$

where e_k is the chain modulus, g the modulus for shear between the chains and $\langle\sin^2\phi\rangle$ the parameter characterizing the orientation distribution of the crystallites, or more particularly the chain axes. Hence, the concave shape beyond

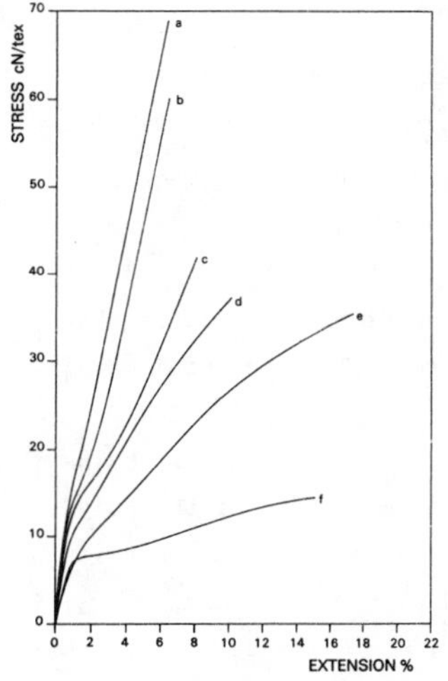

Fig. 1a
Tensile curves of various filaments of regenerated cellulose fibres measured with a gauge length of 2.5 cm and a strain rate of 20%/min., 1 cN/tex = $\rho.10^{-2}$ GNm^{-2} with ρ in gcm^{-3}: a) Cordenka EHM , b) Fortisan , c) exp. fibre, d) Lilienfeld Sedura, e) Cordenka 700, f) Bemberg.

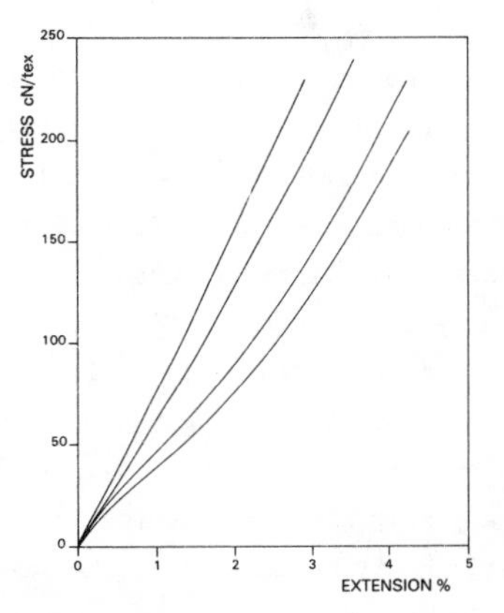

Fig. 1b
Tensile curves of PpPTA fibres with different degrees of orientation. Conditions as in fig. 1a.
Note the different scale of the figures.

the yield point of the tensile curve is caused by progressive contraction of the orientation distribution of the chains in the fibrils. Subsequently, it can be shown that for small elastic extensions the relation between stress and strain is well approximated by the expression

$$\varepsilon = \frac{\sigma}{e_k} + \frac{1}{2} \langle \sin^2\phi_0 \rangle \left[1 - \exp(-\frac{\sigma}{g})\right] \qquad (2)$$

where $\langle \sin^2\phi \rangle$ is the initial value of the orientation parameter.

The hysteresis or the viscoelastic effect during repeated loading of aramid fibres is small and the experimental σ - ε curve, measured during repeated loading, is well-approximated by eq. 2. In the case of cellulose fibres, however, the viscoelastic effect is not negligible. This is probably due to the viscoelastic rotation of the crystallites concentrated in the defect domains. Still it can be shown that the one-phase series model yields a relation between the strain and the immediate elastic response E, which also holds for the cellulose fibres

$$\varepsilon = g(E_0^{-1} - E^{-1}) + \frac{g}{e_k} \log\left[\frac{E_0^{-1} - e_k^{-1}}{E^{-1} - e_k^{-1}}\right] \qquad (3)$$

Earlier experiments, given in ref. 12, have already shown that for cellulose fibres the dynamic modulus E depends only on the extension of the fibre. Figs. 2a and 2b depict the theoretical curves calculated with eq. 3 and the experimental results obtained for the cellulose and aramid fibres. It should be noted here that little is known yet about the molecular mechanism causing the viscoelasticity in these fibres.

For a polymer fibre consisting of well-oriented non-flexible chains the axial extension is thus brought about by an elastic extension of the polymer chain and by a rotation of the chain in the stress direction. The rotation consists of a purely elastic contribution determined by the shear modulus g, of a viscoelastic contribution and in the first extension of a plastic contribution. Consequently impurities and inhomogeneities in the fibre will hamper the rotation process, which leads to stress concentrations and thus to breakage of the fibre.

Due to the progressive increase of the dynamic modulus with extension, caused by the contraction of the orientation distribution, the fibre will become increasingly brittle and thus be even more sensitive to impurities. As shown by eq. 1, the contribution of the chain modulus to the dynamic

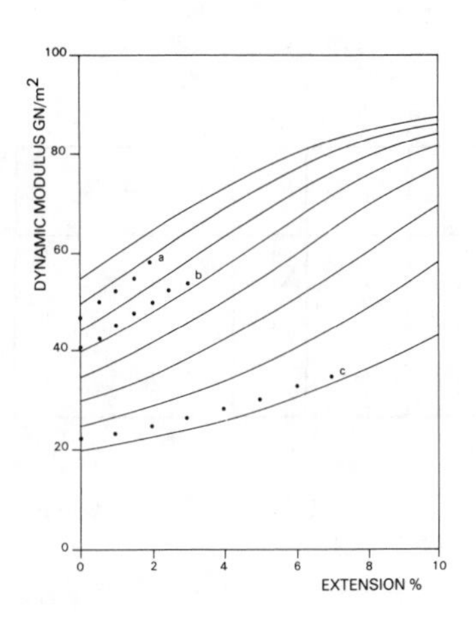

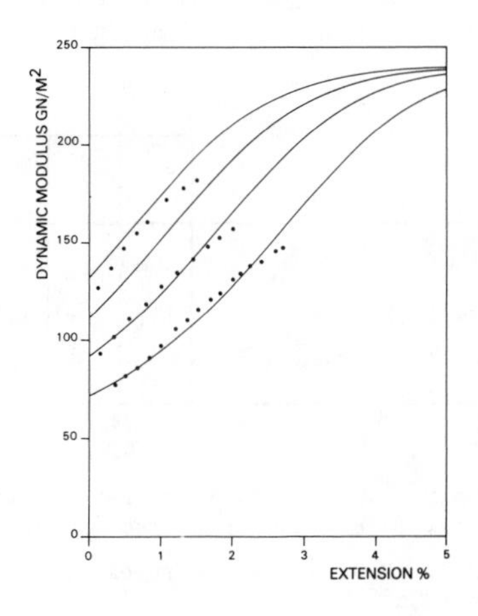

Fig. 2a

Dynamic modulus as a function of the extension for cellulose fibres. Measurements are indicated by dots: a) Cordenka EHM, b) Fortisan and c) Cordenka 700. Solid curves were calculated with eq. 3 for e_k=90 GNm^{-2} and g=2.5 GNm^{-2}.

Fig. 2b

Dynamic modulus as a function of the extension for PpPTA fibres. Measurements are indicated by dots and solid curves were calculated with eq.3 for e_k=240 GNm^{-2} and g= 2 GNm^{-2}. Note the different scale of the figures.

modulus increases during this process. This suggests a comparison between the cellulose and aramid fibres of the modulus and strength values normalized by the chain modulus e_k. The data are presented in Table 1 and show a remarkable agreement, emphasizing the close similarity between the structure and process of axial extension of well-oriented cellulose fibres and of aramid fibres. Table 1 also indicates that for cellulose fibres an appreciable larger tensile modulus and strength would be attained if the chains have the cellulose I conformation in the fibre; for further details the reader is referred to refs. 1 and 2.

Table 1:

Chain moduli e_k in GNm^{-2} compared with the initial fibre modulus E_f and the highest dynamic modulus observed during extension of the fibre E_m. The highest observed filament strength is σ_m expressed in GNm^{-2}.

	e_k	E_f	E_m	E_f/e_k	E_m/e_k	σ_m	$100\ \frac{\sigma_m}{e_k}$
Aramid (PpPTA)	240	90	180	0.38	0.75	4.3	1.8
Cellulose I (Manila)	137	--	118	---	0.86	---	---
Cellulose II (Cordenka EHM)	80	34	60	0.38	0.75	1.3	1.6

REFERENCES

1. Northolt, M.G., Conf. Proc. 23rd. Intern. Man-Made Fibre Congress, Dornbirn, Austria. To be published also in Lenzinger Ber.
2. Northolt, M.G. and H. de Vries, submitted for publication in Colloid & Polymer Sci.
3. Northolt, M.G., Polymer 21, 1199 (1980).
4. Northolt M.G. and R. v.d. Hout, Polymer (1985), in press.
5. Yachi T., J. Hayashi, M. Takai and Y. Shimizu, J. Appl. Pol. Sci., Appl. Pol. Symp. 37, 325 (1983).
6. Sarko A., J. Appl. Pol. Sci.,Appl. Pol. Symp. 28, 729 (1976).
7. Ballou J.W., ACS. Polymer Preprints 17, nr. 1, 75 (1976).
8. Dobb M.G., D.J. Johnson and B.P. Saville, J. Pol. Sci. Pol. Phys. Ed. 15, 2201 (1977).
9. Warner S.B., Macromolecules 16, 1546 (1983).
10. Hermans P.H., "Physics and Chemistry of Cellulose Fibres", Elsevier, Amsterdam (1949).
11. Kast W., Kolloid Z. 125, 45 (1952).
12. De Vries H., Appl. Sci. Res. A3, 111 (1952).

NOTATION

e_k	elasticity modulus of the chain
g	modulus for shear between chains
E	dynamic modulus
$\langle \sin^2\phi \rangle$	orientation distribution parameter
ε	strain
σ	stress

CRYSTALLINE ORDER IN NYLON 4,6

R.J. GAYMANS, D.K. DOEKSEN AND S. HARKEMA

Twente University of Technology, Enschede, The Netherlands

SYNOPSIS

The unit cell of nylon 4,6 was studied with WAXS and was found to be triclinic, a = 4.9, b = 5.5, c = 14.8 Å, α = 46, β = 78, γ = 64 degrees and ρ_c = 1.28 g/cm^{-3}. The long periods were studied with SAXS and correlated with the melting temperatures of the samples. Extrapolating results to infinitely large lamella the T_m^o was found to be 350°C, the ΔH_f 210 J/g and the fold surface free energy 75 erg/cm^2.

1. INTRODUCTION

Nylon 4,6 is a new polyamide, which can be prepared from 1,4 diaminobutane and adipic acid. It has a high melting transition temperature (290°C) and excellent mechanical properties both a room temperature as at higher temperatures. DSM is marketing this polymer.

$$\left(\overset{H}{N} - (CH_2)_4 - \underset{H}{N} - \overset{O}{\overset{\|}{C}} - (CH_2)_4 - \underset{O}{\underset{\|}{C}} \right)_n$$

In polymers the crystalline order has a profound effect on the mechanical properties particular in the region between the glass transition and the melting temperature (ref. 1). Crystalline polymers have three levels of order: Unit cell, lamelar and sferulitic. In this preliminary study the unit cell and lamellar structures have been looked at.

2. UNIT CELL

Nylon 4,6 has a high amide content and belongs to the group of even-even polyamides. The even-even polyamides have two centers of symmetry, in the chain. When centers of symmetry are present the chain has no resulting dipole moment. The chains crystallize in a parallel fashion in a triclinic cell (α-structure). The equal spacing of the amide groups in nylon 4,6 makes for this polymer a second chain packing feasable (β-structure) (fig. 1).

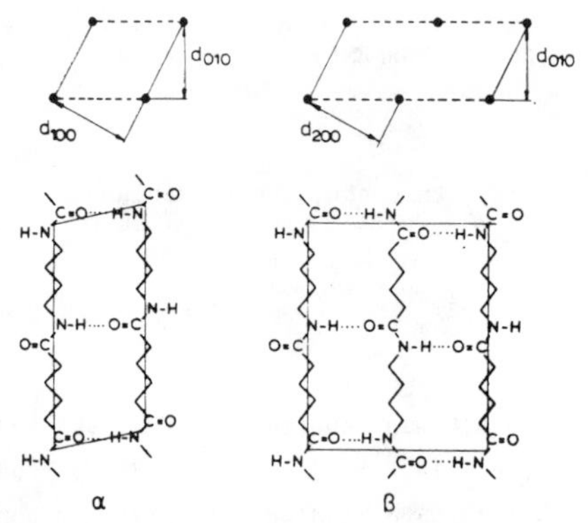

Fig. 1. Cell structures

In both the α and β-structure all the amide groups take part in the hydrogen bonding with the chains fully extended. The β-phase probably being monoclinic will have its WAXS diffraction spots 001 of an oriented sample (fig. 2) on the meridian.

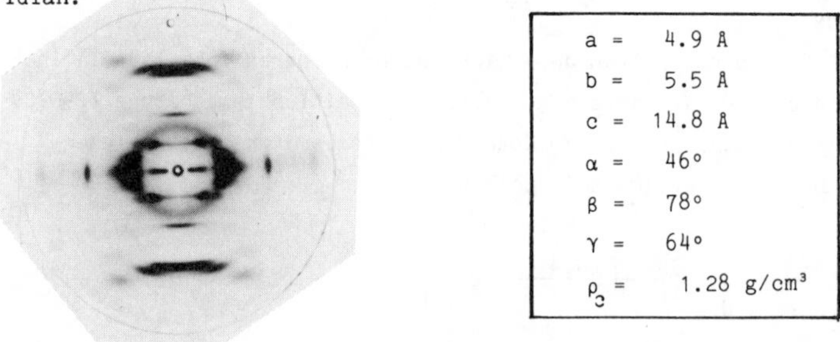

a =	4.9 Å
b =	5.5 Å
c =	14.8 Å
α =	46°
β =	78°
γ =	64°
ρ_c =	1.28 g/cm³

Fig. 2. WAXS-diagram of a fibre

Table 1. Cell constants α-structure

A number of samples have been analysed and the β-phase could not be detected. In other polyamides where β-type packing can be expected (e.g. nylon 7.9) it was not observed either (ref. 2). The direction of the dipole moment might play a role in this.

The cell dimensions were studied on double oriented samples and the results are collected in Table 1.

3. LAMELLA

The SAXS diffraction pattern of nylon 4,6 fibers showed a two point diagram. With annealing both the lamellar thickness and the width increased. The lammellar thickness of a melt quenched sample was 65 Å and an annealed sample (1 h at 280°C) 120 Å. The lamelae are the crystallites in the polymer and their small size make that the surface effects influence the properties. Hoffman and Weeks have given the relationship between the lamellar dimensions and the melting temperatures (ref. 3).

$$T_m = T_m^o \left(1 - \frac{1}{\Delta H_f}\left(\frac{2\sigma_e}{l} + \frac{4\sigma}{x}\right)\right)$$

where T_m^o is the melting temperature of an infinitly large lamella, ΔH_f the heat of fusion of ideal crystal, σ_e and σ the surface free energies of the folded surfaces and the side surfaces and l and x the thickness and the width of the lamella.

The melting temperatures and the heats of fusion have been studied with DSC. Nylon 4,6 has a double melting temperature- like most semicrystalline polymers- and the lower temperature was found to increase with annealing.

The surface free energies of the folded surfaces are much larger than those of the side surfaces. Combine this with the fact that the thickness l is usually much smaller than the width it can be concluded that the effect of the side surfaces is small compared to that of the folded surface and hence can be neglected.

In Figure 3 the reciprocal lamellar thickness with the corresponding melting temperature are given. Extrapolating the data to infinitly large lamellar thickness gives T_m^o, which was found to be 350°C. The ΔH_f of this ideal structure was obtained by extrapolating the ΔH_f versus T_m data to T_m^o (350°C) (fig. 4) and was found to be 210 J/g. From these data the surface free energy of the folded surfaces was calculated and found to be 75 erg/cm².

REFERENCES

1. D.W. van Krevelen, Properties of Polymers, Elsevier, Amsterdam, 1976, Chapter V.
2. Y. Kinoshita, Makromol. Chem, 1, 33 (1959).
3. J.D. Hoffman, J.J. Weeks, J. Res. Nat. Bur. Stand., Sect. A 1962, 66, 13.

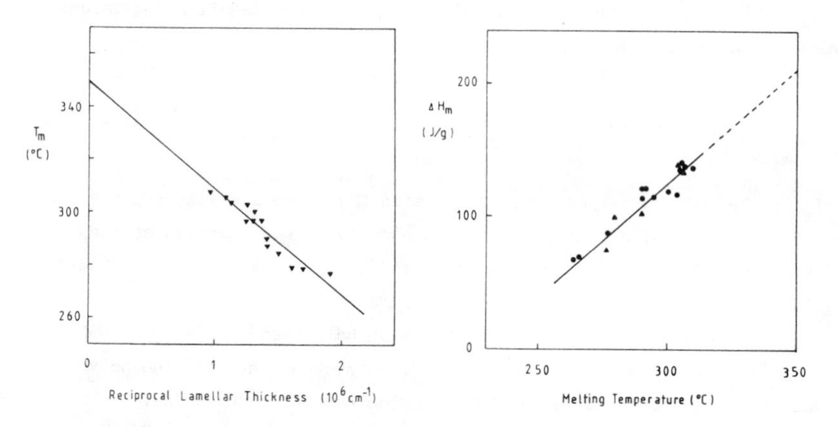

Fig. 3 Influence of reciprocal lamellar thickness on the melting temperature

Fig. 4 Meltingtemperature v.s. heats of melting: ● films, ▼ fibres

PULSED EPR STUDY OF THE TRAPPING PROCESS OF RADICALS IN POLYETHYLENE

C.P. KEIJZERS, H.C. van LIESHOUT, A.A.K. KLAASSEN and E. de BOER

Department of Molecular Spectroscopy, Faculty of Science, University of Nijmegen, Toernooiveld, 6525 ED Nijmegen, The Netherlands

INTRODUCTION

It has been suggested that irradiation of polyethylene (PE), with γ or X-rays, creates radicals in spurs, i.e. regions with high local concentrations. Both, Nunome c.s.[1] and Shimada c.s.[2] applied microwave power satuaration experiments in order to verify this model. This technique results in the determination of a (not absolute) value for the product of the electron spin-lattice (T_1) and the electron spin-spin (T_2) relaxation times. In order to draw conclusions from changes in the magnitude of this product (by changing the irradiation dose, the irradiation temperature or by heat treatment after the irradiation) both groups of authors assumed that T_2 is a measure of the local radical concentration and that T_1 is independent of the history before the time of measurement. In order to test these assumptions, it was decided to measure the effect of heat treatments on T_1 and T_2 by using Electron Spin Echo spectroscopy.

EXPERIMENTAL

The two samples were obtained from Dutch State Mines, Geleen, The Netherlands. Sample one was a stretched (degree of stretching = λ = 10), melt-spun PE fiber, sample two a highly stretched (λ = 60), solution-spun PE tape. After degassing for 12 hours at 10^{-2} Torr, the sampletubes were sealed and irradiated at liquid nitrogen temperature for 12 - 18 hours with a ^{137}Cs γ-source. The total radiation dose was 3 - 5 Mrad. After irradiation and subsequent annealing of the sample tube, the sample was transferred to the TE102-cavity (Q-factor ~ 5000) of the spectrometer without any increase of the sample temperature. The pulsed EPR spectrometer has been described in some detail in ref. 3.

In order to characterize the created radicals, the echo-induced EPR spectrum was measured at 40 K by monitoring the echo intensity while sweeping the DC-magnetic field (Figs. 1,2). This field was oriented perpendicular to the fiber/tape axis.

The relaxation times T_1 and T_2 were measured on the highest intensity peak in the spectrum. In order to minimize spin-diffusion influences on the measured relaxation data, the pulse sequence consisted of a long high-power saturation pulse, followed by a $\pi/2$ and a π-pulse of 100 and 200 nanosec duration respectively. For the T_1-measurements the time between the saturation and the $\pi/2$ pulse was varied; T_2 was measured by varying the time between the $\pi/2$ and the π pulses. After T_1, T_2 measurement at 40 K, the sample was heated for a fixed period of time (15 or 30 minutes, Figs. 3 and 4) to increasingly higher temperatures, after which T_1 and T_2 were measured again at 40 K. The heating times of 15/30 minutes were chosen such that after that period the relaxation times at that temperature did not change markedly.

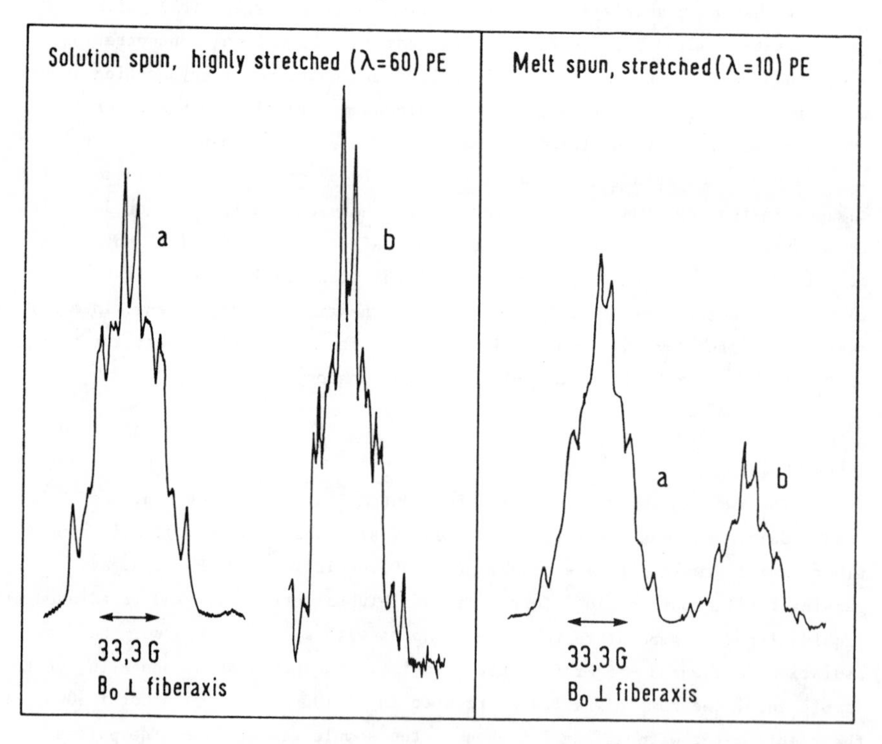

Echo induced EPR spectra before (a) and after (b) heat treatment measured at 40 K. *Fig. 1* Solution spun PE, 5 Mrad at 77 K. *Fig. 2* Melt spun PE, 3.6 Mrad at 77 K.

RESULTS AND DISCUSSION

The echo-induced EPR spectra (Figs. 1,2) have the appearance of CW-EPR-spectra in absorption mode. For both samples ten peaks can be distinguished. This is typical for alkyl radicals when measured perpendicular to the polymer chain axis[1]. The resolution is much better for the highly stretched, solution spun, PE. This corresponds with the higher degree of crystallinity of this sample which in turn is responsible for the very great strength of this material. After the heat treatment, the resolution of both samples increases. This concurs with the observation of Shimada et al.[1] in CW-EPR experiments. These authors attributed the resolution increase to a disappearance of an alkyl radical with a large linewidth which radical was thought to be created in the amorphous parts of the sample and which is, therefore, more reactive. The remaining, more stable, radicals were assumed to be located in the crystalline part and to possess, therefore, a smaller linewidth.

The measured relaxation times are depicted in Figs. 3 and 4. Roughly speaking T_1 is 10^{-3} sec immediately after the irradiation and increases by a factor 3.5 during the heat treatments. T_2, on the other hand, is much faster (~ 10^{-6} sec) and

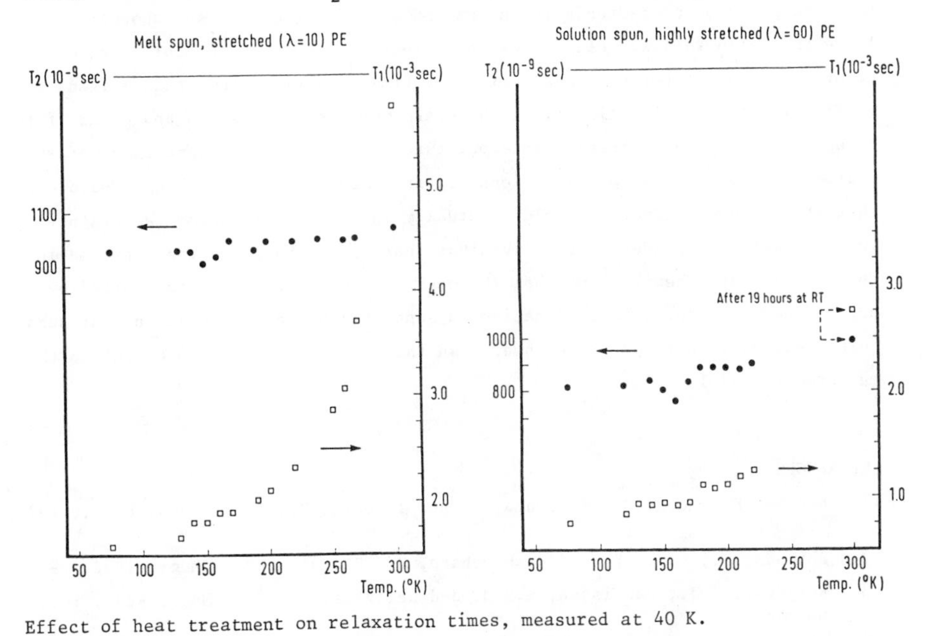

Effect of heat treatment on relaxation times, measured at 40 K.

Fig. 3 Melt spun PE, 3.6 Mrad at 77 K. 15 min. at every temperature.

Fig. 4 Solution spun PE, 5 Mrad at 77 K. 30 min. at every temperature.

is almost constant during the treatment. These observations agree very nicely with the results of Shimada et al.[2]. They found with saturation experiments an increase by a factor 3.3 - 3.6 in the product T_1T_2. Assuming that this increase was due to an increase in T_2, and observing that the increase was almost independent of the irradiation dose, they suggested that the radicals are created in spurs with a high local concentration. The heat treatment would cause a diffusion out of the spurs, leading to a lower local concentration and a longer T_2 relaxation time.

Our experiment, however, shows that this suggestion is wrong. The increase in T_1T_2 is due to an increase in T_1 and T_2 is constant. This implies that we see no reason to assume that the radicals are created in regions with a high local concentration, at least not at an irradiation temperature of 77 K. It might still be true that irradiation at much lower temperatures does produce radicals in spurs. In fact, Nunome c.s.[1] did find large differences between the saturation behaviour of PE irradiated at 1.5 and 4.2 K on one hand and 77 K on the other, but an experiment like ours is necessary to draw any conclusions.

One might wonder whether the increase in T_1 can be explained by the suggested disappearance of radicals in the amorphous phase (ref. 1, see above). We believe that also this suggestion must be rejected because of the following observation: heat treatment for longer periods of time at a fixed temperature leads asymptotically to the relaxation times which are depicted in Figs. 3 and 4 and if the change in T_1 with heat treatment-temperature were due to a disappearance of the radicals out of the amorphous regions (by diffusion and/or decay) one would expect that at any temperature, T_1 would eventually increase to the room-temperature value. A possible explanation could be that there is a change in the conformation of the radicals upon heating and that the magnitude of that change is limited by the applied temperature. This explanation is supported by the observation that subsequent heating to a temperature lower than the previous one has no effect on the measured T_1 and T_2.

REFERENCES

1. Nunome, K., Muto, H., Toriyama, K. and Iwasaki, M., Chem. Phys. Lett., 1976, 39, 542.
2. Shimada, S., Hori, Y. and Kashiwabara, H., Radiat. Phys. Chem., 1982, 19, 33.
3. Reijerse, E.J., Paulissen, M.L.H. and Keijzers, C.P., J. Magn. Res., 1984, 60, 66.
4. Shimada, S., Maeda, M., Hori, Y. and Kashiwabara, H., Polymer, 1977, 18, 19.

ANALYSIS OF FILLED RUBBERS USING SAXS

R. J. YOUNG*, D. AL-KHUDHAIRY* and A. G. THOMAS+

*Department of Materials, Queen Mary College, Mile End Road, London, E1 4NS, U. K.
+Tun Abdul Razak Laboratory, Malaysian Rubber Producers Research Association, Brickendonbury, Hertford, SG13 8NL, U. K.

SYNOPSIS

It is demonstrated that small-angle X-ray (SAXS) scattering can be used to characterize the structure of fillers such as carbon black and silica both before and after their incorporation into natural rubber. The results are shown to be in good agreement with conventional characterization techniques.

1 INTRODUCTION

It is well established that the addition of certain types of fillers to rubbers can greatly increase their stiffness, strength, abrasion and tear resistance and other important physical properties[1,2]. There are several different types of filler that can be employed and the structure of such engineering rubber systems is generally characterized using transmission electron microscopy[3,4]. SAXS is a technique that is used widely to characterize multiphase systems such as filled rubbers which have structure on the 1-100nm scale[5]. In this present study, the SAXS data have been analysed using a SAXS computer analysis package developed by Dr. C. G. Vonk[6] of DSM.

2 EXPERIMENTAL

The polymer used was SMRL natural rubber and the fillers employed were HAF carbon black (Cabot 330, ASTM No. N330) and fine particle silica powder (VN3). Four different loadings of the two types of filler were used which were nominally 10, 20, 30 and 40 phr. The compositions were checked by both density measurements and thermogravimetric analysis (only for silica fillers). The structure of the filler particles was examined by transmission electron microscopy. The individual particles of filler were prepared for microscopic examination by dispersing a few milligrams of the powders in about 5ml of butanone in a glass container using ultrasonic energy. A single drop of the suspension was then placed on a standard copper microscope grid covered with a carbon support film and the solvent allowed to evaporate. The SAXS experiments were performed on a Kratky small-angle X-ray scattering camera used in the step-scan mode with a proportional counter. The SAXS data were processed with the aid of the computer program FFSAXS5 described previously by Vonk[6]. Volume distribution function $D_V(d)$ were determined as a function of the particle diameter, d by the two independent methods available in FFSAXS5 and described by Vonk[6].

3 RESULTS AND DISCUSSION

3.1 Characterization of Fillers

The particle size of fillers is generally determined using either electron microscopy or gas adsorption methods where good agreement between the different techniques is generally obtained[3]. In this present study the SAXS particle size distribution function $D_V(d)$ has been compared with the size distribution measured using electron microscopy. The aggregation of particles makes the determination of the size distribution of the individual particles rather difficult and prone to error. Nevertheless, an attempt has been made to determine the size of the particles by measuring manually the diameters of a large number of individual particles and the results are presented in the form of normalised histograms in Fig. 1. The normalized particle-size distribution functions determined from SAXS data are plotted as curves in Fig. 1 along with the histograms derived from direct particle size measurement. It can be seen that the good agreement gives considerable confidence in the use of SAXS to determine the dimensions of filler particles. The average particle size $\bar{d}$ can also be determined for both methods and the results are summarized in Table 1. The results of such a calculation yield a value of $\bar{d}$ = 33.5nm for the carbon black and $\bar{d}$ = 19.7nm for the silica. The value for the carbon black is close to the 31.7nm quoted[3] for a similar black with the same ASTM number (N330). Similarly the value of $\bar{d}$ for the VN3 silica is within the quoted[4] range of 15-20nm. The program FFSAXS5 also generates average particle sizes directly from the particle size distribution functions. Values of 24.6nm for the carbon black and 21.5nm for the silica are produced from $D_V(d)$.

It is possible also to calculate the specific surface area, s, of the filler particles if it is assumed that they are in the form of non-porous spheres. The values of s determined using SAXS are also given in Table 1. These values of specific surface area are in general agreement with values of s determined from adsorption methods[3,4].

3.2 Characterization of Filled Rubbers

The volume distribution functions $D_V(d)$ are plotted in Fig.2 for the carbon black filler and two samples of black-filled rubber. It can be seen that the distribution curves for the filled rubbers are similar to each other but differ significantly from that of the isolated carbon black. The average particle diameters determined using X-rays are given in Table 1 and it can be seen that the values of $\bar{d}$ remain approximately constant as the amount of filler in the rubber increases whereas $\bar{d}$ for the carbon black is considerably higher when it is in the rubber. It appears therefore that the structure of the carbon black has been significantly affected by incorporation in the rubber. The volume distribution functions $D_V(d)$ are plotted in Fig.2 for the silica filler and two samples of silica-filled rubber. The shape of all three curves is very similar showing that the particle size distributions are very similar for both the filler and filled rubbers. It seems that unlike the carbon black, mixing and incorporation of the silica in the rubber does not significantly affect the particle size distribution.

The volume fraction of scatterers, Φ, in the filled rubbers can be readily determined from the total integrated intensity of the radiation scattered at small angles[5] and the results of such calculations are listed in Table 1. It can be seen that there is reasonably good agreement with the volume fraction determined from density and thermogravimetric measurements.

The specific surface areas, s, of the filler particles in the rubbers can be calculated using SAXS and it can be seen that although there is considerable spread in the data the specific surface areas of the carbon black particles in the rubber are generally higher than the accepted value[3] and the values of s determined from TEM assuming the particles to be spherical. On the other hand the values of s for the silica-filled rubbers are in the range 150-200m^2/g and so very similar to the value of 156m^2/g determined from TEM[4].

REFERENCES

1. 'Reinforcement of Elastomers', Gerard Kraus (Ed), Interscience, 1965.
2. 'Chemistry and Physics of Rubber-like Substances', L. Batener (Ed) Maclaren, 1963.
3. Medalia,A.I., in 'Carbon Black-Polymer Composites' E.K.Sichel (Ed) Marcel Dekker, 1982.
4. Bachmann,J.H.,Sellers,F.W.,Wagner,M.P. and Wolf.R.F.,Rubb.Chem.Tech. 32 (1952) 1286.
5. Glatter,O. and Kratky,O. 'Small-angle X-ray Scattering',Academic Press,1982.
6. Vonk,C.G., J. Appl. Cryst. 9(1976)433.

Sample	Black	A1	A2	A3	A4	S_iO_2	B1	B2	B3	B4
$\bar{d}$ (nm)-TEM	33.5	-	-	-	-	19.7	-	-	-	-
$\bar{d}$(nm)-SAXS	24.6	40.7	40.1	41.2	39.4	21.5	20.3	23.5	18.9	19.0
s(m^2/g)-TEM	99.5	-	-	-	-	156.2	-	-	-	-
phr-Measured	-	10	20	30	40	-	9.74	15.35	28.65	35.49
Specific gravity	1.80	0.961	0.998	1.035	1.058	1.95	0.958	1.009	1.078	1.099
Φ(%)-From phr	-	4.8	9.2	13.2	16.7	-	4.4	6.7	11.8	14.2
Φ(%)-From density	-	4.7	8.9	13.1	15.7	-	3.7	8.6	15.3	17.4
Φ(%)-From SAXS	-	4.3	8.2	9.8	13.9	-	3.5	7.6	8.5	10.3
s(m^2/g(sample))	-	18.5	29.2	24.7	40.2	-	19.3	31.5	42.5	52.0
s(m^2/g(filler))	-	185	146	82.3	100	-	198	205	148	146

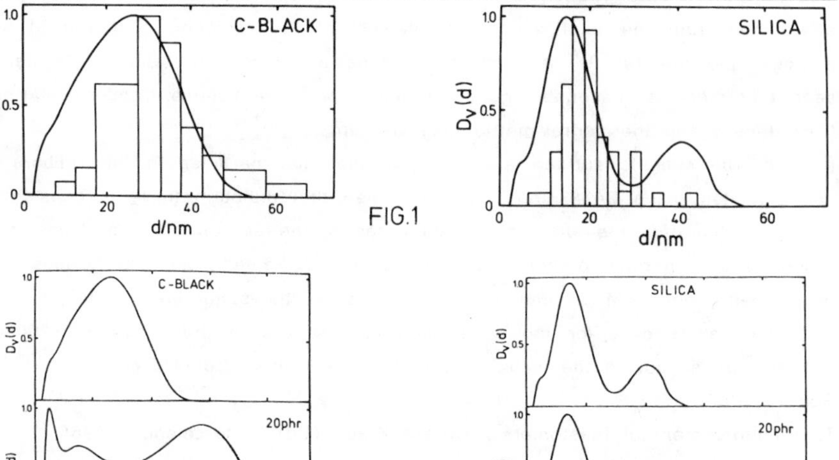

FIG.1

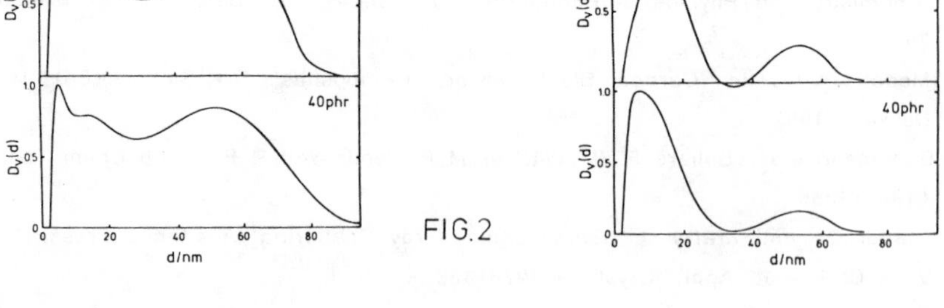

FIG.2

SAXS STUDIES OF SEMI-CRYSTALLINE POLYMER BLENDS USING SYNCHROTRON RADIATION

M. VANDERMARLIERE[a], C.RIEKEL[b], M. KOCH[c], G. GROENINCKX[a] and H. REYNAERS[a]

(a) University of Leuven, Dept Chemistry, Belgium. (b) University of Hamburg,FRG.
(c) EMBL Outstation, Hamburg, FRG.

SYNOPSIS

The melting of polycaprolactone (PCL) in blends of PCL and polycarbonate was followed using differential scanning calorimetry and dynamic small-angle X-ray scattering using synchrotron radiation.

1 INTRODUCTION

Physical mixtures of poly-ε-caprolactone (PCL) with polycarbonate of bisphenol A (PC), prepared by melt mixing, form miscible blends over the whole composition range[1]. Due to the plasticizing action of PCL, polycarbonate, which as a homopolymer is normally amorphous, can crystallize in the mixtures under well defined conditions of temperature. Hence we are dealing with a blend with two crystallizable components in which the amorphous phase is a compatible mixture of the two constituents.[4]

The morphology of PCL/PC blends was studied by time-resolved small-angle X-ray scattering.

2 MATERIALS

Polycaprolactone flakes (Aldrich Europe) having Tg = -52°C, T_m° = 64°C, $\overline{M}_w$ = 30.500 and $\overline{M}_n$ = 17.000 were mixed with PC pellets (Lexan) having Tg = 145°C, T_m° = 295°C, $\overline{M}_w$ = 42.500 and $\overline{M}_n$ = 16.000 in the following compositions by weight : 80/20, 60/40, 40/60, 20/80.

The blends were prepared by melt mixing in a Brabender Plasticorder (55 g of the dried components were mixed for 10 min at 100 rpm and at ± 260°C).

3 EXPERIMENTAL TECHNIQUES

The SAXS experiments were performed at the X-13 camera of the European Molecular Biology Laboratory outstation at the storage ring DORIS of the Deutsches Elektronen Synchrotron in Hamburg, Federal Republic of Germany.

The X-13 bench is a general purpose, double focussing, single wavelength X-ray camera. A 1-dimensional position sensitive detector was used. The time resolution was 20 s, the angle covered about 2° and 1' per channel. The oven was fixed inside a vacuum tube and the heating rate used was 10°C/min[2].

The melting behaviour was also followed in a Perkin-Elmer differential scanning calorimeter (DSC-2C) under a nitrogen atmosphere.

4 RESULTS AND DISCUSSION

4.1 PCL homopolymer

In the DSC measurements one melting endotherm is observed, which starts at about 39°C, and ends at 65°C, the maximum being at 62.5°C (fig.1). An increase in the long spacing (15.1 nm) is seen as soon as the melting domain of the sample is reached. This is accompanied by a sharpening of the diffraction peak, indicating the melting of the smaller crystals within the lamellar structure.

The SAXS invariant, Q, is given by

$$Q = \int_0^{\infty} I(s)\, s^2\, ds \qquad (1)$$

where I(s) is the scattered intensity measured at the scattering angle $s = 2\sin\theta/\lambda$, λ the wavelength of the radiation, and θ the half of the scattering angle. The invariant of a semi-crystalline two-phase system reflects the degree of crystallinity since

$$Q = K\, \phi_c\, (1-\phi_c)\, (\rho_c-\rho_a)^2 = K\langle\eta^2\rangle \qquad (2)$$

Here, ϕ_c is the volume fraction of the crystalline phase, $\rho_c-\rho_a$ is the electron density difference between crystalline and amorphous phases, K is a constant depending on the geometry of the scattering instrument and $\langle\eta^2\rangle$ is the averaged squared electron density difference.

The evolution of the scattering power of PCL as measured by the invariant exhibits an increase as soon as melting starts of the less stable crystals. This observation is in agreement with theory which predicts an increase of Q for a degree of crystallinity above 50 %. The maximum value is reached at 58°C, where the observed periodicity also is at its maximum. At this point a decrease in the invariant is observed and at 60°C the SAXS pattern has faded away completely. It is clear from these results that considerable crystallinity still persists in the sample after the disappearance of the superstructural order as observed by SAXS. The same observation was made by Russell and Koberstein for the melting behaviour of polyethylene[3].

4.2 PCL/PC blends

For the dynamically crystallized PCL/PC blends one observes a double PCL melting peak (fig.2). If the PCL content is increased the first peak decreases and the second one increases. On further heating of samples with less than 80 % PCL, exothermic peaks are observed which must be attributed to the dynamic crystallization of PC.

During the melting of PCL one observes a sudden increase in the long spacing (fig.2). The magnitude of the increase depends on the amount of PCL in the blend. When the temperature is raised further the long spacing remains constant until about 100°C and then gradually increases again. The invariant (fig.3) increases slightly with temperature, but passes through a minimum between 50 and 60°C as a consequence of the melting of PCL. The increase between 60 and 70°C for the 60/40 blend corresponds to a crystallization exotherm. The presence of crystallizable PC accounts for the SAXS profile after the melting of PCL.

This study clearly demonstrates the potential of this new-experimental approach for the investigation of phase transitions in semi-crystalline polymer blends.

ACKNOWLEDGEMENTS

Financial support of the Belgian Science Foundation (NFWO) is gratefully acknowledged. M. Vandermarliere thanks the Phillips Petroleum Chemicals Company for its research fellowship.

REFERENCES

1. C.A. Cruz, D.R. Paul and J.W. Barlow, J. Appl. Polym. Sci., 23, 589 (1979).
2. G. Elsner, C. Riekel and H.G. Zachmann in "Advances in Polymer Science 67", p.35-36, Springer Verlag, Berlin, Heidelberg (1985) .
3. T.P. Russell and J.T. Koberstein, personal communication.
4. G. Groeninckx and M. Vandermarliere, in "Polymer Alloys : Structure and Properties", Proceedings of 16th Europhysics Conference on Macromolecular Physics, Brugge, Belgium, 1984, p.70

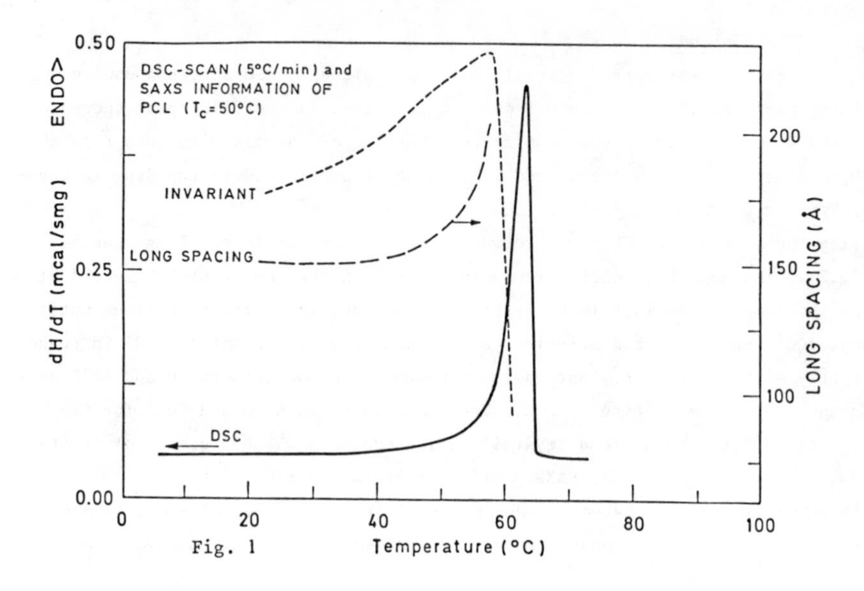

Fig. 1

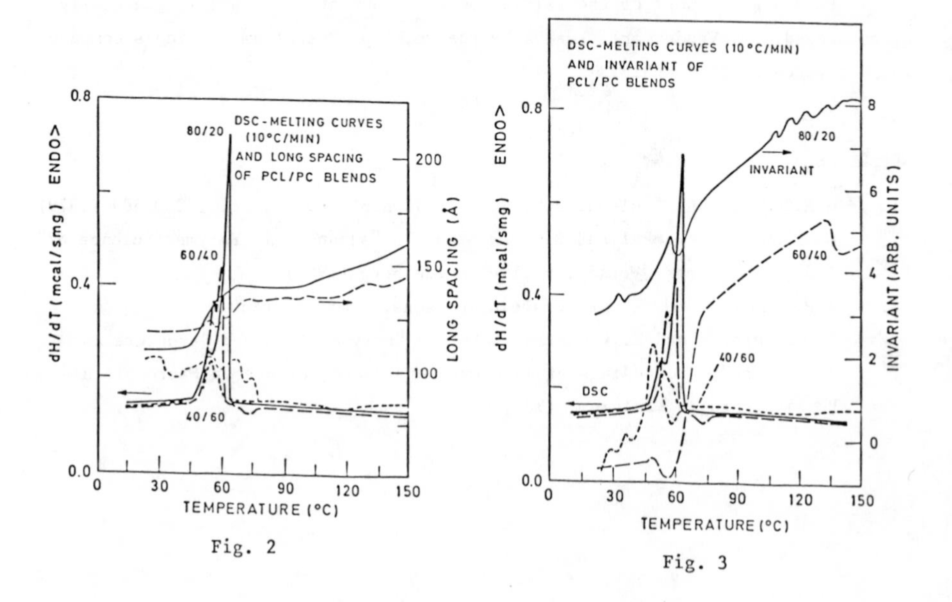

Fig. 2

Fig. 3

ULTRA-DRAWING OF POLYPROPYLENE

A. PEGUY and R. ST-JOHN MANLEY*

Centre de Recherches sur les Macromolécules Végétales (C.N.R.S.)
B.P. 68, 38402 Saint Martin d'Hères (France)
*P.P.R.I.C. Department of Chemistry, McGill University, Montreal, Québec, Canada H3A 2A7

SYNOPSIS

Ultra-high-modulus films of high molecular weight isotactic polypropylene were produced by drawing dry gels made from solution according to the method of Smith and Lemstra. The maximum draw ratio attained was 57. For a draw ratio of 47.5 the mechanical properties were a Young's modulus of 36 GPa, and a tensile strength of 1.03 GPa. These values may be compared with the theoritical values of 41.2 GPa and 1.23 GPa, respectively.

I INTRODUCTION

During the last 20 years research on the preparation of fibers and films with high modulus and high strength has become a topic of increasing interest (1). A variety of technique are known to produce orientation in polymers, such as spinning of fibers from liquid crystals (2), super-drawing (3), high-pressure extrusion (4), hot-drawing (5), gel-state spinning (6), flowing polymer solutions (7), zone-annealing method (8) and solution spinning/drawing (9). This last technique has been employed by Smith and Lemstra to achieve ultra-high strength and modulus in fibres of polyethylene. They have found that gels exhibited remarkably good drawability at elevated temperatures even after complete removal of solvent. This phenomenon was attributed to a reduced number of entanglements per macromolecule in solution cast polymer.

Morphological and X-ray studies revealed that the macromolecules were well oriented in the fibre direction of the highly drawn polyethylene, thus supporting the view that there is a strong correlation between the Young's modulus and the degree of molecular orientation (10).

Over the years the drawing behaviour of melt spun polypropylene has been extensively investigated. Our present paper describes a study of the ultra-drawing of dry gel films of high molecular weight polypropylene produced by gelation/crystallization from dilute solutions according to the method of Smith and Lemstra (9).

2 EXPERIMENTAL

The high molecular weight polymer used was isotactic polypropylene Profax Flake 7862-23 from Hercules. The intrinsic viscosity of this product was 16.8 which corresponds to a molecular weight of approximatively 3.4×10^6. Before dissolution the solvent (decahydronaphtalene) was degassed for one hour under vacuum. Di-t-butyl-p-cresol was added as a stabilizer (0.5 % w/w on polymer). The polymer was then dissolved in hot solvent under nitrogen to form 0.75 to 1.5 % w/v solutions. Temperatures close to the solvent boiling point were necessary to dissolve all polymer particles. Gels were generated when the hot solutions were quenched by pouring them into an aluminium tray cooled to about -25°C. The bulk of the solvent was allowed to evaporate in a current of air under ambient conditions. After washing with methanol a transparent and brittle film was obtained which could only be drawn at elevated temperatures (ranging from 110°C to 160°C). The draw ratio (λ) was determined in the usual way.

3 RESULTS

The maximum elongation at break (λ max) was about 5700 % which is much higher than the so-called "natural draw ratio" which is about 7 (11).

WAXS patterns of dry polypropylene gel films drawn at 140°C to a draw ratio of 47 leads to a highly oriented X-ray fibre pattern with the c-axis (chain direction) oriented parallel to the filament axis. For draw ratios greater than 30, the WAXS patterns displayed spots.

For a draw ratio of 47.5 the mechanical properties were a Young's modulus of 36 GPa and a tensile strength of 1.03 GPa which are substantially higher than those reported by other investigators using different techniques for the production of high-modulus PP fibres. These values represent, respectively 87 % and 84 % of the theoritical values.

From these results it's clear that gelation/crystallization is advantageous to improve the drawability of HMWPP, hence the modulus and the tenacity are close to the theoritical values.

A study about the structure and morphology of PP gels was investigated. X-ray examination showed that the undrawn films are essentially composed of preferentially oriented lamellaes, analogous to sedimented mats of single crystals. In that case the drawability is excellent. A scanning electron micrograph study showed that the presence of spherulitic crystallization induced a poor drawability.

4 CONCLUSION

The results of this work have led to a several number of interesting observations. We guess that the reduction of the number of trapped entanglements in the solid polymer is not only the crucial point which explains the enhancement of the drawability of HWMPP through casting from semi-dilute solution. To explain the high draw ratio found for PP or PE (9,12) the contribution of the unfolding of stacked lamellaes on the drawing behavior cannot be ignored.

Our results on polypropylene, even if this polymer is not of primary technological importance for high stiffness products, are particularly interesting because they enable us to compare PE with a crystalline polymer having completely different molecular structure and morphological features. Further details on the morphology of drawn films and on the thermal behaviour of these samples will be reported elsewhere.

REFERENCES

1. Cifferri, A. and Ward, I.M., Ultra-high modulus Polymers, Applied Science, London, (1979).
2. Kwolek, S.L., U.S. Pat., 3 671 542 (1972)
3. Capiati, N.J. and Porter, R.S., J. Polym. Sci., Polym., Phys. Ed., 13, 1177 (1975).
4. Gibson, A.G. and Ward, I.M., J. Polym. Sci., Polym. Phys., Ed., 16, 2015 (1978).
5. Wu, W. and Black, W.B., Polym. Eng. Sci., 19, 1163 (1979).
6. Kalb, B. and Pennings, A.J., Colloid Polym. Sci., 254, 868 (1976).
7. Barham, P.J. and Keller, A., J. Mater. Sci., 15, 2229 (1980).
8. Kumugi, T., Ito T., Hashimoto, M. and Doishi, M., J. Appl. Polym. Sci., 28, 179 (1983).
9. Smith, P., Lemstra, P.J. and Pijpers J.P.L., J. Polym. Sci., Polym. Phys. Ed., 20, 2229 (1982) and references inthere of same authors.
10. Peterlin, A., Polym. Eng. Sci., 19, 118 (1979).
11. Taylor, W.N. and Clark, E.S., Polym. Eng. Sci., 18, 518 (1978).
12. Peguy, A. and St-John Manley, R., Polym. Com., 25, 39 (1984).

SPINNING OF FIBERS FROM CELLULOSE SOLUTIONS IN AMINE OXIDES

I. QUENIN, H. CHANZY, M. PAILLET and A. PEGUY

Centre de Recherches sur les Macromolécules Végétales (C.N.R.S.)
BP 68, 38402 Saint Martin d'Hères (France)

SYNOPSIS

The thermoplastic properties of cellulose solutions in tertiary amine oxides have been used for the elaboration of a new spinning system where post-stretching is possible. This is achieved at a temperature slightly above the Tg of the solution, i.e. when the relaxation of the cellulose chains is restricted. Preliminary results indicate that this new process may prove to be an attractive route to the production of cellulosic fibers with enhanced mechanical properties.

Our laboratory is involved in the study of cellulose solutions in tertiary amine oxides (1-8), namely N-methylmorpholine N-oxide (MMO) and N,N-dimethyl ethanolamine N-oxide (DMEAO). An ultimate goal of this project is the development of new systems for producing cellulose fibers (9). In the present work, two spinning systems have been used : a standard dry-jet, wet spinning system as a reference and a low temperature system. In the former, precipitation is achieved with a polar solvent (usually water) which induces intermolecular hydrogen bonding. Under optimal experimental conditions, the resulting fibers exhibit maxima in Young's modulus of 35 Gpa and in tenacity of 0.9 Gpa. These values are approximately the same as those obtained from the best viscose fibers, but far from the theoretical values predicted for cellulose. The values found in the literature are not in full agreement ; Young's modulus is estimated to range from 90 to 250 Gpa, according to different authors (10-12). Consequently, the theoretical value of tenacity, estimated as one-tenth of the magnitude of the Young's modulus, ranges from 9 to 25 Gpa.

The dry-jet, wet-spinning system does not yield fibers with better mechanical properties, presumably because post-stretching cannot be performed following coagulation. D.S.C. analysis shows that, from -100°C to +10°C, cellulose solutions in amine oxides behave like thermoplastic materials, as illustrated in fig. 1 and 2. To evaluate the effect of post-stretching of cellulose fibers drawn from amine oxide solutions, a modified spinning system including a cold stage has been developed.

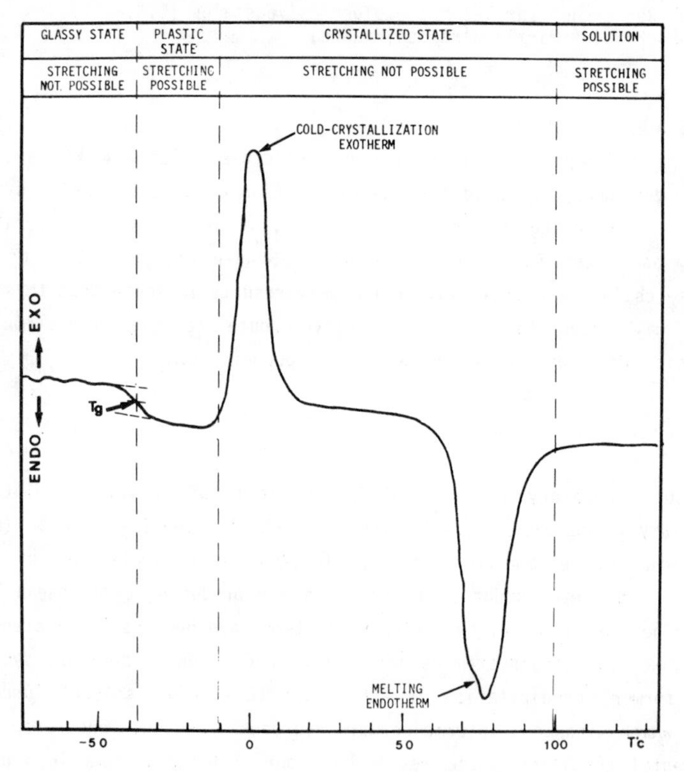

Fig.1. D.S.C. curve of a solution of cellulose in MMO quenched in liquid nitrogen. Scanning rate, 20°C/min.

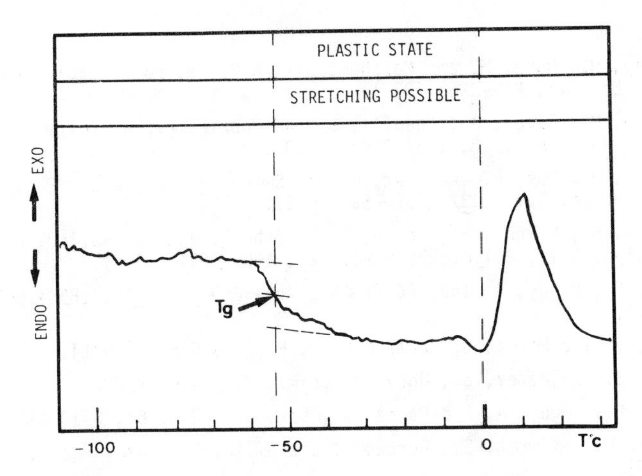

Fig.2. D.S.C. curve of a solution of cellulose in DMEAO quenched in liquid nitrogen. Scanning rate, 20°C/min.

The process can be summarized in the following 3 steps :

- The solution is extruded into a cold aprotic solvent (n-hexane, $CFCl_3$,...) where the cellulose-amine oxide mixture is vitrified.
- The bath temperature is increased to a temperature slightly above the material's Tg and then the fiber is stretched.
- The fiber is then coagulated in a polar solvent (water, ethanol, methanol,...) at a temperature lower than Tg.

These different steps must be carried out continuously to prevent the relaxation of the cellulose chains. For this purpose, a special laboratory spinning device, which approximates the temperature values obtained in D.S.C. analysis, has been built.

In preliminary trials, 500 % draw ratios have been performed and the resulting fibers have mechanical properties comparable to those obtained with the reference process. However, possible variations in post-stretching conditions, make this second process a better prospect for the future improvement of mechanical properties in cellulose fibers.

REFERENCES

1. Chanzy, H., Dubé, M. and Marchessault, R.H., J. Polym. Sci., Polym. Lett. Ed., 17, 219-226 (1979).
2. Chanzy, H., Péguy, A., Chaunis, S. and Monzie, P., J. Polym. Sci., Polym. Phys. Ed., 18, 1137-1144 (1980).
3. Chanzy, H., Noé, P., Paillet, M. and Smith, P., J. Appl. Polym. Sci., Appl. Polym. Symp., 37, 239-259 (1983).
4. Chanzy, H., Nawrot, S., Péguy, A., Smith, P. and Chevalier, J., J. Polym. Sci., Polym. Phys. Ed., 20, 1909-1924 (1982).
5. Maia, E., Péguy, A. and Pérez, S., Acta Cryst., B 37, 1858-1862 (1981).
6. Maia, E. and Pérez, S., Acta Cryst., B 38, 849-852 (1982).
7. Maia, E. and Pérez, S., Nouv. J. Chim., 7, 89-100 (1983).
8. Maia, E., Péguy, A. and Pérez, S., Can. J. Chem., 62, 6-10 (1984).
9. Chanzy, H., Nawrot, S., Pérez, S. and Smith, P., TAPPI Proc., 127-132 (1983)
10. Sakurada, I, Nakushina, Y. and Ito, T., J. Polym. Sci., 57, 651-660 (1962).
11. Sakurada, I., Ito, T. and Nakamae, K., Makrom. Chem., 75, 1-10 (1964).
12. Gillis, P., J. Polym. Sci., A2, 7, 783-794 (1969).

Part IX

NEW DEVELOPMENTS

FUTURE TRENDS IN POLYMER CHEMISTRY

G. SMETS

KULeuven, Laboratorium voor Macromoleculaire en Organische Scheikunde,
Celestijnenlaan 200 F, B-3030 Heverlee-Leuven, Belgium

SYNOPSIS

Several research orientations in polymer chemistry offering application possibilities are discussed successively, namely i) new methods of polymerization ii) transformation reactions on polymer iii) synthesis of graft- and block copolymers iv) organic photochemistry v) intermacromolecular complexation. These different items are discussed on the basis of recent data of the literature.

1 NEW METHODS OF POLYMERIZATION

The elaboration of new methods of polymerization is an important aspect of polymer chemistry. Promising examples are the group transfer polymerization of α.β.unsaturated esters, ketones, nitriles and carboxamides[1,2], the free radical ring opening polymerization of methylene-heterocyclic monomers[3,4] and the zwitterion polymerization, so called no-catalyst polymerization.
In the group transfer polymerization trimethylsilyl ketene acetals are used as initiator in the presence of a suitable catalyst, usually a bifluoride nucleophile. The trimethylsilyl group is transferred to the incoming monomer at each polymerization step by a living type mechanism producing at room temperature and almost in quantitative yields narrow molecular distribution polymers. Moreover the reactive end group of these polymers can easily be converted into other functional group giving telechelics or semi-telechelics. The reaction scheme can be illustrated in reaction scheme 1.

The driving force of the free radical ring-opening polymerization is the higher stability of a carbonyl C=O bond as compared to a carbon-carbon double bond. This polymerization proceeds easily with cyclic ketene acetals and aminals,

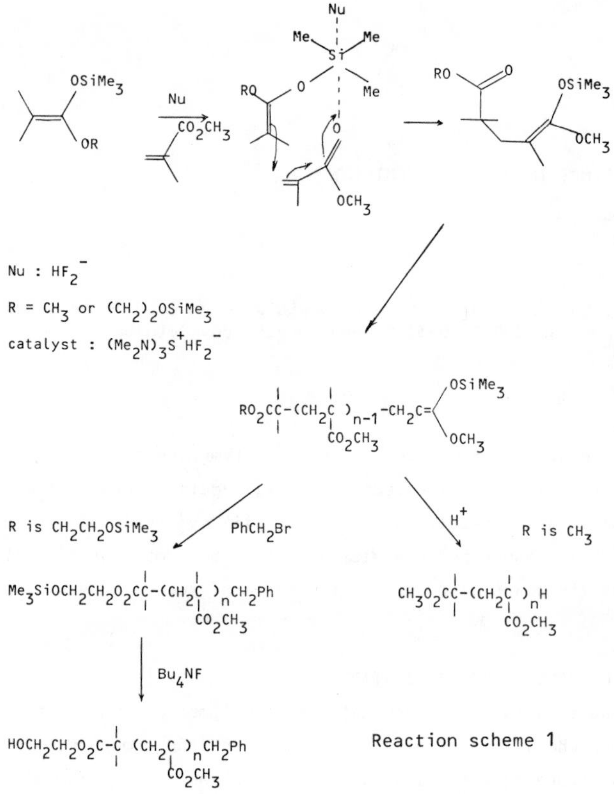

Nu : HF_2^-

R = CH_3 or $(CH_2)_2OSiMe_3$

catalyst : $(Me_2N)_3S^+HF_2^-$

R is $CH_2CH_2OSiMe_3$ $PhCH_2Br$

$Me_3SiOCH_2CH_2O_2CC-(CH_2C)_nCH_2Ph$

Bu_4NF

$HOCH_2CH_2O_2C-C(CH_2C)_nCH_2Ph$

H^+ R is CH_3

$CH_3O_2CC-(CH_2C)_nH$

Reaction scheme 1

with cyclic vinyl ethers, with unsaturated spiro orthocarbonates and orthoesters. In the case of the ketene acetal of tetramethylene glycol it can be schematized as follows :

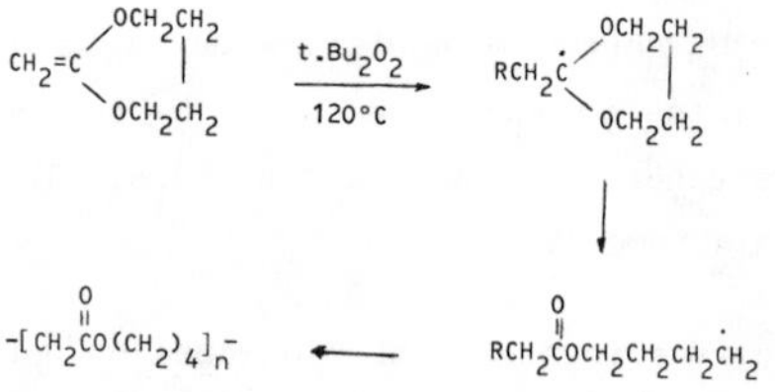

$$CH_2{=}C(OCH_2CH_2)_2 \xrightarrow[120°C]{t.Bu_2O_2} RCH_2\dot{C}(OCH_2CH_2)_2 \rightarrow RCH_2\overset{O}{\overset{\|}{C}}OCH_2CH_2CH_2\dot{C}H_2 \rightarrow -[CH_2\overset{O}{\overset{\|}{C}}O(CH_2)_4]_n- \quad \text{(eq. 2)}$$

By copolymerization of these compounds with vinyl monomers, incorporation of ester/amide groups into the backbone of the resulting copolymer becomes possible. Such copolymers are biodegradable; moreover they offer the possibility of synthesis of telechelics.

$$CH_2{=}C\begin{matrix}OCH_2CH_2\\ |\\ OCH_2CH_2\end{matrix} + CH_2{=}\underset{\phi}{CH} \longrightarrow -CH_2\overset{O}{\overset{\|}{C}}O(CH_2)_4-[CH_2\underset{\phi}{CH}]_nCH_2\overset{O}{\overset{\|}{C}}-O(CH_2)_4CH_2\underset{\phi}{CH}--$$

$$\downarrow H_3O^+$$

(eq. 3)

$$HO(CH_2)_4-[CH_2\underset{\phi}{CH}]_nCH_2COOH$$

These two new methods may have a rather limited but very specific domain of applications. Both methods illustrate however the still existing large possibilities offered by a basic knowledge of organic reaction mechanism. Such a conclusion arose also from the study of the already well described no-catalyst copolymerization [5,6,7]. When phenyl cyclic phosphonite reacts with β-propiolactone or with an equimolecular mixture of methyl acrylate and carbon dioxide, ring-opening of the phosphonium ring intermediates proceeds with an Arbusov rearrangement.

$$\text{(cyclic phosphonite, P–Ph)} + \beta\text{-propiolactone} \longrightarrow \text{(cyclic } \overset{+}{P}(Ph)CH_2CH_2CO_2^-) \rightleftharpoons -[CH_2CH_2O\overset{O}{\overset{\|}{P}}(Ph)CH_2CH_2COO]_n^-$$

$$\text{(cyclic phosphonite, P–Ph)} \xrightarrow{\text{Me-acrylate} + CO_2} \text{(cyclic } \overset{+}{P}(Ph)CH_2-\underset{CO_2CH_3}{CH}-CO_2^-) \rightleftharpoons -[CH_2-CH_2O\overset{O}{\overset{\|}{P}}(Ph)CH_2\underset{CO_2CH_3}{CH}COO]_n^-$$

(eq. 4)

This redox copolymerization was then extended to a new polymerization method, the so-called deoxy-polymerization in which an α-keto-acid is polymerized into a polyester in the presence of an equimolecular amount of sterically hindered cyclic phosphite .

n Ph-COCOOH ⟶ -[CH C O-]$_n$ (Ph, O)

P-OPh ⟶ (cyclic phosphate, O, OH)

(eq. 5)

2 TRANSFORMATION REACTIONS OF POLYMERS

Chemical modification of polymers remains an unlimited domain of application. It will be illustrated by few recent and typical examples.

Polyacetylenes prepared from acetylenic monomers are insoluble and intractable; consequently their purification and further fabrication are very difficult. It was shown recently that polyacetylene synthesis can also be obtained in a two-stage route via soluble precursor polymers[8,9,10]. Thus 7.8 substituted tricyclo [4.2.2.0] deca 3.7.9.dienes were polymerized in the presence of a metathesis catalyst. The resulting polydecatrienes are soluble and can be easily purified; they undergo a symmetry allowed thermal elimination reaction and are converted to polyacetylenes with adequate morphologies unaccessible by direct polymerization of acetylene monomer. The synthesis is represented in eq. 6 for the bistrifluoromethyl and benzo substituted compounds.

WCl_6 / Me_4Sn ; Δt, 100-150° ⟶ naphthalene + polyacetylene (cis) ; (tr.)

(eq. 6)

CF_3 CF_3 ; WCl_6 / Me_4Sn ; Δt ⟶ bistrifluoromethyl benzene + polyacetylene

The formation of bistrifluoromethyl benzene and of naphthalene in these retrocycloadditions constitutes the driving force of these reactions.

Reductive polyheterocyclisation is a new approach to the synthesis of polyheteroarylenes[11,12]. It is based on the reduction followed by cyclization of precursor o.nitrosubstituted heterochain polymers. The method presents several advantages when compared to the conventional high temperature polycondensation technique, namely the high stability of o.o' dinitrosubstituted aromatics when compared to the corresponding o.o' diamino or dihydroxy ones, the possibility of preparing the precursor polymers at much lower temperature and in milder conditions allowing the use of dicarboxylic acid dichlorides, and possibility of using the hydrochloric acid, which is produced, together with metals (usually iron) as reducing agent and at the same time as catalyst for heterocyclization. Comparison between the conventional high temperature method and the reductive heteropolyheterocyclization is shown below :

(a) H_2N, NH_2, HX, XH benzene + YCOArCOY $\xrightarrow{-HY}$ NHCOArCONH (XH, HX) $\xrightarrow[-2H_2O]{\Delta t}$ {–N=C(X)–Ar–C(X)=N–}$_n$

(b) HX, XH, NO_2, NO_2 benzene + YCOArCOY → XCOArCOX (NO_2, O_2N) $\xrightarrow[[H]]{metal}$ XCOArCOX (NH_2, H_2N) $\xrightarrow[HCl]{-2\ H_2O}$ polymer

(eq. 7)

Another interesting example of polymer modification is given by the functionalization of an unsaturated linear polyacetal obtained by cationic ring opening polymerization of 4H, 7H, 1.3 dioxepin[13,14]. Various addition reactions were carried out on the C=C double bonds of the repeat unit, e.g. with iodine isocyanate, and sulfenyl chloride. With peracid a quantitative epoxydation is obtained, and as expected the resulting epoxydized polyacetal can be cured and crosslinked.

$$\text{dioxepin} \xrightarrow{BF_3OEt_2} -OCH_2OCH_2(H)C{=}C(H)CH_2O--- \xrightarrow{mClBzO_3H} -OCH_2OCH_2CH\overset{O}{—}CH\text{-}CH_2--- \rightarrow \text{C.L.}$$

(eq. 8)

3 GRAFT- AND BLOCKPOLYMERS

The remarkable properties and several application possibilities of these materials (coating, adhesion, microemulsion, compatibilization, biopolymers, etc) justify the intensite research efforts spent for the synthesis and chemistry of macromers, telechelics and hemitelechelics. Macromers are linear polymers or oligomers terminating usually with an unsaturated group, which enables them to polymerize or copolymerize with other monomers. Their syntheses were discussed recently in a critical review to which the reader should refer.[15] Some general considerations seem however worthwhile to be done.

Homopolymer macromers must be very compact systems of which the hydrodynamic volume should be much smaller than that of a linear macromolecule of same molecular weight; it depends evidently on the length of the side chains and the overall degree of polymerization. Equation 9 shows the structure of the polymacromer of β-polystyrylethyl methacrylate.

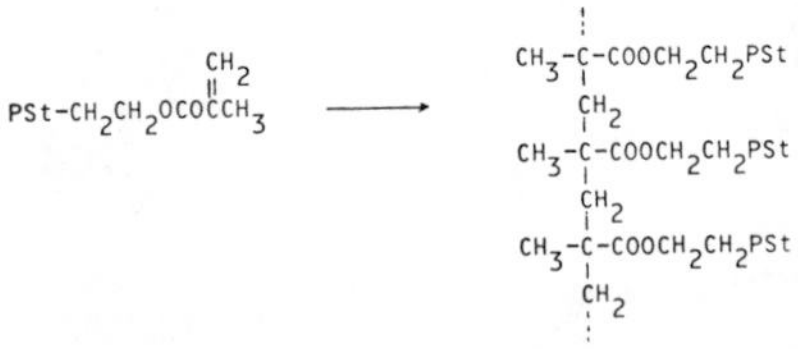

$$PSt{-}CH_2CH_2OCOC(=CH_2)CH_3 \longrightarrow \begin{array}{c} \vdots \\ CH_3{-}C{-}COOCH_2CH_2PSt \\ | \\ CH_2 \\ | \\ CH_3{-}C{-}COOCH_2CH_2PSt \\ | \\ CH_2 \\ | \\ CH_3{-}C{-}COOCH_2CH_2PSt \\ | \\ CH_2 \\ \vdots \end{array} \qquad \text{(Eq. 9)}$$

Only few examples are described; the applicability of gel permeation chromatography to these systems is however questionable. Macromers (molecular weight from 500 to 20.000) are however mostly used in free radical copolymerization for the synthesis of graft copolymers of well-defined structure concerning the number and length of the side-chains. Experimentally it has been shown that the copolymerization reactivity parameters of macromers are equal[16,17,18] or slightly lower[19,20] than those of the corresponding monomers, i.e. their reactivity is only little affected by their chain length. Thus the copolymerization of a styryl terminated polyamine with styrene furnishes comb-graft copolymers of following structure[21] :

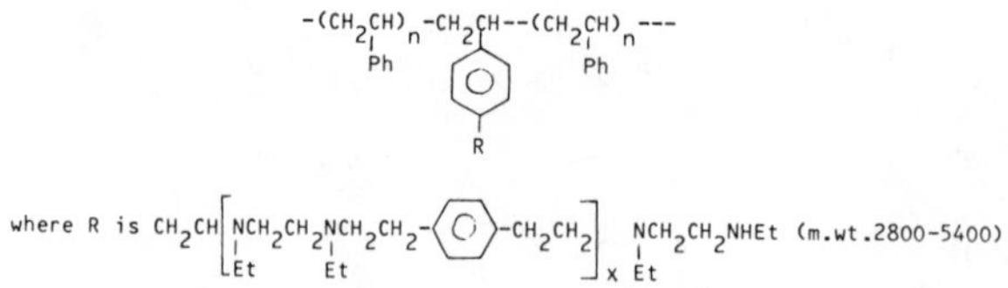

$$-(CH_2\underset{Ph}{CH})_n{-}CH_2\underset{C_6H_4-R}{CH}{-}{-}(CH_2\underset{Ph}{CH})_n{-}{-}{-} \qquad \text{(Eq. 10)}$$

where R is $CH_2CH\left[\underset{Et}{N}CH_2CH_2\underset{Et}{N}CH_2CH_2{-}C_6H_4{-}CH_2CH_2\right]_x\underset{Et}{N}CH_2CH_2NHEt$ (m.wt.2800-5400)

The number of grafts is roughly proportional to the initial concentration of macromer, although regularly lower when the mole percent macromer in the monomer mixture exceeds one mole % (around 20 wt.%).

From kinetic point of view several problems have to be taken into account. Indeed the low concentration of active sites causes a low reaction rate and often a low degree of polymerization, a low accessibility of radicals and unsaturated end groups may be responsible for a lower reactivity of the macromer, uncomplete conversion requires a difficult separation of unreacted macromer from the graft copolymer, transfer reaction with chain segments of the macromer may occur intramolecularily and cause supplementary branching.

The synthesis of block copolymers of well-defined structure is often based on the condensation of hemitelechelics and telechelics containing respectively one and two reactive endgroups. Structure A_xB_y results from the condensation of two hemitelechelics A_xB_y, structure $A_xB_yA_x$ from a bifunctional B_y and two monofunctional A_x while $(A_xB_y)_n$ multiblock is obtained from two bifunctionals A_x and B_y. Analogous difficulties as for the graft copolymers are again encountered. Special attention has to be given here to the correct functionality of the partners to be condensed. Moreover the degree of conversion of a condensation reaction is exceptionnally equal to the unity on account of the high dilution of the reactive functions. Consequently the degree of conversion can only be evaluated on the basis of the residual functionality and from the comparison of the calculated and experimentally determined molecular weights $\overline{M}_n$.[22] In most cases fractionation of the reaction product will be required.
In order to improve the yields of the condensation reaction it may be desirable to convert an end group into a more reactive one, e.g. the hydroxyl end groups of polyethylene oxide into amino ones in order to increase the nucleophilicity of the reagent[23]

$$HO(CH_2CH_2O)_xCH_2CH_2OH \xrightarrow{TosCl} TosO(CH_2CH_2O)_xCH_2CH_2OTos$$

(Eq. 11)

$$\xrightarrow[KOBu^t]{HSCH_2CH_2NH_2;} H_2NCH_2CH_2S(CH_2CH_2O)_xCH_2CH_2SCH_2CH_2NH_2$$

Sometimes a two-block system can be obtained directly with one condensation catalyst in a two-stage living polymerization, e.g. the use of tetraphenylporphinatoaluminum chloride for the synthesis of a polyester-polyether diblock system of which both sequences have a narrow molecular weight distribution[24].

$$\beta\text{-lactone} \xrightarrow{(TPP)AlCl} (TPP)Al\text{-}(OCOCH_2\overset{R}{C}H)_nCl \quad \text{living P, narrow MD, via TPPAl-carboxylate}$$

$$\xrightarrow{m \text{ epoxide } (R^1)} (TPP)Al\text{-}(O\underset{R^1}{C}HCH_2)_m\text{-}(OCOCH_2\overset{R}{C}H)_nCl \quad \text{via TPPAl-alkoxide} \qquad \text{(eq. 12)}$$

(TPP)AlCl : tetraphenylporphinato-Al-chloride

On the basis of their micellization and emulsifying properties it is to foresee an increasing use of amphiphilic block and graft copolymers in emulsion polymerization systems especially in non-aqueous media and for the polymerization in structured media as bilayers, vesicles, membranes, etc.[25,26]

4 ORGANIC PHOTOCHEMISTRY

A better understanding of the physical and chemical processes occurring on irradiation of polymers (excited state lifetime, energy transfer, electron exchange, etc.) and the relations between these processes and the organic structure of the polymers, tacticity and stereochemistry included, have permitted considerable progresses in organic polymer photochemistry. The primary processes have been however usually studied in dilute, sometimes very dilute, systems while on the other hand most photochemistry applications refer to bulk polymers. Large discrepancies are observed between these two states of matter; they originate from the intimate dependence of polymer photochemistry on polymer properties. In many cases bulk properties of the solid polymer (α and β transitions, segmental mobility) may superseed completely the primary photophysical and chemical processes, which then obey the classical WLF equation[27]. As a consequence the incorporation of photosensitive probes in a polymer has become a source of information about chain mobility, conformation, entanglement, etc.. Free volume distribution and distribution of reactive sites influence strongly the quantum yield of photochemical reactions. Thus photocrosslinking reactions not only depend on the short lifetime of the excited states, but also on the imperfect orientation of the reaction sites and the lack of back-bone deformation and conformational mobility[28]. With the same reactive groups, segment mobility is usually more important than the number of photocrosslinkable groups. Interesting are therefore the light-sensitive polymers based on unsaturated elastomers and unsaturated polyamide-imide containing dimethyl meleimide side- and end groups respectively.

(eq. 13)

+ ClS– · n · R · Cl · hν

cyclodimerization and photocrosslinking

–S–Ar–N · N–Ar–S–

In the solid state cage effects and diffusion restrictions limit severely the application of dissociation reactions while on the contrary they may be beneficial for electron donor-acceptor systems[30,31]. These considerations about interrelations polymer photochemistry and polymer structure and properties remain valid in the cases of organized systems, in mono- and multilayer assemblies, in membranes.

The photochemical synthesis of block polymers is another domain of interest on account of its applicability to most vinyl monomers, its selectivity of initiation by an adequate choice of the photosensitive groups, and the low temperature of reaction. The block copolymers can be obtained either by incorporation of light sensitive groups in a prepolymer (condensation polymer) either by putting photosensitive end groups on a preexisting polymer; in both cases the photolysis is carried out in the presence of a polymerizable monomer. A recent example is given by the photolysis in the presence of methyl methacrylate of polybisphenol-A-carbonate containing incorporated benzoin ether groups.

$$H(OC_6H_4\overset{Me}{\underset{Me}{C}}C_6H_4O\overset{O}{C})_mOC_6H_4\overset{O}{C}\underset{OMe}{C}H\,C_6H_4O(\overset{O}{C}OC_6H_4\overset{Me}{\underset{Me}{C}}C_6H_4O)_nH$$

hν, methyl methacrylate

(eq. 14)

$$\text{poly(bisphenol A carbonate)}_mOC_6H_4O\overset{O}{C}\,(CH_2\overset{Me}{\underset{CO_2Me}{C}}-)_xH$$

block copolymer

5 INTERMACROMOLECULAR COMPLEXES

Macromolecules with complementary binding sites may associate into interpolymer complexes[33,34]. The origin of these interactions may be i) Coulombic forces on mixing oppositely charge polyelectrolytes, e.g. poly-N-ethyl-vinylpyridinium salts with sodium polymethacrylate or sodium polyethylenesulfonate,and polydimethylamino ethyl methacrylate hydrochloride with sodium polyphosphate[35,36]. ii) hydrogen-bonding between H-donor and H-acceptor groups, e.g. poly(meth)acrylic acid and polyethyleneoxide, polyvinylalcohol and poly-N-vinylpyrrolidone[37]. iii) charge transfer between electron-donor and -acceptor polymers, e.g. mixture of polymers carrying N-substituted carbazol (donor) and tri-nitrofluorenone (acceptor) side groups[38]. iv) van der Waals' forces resulting in stereocomplexes, as isotactic polymethyl methacrylate and syndiotactic polymethacrylic acid or synd. polymethyl methacrylate[39,40].

The growing interest in these interpolymer complexes results partially from their potential uses, as biopolymers, ultrafiltration and dialysis membranes, coating with antistatic properties, etc... From scientific point of view, the formation of interpolymer complexes introduced the concept of template (matrix) polymerization in which monomers or/and propagating chains are adsorbed on the matrix polymer. Presumably such adsorption (binding) will affect the polymerization kinetics, and possibly the internal structure and molecular weight of the newly formed polymer. In this respect, interesting is the free radical polymerization of N-2-methacryloyloxyethyl derivative of adenine (MAOA) in the presence of poly-methacryloyloxyethyl thymine(poly MAOT). The initial rate of poly-

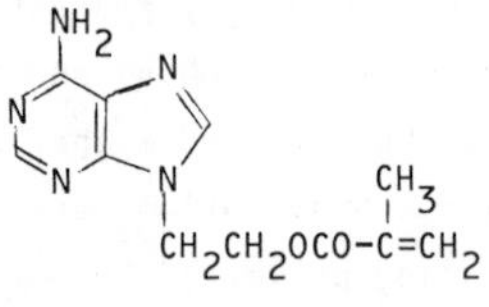

NH_2 … $CH_2CH_2OCO\text{-}C(CH_3){=}CH_2$ O … NH … R … $CH_2CH_2OCO\text{-}\!\left[\!\text{-}CH_2\text{-}C(CH_3)\text{-}\right]_n$ (eq. 15)

N-β-methacryloyloxyethyl adenine (MAOA)

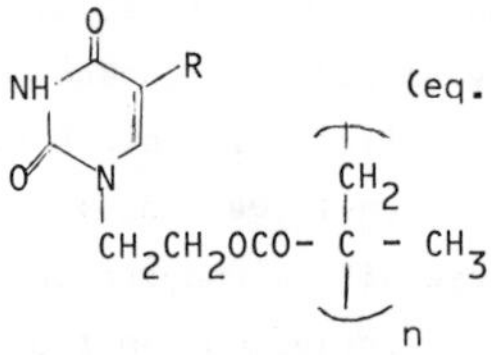

R = CH_3 poly MAOT thymine

R = H poly MAOU uracil

merization is about four times greater in the presence of poly MAOT than in its absence and maximum rate is obtained at base-base pairing stoichiometry. However the influence of the molecular weight was not observed; on the contrary the rate of polymerization depends on the stereoregularity of the matrix polymer. The polymerization of MAOA in the presence of poly MAOU proceeds in following order isot. > syndio. > atactic[41,42]. Specific interaction between complementary nucleic acid bases seem also to play an important role in the permeation through hydrogel membrane[43].

On the other hand non-stoichiometric polyelectrolyte complexes participate in interpolymer exchange and substitution reactions. The knowledge of the reaction mechanism should contribute considerably to a better understanding of biological processes[36].

Acknowledgement - Financial support of the Ministry of Science and Scientific Programmation is gratefully acknowledged.

REFERENCES

1. O.W. Webster, W.Z. Hertler, D.Y. Sogah, W.B. Farnham & T.V. Rajanbabu, J.Am. Chem.Soc. 105, 5706 (1983)
2. D.Y. Sogah, O.W. Webster, J.Polym.Sci. Polym.Lett.Edit. 21, 927 (1983)
3. W.J. Bailey, Polymer Prepr. A.C.S. 25, 1, 210 (1984) and references therein
4. W.J. Bailey, Makromol.Chem.Suppl. (1985) in press, Polymer Chemistry Meeting, Leuven, September 1984.
5. T. Saegusa, Pure and Applied Chemistry 53, 691 (1981); Proceedings of the Robert A. Welch Foundation Conferences on Synthetic Polymers, Nov.15 (1982) Houston, Texas, and references therein.
6. T. Saegusa, S. Kobayashi and T. Kobayashi, Macromolecules, 14, 463 (1981)
7. S. Kobayashi, T. Yokoyama, K. Kawabe and T. Saegusa, Polymer Bull. 3, 585 (1980)
8. M.M. Ahmad, A.B. Alimuniar, J.H. Edwards, W.J. Feast, I.S. Millichamp and S. Spanomanolis, Polymer Prepr. A.C.S., 25, 2, 217 (1984)
9. J.H. Edwards, W.J. Feast and D.C. Bott, Polymer 25, 395 (1984)
10. D.C. Bott, Polymer Prepr. A.C.S., 25, 2, 219 (1984)
11. V.V. Korshak, A.L. Rusanov and D.S. Tugushi, Polymer, 25, 1539 (1984)
12. V.V. Korshak and A.L. Rusanov, Vysokomolek.Soedinenia, 26, 1 (1984)
13. H. Hellerman and R.C. Schulz, Makromol.Chem. Rapid Commun., 2, 585 (1981)
14. R.C. Schulz, Makromol.Chem.Suppl. (1985) in press, Polymer Chemistry Meeting, Leuven, September 1984.
15. P. Rempp and E. Franta, Adv.Polym.Sci. 58, 3 (1984)
16. K. Ito, N. Usami and Y. Yamashita, Macromolecules, 13, 216 (1980)
17. Y. Yamashita, K. Ito, H. Mizuno and K. Okada, Polymer J., 14, 255 (1982)

18. Y. Yamashita, Y. Tsukahara, K. Ito, K. Okada and Y. Tajima, Polym.Bull., 5, 335 (1981)
19. R. Asami, M. Takaki and T. Matsus, Prepr.1st SPSJ Internat.Polym.Confer., Kyoto, Aug. 1984, p.76.
20. K. Ito, K. Uchida, A. Hayashi, H. Tsuchida, T. Kitano and E. Yamada, Prepr. 1st SPSJ Intern.Polym.Confer., Kyoto, Aug. 1984, p.77.
21. T. Nishimura, M. Maeda, Y. Nitadori, T. Tsuruta, Makromol.Chem. Rapid Commun., 1, 573 (1980)
22. Y. Chujo, A. Hiraiwa and Y. Yamashita, Makromol.Chem., 185, 2077 (1984)
23. G. Ziegast and B. Pfannemüller, Makromol.Chem. Rapid Commun., 5, 363 (1984)
24. T. Yasuda, T. Aida and S. Inoue, Macromolecules, 17, 2217 (1984)
25. G. Riess and D. Rogez, ACS Polymer Preprints, 23, 19 (1982)
26. G. Riess, Makromol.Chem.Suppl. (1985) in press; Polymer Chemistry Meeting, Leuven, Sept. 1984.
27. L. Bokobza, E. Pajot-Augy, L. Monnerie and A. Castellan, H. Bouas-Laurent, Polym.Photochem., 5, 191 (1984)
28. E. Pitts and A. Reiser, J.Am.Chem.Soc., 105, 5590 (1983)
29. J. Berger, Polymer, 25, 1629 (1984); Polymer, 26, Commun. 11 (1985)
30. G. Smets, 1st Internat.Conference of SPSJ, Kyoto, Aug. 1984, Polymer J., in press, Jan. 1985
31. G. Smets, S.N. El Hamouly and T.J. Oh, Pure and Appl.Chem., 56, 439 (1984)
32. G. Smets and T. Doi, Intern.Symposium on New Trends in the Photochemistry of Polymers, Stockholm, Aug.26-29, 1985, in press
33. E. Tsuchida and K. Abe, Adv.Polym.Sci., 45, 3 (1982) and ref. therein.
34. E.A. Bekturov and L.A. Bimendina, Adv.Polym.Sci., 41, 99 (1981)
35. V.A. Kabanov, A.B. Zezin, Pure Appl.Chem., 56, 343 (1984); Makromol.Chemie Suppl., 6, 259 (1984)
36. V.A. Kabanov, A.B. Zezin, V.A. Izumrudov, T.K. Bronich and K.N. Bakeev, Makromol.Chemie Suppl. (1985) in press, Polymer Chemistry Meeting, Leuven, Sept. 1984.
37. Y. Osada, J.Polym.Sci., Polym.Chem.Ed., 17, 3485 (1979)
38. S. Tazuke and H. Nagahara, Makromol.Chemie, 181, 2217 (1980)
39. A. De Boer, G. Challa, Polymer, 17, 633 (1976)
40. F. Roerdink, G. Challa, Polymer, 21, 509 (1980)
41. M. Akashi, H. Takada, Y. Inaki and K. Takemoto, J.Polym.Sci., Polym.Chem.Ed., 17, 747 (1979)
42. K. Takemoto and Y. Inaki, Adv.Polym.Sci., 41, 1 (1981)
43. M. Miyata, S. Senda, K. Takemoto, Makromol.Chemie, 185, 647 (1984)

RECENT INVESTIGATIONS OF INTERPENETRATING POLYMER NETWORKS

KURT C. FRISCH

Polymer Institute, University of Detroit, Detroit, Michigan 48221

SYNOPSIS

High loss IPNs which dampen mechanical vibrations over a broad temperature range of 5-35°C have been prepared. These IPNs consisted of a 50/50 PU/epoxy ratio. The PU component consisted of a polycaprolactone polyol, a modified MDI and small amounts (0-10%) of 1,4BD. A bis(glycidyl) ether of bisphenol A was used as the epoxy component and dibutyltin dilaurate and BF_3-etherate were used as the urethane and epoxy catalysts, respectively. These IPNs exhibited a tan δ greater than 1.75 over the desired temperature range. A plasticizer (Sundex 740-T) was found to tailor the maximum in the tan δ peak while a study of fillers so far has resulted in no conclusive trends and requires further study.

IPNs based on polyurethanes and (methacrylate) copolymers with opposite charge groups (tertiary amine and carboxyl groups, respectively) have been prepared and their properties and morphology studied. The mechanical properties and compatibility between the component networks were significantly improved presumably due to the interaction of the opposite charge groups in these systems.

1 INTRODUCTION

Interpenetrating polymer networks are unique polymer alloys which consist of two or more crosslinked polymer networks held together by permanent entanglements with only accidental covalent bonds between them. They are polymeric catenanes produced by homocrosslinking of two or mor polymer systems. Formation of IPNs is the only way of combining crosslinked polymers with the resulting mixture exhibiting (at worst) only limited phase separation.

A significant feature of IPNs is the fact that the combination of varied chemical networks, resulting often in controlled different morphologies, frequently results in synergistic behavior. For instance, if one polymer is elastomeric and the other polymer is glassy at room temperature, one obtains either a reinforced rubber or a high impact plastic depending upon which phase is continuous.[1]

IPNs synthesized to date exhibit varying degrees of phase separation which depends primarily on the compatibility of the respective polymers. Complete compatibility is not necessary to achieve complete phase mixing since the permanent entanglements (catenation) can effectively prevent phase separation. With intermediate conditions of compatibility, intermediate and complex behavior results. Thus, IPNs with dispersed phase domains ranging from a few micrometers (incompatible) to a few tens of nanometers (intermediate)[2] and finally, to those with no resolvable domain structure (complete mixing)[1] have been reported.

This paper deals specifically with two different aspects of IPNs, one aimed at vibration attenuation and the other with the effect of charge groups in two component IPNs.

2 IPNS FOR VIBRATION ATTENUATION

2.1 General

In this study IPNs of low and high Tg polymers were prepared in order to obtain "semi-compatible" behavior, to thus attain the desired broad range of mechanical energy absorption.[3] Polyurethane/epoxy IPNs were selected because previous investigations of IPNs based on similar polymers resulted in homogenous morphologies (an unusual phenomenon) and the simplest approach seemed to be "decompatibilize" the polymers.[4,5] Emphasis was placed on maximizing the dynamic mechanical properties of the resulting IPNs, i.e. synthesis of IPNs with a high and broad tan δ (tan $\delta > 1.75$ in the temperature range of 5°-35°C) as well as a Young's modulus (E') of $10^8 - 10^9$ dynes/cm^2 in the rubbery plateau.

2.2 Preparation and Characterization of IPNs

The IPNs were parepared by the one-shot simultaneous polymerization technique (SIN).[6] They were composed of an elastomeric polyurethane component (the low Tg component) and a glassy epoxy component (the high Tg component). The polyurethanes were prepared from a carbodiimide-modified diphenylmethane diisocyanate (MDI) (Isonate 143-L, The Upjohn Co.), two different polyols an acrylonitrile-containing graft polyether polyol (Niax polyol 31-28, Union Carbide) and a polycaprolactone polyol (Tone 0230, Union Carbide), and 1,4-butanediol (1,4BD,GAF Corp.) chain extender,employing a dibutyltin dilaurate catalyst.

Three different types of epoxy resins were used: a bisglycidyl ether of bisphenol A (DER 330, Dow Chemical Co.), an epoxy novolac (DEN 431, Dow Chemical Co.), and a tetraglycidyl ether of methylene bis(aniline) (Araldite MY 720, Ciba-Geigy Co.). Three epoxy catalysts were used: a Lewis acid etherate (BF_3-etherate, Eastman Chemical), a BCl_3-amine complex (XU-213, Ciba-Geigy) and a tertiary amine catalyst (DMP-30, Rohm & Haas). These catalysts were selected because they cause crosslinking of the epoxy resins with a minimum of side reactions with any of the polyurethane components.

In order to obtain a broader and higher tan δ as well as a higher rubbery modulus, a number of fillers having varying geometries as well as a plasticizer were added in some formulations. They included mica, Wollastonite, alumina trihydrate, carbon black and glass fibers (with and withouth silane sizing). The plasticizer used was an oil based mostly an aromatic and naphthenic fractions (Sundex 740T, Sun Oil Co.).

The preparation of the IPNs was carried out by mixing of two components. One component contained the isocyanate and epoxy resin and the other component contained the respective polyol, 1,4-butanediol and catalyst (when fillers or plasticizers were used, they were blended in with this component). The two components were mixed for 30 seconds at room temperature and the mixture was then quickly poured into a pre-heated mold and pressed on a laboratory platen press at 100°C. The sample (approximately 4 minutes gel time) was demolded after 30 minutes and was then post cured for 5 hours at 120°C. All dynamic mechanical measurements were made on a Rheovibron dynamic viscoelastometer, DDVII (Toyo Mfg. Co.) at 110 Hz and a scanning rate of 1-2°C per minute in the glass transition region or 3-5°C per minute in the non-transition region. The specimens were in the form of rectangular films (2.4x0.2x0.05cm). All tests were carried out at a frequency of 110 Hz.

2.3 Results and Discussion

2.3.1 Polyurethane Behavior: At first pure polyurethanes based on Isonate 143-L and various polyols without chain extender, using T-12 catalyst were evaluated on a Rheovibron to determine which type of polyol gave the highest tanδ values. The results were poly(oxypropylene) glycol > poly(oxypropylene-oxyethylene) triol with 21% graft acrylonitrile (Niax 31-28, Union Carbide) > poly(1,4-oxybutylene) glycol > poly(caprolactone) glycol. Mixtures of these polyols in the resulting polyurethanes indicated that there was no effect on the transition range (breadth) and the tan δ values exhibited merely values intermediate between those of the polyurethanes based on the individual polyols.

However, the final polyol selection was made not only just on the tan δ values since a fast curing reaction inject molding (RIM) IPN was the desired method of synthesis. Therefore all subsequent systems were based on polyurethanes based on Isonate 143-L and either a poly(caprolactone) glycol (Niax Tone 0230, Union Carbide) or a graft acrylonitrile polyether triol (Niax 31-28, Union Carbide).

2.3.2 Effect of Chain Extenders: Various weight fractions of 1,4BD chain extender were introduced into the PU/epoxy IPNs. The Rheovibron data of these IPNs indicated that the degree of phase separation increased (i.e. E" and tan δ peaks broadened) as the 1,4BD content increased from 2-15% based on polyols. These results were independent of the type of epoxy used and was most evident when the PU content in the IPN was 50% or greater. However, despite an increase in the height of the tan δ and the broadening of tanδ and E" peaks which indicated a broader temperature range for vibration attenuation, a simultaneous shifting of these peaks to a higher temperature occurred. Since the object of this study was to prepare IPNs of high damping from approximately 5 to 35°C, only small amounts of 1,4BD (0-10%) were used for the best performing systems.

Addition of trimethylolpropane (TMP) and N,N-bis(2-hydroxypropyl) aniline (Isonol N-100, Upjohn Co.) as chain extenders for the PU greatly broadened the tan δ and E" peaks in the following order: Isonol N-100 > TMP > 1,4BD (60/40 PU/epoxy IPN). However, both the Isonol N-100 and TMP caused a shift to a higher temperature (out of the desired range) as well as drastically lowering the tan δ peaks due to the increase in stiffness (E') as a result of the increase in the hard segment content of the PU.

2.3.3 Effect of Epoxy Type: Different types of epoxies were investigated, among them a) a bis(glycidyl) ether of bisphenol A, (DER 330, Dow Chemical Co.), b) an epoxy novolac (DEN 431, Dow Chemical Co.), and c) a tetraglycidyl ether of 4,4'-methylene bis(aniline) (Araldite MY 720, Ciba-Geigy Co.). DER 330 was selected because it gave a higher though narrower tan δ peak than DEN 431 in the desired temperature range and Araldite MY 720 caused too much stiffness, and hence, low damping.

2.3.4 EFfect of Epoxy Catalyst: Different types of epoxy catalysts were used but only BF_3-etherate alone produced IPNs in which a high tanδ peak resulted presumably due to the fact that the epoxy with it functions somewhat like a highly crosslinked plasticizer instead of a rigid plastic. Hence BF_3-etherate was used in the subsequent formulations.

2.3.5 Effect of PU/Epoxy Ratio: The PU/epoxy ratio was varied to take advantage of the lightly crosslinked epoxy. From an initial value of 60/40 PU/epoxy, systems were prepared at ratios of 50/50, 40/60 and 30/70. The tan δ values kept climbing from a maximum of 1.6 in the 50/50 IPN to off-scale (tanδ >1.75) in the 40/60 and 30/70 blends. The tanδ at 30/70 PU/epoxy went off scale and did not return back on scale because the material became too soft to obtain dynamic mechanical measurements as the temperature reached 40°C. Hence, the 50/50 and 40/60 PU/epoxy mixtures consisting of Tone 0230-Isonate 143-L (PU) and DER 330 (epoxy) with the respective catalysts were deemed to be the best systems (see Figs. 1 & 2).

2.3.6 Effect of Plasticizer and Fillers: To a system containing Niax Tone 0230, 10% 1,4BD, Isonate 143-L, and DER 330, was added a plasticizer consisting essentially of aromatic and naphthenic hydrocarbons (Sundex Oil 740T, Sun Oil Co.), in amounts ranging from 10-20%.

It was shown that as more plasticizer was incorporated, a decrease in the temperature at which maximum damping occurs resulted without affecting the other system variables (i.e. tanδ peak area or E'). Thus, the plasticizer can be used to tailor the location of the maximum in the tan δ peak.

A series of fillers was also evaluated including alumina trihydrate, chopped glass fibers, mica, Wollastonite, and carbon black. A modest increase in the tanδ was observed for alumina trihydrate and 10% untreated chopped glass fibers while mica and Wollastonite (10% concentration) had no effect on the tanδ of the IPNs although both increased the storage modulus E' slightly while 10% carbon black decreased the maximum tan δ significantly.

3 IPNS WITH OPPOSITE CHARGE GROUPS

3.1 General

Few studies have been reported in literature with regard to improving the compatibility between polymers by introducing opposite charge groups into the respective polymer backbones. We have recently prepared two component IPNs with opposite charge groups using polyurethanes and methacrylate copolymers as well as polyurethanes and epoxies. Three component IPNs with opposite charge groups based on polyurethanes, epoxies and acrylics have also been prepared.[5-7] For brevity sake, only two component IPNs based on polyurethanes and methacrylate copolymers will be described in this paper.

3.2 Preparation and Characterization of IPNs

The composition of the polyurethanes, polymethacrylates and IPNs are shown in Table 1. The prepolymer of MMA and the precopolymer of MMA and MAA were prepared by solution polymerization at different mole ratios of MMA/MMA (4/1, 6/1, 8/1 and 10/1) in the presence of a chain transfer agent (CCl_4), to obtain low molecular weight prepolymers, which were stored in the refrigerator prior to use. The solids concentration in these prepolymer solutions was determined by removal of the solvent and monomer.

Polyurethane prepolymers made from MDI and PTMO 1000 at a NCO/OH ratio of 2/1 were dissolved in dioxane which resulted in a 50-60% (by wt.) solution and were mixed with a MBCA solution (50% MBCA in cyclohexanone) to prepare film samples by evaporation of the solvent.[6] PTMO 2000 and BHPA were mixed to obtain a MW of 1000 and reacted with MDI at a NCO/OH=2/1 to yield a PU prepolymer that contained tertiary amine groups (NPU). This prepolymer was mixed with MBCA as described above.

Two types of IPN were prepared: IPN-I. The PU prepolymer solution from MDI and PTMO 1000 was mixed with the prepolymer solution of MMA at a

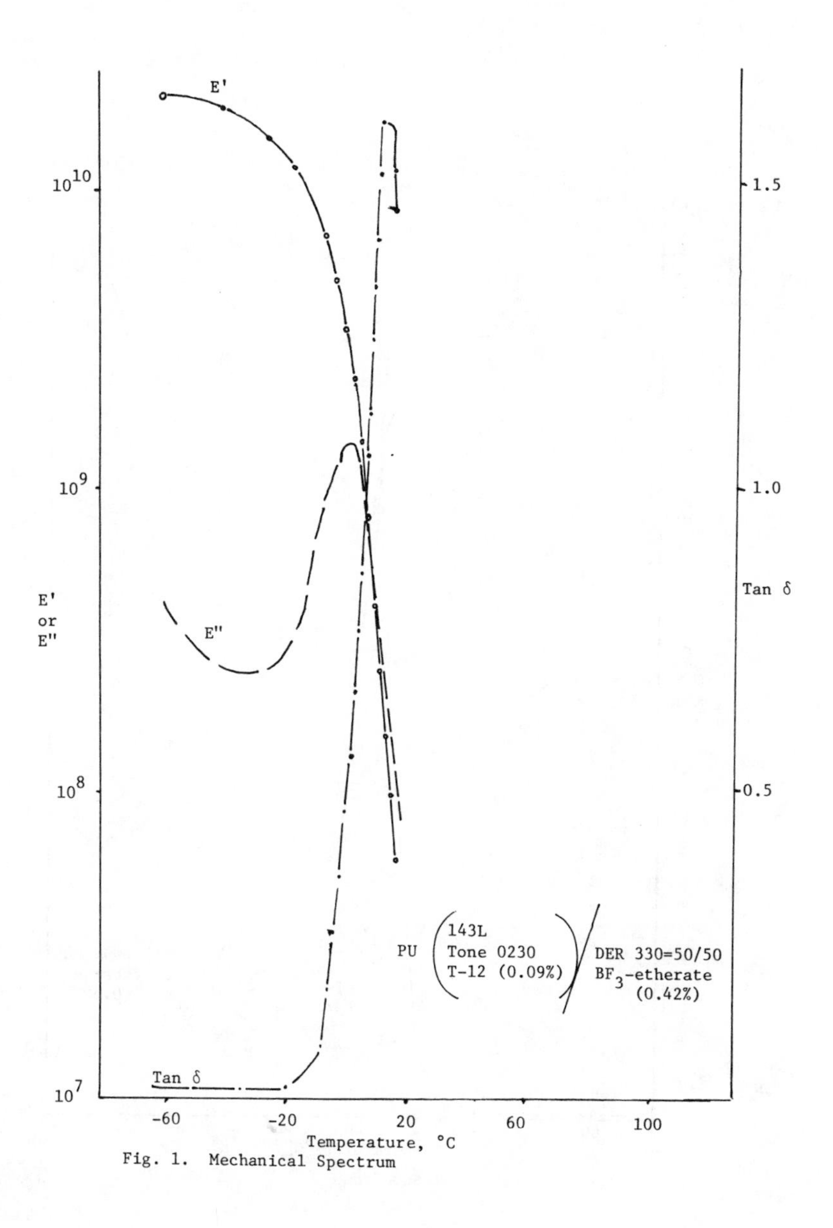

Fig. 1. Mechanical Spectrum

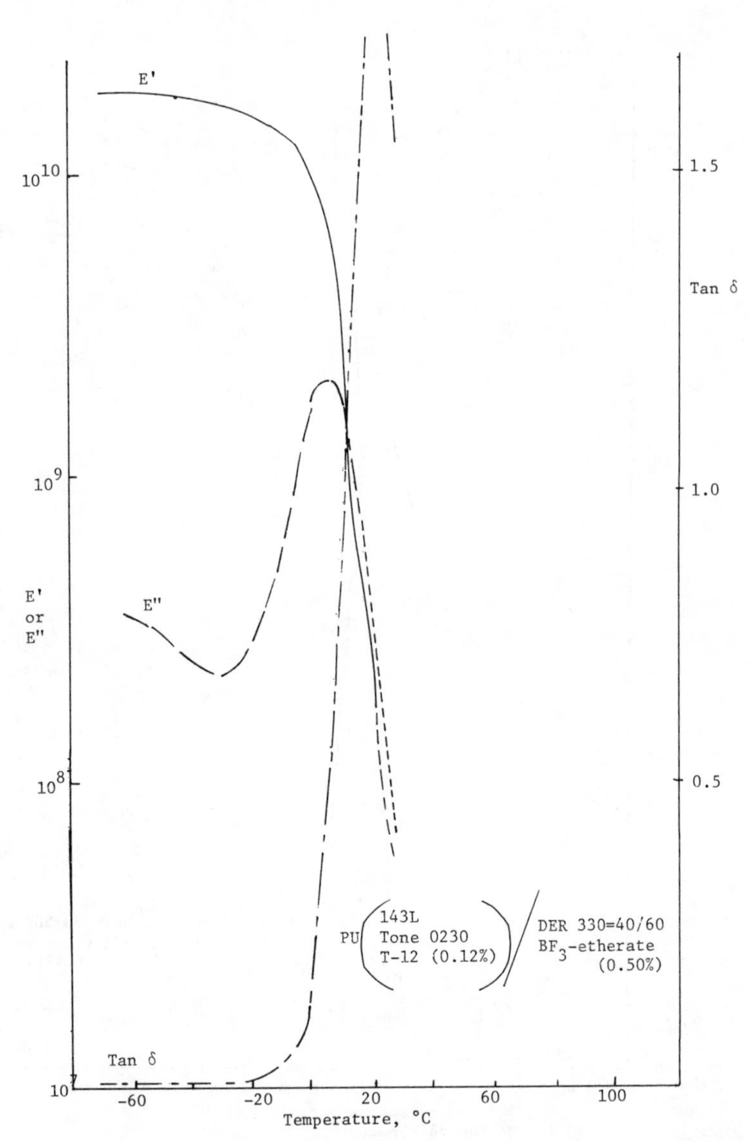

Fig. 2. Mechanical Spectrum

weight ratio of PU/MMA=80/20. It was then mixed with MBCA (NCO/NH_2=1.05), ethylene glycol dimethacrylate (EGDMA), and benzoyl peroxide (BPO). This is an IPN without a charge group (Table 1).

IPN-II. The NPU prepolymer solution was mixed with the prepolymer solution of MMA and MAA which had different mole ratios of MMA/MAA at a weight ratio of 80/20. MBCA, EGDMA, and BPO were then added to this solution to prepare film samples. These IPNs contained charge groups in both polymer components (Table 1). The IPNs were characterized by their mechanical (Instron) and adhesive properties (lap shear and 180° peel strength) as well as by their Tg's (DSC) and scanning electron microscopy (SEM).

3.3 Results and Discussion

The effects of charge groups on the mechanical properties of IPNs are seen in Table 2. It is evident that the IPN-II series with opposite charge groups exhibits the best balance of properties compared with pure PU. The improvement in the IPN-II series is due to the compatibility between the two polymers because of the interaction between opposite charge groups (tertiary amine and carboxyl groups). This implies that this specific interaction gave rise to a negative enthalpy of mixing, and as a result the free energy of mixing, ΔG_{mix} could perhaps be zero or negative.[8] The last sample in Table 2 which was prepared without a chain transfer agent, exhibited lower physical properties than those in the preceding sample which was prepared in the presence of CCl_4. It is interesting to note the effect of the concentration of charge groups on the mechanical properties of IPNs as shown in Table 3. As the concentration of carboxyl groups increases (by decreasing the MMA/MAA ratio), the compatibility between the two polymers will improve because of the interaction between the respective charge groups.[9,10] IPN-I samples without charge groups exhibited lower properties than the IPN-II samples with charge groups, due to the absence of interaction between molecular chains.

The adhesive properties of these IPNs on aluminum plates are shown in Table 4. The increase in the lap shear and peel strength is remarkable with increasing concentration of carboxyl groups in the IPN-II series. The increase in the lap shear strength in the P(MMA-MAA) with increasing carboxyl group concentration is also noteworthy.

The effect of opposite charge groups on the morphology of these systems was apparent in their Tg measurements as well as the SEM micrographs. Two Tg's were observed in the IPN-I series (no charge groups) while only one Tg was found in the IPN-II series. Likewise, the SEM micrographs (not shown here) clearly indicated the effect of no charge groups as well as the effect of increasing concentration of carboxyl groups in the IPN-II series. The particle sizes of the latter decreased with increasing charge groups concentration of carboxyl groups in the IPN-II series while the IPN-I series exhibited by far the largest particle sizes.

ACKNOWLEDGEMENTS

The author wishes to acknowledge the valuable assistance of Dr. H.L. Frisch of SUNY, Albany, Drs. D. Klempner and H.X. Xiao of the Polymer Institute, University of Detroit, and Dr. R. Ting of the U.S. Naval Research Lab, Orlando, FL in these investigations.

TABLE 1

Composition of Polyurethanes, Polymethacrylates, and IPNs

Description	Materials	Composition
PU	Polyurethane	PTMO[1] 1000, MDI and MBCA,[2] NCO/OH = 2/1, NCO/NH$_2$ = 1.05
NPU	Tertiary amine-containing polyurethane	PTMO 2000, MDI, BHPA,[3] and MBCA, NCO/OH = 2/1, NCO/NH$_2$ = 1.05
PMMA	Poly(methyl methacrylate)	Prepolymer solution of MMA
P(MMA-MAA)	Poly(methyl methacrylate-methacrylic acid)	Prepolymer solution of MMA and MAA at different ratios of MMA/MAA (4/1, 6/1, 8/1, and 10/1)
IPN-I	PU/PMMA	PU/PMMA = 80/20
IPN-II	NPU/P(MMA-MAA)	NPU/P(MMA-MAA) = 80/20

1 1,4-Poly(oxybutylene) glycol

2 Methylene bis(2-chloroaniline)

3 Bis(2-hydroxypropyl) aniline

TABLE 2

Effect of Charge Groups on Mechanical Properties of IPNs

Composition	MMA/MAA (mol)	Tensile strength (psi)	Modulus 100% (psi)	Modulus 300% (psi)	Elongation (%)	Hardness shore A
PU (100%)	—	4401	1084	1574	910	90
NPU (100%)	—	5533	1589	2457	650	92
IPN-I (80/20)	1/0	5541	1571	2716	557	92
IPN-II (80/20)	4/1	7226	1699	2966	685	91
IPN-II (80/20)	4/1	6781	1610	2906	695	92

TABLE 3

Effect of Charge Groups Concentration in Methacrylate Polymers on Mechanical Properties of IPNs

Composition	MMA/MAA (mol)	—COOH (mol %)	Tensile strength (psi)	Modulus 100% (psi)	Modulus 300% (psi)	Elongation (%)	Hardness Shore A
NPU	—	—	5533	1589	2457	650	92
IPN-II (80/20)	4/1	20	7226	1699	3199	685	91
IPN-II (80/20)	6/1	14.3	6341	2001	3135	633	91
IPN-II (80/20)	8/1	11.1	6419	1850	3076	643	91
IPN-II (80/20)	10/1	9	5925	1900	2966	623	92
IPN-I (80/20)	—	—	5826	1198	2048	720	88

TABLE 4

Effect of Charge Groups of Methacrylate Polymers on Adhesive Strength of IPNs

Composition	MMA/MAA (mol)	—COOH[a] (mol %)	Substrate	Lap shear (psi)	Peel strength (lb in.)	Type of failure
IPN-II (80/20)	4/1	20	A1-A1	744	13	Cohesive
IPN-II (80/20)	6/1	14.3	A1-A1	588	10	Cohesive
IPN-II (80/20)	8/1	11.1	A1-A1	313	8	Cohesive
IPN-II (80/20)	10/1	9	A1-A1	267	7	Cohesive
IPN-I (80/20)	—	—	A1-A1	201	5	Cohesive
P(MMA-MAA)	4/1	20	A1-A1	314	—	Cohesive
P(MMA-MAA)	6/1	14.3	A1-A1	254	—	Cohesive
P(MMA-MAA)	8/1	11.3	A1-A1	209	—	Cohesive
P(MMA-MAA)	10/1	9	A1-A1	92	—	Cohesive
PMMA	—	—	A1-A1	91	—	Cohesive
NPU	—	—	A1-A1	198	8	Adhesive

REFERENCES

1. Frisch, K.C., Frisch, H.L., Klempner, D. and Mukherjee, S.K., J. Appl. Polym. Sci., 18, 689 (1974); Frisch, K.C., Klempner, D., Antczak, T. and Frisch, H.L., ibid, 18, 683 (1974); Frisch, K.C. Klempner, D., Migdal, S. and Frisch, H.L., J. Polym. Sci., (A-1), 12, (4), 885 (1974); Frisch, K.C., Klempner, D., Migdal, S., Frisch, H.L. and Ghiradella, H, Polym. Eng. Sci., 14, 76 (1974); Frisch, H.L., Frisch, K.C., and Klempner, D. ibid, 14, 562 (1974).
2. Sperling, L.H. and Friedman, D.W., J.Polym. Sci., A-2, 7, 425 (1969); Sperling, L.H., Taylor, D.W., Kirkpatrick, M.L., George, H.F. and Bardman, D.R., J. Appl. Polym. Sci., 14, 73 (1970); Sperling, L.H., George, H.F., Huelck, V., and Thomas, D.A., J. Appl. Polym. Sci., 14, 2815 (1970).
3. Grates, J.A., Thomas, D.A., Hickey, E.C. and Sperling, L.H., J. Appl. Polym. Sci., 19, 173 (1975).
4. Frisch, H.L., Frisch, K.C. and Klempner, D., Polym. Eng. Sci., 14, (9), 646 (1974).
5. Frisch, H.L., Cifaratti, J., Palma, R., Schwartz, R., Foreman, R., Yoon, H., Klempner, D. and Frisch, K.C., in "Polymer Alloys", edited by Klempner, D. and Frisch, K.C., Plenum Publishing Co. (1977).
6. Xiao, H.X., Frisch, K.C. and Frisch, H.L., J. Polym. Sci., 21, 2547 (1983); Xiao, H.X., Frisch, K.C., and Frisch, H.L., J. Polym. Sci., 22, 1035 (1984).
7. Cassidy, E.F., Xiao, H.X., Frisch, K.C. and Frisch, H.L., J. Polym. Sci., 22, 1839 (1984); Cassidy, E.F., Xiao, H.X., Frisch, K.C. and Frisch, H.L., J. Polym. Sci., 22, 1851 (1984); Cassidy, E.F., Xiao, H.X., Frisch, K.C., and Frisch, H.L., J. Polym. Sci., 22, 1839 (1984).
8. Paul, D.R. and Newman, S., "Polymer Blends", Academic Press, New York (1978).
9. Prud'homme, R.E., Polym. Eng. Sci., 22 (2), 90 (1982).
10. Eisenberg, A., Smith, P. and Zhou, Z.L., Polym. Eng. Sci., 22 (17), 1117 (1982).

POLYMERS WITH METAL-LIKE CONDUCTIVITY; STRUCTURE, PROPERTIES AND APPLICATIONS

G. WEGNER
Max-Planck-Institut für Polymerforschung,
D-6500 Mainz, Germany

SYNOPSIS

Metal-like conductivity in polymers is linked to a salt-like structure in which ionic charges reside on the segments of the polymer chain and are balanced by gegenions. These form a sublattice of their own. In general terms these materials may be considered as ion-radical salts of polymers. The similarity of all electronic phenomena seen in organic metals, independently whether they contain polymeric or low-molecular weight substructures, suggests the importance of inter-chain mechanisms of charge transport as compared to on-chain transport of carriers. Possible applications of polymeric organic metals are still hampered by the limited environmental stability of these materials although some progress has been made especially in the area of poly(pyrrole) salts.

1 INTRODUCTION

Metal-like conductivity in polymers is a subject of worldwide intensive research since the seminal work of the Philadelphia group[1] has appeared in 1977. Although still controversial with regard to details of the chemistry and physics involved, a number of basic principles how to achieve, maintain and understand conductivity and related electromagnetic properties in these materials have been elucidated.

It is important to realize that work on the electrical properties of polymers has to be compared against progress made in the area of low molecular weight so-called organic metals as well[2]. This is specifically true with regard to the relevance of different possible mechanisms of charge transport in such materials.

Another question which has been barely discussed in the open literature is the problem of stability of conducting polymers towards environmental or internal mechanisms of deterioration. The materials known so far do not seem to meet the required standards for technical application which does not limit their importance for academic studies but rather asks for more intense work from the chemistry side.

2 STRUCTURAL PRINCIPLES

One of the principle requirements to gain intrinsically conducting polymeric systems is that segments of initially insulating polymers are oxidized or reduced to form ion-radical states able to interact with neutral segments in a manner described generally as charge-transfer-interaction. The required interaction imposes steric restraints with regard to possible chemical structures of the polymer segments, e.g. planarity is a prerequisite and bulky substituents preventing mutual interaction of segments of different chains are likely to prevent the formation of the CT-complex.

Further, high conductivity is linked to a more or less well ordered salt-like structure of the polymeric organic metal; in other words, the charges residing on the segments of the polymer chain have to be balanced by suitable gegenions which form a sublattice of their own. The gegenions are - in a first approach - merely spectators to what happens in terms of electronic interactions in the other sublattice build-up by the charged polymer segments.

By the same token any conducting polymer behaves naturally as an ion-exchange material or redox-polymer.

The gegenions have to be selected as to show the minimum possible chemical reactivity toward the charged polymer segment. Thus "closed-shell", non-complexing ions with little or no electro- or nucleophilicity are the best choice.

The stability of individual conducting polymers can be easily envisaged from the known redox-potential and - chemistry of model compounds. The most important factor is the redox potential of the polymer, second the nucleo- and electrophilicity of the charged segments.

2.1 Radical Cation-salts as models for conducting polymers

The simplest organic metals can be produced from naphthalene[3] and similar arenes such as pyrene, perylene, fluoranthene or triphenylene[4-5]. Crystals of the general stoichiometry $(arene)_2^{+}X^{-}$ ($X=ClO_4$, BF_4, PF_6, SbF_6, AsF_6 etc.) are obtained, if these compounds are electrochemically oxidized in an inert solvent containing e.g. $(C_4H_9)_4N^{+}X^{-}$ as the supporting electrolyte. The counterion X^{-} serves to provide electroneutrality to the salt crystals growing from the anode. They balance the charge located on the aromatic cation-radical stack (Fig. 1). These materials show conductivities σ of typically 1000 ($S\ cm^{-1}$) in stack direction and of the order of 1-10 ($S\ cm^{-1}$) normal to stack direction.

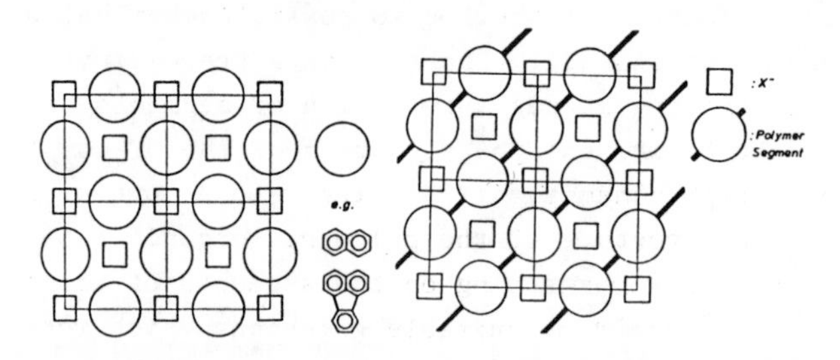

Figure 1 Analogy of the packing in radical cation salts (left) and conducting polymers (right).

The interaction found between the aromatic rings in the conducting stacks in radical cation salts can be regarded as a model for interchain interactions in conducting polymers. As a consequence many of the models which have been proposed for the structure of highly conductive polymer phases are derived from the structural principles of radical cation salts. This concept is schematically illustrated in Fig. 1.

All radical cation salts can be characterized as columnar structures. The aromatic rings are arranged in stacks leaving channels in which the counterions and in some cases additional molecules like solvent or neutral arenes are situated. Within the stacks extremely small interplanar spacings of 3.2 to 3.3 Å are observed. The rings are usually oriented perpendicular to the stacking direction. Consideration of these facts indicates that the term "doping" is totally inadequate to describe the structural and

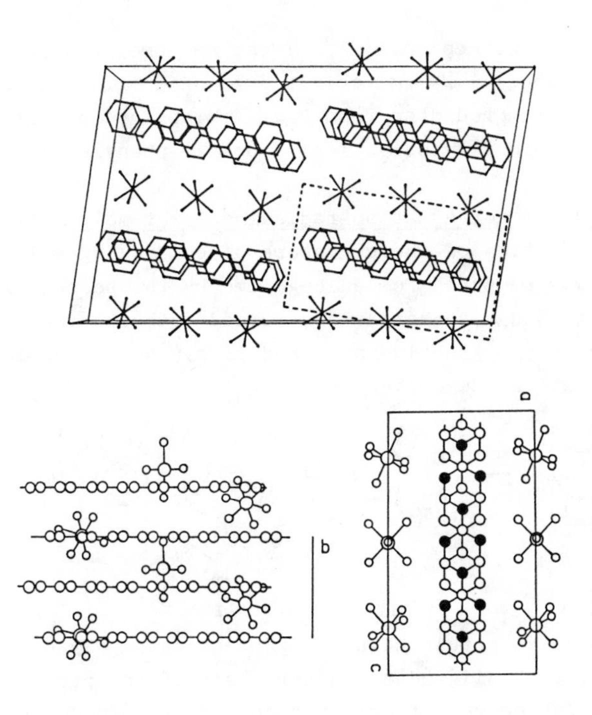

Figure 2 Comparison of the quaterphenyl cation radical salt with SbF_6-counterions (top) and the structural model for conducting salts of poly-p-phenylene (Enkelmann 1985[5])

chemical changes required to prepare an organic metal. We rather deal with oxidation of the arene or polymer segment resp. and with the formation of a new structure which has nothing in common with the initial structure. In addition, the electronic properties of the organic metal arise from intermolecular interaction between the units within the stack; thus the packing of the molecules is at the origin of all considerations relating to the electronic properties.

An immediate prediction from this model is that the conductivity in polymeric systems should be rather independent of chain length, since the electronic properties should arise because of the same principles as in the low-molecular weight analogues. The missing link between the low-molecular weight organic metals and the polymers has in fact been found in the case of the cation-radical salts of ter- and quaterphenyl [5,6]. These are prepared as described for the other $(arene)_2^+ \cdot X^-$ materials by anodic oxidation and exhibit the structure predicted from Fig. 1. They are models for the organic metal obtained by oxidation of poly(p-phenylene)[5,7].

2.2 The structure of a cation-radical salt of polyacetylene

The elucidation of the structure of conducting salts of polyacetylene was hampered for a long time by the poor quality of the samples obtained via Shirakawa's method[1] considering orientation and homogeneity. Adopting a method of synthesis developed by Edwards and Feast[8] as described by

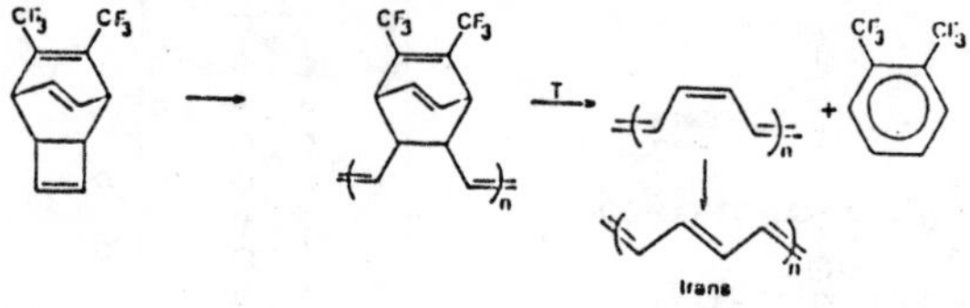

I II III

we have been able to obtain highly oriented trans-polyacetylene by simultaneous stretch alignment and pyrolysis of the precursor polymer II [9]. The high degree of orientation is retained during oxidation with iodine and other oxidative treatments from the gas or solution phase including electrochemical oxidation.

Electrochemical oxidation in the presence of tetrabutylammonium hexafluoroantimonate gives rise to a stable composition of the formula $|(CH)(SbF_6)_{0.06}|_x$. Excellent fibre diagrams could be obtained and analyzed (2,10). The crystal structure was solved in terms of a projection along the chain axes which is compatible with all available X-ray and neutron diffraction data and other experimental evidences, Fig. 3. It consists of two mutually incommensurate sublattices of polymer chains and counterions. The latter form a hexagonal lattice which is disordered in the polymer chain direction. The structural model is in close analogy to the building principles of radical cation salts of arenes. In this structure an important contribution to the conductivity is expected to come from interchain charge transfer interactions.

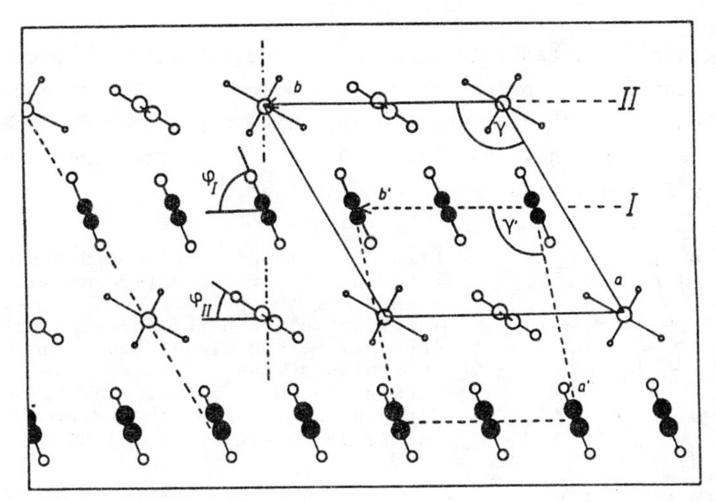

Figure 3 Crystal structure of $|(CH)(SbF_6)_{0.06}|_x$, projection on the a-b plane (along the polymer chain axes).

The pertinent data of the structure described by Fig. 3 are Sublattice I: a = 8.66 Å, b = 7.36 Å, c = 2.4 Å, $\alpha = \beta = 90^\circ$, $\gamma = 101^\circ$, $\phi_I = 69^\circ$, $\phi_{II} = 27^\circ$
Sublattice II: a = b = 9.8 Å, c = 5.4 Å, $\alpha = \beta = 90^\circ$, $\gamma = 120^\circ$
The incommensurability of the two sublattices can be removed by chosing an unusually large unit cell; in the light of the inherent disorder present in these samples it does not seem reasonable to address too much importance to this incompatibility for the time beeing.

The present structure differs qualitatively and quantitatively from previous proposals which simply assumed that salts of polyacetylene can be characterized as intercalation compounds consisting of alternating layers of polymer chains and counterions[2,11].

2.3 Salts of polypyrrole with layered structure

Symmetrical gegenions discussed so far can be replaced by anisometric ions such as the ions of detergent molecules without changing the stacking of the segments of the polymer.

We have realized this idea in the case of poly(pyrrole) salts prepared by electrochemical oxidation of pyrrol in water in the presence of a supporting electrolyte which contains simple detergent ions[12]. Suitable detergents are n-alkyl sulfates, -sulfonates, -phosphates and phosphonates. The polymeric salts give rise to a small-angle X-ray peak which scales with the number n of carbon atoms per alkyl chain as $d(n)=0.19\cdot n+1.2$ (nm). The structure of these salts is schematically depicted by figure 4.

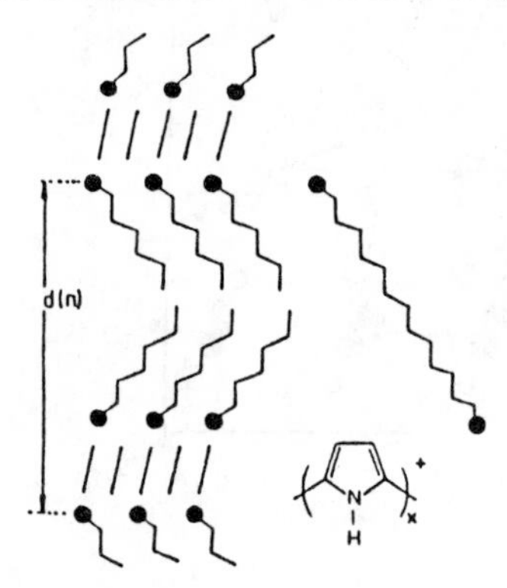

Figure 4 Short range order structure in polypyrrole detergent salts, seen along the polymer chain axis. The short black bars represent polymer chains, the zigzag lines the alkylchains of the detergent, the black dots stand for the ionic head groups. Note the stacking of the polymer chains as in figure 1b and 3.

Typical data on composition and stoichiometry are compiled in Tab. 1.

Table 1

Composition and conductivity (25°C) of salts of polypyrrole with tenside anions

anion	σ (S cm^{-1})	stoichiometry[a]	anion weight fraction
$H(CH_2)_{12}SO_4^-$	80	4:1	0.51
$H(CH_2)_{16}SO_4^-$	5	3.8:1	0.56
$H(CH_2)_4SO_3^-$	10	3.5:1	0.37
$H(CH_2)_8SO_3^-$	160	3:1	0.50
$H(CH_2)_8PO_4H^-$	12	4:1	0.43

a) ratio of pyrrole units to counterions

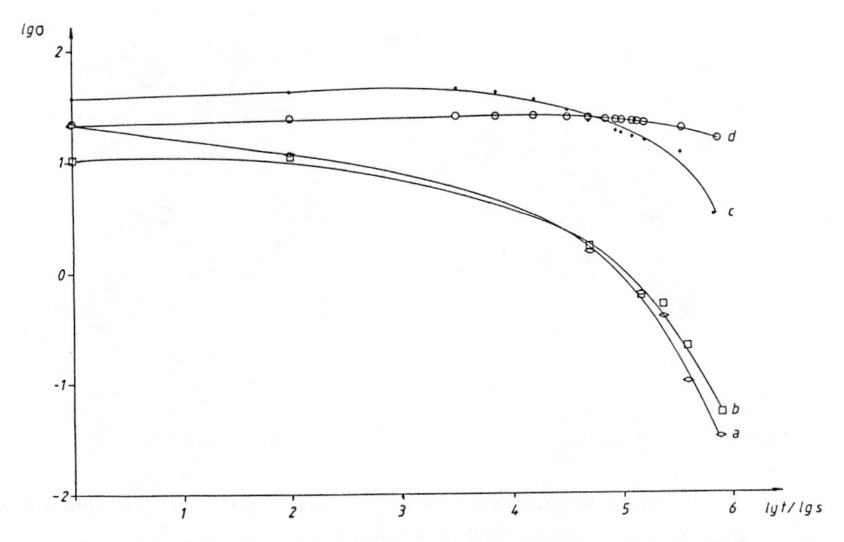

Figure 5 Conductivity vs. storage time for various polypyrrole tenside salts; a,b: films of salt with $H(CH_2)_{12}SO_4^-$ in pure O_2 at 100°C; c: film with $H(CH_2)_{12}SO_3^-$ in air at 100°C; d: film with $^-O_3S-(CH_2)_{10}-SO_3^-$ in air at 100°C. All data from Wernet 1985 (13).

The temperature dependence of σ is similar to that of other conducting polymers. It can be described by the Mott-variable-range hopping model. Samples with a ratio σ (300K)/ σ(10K) = 1.4 could be obtained.

Conducting polypyrrole salts are quite stable toward environmental exposure and temperature. This is indicated by Fig. 5. They are also surprisingly stable in electrochemical redox-reactions so that their use as half-cell component in applications for battery purposes seems to feasible.

2.4 Organic metals based on phthalocyaninatopolysiloxanes

The necessary organization of structural elements into stacks is achieved quite differently in the case of phthalocyaninatopolysiloxanes[14]. Here, metallo macrocycles, namely phthalocyanines are connected with strong covalent links perpendicular to the planes. The synthesis works by polycondensation as shown in Fig. 6.

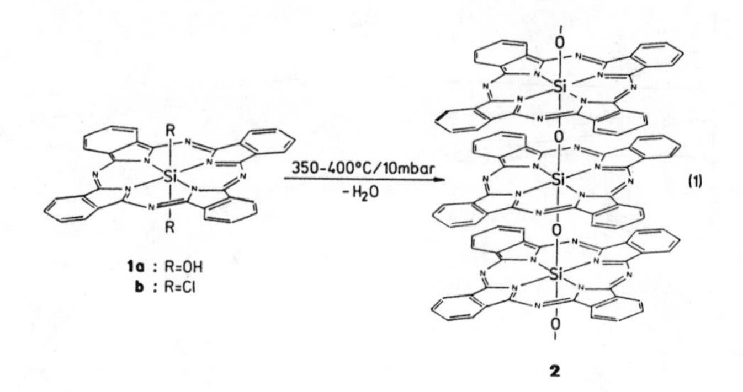

Figure 6   Synthesis of phthalocyaninatopolysiloxane[14]

By this method highly crystalline polymer powders of considerable thermal stability are obtained. The average degree of polymerization $\bar{P}_n$ is in the range between 50 and 200. Doping of these materials with I_2, Br_2, substituted quinones, nitrosyl salts such as $NOBF_4$ or $NOPF_6$ leads to partially oxidized samples with specifi conductivities of 10^{-2} to 1,0 Ω^{-1} cm^{-1} (15).

The insulating powder-like polymer can be oxidized electrochemically as well[16]. If the counterion supposed to balance the positive charge on the stacks is small enough to enter the interstitial channels between the stacks in the solid material, the sample is effectively oxidized and thus becomes conducting.

Data obtained[16] with various counterions at an approximate composition $|(PcSiO)X_{0.5}|_n$ where Pc = phthalocyaninering and x = counterion are $\sigma(BF_4) = 0.28$, $\sigma(PF_6) = 0.09$, $\sigma(HSO_4) = 0.053$, $\sigma(AsF_6) = 0.19$ ($S\ cm^{-1}$).

The conductivity of electrochemically oxidized samples is found independent of $\bar{P}_n$ for $\bar{P}_n$ 7. Thus, the function of the polymer backbone is only to bring about the desired and necessary packing of the macrocycles in stacks, but the actual chain length is of little consequence to the electric properties.

This is in agreement with findings in other systems such as poly-p-phenylene[2,6] and polyacetylene[17].

REFERENCES

1. H. Shirawaka, E.J. Louis,A.G. Mac Diarmid, C.K. Chiang, A.J. Heeger, J. Chem. Soc. Chem. Commun 1977, 578

   A.G. Mac Diarmid, A.J. Heeger in W.E. Hatfield, Ed. Molecular Metals, Plenum Press, New York 1979, p. 161

2. G. Wegner, Angew. Chem. 93, 352 (1981)
   G. Wegner, M. Monkenbusch, G. Wieners, R. Weizenhöfer, G. Lieser, W. Wernet, Mol.Cryst.Liq.Cryst. 118, 85 (1985)

3. H.P. Fritz, H. Gebauer, P. Friedrich, P. Ecker, R. Artes, U. Schubert, Z. Naturforsch. B 33, 498 (1978)

4. V. Enkelmann, B.S. Morra, Ch. Kröhnke, G. Wegner, J. Heinze, Chem. Phys. 66, 303 (1982)

5. V. Enkelmann, K. Göckelmann, G. Wieners, M. Monkenbusch, Mol.Cryst.Liq.Cryst. 120, 195 (1985)

6. V. Enkelmann, G. Wieners, J. Eiffler, Makromol.Chem. Rapid Commun. 4, 337 (1983)

7. L.W. Shaklette et al. Synth. Metals 1, 307 (1979)

8. J.H. Edwards, W.J. Feast, Polymer 21, 595 (1980)

9. G. Wegner, Mol.Cryst.Liq.Cryst. 106, 269 (1984)
   G. Lieser et al. Polymer Preprints ACS Div. Polym. Chem. 25 (2), 221 (1984)

10. W. Wieners, R. Weizenhöfer, M. Monkenbusch, M. Stamm, G. Lieser, V. Enkelmann, G. Wegner, Makromol.Chem. Rapid Commun. 1985 in press

11. K. Shimamura et al. Makromol.Chem.Rapid Commun. 3, 269 (1982)

12. W. Wernet, M. Monkenbusch, G. Wegner, Mol.Cryst.Liq.Cryst. 118, 193 (1985)

13. W. Wernet, Ph.D.-Thesis, Freiburg 1985

14. R.D. Joyner, M.E. Kenney, Inorg.Chem. 1, 717 (1962)

15. C.W. Dirk et al. J.Amer.Chem.Soc. 105, 1539 (1983)

16. E.A. Orthmann, V. Enkelmann, G. Wegner, Makromol.Chem.Rapid Commun. 4, 687 (1983)

17. M.A.Schen, J.C.W. Chien, F.E. Karasz, Polymer Preprints ACS Div.Polym.Chem. 25 (2), 227 (1984)

THE MECHANICAL PROPERTIES OF POLYPYRROLE PLATES

D. BLOOR and R.D. HERCLIFFE (Physics Department) and
C.G. GALIOTIS and R.J. YOUNG (Materials Department)

Queen Mary College, Mile End Road, London E1 4NS, United Kingdom.

1 INTRODUCTION

Conductive organic polymers have been intensively investigated during the last decade. Polyacetylene has been the most thoroughly examined but increasing attention is now being paid to nitrogen containing polymers such as polypyrrole and polyaniline which display superior stability to ambient conditions when in the conductive state (1).

To date, however, little attention has been paid to the mechanical properties of the polypyrroles. Diaz and Hall have reported conditions under which thin films of polypyrrole p-toluenesulphonate can be prepared with high tensile strenths and that the tensile strength varies in an, as yet, unexplained manner with the solvent system (2).

In the present work, the preparation of relatively thick (ca 1 mm) plates of polypyrroles is described which, in some cases, exceed the strengths reported by Diaz and Hall.

2 PREPARATION

The plates were prepared in single chamber cells which contained vertical platinised electrodes 3.5 x 3.5 cm in area and 4 cm apart. The imposed current was controlled galvanostatically to within 0.02% by shorting a Thompson Electrochem type 401 potentiostat through a resistor. Electrolyte homogeneity was maintained by a magnetic follower. Oxygen was excluded from the reactors by a slow argon purge.

The solvent (propylene carbonate, 220 ± 16 g) was dried and distilled prior to use. The pyrrole (0.14 moles, an excess) and electrolyte salts (45 ± 6 mmoles, an excess) were dried before use.

3 MECHANICAL PROPERTIES

The polypyrrole plates were removed from the anode with a scalpel and entrained solvent and electrolyte were removed by soxhlet extraction with acetonitrile followed by drying in a dynamic vacuum. Samples were prepared for tensile testing by cutting into strips 8 x 30 mm to which aluminium grips were attached using epoxy resin cured overnight at 50°C. Stress/strain curves were measured by deforming specimens to failure using an Instron machine.

Figure (2) is a plot of ultimate tensile strengths as a function of current density. Most of the experiments were performed using dry solvent and the results, represented by circles, indicate that there is a strong inverse relationship between the current density and the ultimate tensile strength of the samples. One point has been disregarded in drawing the line because that particular sample failed at the grips and hence the value may be erroneously low. Even if this is not the case, the same broad conclusions are made.

Two points in Fig. (2), represented by triangles, refer to plates made in solvent to which 1% of water has been added. The effect of added water is to increase the tensile strength and, as indicated by the figures in Table (1), to cause dramatic increases in the ductility. This result is, as first sight, surprising since Diaz and Hall found that increases in the water content of mixed solvent systems from 1% upwards almost always resulted in a decrease in the observed strength.

The effect of water can be understood if it is assumed that it can catalyse a slow step in the reaction mechanism. A scheme is suggested in Fig. (1) in which water, by acting as a base, increases the rate of a deprotonation reaction. Such an effect could increase the kinetic chain length of the reaction, leading to a product with a higher molecular weight. Alternatively, water may act merely to prevent the reduction of tetraalkylammonium at the cathode, although it is not known at present in what manner the products of such a reaction might interfere with the polymerisation kinetics. Further work is in hand to investigate these mechanisms.

Diaz and Hall suggested that the effect of excess water on both the conductivity and mechanical strength of this material may be related to a reduction in conjugation length caused by an excessively polar solvent system (2).

The effect of current density on mechanical strength could be due to variations in the degree of polymerisation or chain regularity or to morphological effects. Little variation has been detected in the internal bulk morphology of samples using scanning electron microscopy down to a resolution approaching 1 μm and the samples were space-filling and void free with no evidence of anything equivalent to intergranular fracture. Similarly, surface morphologies could affect the mechanical strength by the absence or presence of stress concentrating irregularities (3) but, within the investigated range of current densities, there was only a rough correlation with surface morphology.

The observation of rough furrowed fracture surfaces and the onset of yielding before failure, as observed in the stress/strain curves of p-toluenesulphonate samples, is typical of the behaviour exhibited by amorphous semi-ductile polymers such as polycarbonates (3).

Interestingly, polypyrrole perchlorate and hydrogen sulphate plates prepared from the tetrabutylammonium salts under comparable conditions displayed rough dendritic surfaces and a brittle failure characterised by elastic deformation to a low ultimate stress. The origin of this effect is not known, but Diaz and Hall made similar observations of low ultimate tensile strengths for perchlorate and tetrafluoroborate samples (2).

4 CONDUCTIVITY

The conductivities in Fig. (3) were measured over a range of temperatures using four in-line, silver paste on nickel electrodes under a dry argon atmosphere. They are typical of values reported by previous authors and display only weakly activated behaviour in the region of room temperature.

ACKNOWLEDGEMENTS

This work was financially supported by the SERC and Lucas Group Services to whom the authors wish to express their gratitude. We are also grateful to Marston Palmer Ltd. for the gift of electrode materials.

REFERENCES

1. Proc. Int. Conf. Low Dimensional Conductors, Les Arcs, France, Dec. 1982, J. de Physique, Coll. C 3, (1983)
2. A.F. Diaz and B. Hall, IBM J. Res. Dev., 27, 342-7 (1983)
3. A.J. Kinloch and R.J. Young, Fracture Behaviour of Polymers (Applied Science, London 1983)

Table 1: Mechanical properties (see Fig 4 also)

Sample	c.d.[1] (mA/cm^2)	Thickness (mm)	Young's Modulus (GPa)	UTS[2] (MPa)	SF[3] (%)	Anion
1	3.2	1.04	3.4	34.3	-	pTS[4]
2	3.2	0.98	-	36.6	-	pTS
3	1.9	0.88	3.9	49.7	2.6	pTS
4	0.7	1.00	4.1	64.8	-	pTS
5	0.3	0.38	3.8	>50.4[5]	2.9	pTS
6[6]	1.7	0.80	3.3	65.1	>6.0[7]	pTS
7[6]	1.7	0.85	3.3	54.7	5.0	pTS
8	1.4	0.82	3.4	24.9	0.7	HSO_4
9	1.4	0.75	3.1	22.2	0.6	HSO_4
10	3.2	1.20[8]	3.2[8]	11.0[8]	0.4[8]	ClO_4

1 c.d. = current density
2 UTS = ultimate tensile strength
3 SF = strain at failure
4 pTS = p-toluenesulphonate
5 This sample failed at the grips.
6 1% of water was added to the solvent.
7 The strain gauge became detached at 6%.
8 These values are approximations because the sample surface was rough.

Figure 1 catalysis by water

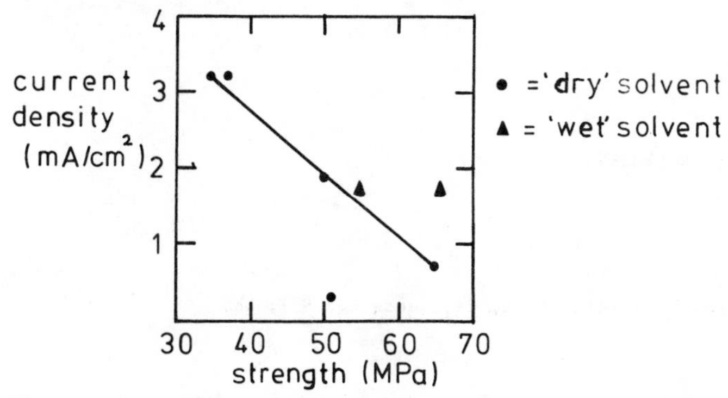

Figure 2 UTS of polypyrrole p-toluenesulphonate vs. c.d.

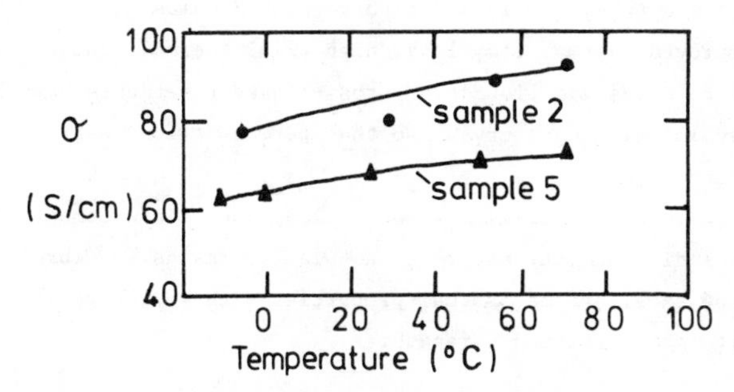

Figure 3 Conductivities (σ) vs. Temperature

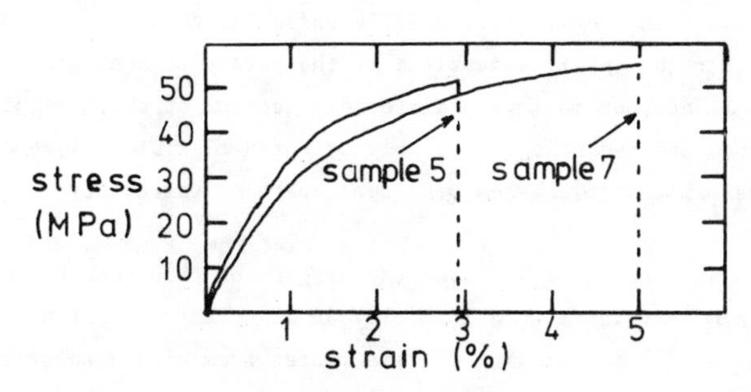

Figure 4 Mechanical properties

HIGH MODULUS FLEXIBLE POLYMERS

IAN M. WARD

Department of Physics, University of Leeds, Leeds LS2 9JT, UK.

SYNOPSIS

At Leeds University processes have been developed for the production of high modulus flexible polymers by tensile drawing (monofilaments, multifilament yarns, tape and net) hydrostatic extrusion and die drawing (solid sections, including rods, sheet and tubes). In addition to high modulus and strength, the products show improved thermal stability, high axial thermal conductivity and low permeability to gases and liquids. Cross-linked treatments have been developed which virtually eliminate creep, so that permanent load bearing applications can be envisaged.

Applications which are under evaluation include the reinforcement of brittle matrices, including cement, concrete and plastic resins. Fibre reinforced resins show excellent mechanical properties with very high impact energy as well as high stiffness and strength.

1 INTRODUCTION

In presenting a paper to a conference which aims to bring together Polymer Science and Technology, it is clearly essential to adopt a different approach from either the purely scientific or the purely technological. In this paper, I shall attempt to give a historical account of the interplay between the science and technology of highly oriented flexible polymers produced by solid phase deformation processes developed at Leeds University in the last fifteen years.

It is important to recognize that the origins of this research were purely academic and that there were initially at least three apparently different approaches (1) the drawing of fine fibres from dilute solutions, which originated in research by Pennings[1] at the DSM laboratories (2) the extrusion of molten polyethylene under high pressure and subsequent solidification, which was pioneered by Porter[2] in USA (3) the Leeds approach of solid phase deformation by tensile drawing[3] and hydrostatic extrusion[4]. The gap between the predicted theoretical chain modulus of polyethylene and the actual values obtained prior to the achievements of the early seventies had been appreciated from the classic calculations of Treloar[5] and Simanouchi et al[6] and the crystal strain measurements of Sakurada and his colleagues[7]. An extremely prescient paper by Frank[8] in 1970, foresaw these recent developments and emphasised the significance of the earlier work by Pennings[1] and Schmidt and Van der Vegt[9] on polymer solutions and melts, and the initial tensile drawing studies of Andrews and Ward[10].

Full recognition must be given to the generality of the solution, melt and solid phase deformation routes to high modulus materials, and their interplay has been beautifully appreciated by the gel spinning and drawing route to high modulus fibres and films pioneered by Smith and Lemstra in the DSM laboratories[11]. The purpose of the present review is to consider only solid phase deformation processes, and products made by these processes.

2 TENSILE DRAWING

The guide lines for producing high modulus polyethylene by tensile drawing were established by small-scale drawing experiments on 2cm gauge length dumbbell specimens cut from compression moulded sheets[12]. The previous studies of Andrews and Ward[10] showed that the modulus was uniquely related to the draw ratio, indicating that it was essential to achieve high draw ratios if high moduli were to be obtained. This observation is the corner-stone of these developments. Looked at another way, the measurement of modulus provides a sound operational guide to the effectiveness of the drawing process. A feature which had not been appreciated was the time-dependent nature of the plastic deformation. The dumbbell specimens were drawn in an Instron tensile testing machine for different times in the range from a few seconds up to several minutes. In this way a series of specimens was produced and the maximum draw ratio (i.e. the draw ratio in the centre of the deformed specimen) shown to be a function of time[13]. Plots of draw-ratio versus time were then constructed for samples of different molecular weight characteristics with different initial thermal treatments e.g. quenching into cold water from compression moulding at

160°C or slow cooling from 160°C to 110°C before quenching. The conditions of drawing were also varied, most usually by changing temperature but drawing at comparatively low rates of cross-head extension (e.g. 2cm/minutes). From many such experiments on polyethylenes the following conclusions were made[13,14].

(1) High draw ratios are most readily achieved for samples of low molecular weight, especially low weight average molecular weight.

(2) The initial thermal treatment affects the drawing behaviour markedly for low molecular weight samples.

(3) The optimum draw temperature increases with increasing molecular weight, typically from 75°C to 115°C for samples with $\bar{M}_w$ = 100,000 and 300,000 respectively for these low draw rates.

(4) The modulus is a unique function of the draw ratio, irrespective of molecular weight, initial thermal treatment and drawing conditions, provided that the latter have been optimised for most effective drawing. (Fig.1). The modulus-draw ratio curve can therefore be regarded as an upper limit boundary.

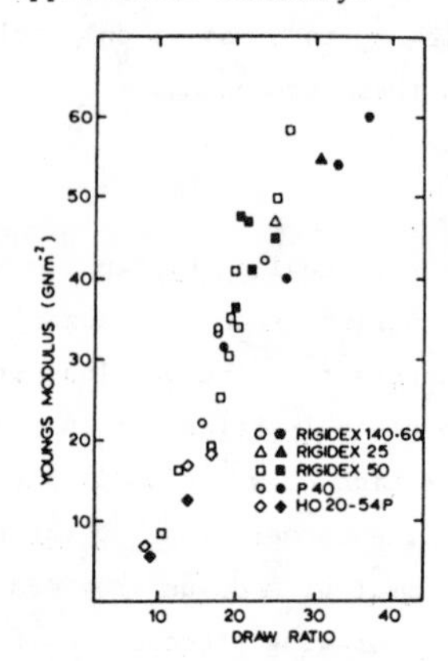

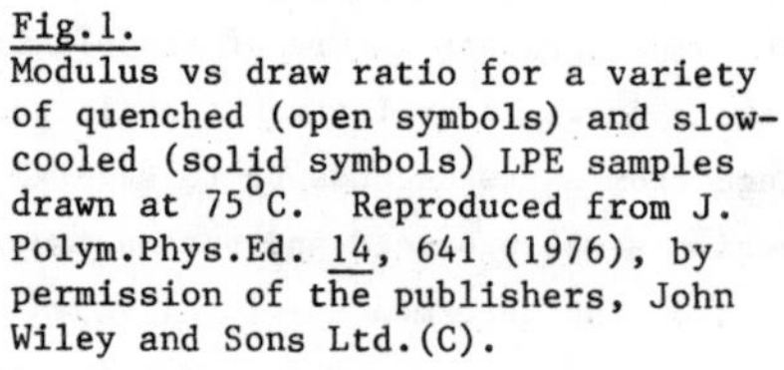

Fig.1.
Modulus vs draw ratio for a variety of quenched (open symbols) and slow-cooled (solid symbols) LPE samples drawn at 75°C. Reproduced from J. Polym.Phys.Ed. 14, 641 (1976), by permission of the publishers, John Wiley and Sons Ltd.(C).

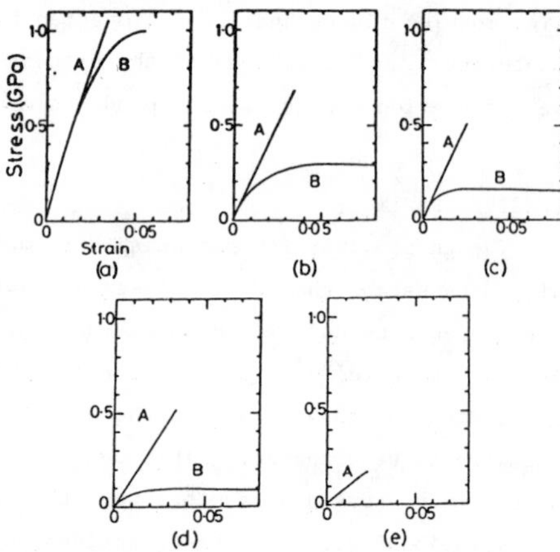

Fig.2.
Stress-strain curves for (A) electron irradiated fibres (200 kGy in acetylene); (B) unirradiated fibres. Strain rate (s^{-1}):$8.3x10^{-3}$ (a) $8.3x10^{-5}$ (b),(d),(e) $2.1x10^{-6}$ (c) Temperature (°C):23°C (a),(b), (c) 70°C (d) 130°C (e).Reproduced from Polymer Communication 25,298 (1984) by permission of the publishers Butterworth & Co (Publishers) Ltd. (C).

Related structural studies identified differences in morphology due to the different thermal treatments and have underlined our understanding of the drawing process. The key idea is the stretching of a molecular network whose network junction points are either physical entanglements (which predominate at high molecular weight) or the incorporation of different chains in the crystalline lamellae (which is important at low molecular weights). The essence of the drawing process is to find conditions which permit effective draw. Although this is easiest for low molecular weight samples, because of the low melt viscosity, it is still possible to find conditions for drawing higher molecular weight polymers, although these may be more restricted. Even polymers with $\bar{M}_w$ ~ 800,000 were drawn to comparatively high draw ratios by Capaccio et al[14]. Moreover, although the modulus-draw ratio relationship is not changed the subsequent products will probably show higher strengths and better creep behaviour.

The Leeds research on tensile drawing soon led to a practical continuous drawing process, where monofilaments were drawn 30x in a glycerol bath at 120°C at rates up to 500m/min. Further work led to a drawing process for multifilament yarns, again at realistic production rates, albeit much slower than those now available for conventional textile fibres of polyester or nylon[15]. The developments at Leeds and ICI Fibres were the forerunner of a pilot plant production by the Celanese Research Company, which has permitted the availability of material in several kilogram quantities for development applications trials.

2.1 Recent developments on high modulus polyethylene fibres

2.1.1 Irradiation-cross-linking

A major disadvantage of high modulus polyethylene fibres is their poor creep resistance. Although some advantages can be obtained by increasing polymer molecular weight, by light cross-linking prior to drawing or by the use of co-polymers, these improvements are comparatively marginal and only permit a small fraction of the short time strength of ~ 1GPa to be utilized in load bearing situations. It is interesting to note in passing that such comparatively low levels of permanent load (corresponding to ~ 0.1 GPa) can be quite adequate for some Civil Engineering applications such as Geo-grids.

Fortunately, creep in polyethylene fibres can be virtually eliminated by appropriate electron-irradiation of the final drawn fibres[16,17]. Some preliminary results obtained by irradiating a typical multifilament yarn are shown in Fig.2. It can be seen that the effect of irradiation is to change the stress-strain curves from yield and plastic deformation to essentially elastic-brittle behaviour. Measurements of gel content and the rubber elasticity behaviour above the melt point show that the primary effect of the irradiation, which must be under carefully controlled conditions, is to produce a cross-linked structure with an average chain length between cross links corresponding to a molecular weight of about 5,000. The use of acetylene as a sensitiser enables the irradiation times to be significantly reduced. More recent studies with commercial electron irradiation facilities have demonstrated the practicality of this treatment, with only a marginal projected increase in the cost of the fibres.

Cross-linking also decreases the temperature sensitivity of the polyethylene fibres, so that they can withstand temperatures at least up to the polymer melting point (130°C) without complete loss of properties. Furthermore their behaviour in cyclic loading is vastly improved as shown in Fig.3, where the loading programme corresponds to that required for assessment of ropes and cords. It can be seen that untreated fibres fail after about 15 minutes, whereas the treated fibres show substantially unchanged properties after one week. It is interesting to compare these results with those for a standard polypropylene rope fibre. Not only is the cross-linked polyethylene much more stable, it also shows very clearly its much higher modulus.

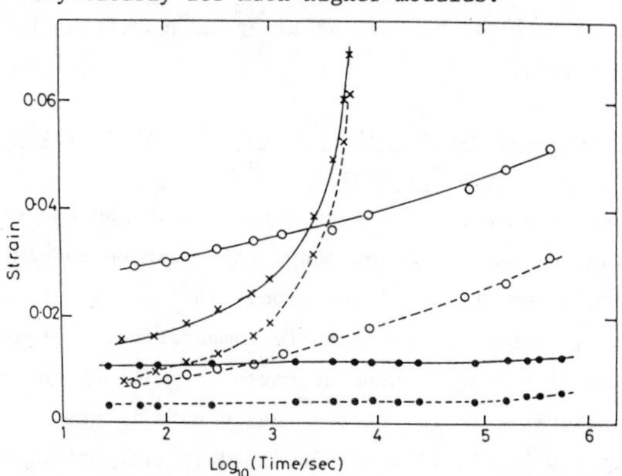

Fig.3.
Strain at 0.30 GPa (——), and at 0.01 GPa (----) in fatigue cycling tests (x) Control unirradiated polyethylene fibre (cycle time 40 s). (●) Polyethylene fibre irradiated 200 kGy in acetylene (cycle time 28 s). (O) Polypropylene monofil (cycle time 26 s). Reproduced from Polymer Communications 25, 298 (1984) by permission of the publishers Butterworth & Co (Publishers) Ltd. (C).

2.1.2 Surface treatment

The chemical stability of polyethylene is a great advantage in many respects. For example, the cross-linked low creep fibres can be envisaged as reinforcements for cementitious matrices. In one respect, however, this inertness is a disadvantage in that it does not readily permit the creation of a good bond between the fibres and a polymeric matrix. The recognition of this potential disadvantage has led to the development of surface treatments to give a greatly improved bond between ultra high modulus polyethylene fibres and any polymeric matrix. In practice, most of our studies at Leeds have involved epoxy resins, but comparable limited data have also been obtained for polyester resins.

The research was initiated by determining the pull-out adhesion of monofilaments subjected to a range of different surface treatments[18]. A low-viscosity epoxy resin (Ciby-Geigy XD927) intended for high-strengths composite structures was used. This resin could be cured initially at room temperature and post-cured at 80°C, well below temperatures likely to cause any deterioration of fibre properties. Table 1 summarises the results obtained, and it can be seen that there is significant improvement for both plasma and acid treatments, although the mechanism of failure appeared to be different in the two cases. Untreated and acid-treated monofilaments showed a fairly smooth surface and failure of the pull-out samples involved sliding along the monofilament/resin interface. Plasma treatment, on the other hand, produces a cellular surface on the monofilament surface into which the resin penetrates to give mechanical keying between resin and monofilament. In the pull-out test the surface layer of the monofilament peels off and remains attached to the resin.

Table 1:
Summary of pull-out adhesion for UHMPE monofilaments (room temperature)

ADHESION LEVEL	RANGE (MPa)	TREATMENTS
LOW	0.4 - 0.7	UNTREATED CERIC SOLUTIONS
MEDIUM	0.8 - 1.7	CHROMIC ACID ALL PLASMA WITH CF_4 GAS WEAK PLASMA WITH Ar GAS
HIGH	1.8 - 3.0	STRONGEST PLASMA WITH Ar GAS ALL PLASMA WITH He GAS WEAK PLASMA WITH O_2 GAS
VERY HIGH	4.0 - 5.0	STRONGEST PLASMA WITH O_2 GAS

Subsequent tests on unidirectional fibre composites reinforced with surface treated UHMPE fibres showed an increase in the interlaminar shear strength of the composite to about 30 MPa compared with 15 MPa for untreated fibres.

2.1.3 Strength

It will be clear from the previous discussion that the definition of the strength of the UHMPE fibres must include a prescription for the test conditions, especially strain rate (and temperature). It was shown by Cansfield et al[19] that the standard UHMPE fibres show very high strain rate sensitivity so that their strengths are well above 1 GPa for high test rates. This is a valuable attribute with regard to potential ballistic applications, and has led to a recent research project to examine in detail the increases in tensile strength which can be obtained for the melt spinning and drawing route.

It is, of course, well known that molecular weight is a key variable in determining the tensile strength of textile fibres. It is also evident from much previous work that increasing draw ratio gives increases in both modulus and strength, provided that effective drawing occurs. It is therefore necessary to examine the drawing behaviour of samples with different molecular weights with the specific aim of achieving high strengths in addition to high modulus, recognising that although increasing molecular weight should lead to an increased strength for a given level of modulus, it may also limit the maximum draw ratio and hence the maximum modulus which can be obtained. For this study, small quantities of fibres differing primarily in $\bar{M}_n$, but also with $\bar{M}_w$ ca 10^5, $2x10^5$ and $3x10^5$ respectively were prepared using a micro-spinning facility described in a previous publication[10]. Fibres were then drawn to draw ratios of 15 and 20, taking care to optimise the drawing process so as to achieve high modulus materials. It was found that the fibre strengths increased with increasing $\bar{M}_n$, and also that fibres of similar $\bar{M}_n$ but higher $\bar{M}_w$ showed higher strengths. Polymers were then selected for a further study in which fibres were drawn to their maximum practical draw ratio on the small scale continuous drawing facility. It was found that fibre strengths at high rates (or low temperatures) can readily be achieved in the range 1.5 GPa for a comparatively modest increase in polymer molecular weight. Figure 4 shows results for two selected polymers, in terms of tenacity vs draw ratio, where the drawing temperature has been adjusted as indicated to achieve maximum effective draw.

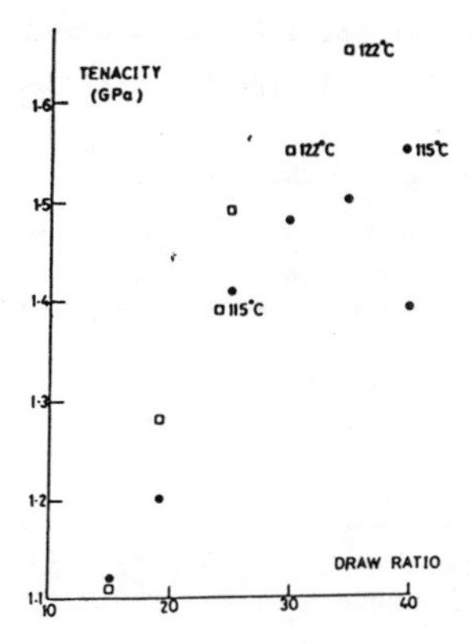

Fig.4.

Tenacity vs draw ratio; Temperature =-55°C. Alathon 7030 and Unifos 2912 drawn at 100°C, unless otherwise stated.

2.1.4 Applications

A number of potential application areas are now being actively examined. The improvements in creep behaviour suggest that the possible use of the UHMPE fibres in ropes and cords need no longer be discounted. A comparatively low cost fibre is available with high modulus and high strength, especially if these quantities are considered in terms of specific properties where weight is also taken into account. Table 2 illustrates the comparison with carbon, glass and Kevlar 49 fibres, and it can be seen that the specific properties of UHMPE are in the same range. Taking into account the recent developments described above which suggest that 50% improvements in modulus and strength should be available, UHMPE stands up to this comparison very well. Other advantages of UHMPE fibres such as high chemical resistance, excellent sunlight and weathering properties, buoyancy in water and easy handleability due to the high extension to break, are also noteworthy.

Although cement reinforcement is a possibility, especially in view of the known good performance of high modulus polyethylene net in cement[21], research at Leeds has concentrated on resin composites[22]. The particular advantage of UHMPE fibres comes from their high energy absorbing characteristics. Results for unidirectional fibre composites, made from commercially prepared pre-pregs, are shown in Table 3, where the results are presented in terms of specific properties. It can be seen that for the wholly UHMPE composites, the tensile modulus and strength reflect very well the 50% fibre concentration, and there

Table 2.
Properties of reinforcing fibres (room temperature)

Property / Fibre	Tensile Modulus (GPa)	Tensile Strength (GPa)	Elongation at Break %	Density ρ g/cm^3	Specific Modulus GPa/ρ	Specific Strength GPa/ρ	Maximum Working Temperature °C
Carbon	230	1.5	1.5	1.8	128	0.8	>1500
Glass	75	2.0	2.5	2.5	30	0.8	250
Kevlar 49	125	3.0	3.0	1.45	86	2.1	~ 180
Polyethylene	50	1.2	4-18	0.96	52	1.2	120

Table 3.
Specific mechanical properties of pre-preg composite systems - various continuous reinforcement/code 91 epoxy resin.

Reinforcement	ILSS MPa/ρ (x10^3)	UFS MPa/ρ (x10^3)	FM GPa/ρ (x10^3)	TM GPa/ρ (x10^3)	TS GPa/ρ (x10^3)	CS MPa/ρ (x10^3)	"Flat" Charpy Test-Energy Absorption at 1st Impact kJm^{-2}/ρ (x10^3)
UHMPE (untreated)	14	155	38	38	0.40	70	125
Kevlar 49	39	394	49	56	0.96	205	130
Carbon EXAS	42	1025	67	88	1.25	675	50
E-Type Glass	34	585	21	29	0.80	500	155
UHMPE (untreated) - Carbon EXAS	27	385	38	66	-	320	120
UHMPE (untreated) - E-Type Glass	19	170	29	32	-	175	165
UHMPE (untreated) - Kevlar 49	18	240	34	48	-	135	115

ρ : Density as given in Table 2

are no unexpected features. The impact energy of this composite is high, and moreover it does not shatter on impact because of the very high extensibility of the fibres. This characteristic can probably be used to better advantage in hybrid composites, as shown in the last three rows of this slide. Such composites can retain the superior strength and stiffness of the pure carbon or glass composite, but gain appreciably in energy absorption with very little loss of strength or stiffness because of the near comparability of the UHMPE fibres in these respects.

3 HYDROSTATIC EXTRUSION

The production of high modulus polyethylene in the form of circular rods of large section (~1cm) was first achieved by hydrostatic extrusion[4,23]. In this process a solid billet of polymer is plastically deformed in the solid phase by forcing it through a converging die by application of pressure to a fluid which surrounds the billet at the entrance to the die. The deformation ratio is defined as the ratio of the initial to the final cross-sectional area of the billet (for deformation at constant volume) and is equivalent to the draw ratio of the tensile drawing process. It was shown that the modulus/deformation ratio relationship was identical to the modulus/draw ratio relationship, and in polyethylene, products with Young's moduli up to about 60 GPa could be obtained[4,23].

It was soon found that the rates of extrusion were comparatively slow (~ mm/min) for high modulus products, and this led us to make a detailed analysis of the mechanics of the hydrostatic extrusion process[24,25]. This analysis was based on the observation that the deformation within the die is tensile, so that it is only necessary to consider the axial equilibrium of thin disc-shaped elements within the die. Moreover the strain and strain rate within the die can be defined by the Avitzur analysis[26]. Essentially the treatment involves an extension of the Hoffman and Sachs' analysis of flow of metals through a conical die[27], to include the dependence of the flow stress on plastic strain, on strain rate and on the hydrostatic pressure. It was found that the sensitivity of the flow stress to these three factors readily explains the comparatively low deformation rates which can be achieved. For high deformation ratios, the flow stress shows a rapidly increasing dependence on strain rate as well as strain, and attempting to increase the extrusion rate by increasing the applied hydrostatic pressure is self-limiting because the flow stress increases quite rapidly with increasing pressure. These considerations made us recognise the fundamental advantages of tensile drawing compared with hydrostatic extrusion in these respects, and led to the development of the die-drawing process which is described below.

There are, however, two important points to make about the hydrostatic extrusion process. First, the limitations highlighted by the Hoffman-Sachs analysis apply to an isothermal process. For large sections it is also possible to find a stable adiabatic regime, where the heat generated reduces the flow stress so that extrusion velocities ~50cm/min can be obtained[28]. Secondly, hydrostatic extrusion can be applied to short glass fibre reinforced polymers, to achieve a product with a high degree of both fibre and matrix orientation[29].

4 DIE-DRAWING

Figure 5 is a schematic diagram of the die-drawing facility[30,31], where three regimes of deformation can be envisaged (I) conical die flow (II) tensile deformation within the heated die and (III) tensile deformation outside the die. There are two major advantages of die-drawing over hydrostatic extrusion.

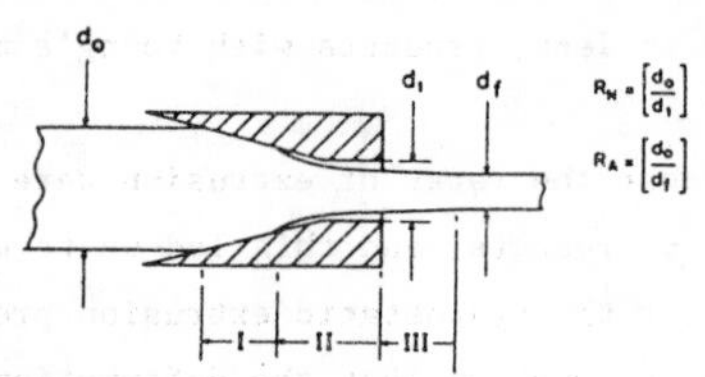

Fig.5.
Schematic representation of the die-drawing process with heated die and stable post-die neck region. Reproduced from Polym.Eng.Sci., 20, 1229 (1980) by permission of the publishers, Society of Plastics Engineers.(C).

First, there is only a hydrostatic component of stress in the first stage of the deformation when the flow stress is comparatively low. Secondly, the strain rate field is similar to that in tensile drawing in that the highest strain rates occur in the neck where the degree of plastic deformation is comparatively low. The very high flow stresses encountered in hydrostatic extrusion are therefore not reached.

For low haul-off velocities the billet remains in contact with the die wall until it reaches the die exit and the total deformation ratio is defined by the ratio of the cross-sections of the initial billet and the die exit, as in hydrostatic extrusion. As the haul-off velocity is increased, the billet comes away from the die wall and free necking occurs within the die. The deformation ratio therefore increases with increasing haul-off velocity, which means that the highest modulus products are produced at comparatively fast rates, typically (~50cm/min). It is important to note that again there is a unique relationship between modulus and final deformation ratio. The die-drawing process has been shown to be a very flexible process. For example, in polyethylene it was possible to produce comparable high modulus products for a range of reduction ratios within the die, with the die heated to a range of temperatures 80-110°C[32]. (Fig.6).

The versatility of the die-drawing process led us to explore its potential for many polymers in addition to the polyolefines. The polymers studied have included polyoxymethylene[29], polyvinylidene fluoride[33], polyetheretherketone[34], polyvinylchloride, nylon and polyester[35]. Because the properties of the polymer are available in 100% concentration as distinct from

50% concentration for a high modulus fibre reinforcement, it is possible to reduce the requirement of very high draw ratio to medium to high draw ratio and still obtain valuable property enhancement. Not only does this reduction in

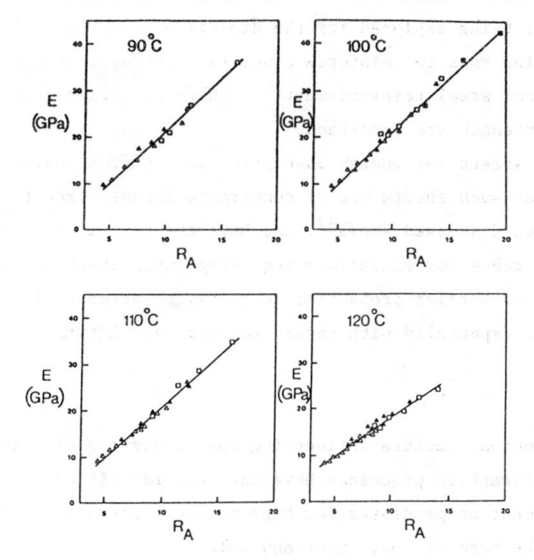

Fig.6.
Die drawing of HDPE. Variation of static three point bend modulus (E) with draw ratio (R_A). Drawing temperatures as indicated (Δ) R_N=4; (▲) R_N=6; (□) R_N=8; (■) R_N=10.

the draw ratio generally enable a much more flexible process to be devised with correspondingly higher production rates, it also generally produces a more acceptable portfolio of properties e.g. appreciable retention of lateral strength but still permitting a considerable increase in axial strength and stiffness.

There are several additional advantages of the die-drawing process. First, it is capable of continuous operation, by direct extrusion of the initial billet into the heated die-drawing chamber. Secondly, it is possible to die-draw tubes. Moreover the mandrel can be arranged to give hoop expansion so that a biaxially oriented tube can be produced. In polyester this product can be envisaged as forming the basis for a transparent bottle, similar to those now made by alternative injection moulding and expansion processes.

Finally it is important to note that the die-drawn products can be post-formed into more suitable shapes without loss of properties. For example, the biaxially oriented polyester tube can be slit and passed through rollers to give a thick and very tough transparent sheet.

It is only possible at this stage to indicate some of the application areas which are being explored for the die-drawn products. These include

(1) Polyolefine rods to reinforce concrete structures in place of traditional steel reinforcement[36]. Here corrosion resistance and impact strength are important.

(2) Oriented sheets for energy absorption in impact situations. It is to be noted that such sheets are of comparable impact strength to the UHMPE composites discussed above[37], and have the advantage of post-formability.

(3) Oriented tubes for situations requiring improvements in chemical resistance, barrier properties (e.g. oxygen permeability) and hoop strength, especially with regard to long term internal pressure[38].

5 CONCLUSIONS

The principal factors influencing the production of oriented polymers by solid phase deformation processes have now been identified, and are being used in the development of processes for high modulus fibres and solid section materials in the form of rod, sheet and tube.

In addition to high modulus and strength, there is considerable interest in the improvement of other physical properties, including chemical resistance, gas barrier and thermal stability. A number of development application areas are now being actively explored.

6 REFERENCES

1. A.J.Pennings & A.M.Kiel, Kolloid Z., 205, 160 (1965)
2. J.H.Southern & R.S.Porter, J.Appl.Polym.Sci., 14, 2305 (1970)
3. G.Capaccio & I.M.Ward, Nature, Physical Sci., 243, 143 (1973)
4. A.G.Gibson, I.M.Ward, B.N.Cole & B.Parsons, J.Mater.Sci., 9, 1193 (1974)
5. L.R.G.Treloar, Polymer, 1, 95 (1960)
6. T. Shimanouchi, M.Asahina & S. Enomoto, J.Polymer Sci., 59, 93 (1962)
7. I.Sakurada, T.Ito & K.Nakumae, J.Polymer Sci., C15, 75 (1966)
8. F.C.Frank, Proc. Roy. Soc. A 319, 127 (1970)
9. A.K.Van der Vegt & P.P.A.Smit, Advances in Polymer Science, Monograph 26, Soc.Chem.Ind. London 1967 p.313.
10. J.M.Andrews & I.M.Ward, J.Mater.Sci., 5, 411 (1970)

11. P.Smith & P.J.Lemstra, J.Mater.Sci., 15, 505 (1980)

12. G.Capaccio & I.M.Ward, Polymer 15, 233 (1974), 16, 239 (1975)

13. G.Capaccio, T.A.Crompton & I.M.Ward, J.Polymer Sci., Polym.Phys.Edn., 14, 1641 (1976)

14. G.Capaccio, T.A.Crompton & I.M.Ward, J.Polymer Sci., Polym. Phys. Edn., 18, 301 (1980)

15. G.Capaccio, F.S.Smith & I.M.Ward, British Patent No. 1506565 (filed 5 March 1974)

16. D.W.Woods, W.K. Busfield & I.M.Ward, Polymer Comm. 25, 298 (1984)

17. D.W.Woods, W.K.Busfield & I.M.Ward, Plast.Rubb. Process. Appln., 2, 157 (1985)

18. N.H. Ladizesky & I.M.Ward, J.Mater.Sci., 18, 533 (1983)

19. D.L.M.Cansfield, I.M.Ward, D.W.Woods, A.Buckley, J.M. Pierce & J.L.Wesley, Polymer Comm. 24, 130 (1983)

20. M. Hallam (unpublished work)

21. D.J.Hannant, Symposium 'Polymers in Civil Engineering' Plastics & Rubber Institute, London 20th May, 1980

22. N.H.Ladizesky & I.M.Ward (unpublished work)

23. A.G.Gibson & I.M.Ward, J. Polymer Sci., Polym. Phys.Edn., 16, 2031 (1978)

24. P.D.Coates, A.G.Gibson & I.M.Ward, J. Mater. Sci., 15, 359 (1980)

25. P.S.Hope & I.M.Ward, J.Mater.Sci., 16, 1511 (1981)

26. B.Avitzur, "Metal Forming: Processes and Analysis", McGraw-Hill, New York, 1968

27. O. Hoffman & G.Sachs, "Introduction to the Theory of Plasticity for Engineers", McGraw-Hill, New York, 1953

28. B.Parsons & I.M.Ward, Plast. Rubb. Process. Appln. 2, 224 (1982)

29. P.S.Hope, A. Richardson & I.M.Ward, Polym. Eng. & Sci., 22, 138 (1982)

30. P.D.Coates & I.M.Ward, Polymer 20, 1553 (1979)

31. A.G.Gibson & I.M.Ward, J.Mater.Sci., 15, 979 (1980)

32. A.Richardson, B.Parsons & I.M.Ward (to be published)

33. A.Richardson, P.S.Hope & I.M.Ward, J. Polymer Sci., Polym. Phys. Edn., 21, 2525 (1983)

34. A.Richardson, F. Ania, D.R.Rueda, I.M.Ward & F.J. Balta Calleja, Polym. Eng. & Sci., (in press)

35. A.Selwood & I.M.Ward, Brit. Pat. Appln. No.8418996 (filed 25.7.84)

36. M.M. Kamal Ph.D. thesis Leeds 1983

37. N.H. Ladizesky & I.M.Ward (to be published)

38. I.M.Ward, A. Selwood, B. Parsons & A.Gray, Plastics Pipes VI Conference York 1985

RADIATION TREATMENT OF POLYMERS

A. Charlesby, Watchfield SN6 8TF, UK

INTRODUCTION

The effect of high energy radiation on polymers has grown from a subject of scientific curiosity, of interest to a limited number of specialist scientists, to a large scale highly-technical industry. Apart from its technological value, numerous developments and applications are being sought in a very wide range of other radiation possibilities. However it must be stated that at present, the major uses of large radiation sources have been with polymer modification, and with sterilisation of medical disposables. This is not because the chemical changes are very different in these two fields, but merely that in these high molecular weight systems, a very small chemical modification at the right place can provide a very considerable change in physical or biological properties.

To understand the behaviour of irradiated polymers, it is therefore necessary to understand why certain, very minor chemical changes can have such large effect, and why radiation, absorbed at random in the irradiated polymer, nevertheless produces a highly selective reaction.

High energy radiation can take the form of any one of a series of particles - protons, electrons, neutrons, alpha particles (He nuclei), gamma or X rays etc. The energy is measured in electron volts (eV) and is usually of several million eV (MeV) down to a fraction of a MeV. This compares with energies of visible photons 1.5 to 3 eV, ultra violet photons up to about 5 eV, and chemical binding energies of a few eV. In penetrating through matter, high energetic particles have sufficient energy to rupture all chemical bonds; the observed selectivity of the nature of radicals or ions produced therefore requires some explanation. One such is 'energy transfer' to certain bonds but this is hardly adequate to explain all the observed effects especially in the solid state.

In the early history of the subject, most research was devoted to the effect of alpha radiation on gases. This choice was due to the low activity of the radioactive sources available at the time, and the low penetration of alpha radiation. An entirely different impulse was given by the growth of interest in radioactivity and nuclear reactions, where one wanted to know the effect of the radiation engendered by nuclear fission on the materials in the vicinity. All types of high energy radiations were involved, and lengthly exposures were studied to determine the suitability of different materials for use in a nuclear environment. In general one was searching for structures which showed little or no effect. A third approach was the more academic study of radiation-induced changes in chemicals, usually liquids. This last subject, radiation chemistry, was intended to study reaction products, processes and reaction kinetics, and still continues in many universities, where it may be considered as a different branch of chemistry, somewhat akin to photochemistry.

Finally there was the search for interesting modifications of physical properties following irradiation. One such was the difference in orientation, melting and flow properties of polyethylene subjected to electron beam radiation, and later to mixed reactor or gamma radiation. There last two approaches have little to do with nuclear energy or nuclear physics, and it is unfortunable that some of the beneficial effects obtained by such studies have become confused in the public mind with nuclear energy, radioactivity and nuclear warfare. An analogy would be to deprecate cooked food on the grounds that the electricity used in its cooking came in part from a nuclear energy plant and is therefore unsafe to eat.

SOURCES OF RADIATION; DOSE AND DOSE RATE

In penetrating matter, high energy radiation may cause changes in nuclear structure; the new nucleus may be unstable (radioactive) and lead to further reactions. A second type of reaction is one whereby the nuclear structure is not affected, buth the nucleus is displaced into new configurations within the lattice or interstitially. A third possibility is when the reaction is with an electron. This can either be expelled, leaving a changed cationic species (ionic reaction) or raised to a higher energy level (excited). The fate of these radical or ionic species will then depend on a variety of conditions such as chemical structure, temperature, molecular configurations, availability of trapping sites for the electron, additives capable of reacting with the radiation-induced species etc.

In the reactions in polymers with which we are concerned, only this last type of interaction is considered. No nuclear transformations, and few if any nuclear displacements are involved. By far the most convenient radiation sources are (radioactive) cobalt 60, emitting gamma radiations of energy 1.33 and 1.17 MeV, and high energy electrons, from linear or other accelerations, of energy usualy 2-4 MeV, but more recently of lower energy, such as 0.3 MeV, for low penetration treatments. Other types of source are currently being examined, such as caesium 137 from radioactive residues, and X rays from the impact of high energy electrons onto a target of high Z value. From the practical point of view the main difference between the effects of gamma radiation from cobalt 60, and electrons from an accelerator is due to their very different dose rates and penetration. In material of unit density, gamma radiation of about 1.25 MeV will penetrate a distance of over 20 cm, as compared with less than 1 cm for a 2 MeV electron. Furthermore the electron beam can be focussed to give a highly intense beam only a mm or less in diameter. With gamma rays, the beam extends equally in all directions. Thus a typical dose rate from a powerful gamma source may amount to several Mrad per hour - the same dose can be accumulated locally in a matter of seconds from an electron beam of the same total power output. The choice of radiation source must be largely determined by object size and shape, especially thickness, and then by other secondary factors.

Doses are expressed in terms of energy absorbed per unit mass. The rad represents an absorption of 100 erg/g, and the more convenient megarad (Mrad = 10^6 rad) an absorption of 10^8 ergs/g or 10 J/g. The more recently-adopted unit is the grey (1 J/kg) or more conveniently the Kgy (= 10^3 J/kg = 0.1 Mrad). For many plastics, the radiation doses used are about 5 to 10 Mrad equivalent to temperature rises of only about 10 °C. This is one of the great advantages of radiation in industry, in that reactions can be induced without the high temperature rise needed to initiate chemical catalysts.

One can proceed directly from power output from a radiation source to product output at a given dose. Thus a power output of 100 kW from an electron accelerator of 4 MeV (penetration of about 1.6 cm in polymer of unit density) can produce 360 x 100 kg* of product at 1 Mrad per hour if fully absorbed or 3.6 tons at 10 Mrad. In practice a lower efficiency of energy absorption will reduce the product output.

---

* (1 kWh = 10^3 x 3600 Joules).

In its passage through matter, each high energy particle such as an electron, gamma photon or α-particle loses energy by a series of collisions, leaving a track of many excited or ionised species in the medium. In gases these species are far apart, and in air for example the average energy lost per ion pair formed is 34.5 eV, i.e. 2.9 ion pairs per 100 eV absorber. This is referred to as the G value. In solids or liquids such as low molecular weight organics, the number of ion pairs produced per 100 eV deposited is considerably lower, and G = 0.1 is typical for alkanes. This is because even with the same number of ions originally produced, the ejected electron loses energy very rapidly in the much denser medium, and reaches low energy (e.g. it is thermalised) while still within the range of its parent cation, to which it migrates to form a highly excited molecule. G = 0.1 represents the very small fraction of electrons which escape the Coulombic field of the parent, or are otherwise trapped.

Highly excited molecules are therefore formed in two ways, either directly from the collision with an energetic particle, or by recapture of its electron. Such excited species are unstable, and the loss of a bonding electron may result in bond fracture together with the formation of radical species which can be observed directly by ESR.

$$-CH_2- \rightsquigarrow -\dot{C}H- \quad + H$$

or

$$-CH_2- \rightsquigarrow -C^+H_2- \quad + e \quad \text{followed by}$$
$$-C^+H_2- \quad + e \quad -\dot{C}H- \quad + H$$

The G value of such reaction is typically about 3. Despite the wide range of bonds present in organic compounds and the high energies available in such excited systems, the subsequent bond fracture tends to occur at selected sites not necessarily the weakest. In polymers such fracture occurs largely in either the main chain, giving chain scission and a reduction in molecular weight (polymer degradation) or in side chains, leaving lateral radicals which can then react in pairs, linking molecules together (crosslinking, with enhancement of molecular weight and ultimately the formation of a network). Other chemical reactions will also occur, such as the formation of unsaturated groups in irradiated polyethylene, their disappearance (in irradiated rubber), the evolution of gaseous compounds formed from the side groups and in suitable systems the trapping of electrons. However the most striking effects arise from the physical changes due to degradation and crosslinking, and these two very different types of reaction will be discussed separately. In both cases a single chemical bond change per molecule with up to 10^5 such bonds may produce a material with very different physical properties.

DEGRADING POLYMERS

Many polymers show a progressive reduction in molecular weight with increasing dose. Included in these are polymethyl methacrylate (PMMA), polyisobutylene (PIB), polytetrafluorethylene (PTFE), cellulose. In the former two polymers of amorphous structure, scissions occur at random along the chain, and in proportion to dose r (in Mrads). Per g of polymer, there will be $0.624 * 10^{18}$ G.r scissions (1 Mrad = 0.624×10^{20} eV/g) and a corresponding increase in the number of molecules. Thus the number average molecular weight M_n decreases from its initial value M_n.

$$1/M_n = 1/M_{n_o} + 1.04 \times 10^{-6}\ G.r$$

with G scissions per 100 eV, usually about 1 to 5. Moreover such scissions are located at random (except possibly at the different chemical structure near chain ends, though there is no good evidence of this). Thus the molecular weight distribution even if not initially random, rapidly becomes so with $M_z/3 = M_w/2 = M_n$. This scission can therefore be used to provide a series of polymer specimens of known molecular weights, and over an extremely wide range. From the radiation chemical point of view one can therefore study, using conventional polymer characterisation techniques, the effect of such factors as temperature or radiation-induced processes in solids (glassy PMMA) and viscous liquids (PMMA above Tg). Moreover by measuring the mechanical properties of a polymer exposed to known radiation doses, one can evaluate the relation between molecular weight and physical properties. The main effect is a reduction in the strain at break.

In polymethyl methacrylate exposed to an electron beam, very fine 'tree' effects can be produced if the electron beam is of inadequate energy to penetrate the specimen completely. A build-up of charge takes place within the bulk of the material until the mutual electrostatic repulsion overcomes the insulating properties, and the discharge, equivalent to a lightning discharge in a solid, leaves a permanent series of traces. This 'tree' effect also occur in many other polymers, and may be decorative, or a serious objection to the use of radiation processes. In any case it can serve as a promosing method for the study of electrical breakdown in insulating solids.

In PMMA the major obvious change is the scission of the main chains, but some sidechains are also released in the form of gases, primarily H_2, CO, CO_2,

CH_4. These are dispersed probably as individual gas molecules, and held within the glassy lattice. On warming the irradiated piece of PMMA, its viscosity greatly decreases, and gas molecules can move together to form bubbles. This might provide interesting data on the diffusion processes of gases, and even of individual molecules throughout the glassy polymer. It is significant that the foam in the heated, irradiated PMMA extends throughout the material, but not into the surface, which remains smooth. It can be surmised that individual molecules can diffuse a limited distance and therefore escape through the surface skin. If this is possible, then diffusion into gaseous conglomerates should be possible within the glassy polymer, but cannot expand into a bubble, because of the surrounding restraints. This novel method of studying the diffusion of molecules in glasses is unexplored.

In two phase systems such as PTFE, partly crystalline and partly amorphous, it is presumed that energy is absorbed equally in both phases, but the reactions must be very different. In the amorphous flexible phase, chain scission and separation of the two fractured ends is readily achievable. In the crystalline phase it is more difficult to visualise such separation when the two fractured ends, in the form of radicals, are kept in close proximity. One would expect then to rejoin unless there are internal strains within a molecule, possibly due to the mutual repulsion of bulky side groups. Another possibility is that energy is transferred from the crystalline to the amorphous regions where it can initiate radical changes and permanent main chain scission. The basic question is still not finally settled - does main chain scission occur in the crystalline regions of such two-phase polymer to a reduced extent, or not at all.

In the case of polytetrafluorethylene (PTFE) this difference in sensitivity of the two components results in a high degree of radiation sensitivity. Even low doses, corresponding a very small percentage of main chain scissions, results in a great reduction in mechanical strength. This is best explained as due to selective scission of tie or linking chains between crystallites, but it is then necessary to explain why so much of the incident energy is concentrated on these relatively few important main chain lengths tying separate crystallites.

The radiation-induced degradation of PTFE has become an industrial process, in which remainds polymer material is degraded to give a much lower molecular weight, then used as a coating or lubricating medium.

A number of other factors can influence the effect of radiation on degrading polymers. Notably of importance are temperature, additives, morphology, solvents. The influence of temperature on the scission of polyisobutylene follows the same pattern as that observed in the activation of a number of bacteria, thereby indicating that in many biological systems, the well-known temperature effect in radiation response is not of some specific biological nature, but derives from a basic physico-chemical reaction in macromolecules.

Many additives can greatly influence these radiation effects notably by reducing the changes produced by a given dose, even at very low concentrations, possibly one percent of less. Thus the energy the additive absorbs must be correspondingly small, and some transfer, either of energy or of reactive groups such as electrons or H atoms must be envisaged. Here again the behaviour of such polymers and additives parallels that of biopolymers with these same additives, pointing to a non-biological origin of radiation protection.

CROSSLINKING POLYMERS

The major effects of radiation on many polymers, notably polyethylene, rubber and polydimethylsiloxane is the fracture of sidechains, resulting in the liberation of such gases of H_2, CH_4 etc., and the formation of radicals on the main polymer chain. By the combination of two such radicals on adjacent chains, a crosslink is formed. Each crosslink involves two crosslinked units - one on each chain, and the crosslink density is expressed in terms of the crosslinking coefficient δ, the number of crosslinked units per weight average molecule M_w. The number of crosslinked units per gram is 0.624×10^{18} G.r, distributed among $6.02 \times 10^{23}/M_w$ molecules, so that $\delta = 1.04 \times 10^{-6}$ G.r.M_w (δ is proportional to dose and G is the number of crosslinked units per 100 eV absorbed).

At low doses ($\delta < 1$) there is an increase in average molecular weight $M_w = M_w^o/(1 - \delta)$ and the formation of molecules of higher molecular weight, primarily branched (X). A vital transition point occurs at a dose r_g (gelation dose) when $\delta = 1$, at which an incipient three dimensional network first begins to form ($r_g = 0.96 \times 10^6/G.M_w$). Thus if G (crosslinked unit) = 1, and $M_w = 4 \times 10^5$, $r_g \sim 2.5$ Mrads.

The relation giving r_g is independent of molecular weight distribution ($\delta = 1$). At higher doses an increasing fraction of the specimen (gel) forms a network, which is insoluble but will swell in any usual solvent. The relation between gel fraction and dose depends on initial molecular weight distribution. In the network itself, higher doses increase the network fraction (g) and reduces the average molecular weight between successive links M_c. Thus the swelling ratio will diminish, at the same time as the elastic modulus increases. There are therefore three methods of assessing the degree of crosslinking - soluble fraction (s = 1-g), swelling and modulus. A fourth technique recently developed is via nuclear magnetic resonance (NMR). The use of NMR for chemical analysis via the T_1 relaxation is very well known. Further studies of the T_2 relaxations reveals that this can be used to evaluate the degree of crystallinity, and also the network fraction, due to either crosslinks or entanglements of a more temporary nature.

The crosslink coefficient δ and the density of crosslinks per g of polymer depends not only on the dose and temperature but also on the morphology of the polymer. For example crosslinking in the crystalline phase of a two-phase polymer may be very different from that in the amorphous regions, or even on or near the crystalline surface. However many experiments to determine the location of links in a polymer such as polyethylene have not yet proved decisive. To measure the density of crosslinking it is necessary to warm the specimen to provide flexibility and they may result in further reaction between radicals trapped in the crystalline regions. At the same dose, and with identical polyethylene material, single crystal specimens show less crosslinking than do more conventionally prepared specimens. On the other hand, irradiation of simple linear alkanes, irradiated in a purely crystalline state, show relatively high degrees of crosslinking. A detailed analysis of their products shows a high degree of internal linkages between adjacent molecules, and not predominantly at or near the crystalline surfaces.

The usual values of G (crosslinked units) = 2 G (crosslinks) ranges from about 2 to 5, roughly comparable with the number of radicals as measured by ESR in samples irradiated at low temperature. It is perhaps surprising that the presence of unsaturation as in rubber does not greatly increase the ability to form crosslinks. On the other hand aromatic groups as in polystyrene greatly reduce the formation of crosslinks; G (crosslink) $\sim$ 0.1. This is ascribed to 'energy transfer' whereby the energy absorbed anywhere in the molecule migrates to the aromatic group, which thanks to its many energy levels, acts as a sink and reduces radition effects. It is then surprising that a similar considerable reduction is not seen in the radiation induced polymerisation of styrene.

The mechanism of crosslinking is still a matter of lively discussion. The main difficulty is that if crosslinks are formed by the reaction of two radicals on adjacent chains, how are they formed in such close proximity? On purely statistical grounds, such pair formation is extremely unlikely ($< 10^{-4}$). Are the radicals formed at random and then migrate, and if so what is the mechanism? - (i) by hole migration - i.e. H-transfer from adjacent bonds, until radicals can meet a partner; (ii) H formed in the initial radiation event carry out a series of addition, abstraction reactions?

$$-CH_2- \rightsquigarrow -\dot{C}H- \quad + H$$

$$H + -CH_2- \longrightarrow -\dot{C}H- \quad + H_2$$

$$H_2 + -\dot{C}H- \longrightarrow H + -CH_2-$$

etc.

so that finally two radical molecules can meet under suitable circumstances to react

$$\begin{matrix} -\dot{C}H- \\ -\dot{C}H- \end{matrix} \longrightarrow \begin{matrix} -CH- \\ | \\ -CH- \end{matrix}$$

The slow loss in radical concentration long after irradiation, and its dependence on hydrogen pressure argues strongly in favour of this hypothesis. The so-called protective effect of many additives, present at low concentration, may be due to their H scavenging ability, which greatly reduces the opportunity for such radicals to move.

The effect of temperature on the reaction must also be considered. At very low temperatures, crosslink formation is reduced by about half, and this may be indicative of two different processes, one of which can be eliminated by additive or temperature, and is a slow process. The other may be immediate, and therefore non-scavengeable and almost independent of temperature.

MEMORY EFFECT

A most important aspect of the radiation process for the modification of polymers is that unlike many chemically initiated catalytic reactions, it varies little with temperature and can therefore be used under a very wide range of conditions, and even in the solid state. This allows temperature to be used as an extra variable at the disposal of the experimenter or technologist, and permits a double process to be developed. Some major industrial developments take advantage of this increased freedom.

One of the earliest economic development is based on the memory effect. Chains in a lightly crosslinked flexible network have considerable freedom between successive crosslinks, and take up configurations determined by thermodynamic properties, and by their molecular arrangement during the crosslinking process. This lies at the heart of the high elasticity shown e.g. by vulcanised rubber. In polyethylene at room temperature, chains are held in position primarily by the crystalline regions, the amorphous chains linking crystallites accounting for its limited flexibility and deformation under stress at room temperature. Above the crystalline melting point, polyethylene becomes a very viscous fluid, held together primarily by molecular entanglements of a dynamic but temporary nature.

However if the polyethylene has been crosslinked by radiation, these chains are held in a permanent flexible network, of modulus determined by the density of crosslinks, i.e. the dose. It then behaves as a rubber, and can be extended or otherwise deformed with full potential recovery. If then cooled to allow recrystallisation, these crystals will hold the polymer in its new shape indefinitely. However on rewarming to remove these cyrstals, this shape is no longer restrained, and due to the network elasticity, recovers the shape it held during irradiation. This memory effect is widely used for electrical and mechanical equipment e.g. for cable junctions, expanding devices and many other products involving a long-term memory facility. No doubt, many new application will be developed.

FOAMED POLYMER

If polyethylene incorporating a blowing compound is lightly crosslinked by radiation at about room temperature, it can then be heated above the melting point, and the blowing agent allowed to foam into this flexible network, which can then expand many times thanks to its highly elastic properties. On cooling crystals reform and hold the expanded network in position.

RUBBER VULCANISATION

In conventional rubber curing with sulphur, the molecular weight between successive links M_c is about 10^4. This density of crosslinking can be achieved with radiation in the absence of a vulcanising agent, but the dose required, about 45 Mrads would not be very economic. In preparing rubber e.g. for tyres, it has been shaped in various forms before vulcanisation, and this is only achieved with difficulty by using the temporary entanglements in high molecular weight polymer. Radiation at low economic doses can be used to provide a lightly crosslinked network of a permanent character, which then allows the operator to shape his tyre components, with no time limitation on manipulation, before it is finally cured by sulphur vulcanisation. Other more purely radiation processes have been developed.

LITHOGRAPHY

In the search for higher resolution in silicon chips, allowing smaller components, more rapid response, one is faced with a natural limitation in the wavelength of light. Electron beam scanning, of much higher resolution can be adopted to etch circuits onto a thin plastic coating on the silicon. With degrading polymer, the irradiated portions are of far lower molecular weight and more readily removed, whereas with crosslinking polymer it is the unirradiated portions which remain soluble. In either case finer circuitry can be developed. The basic researches involve the sharpest resolution (dependent on beam diffusion, molecular weight distribution etc.).

HYDROPHILIC POLYMERS

Irradiation of polymers in aqueous solutons can lead to a crosslinked network, swollen in the water. Such networks can be used for water retention and slow releases purposes. An interesting feature is that the transition to a network occurs very sharply at a dose which decreases as the concentration diminishes (i.e. the molecules are further apart). This apparent contradiction is due to the indirect effect of the radiolytic products of the water, which increase as the polymer concentration diminishes, and there are fewer polymer molecules with which it reacts.

OXYGEN EFFECT

Many radiation events are greatly enhanced in the presence of oxygen, which reacts with the radicals, leading to a chain reaction, often of an undesirable character, somewhat akin to the oxygen reaction in polymers operating at higher temperature. Incorporation of antioxidants can frequently reduce the oxygen effect, and such antioxidants can be considered as antirads or protectors since they reduce the radiation efect. This oxygen enhancement is also seen in radiobiology.

ENHANCED CROSSLINKING

With conventional crosslinking, each link involves at least one distinct ionisation or excitation event, and insofar as the irradiated polymer is uniform in structure, these links are distributed at random. The dose required to form even an incipient network typically amounts to a few Mrads for high molecular weight polymer, and considerably more for much lower molecular weight polymer of the same chemical structure. This high dose requirement may make the process uneconomic for large scale production. However a very different situation often arises, whereby a highly crosslinked system may be obtained at far lower doses, possibly well below a single Mrad. For this to occur a chain reaction is needed, whereby a single radiation event initiates a radical, or less probably an ionic reaction, and this proceeds via a multifunctional monomer or polymer to give rise to a polymerisation reaction extending through a number of polymer chains. The statistics of a network formed in this way are quite different from these for a random distribution of crosslinks. If A is linked to B by a single radiation event there is a high probability of B then being linked to C, then to D etc. Examples of such polymerisation/crosslinking reactions are found in unsaturated polyesters, polyfunctional monomer systems. Polyvinyl chloride is very difficult to crosslink by conventional means, and shows marked dehydrochlorination reactions. However if a polyfunctional monomer such as triallylcyanurate is present, a fully crosslinked structure is readily obtained at a low dose. It may still be queried whether a true copolymer network is formed between these two components, or whether a tight network involving only the polyfunctional monomer is involved, with the PVC trapped within its pores. Such crosslinked PVC is widely used e.g. for telephone cables.

For such multifunctional systems, it is rarely possible to form a truly highly-elastic network, and radiation curing transforms what is often a viscous liquid directly to a glassy or flexible solid. This is because high elasticity requires a very open network, with flexible chains of molecular weight of at least 10^4 between successive crosslinks, and such open networks cannot be formed by linking together shorter molecules.

The main application of such enhanced crosslinking is in the radiation curing of surface films - a rapidly increasing radiation technology. This uses electron accelerators of lower voltage (about 0.3 MeV) since high penetration is not needed, and may in fact be disadvantageous (as in surface curing on wood surfaces). But for this process to be feasible, some very effective form of enhanced crosslinking is necessary to keep dose requirements within reasonable limits.

The use of high energy radiation for surface curing can be compared with the use of heat and solvent evaporation, and with photochemical methods. The need for lengthy heating in dustfree conditions, the release of solvent into the atmosphere, and the high energy consumption (since the whole object must be heated to a relatively high temperature) argue against the conventional technique. Ultraviolet photochemical methods are limited in depth, they need photosensitive groups, and are unable to penetrate coloured films. Curing with high energy electron beams requires a relatively high capital investment, and a reselection of sensitive systems, not necessarily identical with those used in conventional heat and solvent curable points and emulsions. One disadvantage is that the shape of the object whose surface is to be irradiated may not be readily covered by a linear scanned electron beam but reentrant beams are sometimes available.

POLYMERISATION

Detailed studies of polymerisation reactions induced by radiation have been carried out over a number of decades, and have helped in our understanding of the kinetics in both radical and ionic reactions. Radiation plays a major role only in the initiation step, but here it permits considerable control, over an enormous range of intensities i.e. concentration of initiating and propagating chains. The temperature can be chosen without having to consider the temperature dependence of the initiating catalyst, where the exotherm due to the propagation step reacts back on the catalyst, further increasing the propagation rate and possibly leading to an explosive situation. With radiation, no catalyst is necessary, and there are no residues. Polymerisation can also be initiated with monomer in the solid, crystalline state at low temperatures; surprisingly it may occur that solid state polymerisation is more rapid than when the monomer is in the liquid state. In spite of its promise and advantages, the industrial development of this method of polymerisation has been extremely slow; even when a complete process has been developed to the production stage, it has proved extremely difficult to get it adopted on a large scale. Perhaps human reluctance to introduce a very different and novel process into a well-established industry with conventional chemical techniques must be faced.

GRAFTED SYSTEMS

A radiation technique of great promise is in the production of grafted polymer, whereby a polymer A serves as a backbone to side branches from a polymerised monomer B. It can be hoped that by a suitable selection one can produce a copolymer combining many of the advantages of A and B.

Several alternatives are available for this reaction:

(1) Polymer A is irradiated, and then plunged into monomer B.

The radicals trapped in A serve to initiate polymerisation of B as sidechains on A.

(2) Polymer A is allowed to absorb B and both are irradiated together.

(3) Polymer A is irradiated in a solution of B. The difficulty is that in addition to the copolymer A7B, some homopolymer B is formed, which represents a

waste of B, and is difficult to remove from the surface of A-B.
(4) Polymer A is irradiated in air, so that peroxide groups are formed. Subsequently when A is heated in the presence of B, the peroxide groups decompose, leaving radicals to react with B.

All these processes have some disadvantages which can usually be overcome. Nevertheless radiations grafting so far had only commercial success in a few fields: - the formation of grafted films for use as battery separators in a very hostile environment; the surface coating of polymers to render them blood-compatible for use in medical surgery; the surface treatment of hydrophobic textiles to make them more hydrophilic. Numerous other applications have been explored but there are considerable difficulties in taking them to the commercial exploitation stage.

The range of possible grafts which can be achieved is immense, not only in terms of A and B, but also in such basic properties as number and length of side chains B, degree of crystallinity and flexibility, the distribution of B on the surface or interior of A etc. But a major handicap in exploiting the method is the need to discover a major need, widely felt but not yet met. Here as in many aspects of radiation technology, it is the application which has to found, rather than the technique of meeting it. Occasions have arises when radiation techniques have been used to develop a new product, where large scale production is then achieved by more conventional chemical methods. One can understand the attitude of those who have spent all his adult lives in researches involving chemical processes, and who are then reluctant to abandon their great expertise for a very new type of technology, requiring some degree of retraining and another general outlook. So one can expect to see major progress mainly in areas where radiation methods achieve results which cannot be readily met by more conventional methods (such as in solid state reactions) and where a widely-felt need already exists (i.e. where these is a market).

CONCLUSIONS

In this paper it has only proved possible to outline, in somewhat general terms, some aspects of the problems and answers involved in the radiation technology of polymers. One can visualise high energy radiation as a very readily controlled technique for producing radical and ionic species in polymeric systems, over a enormous range of conditions such as temperature, concentrations, environment and physical state. Radiation can therefore serve as a very powerful means of studying polymer characteristics and behaviour under accurately controlled conditions. Furthermore it can be used in the extensions and innovation of production techniques. Finally it can be used to investigate other problems often associated with polymers - the mechanisms of reinforcement, electrical behaviour and discharge phenomena, surface modification, diffusion and retention of gases.

FURTHER READING

A.J. Swallow, Radiation Chemistry, Longman 1973

A. Charlesby, Atomic Radiation and Polymers, Pergamon 1960

A. Chapiro, Radiation Chemistry of Polymeric Systems, Wiley 1962

M. Dole, Radiation Chemistry of Macromolecules, Academic Press 1972

F.A. Makhlis, Radiation Physics and Chemistry of Polymers, Wiley 1975

Radiation Processing: 'Radiation Physics and Chemistry (Pergamon)'

Vol. 9 (1977) Puerto Rico Conference

Vol. 14 (1979) Miami Conference

Vol. 18 (1981) Tokyo Conference

Vol. 23 (1983) Dubrovnik Conference

A. Charlesby, Proc. Phys. Soc. 57 496, 510, 1945

A. Charlesby, Proc. Roy Soc. (London) A215, 187, 1952

HIGH PRECISION REPLICATION OF LASERVISION VIDEO DISCS USING UV-CURABLE COATINGS

G.J.M. LIPPITS and G.P. MELIS

Philips Research Laboratories, P.O.Box 80 000, 5600 JA Eindhoven, The Netherlands

SYNOPSIS

The material requirements are discussed that had to be coped with for the manufacture of optical discs by high-precision replication with UV-curable coatings.

1. INTRODUCTION

Several methods as embossing, compression and injection moulding have been studied for the replication of optical discs from a mould. In this paper casting as the replication method using photo-induced polymerization of polyfunctional acrylate monomers (the Philips 2p process) is discussed. A diagram of the replication process is given in fig. 1, a diagram of the final product in fig. 2. The specifications of disc and LaserVision player are matched to deliver a television signal of excellent quality to the receiver. This means that the disc must satisfy a number of requirements relating to optics, geometry and stability (refs. 1,2).

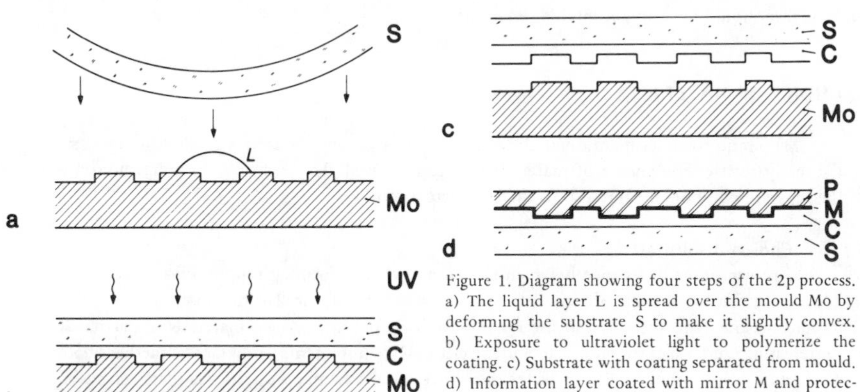

Figure 1. Diagram showing four steps of the 2p process. a) The liquid layer L is spread over the mould Mo by deforming the substrate S to make it slightly convex. b) Exposure to ultraviolet light to polymerize the coating. c) Substrate with coating separated from mould. d) Information layer coated with mirror M and protective layer P.

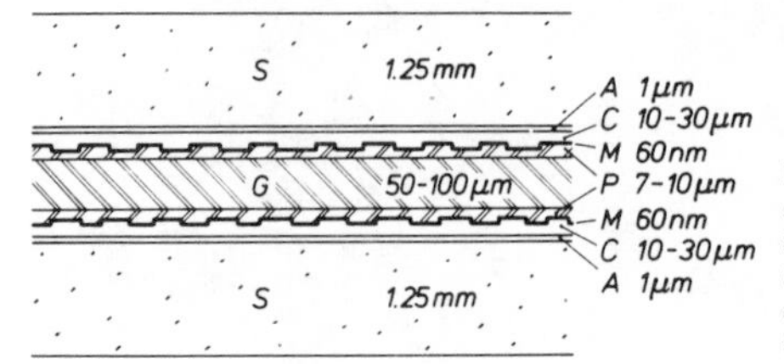

Figure 2. Diagram showing the configuration of a double-sided LaserVision disc. The hole at the center is not shown. S = transparent substrate. A = adhesion layer. C = layer with picture and sound information in the form of pits. M = mirror coating. P = protective layer. G = adhesive layer.

2. REQUIREMENTS

2.1 Optical requirements

For the stored information to be read properly there must be at least 75% optical reflection of the laser beam. Absorption and scattering losses occur as the beam passes twice through the substrate. These losses depend on the type of substrate material and the metal mirror coating. Also the birefringence of the transparent substrate is an important parameter. It must be no greater than 20 degrees.

2.2 Geometrical requirements

The thickness of the substrate is set at 1.25 ± 0.1 mm. Thinner substrates are difficult to handle and to make. Moreover, at this thickness scratches on the outer surface of the disc are so far outside the depth of focus of the lens of the player that no harmful effects are observed. The specification for flatness and parallelism of the two surfaces of the substrate is more critical than that of the thickness. Any departure from flatness of the inner surface of the disc causes vertical movement in the information track. The high rotational speed of the disc imposes restrictions on the tolerable local thickness variations.

2.3 Stability requirements

The disc must meet stability specifications immediately after manufacture and during its use by the consumer. The optical and geometrical characteristics of the disc must remain unaffected under climatic conditions likely to be encountered during transport and storage and the disc should have an acceptable service life.

3. CHOICE OF MATERIALS

The requirements mentioned above are clearly rigorous for a mass production process, and severely restrict the choice of materials that can be used. This restriction applies not only to the separate components but also to the complete assembly.

3.1 Choice of substrate

An important difference between the 2p and thermoforming reproduction methods is that the substrate in the 2p process serves as inert carrier of the 2p information layer. For economic reasons it is necessary to use readily available polymers. Two materials in particular are interesting and meet the optical and mechanical requirements: polymethylmethacrylate (PMMA) and polycarbonate (PC). Besides the above requirements, properties as glass transition temperature and adhesion to the 2p coating have to be considered. They are compared in table 1.

TABLE 1.

Properties of polymethylmethacrylate (PMMA) and polycarbonate (PC) for possible application as substrate material in the 2p process for the manufacture of LaserVision discs.

PROPERTY	PMMA	PC
transmission at 630 nm	about 95%	about 95%
birefringence	acceptable	problematic
glass temperature	100-110°C	140°C
dimensionally stability	sensitive to moisture	good
method of manufacture	extrusion, casting, injection molding	injection molding
price	reasonable	high
compatibility with 2p process	good	good, but tendency to stress cracking in contact with 2p lacquer

3.2 Choice of the 2p coating

The 2p coating determines to a large extent the picture quality of the video disc. The properties of the coating depend on process requirements, e.g. high curing rate, low viscosity, good wetting of the mould (ref. 3), easy release of the cured coating from the metallic mould, excellent adhesion to the substrate and to the metallic mirror coating. High curing rates are obtained with acrylates. The photoinitiator used, a benzoin ether, 1,1-dimethoxy-1-phenyl-acetophenone, is optimal for fluorescent lamps with maximum emission at 350 nm. Details can be found in refs 4, 5. The acrylates are chosen with a view to ultimate properties such as the behaviour of the coating under metallization and the adhesion of the coating to the metallic mirror. We found that these ultimate properties depend on the functionalities, f, of the monomers. Higher f leads to increased crosslink density resulting in increased dimensional stability. However, if the crosslink density becomes too high (e.g. $f \geq 4$), the material will be too stiff. High stiffness coupled to a polymerization shrinkage of about 15% may cause warping of the substrate or cracking of the coating. If f is too low, e.g. $f = 2$, the coating remains too soft. The crosslink density also has a strong effect on the reflectance of the mirror coating in the region of low functionalities as seen in fig. 3. The effect is similar, whether the mirror is applied by evaporation

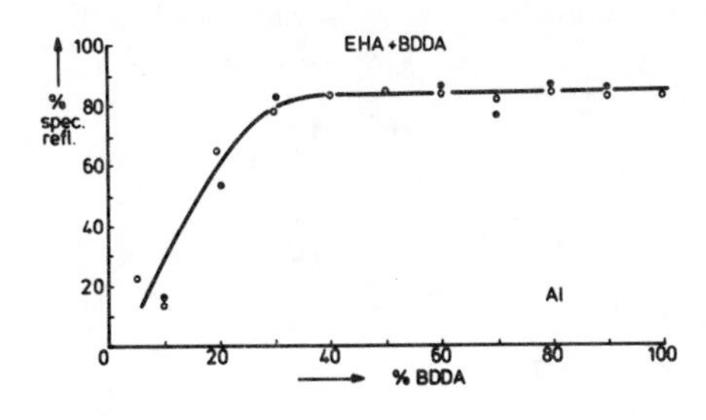

Figure 3. Specular reflectance of aluminium, deposited on fully cured layers of EHA/BDDA. Thickness of metallic coating: 100 nm. ○ : evaporated, 7×10^{-3} Pa. ● : magnetron sputtered in Ar, 3×10^{-1} Pa.

or by sputtering. The explanation is that stresses occurring in the metallic mirror show up as wave patterns owing to shear and deformations in the softer 2p coatings (fig. 4). As a result the

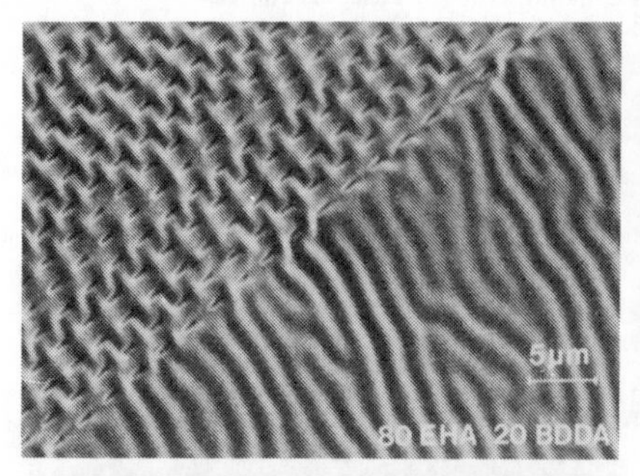

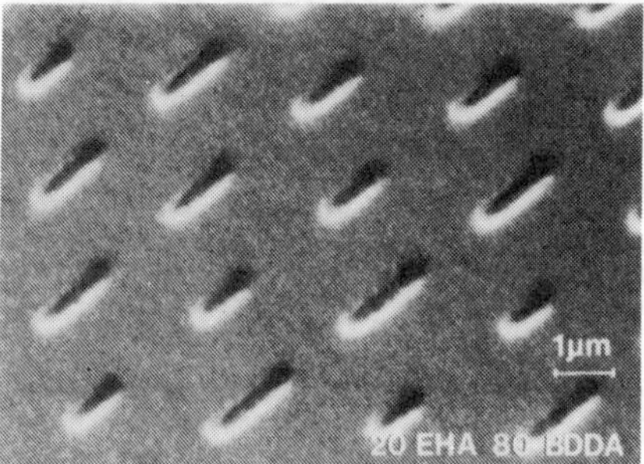

Figure 4. Scanning electron microscope pictures of the LaserVision pattern after metal deposition on a coating of a mixture of 2-ethylhexylacrylate (EHA) and 1,4--butanedioldiacrylate (BDDA). The upper photographs -4a-(magnification 2500x) and the photograph-4b-(magnification 10,000x) were obtained with the composition 80% EHA, 20% BDDA. After metallization, the soft coating appears as a wave pattern. The photograph on the left -4c- (magnification 10,000x) was obtained with the composition 20% EHA, 80% BDDA. Since the coating substrate is sufficiently hard, there is no wave pattern.

reflectance is strongly reduced. The adhesive and reflective properties of the cured monomers have been mapped in a f-k diagram, where k is the weight fraction of unsaturated hydrocarbon groups. Maps of f versus k are shown in fig. 5 for a series of monomers and for mixtures of monomers. Fig. 5a shows the available monomers, fig. 5b also indicates when the adhesion between PMMA and adhesive is good and fig. 5c the same for the adhesion with the metallic mirror. It can be concluded that suitable compositions can be found in the domain with $2 < f < 6$ and $0.15 < k < 0.50$. In this way a number of suitable coatings can be selected (Table 2).

TABLE 2.

Properties of coatings that will give good video discs.

Coating	Composition (weight percent)		Viscosity at 23°C (mPa.s)	Curing time (s) 23°C	80°C
1	57 % TPGDA 10 % TMPTA	29 % NVP 4 % DMPA	7.8	1.5	1.0
2	61 % TPGDA 17.5% TMPTA	17.5% NVP 4 % DMPA	12.3	1.7	1.0
HDDA	96 % HDDA	4 % DMPA	6.7	3.0	1.4

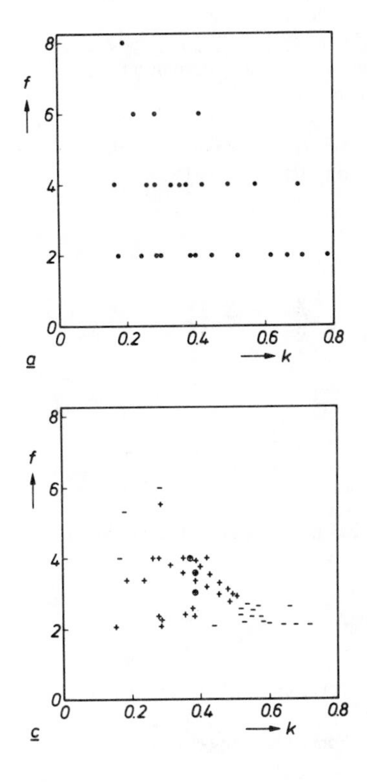

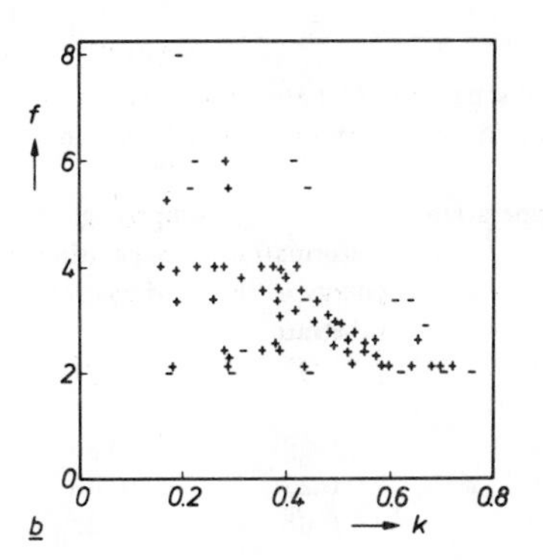

Figure 5. Saturated hydrocarbon fraction k and functionality f of commercially available acrylate monomers (a), and for a number of mixtures composed from them (b and c). In b there is an indication of whether, after curing, good adhesion is obtained at the substrate and whether the information layer separates readily from the mold: + yes, – no. In c there is a further selection, depending on whether the reflective coating adheres to the layer and the material remains bright on metallization (vapour deposition with Al), + yes, – no. The indications for two mixtures and a single component system that will give good video records are circled.

Three of these coatings have been studied with respect to curing rate, extractability of unreacted substances and separation energies, required for release from the mould. Regarding curing rate the mixtures polymerize faster than the individual monomers. When the photopolymerization is carried out at 80°C, the curing rate of the single monomer coating (HDDA) nearly equals that of the mixtures. The extractability is a measure of the quantity and mobility of small molecules present in the material after curing. Extraction curves are determined by liquid chromatography. The one-component system is found to contain less extractable material than the mixtures.

The separation energies are given in Table 3. During release of the substrate with cured information layer from the mould, it is warped in a complicated way. One part of the total separation energy is required for the actual separation (release of adhesion) another part for the (in)elastic deformation of the substrate with information layer. Table 3 shows that the release is easiest for the one-component system, although satisfactory results were obtained in all three cases.

TABLE 3.

Some data of the separation of LaserVision discs from a mould, measured at a constant pulling rate of 500 mm/min, when mixtures 1 and 2 and the single-component system HDDA are used.

Coating	Total separation energy (J)	Comprising: Deformation energy of the substrate (J)	Comprising: Separation energy of coating (J)	Separation time (s)	Maximum pulling force (N)
1	1.27	0.39	0.88	4.2	51
2	0.75	0.27	0.48	3.6	48
HDDA	0.30	0.08	0.22	2.4	31

ACKNOWLEDGMENT

The author is indebted to J.G. Kloosterboer, L.K.H. van Beek and G.E. Thomas for valuable discussions.

REFERENCES

1. A.J.M. van den Broek, H.C. Haverkorn van Rijsewijk, P.E.J. Legierse, G.J.M. Lippits and G.E. Thomas, J. Radiation Curing **11**, (1)2 (1984).
2. H.C. Haverkorn van Rijsewijk, P.E.J. Legierse and G.E. Thomas, Philips Tech. Rev. **40**, 287 (1982).
3. P.E.J. Legierse, J.H.A. Schmitz, M.A.F. van Hoek and S. van Wijngaarden, Plat. and Surf. Fin. **71**, (12), 20 (1984).
4. J.G. Kloosterboer, G.J.M. Lippits and H.C. Meinders, Philips Tech. Rev. **40**, 298 (1982).
5. J.G. Kloosterboer and G.J.M. Lippits, J. Radiation Curing, **11**, (1)10 (1984).

FAST CURING LOW-MODULUS COATINGS FOR HIGH-STRENGTH OPTICAL FIBRES

D.J. Broer and G.N. Mol

Philips Research Laboratories, P.O.Box 80 000, 5600 JA Eindhoven, The Netherlands

SYNOPSIS

UV-curable polyetherurethane acrylates and poly(dimethyl-co-methylphenyl)siloxane acrylates have been investigated for the primary buffer coating of optical fibres. Curing behaviour, E-modulus, glass transition temperature and refractive index have been studied in relation to the molecular structure of both acrylate types.

1. INTRODUCTION

Optical glass fibres are primary coated on line during fibre drawing with a low-modulus buffer coating. Before cabling, a thermoplastic secondary coating is applied by extrusion. This paper deals with the primary buffer coating.

The main requirements of the primary coating materials are related to processing speed, optical functioning and fibre strength. On-line coating during fibre drawing is essential for the protection of the brittle fibre surface[1]. To achieve fibre drawing speeds greater than 5 m/s, the coating must be applied and solidified in a very short time, typically of the order of 0.1 seconds.

The signal-to-noise ratio of the information transported by the optical fibre is enhanced when any cladding light modes which may be present, are removed by the coating. To achieve cladding mode stripping, the refractive index of the coating must be higher than that of the cladding glass, even at the highest operation temperature. For quartz-clad fibres, accounting for the difference in temperature dependency of quartz and the coating, this requirement is fulfilled when the refractive index of the coating at room temperature is higher than 1.475.

Optical losses increase under random external forces due to microbending[2]. Microbending losses are minimized by the application of the low-modulus buffer coating. However, to maintain constant optical properties from − 50°C to + 80°C, thermal transitions affecting the modulus of the coating (e.g. glass transition or crystallization) must be absent in this temperature range.

UV-curable acrylates are one preferred class of optical fibre coatings. Generally, they combine good rheological properties for high speed, low tension extrusion, high curing rates and extended pot lives. Epoxy and urethane acrylates are already widely used. The main drawback of these materials is the temperature of their glass transition. When used as a buffer coating the modulus increases rapidly with decreasing temperature, resulting in an enhanced sensitivity to microbending losses.

Promising materials with respect to the low temperature properties are UV-curing silicone and polybutadiene rubbers[3,4,5]. However, suitable UV-curing silicone rubbers still have to be developed in order to meet the refractive index requirements. The disadvantages of UV-curing polybutadiene rubbers lies in their sensitivity to oxidation. In this paper we present our investigations on UV-curable polyetherurethane rubbers with extended low Tg polyether chains and UV-curable silicone rubbers with aromatic substituents to increase their refractive index.

2. MATERIALS

The investigated oligomeric acrylates are presented in fig. 1. In the case of the polyetherurethane acrylates the material properties were varied by varying the length of the polypropyleneoxide chain. An increasing chain length yields a decreased crosslink density, a decreased volume fraction of

$$CH_2{=}CH\text{-}CO\text{-}O\text{-}CH_2\text{-}CH_2\text{-}O\text{-}CO\text{-}NH\text{-}C_6H_3(CH_3)\text{-}NH\text{-}CO\text{-}O\text{-}[CH(CH_3)\text{-}CH_2\text{-}O\text{-}]_n\text{-}CO\text{-}NH\text{-}C_6H_3(CH_3)\text{-}NH\text{-}CO\text{-}O\text{-}CH_2\text{-}CH_2\text{-}O\text{-}CO\text{-}CH{=}CH_2 \quad a$$

$$CH_2{=}CH\text{-}CO\text{-}O\text{-}CH_2\text{-}CH_2\text{-}CH_2\text{-}Si(CH_3)_2\text{-}O\text{-}[\text{-}Si(CH_3)_2\text{-}O\text{-}]_p\text{-}[\text{-}Si(CH_3)(C_6H_5)\text{-}O\text{-}]_q\text{-}[Si(CH_3)((CH_2)_3\text{-}O\text{-}CO\text{-}CH{=}CH_2)\text{-}O\text{-}]_r\text{-}Si(CH_3)_2\text{-}CH_2\text{-}CH_2\text{-}CH_2\text{-}O\text{-}CO\text{-}CH{=}CH_2 \quad b$$

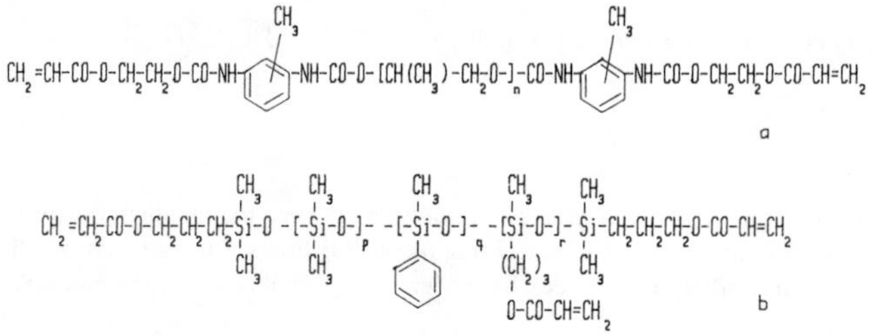

Figure 1. Polyetherurethane acrylate (a) and poly(dimethyl-co-methylphenyl) siloxane acrylate (b)

the polar urethane groups and an increased volume fraction of the low Tg, low refractive index polyether. As a result Tg, refractive index and E-modulus are decreased. Practical application of these oligomeric acrylates with respect to viscosity and polymerization rate is possible with relatively small amounts ($\leq$ 20 wt%) of reactive diluents and slightly increased temperatures ($\leq$ 50°C).

In the case of the poly(dimethyl-co-methylphenyl)siloxane acrylates, the material properties were varied by varying the mole fraction of the methylphenylsiloxane units. An increased number of aromatic rings yields an increased refractive index, Tg and viscosity. To obtain suitable viscosities and mechanical strength, the molecular weight has to be chosen relatively high. In that case crosslinking over $\alpha - \omega$ acrylate groups only, as was done with the polyetherurethane acrylates, results in very low curing rates and gel-like products. Good results were obtained when about 5 mole % acrylate groups are incorporated.

3. PHOTOCHEMICAL CROSSLINKING

For curing, the materials were provided with 0.2 M α, α-dimethoxy-α-phenylacetophenone as a photoinitiator. The rate of conversion of the acrylate groups was measured by low intensity (0.2 mW/cm^2) UV irradiation under nitrogen in a differential scanning calorimeter (DSC), modified for photocuring experiments. Typical examples of polymerization rate vs. conversion are given in fig. 2. As is known[6,7], the conversion of di- and tri-acrylates normally stops far from completion due to increasing steric hindrance and decreasing mobility of the remaining acrylate groups at higher conversion. In contrast, the conversion of polyetherurethane acrylate goes to completion very rapidly, even under the low intensity conditions used. This behaviour is caused by the length and flexibility of the chains connecting the acrylate groups. This prevents vitrification during polymerization at room temperature.

However, in the case of the poly(dimethyl-co-methylphenyl)siloxane acrylates, with both $\alpha - \omega$ and pendant double bonds, complete conversion is more difficult. In spite of the flexibility of the connecting chains, the mobility of the pendant acrylate groups is spatially limited at two sides, which apparently makes them less liable to further reaction. As can be seen from fig. 2, the polymerization rate drops down at a much lower conversion of the acrylate groups than is the case with the polyetherurethane acrylates.

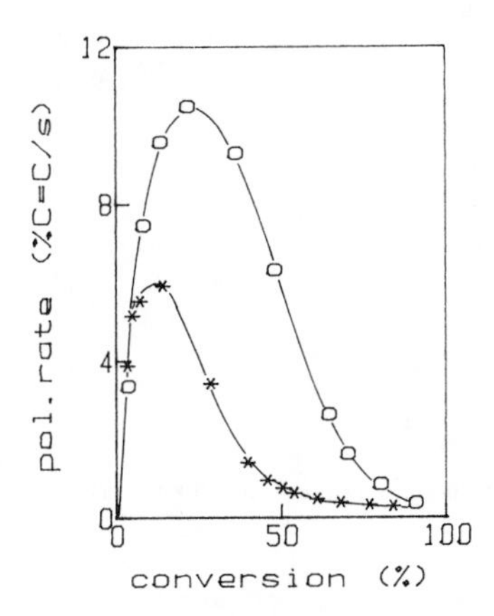

Figure 2.
Polymerization rate vs. the conversion of double bonds for a polyetherurethane acrylate (o) and a polysiloxane acrylate (*)

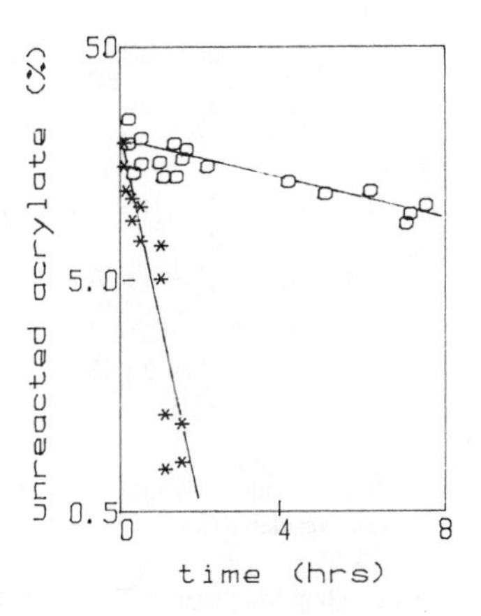

Figure 3.
Dark reaction in air (o) and in nitrogen (*) of a polyetherurethane acrylate after an UV exposure of 0.06 s

Due to the high mobility of the polyether chains, the polyetherurethane acrylates show a remarkable dark reaction after a short flash of high intensity (600 mW/cm^2) UV light. The conversion measured by DSC immediately after 0.06 seconds irradiation is about 80% (fig. 3). When the samples are stored in the absence of oxygen, the reaction proceeds to completion in about 1 hour. In air 95% conversion of a 40 μm film is obtained after 24 hours. In practice, coated fibres are stored in air, but the buffer coating is immediately covered with a topcoating which in turn is freed from oxygen during polymerization.

4. MECHANICAL AND PHYSICAL PROPERTIES

As both Tg and refractive index are affected by changes in molecular structure, they can be plotted against each other as depicted in fig. 4. This figure shows the superiority of the poly(dimethyl--co-methylphenyl)siloxane acrylates with respect of the lowest Tg at appropriate refractive index values. Since no crystallization/melting could be measured by DSC, in contrast to thermally cured silicon rubbers[8], operation temperatures down to − 80°C are possible. However, with polyetherurethane acrylates, too, lower Tg's than normally assumed can be obtained. Moreover, when these materials are modified with a small amount ($\leq$ 20 wt%) of refractive diluent with a high refractive index and a low polarity, some increase in refractive index can be observed without a negative influence on Tg. Minimum Tg values between − 50 and − 60°C were realized in this way. Typical E-moduli are then 2 to 4 MPa for the polyetheruretane acrylates and about 1 MPa for the polysiloxane acrylates.

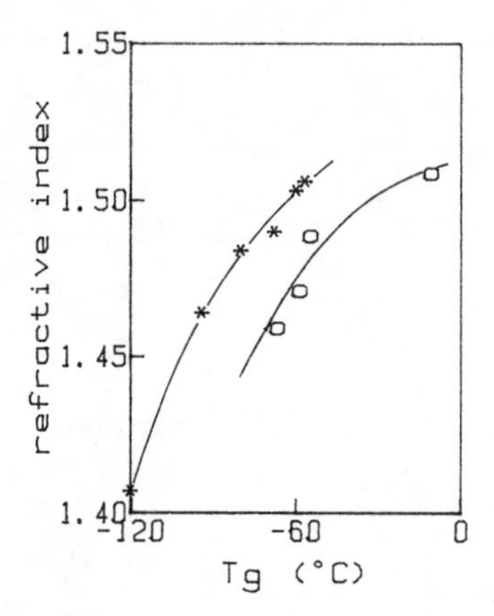

Figure 4. Refractive index (D-line) vs. glass transition temperature for polyetherurethane acrylates (o) and polysiloxane acrylates (x).

Although polyetherurethane acrylates have a higher mechanical strength than poly(dimethyl--co-methylphenyl)siloxane acrylates, both coatings must be provided with a hard top coating to protect the fibre against abrasion during storage, transport or cabling. Suitable materials are among others polyetherurethane acrylates as given in fig. 1, however with smaller polyether chains than used for the buffer coatings.

5. CONCLUSIONS

Both with polyetherurethane acrylates and poly(dimethyl-co-methylphenyl)siloxane acrylates low-modulus, low-Tg coatings can be realized. Within the refractive index requirements, the siloxane acrylates are advantageous with respect to the low temperature properties. With respect to the curing rate, polyetherurethane acrylates are preferred. They cure very rapidly to complete conversion at irradiation times shorter than 0.1 s.

REFERENCES

1. D. Kalish, P.L. Key, C.R. Kurkjian, B.T. Tariyal and T.T. Wang, Optical Fiber Telecommunications, Ch. 12, Academic Press Inc. (1979).
2. D. Gloge, Bell Syst. Tech. J. 54 (2), 245 (1975).
3. H.A. Aulich, N. Douklias and W. Rogler, ECOC83-9th European Conference on Optical Communication (Ed. Melchior and Sollberger), Elsevier Science Publishers B.V. (North-Holland), 377 (1983).
4. T. Kimura and S. Sakaguchi, El. Letters 20 (8), 317 (1984).
5. T. Kimura and S. Yamakawa, El. Letters 20 (5), 202 (1984).
6. J.E. Moore, UV Curing: Science and Technology (Ed. by S.P. Pappas), Technology Marketing Co. (Connecticut, U.S.A.), 146 (1978).
7. J.G. Kloosterboer, G.M.M. van de Hei, R.G. Gossink and G.C.M. Dortant, Polymer Communications **25**, 322 (1984).
8. K.E. Polmanteer and M.J. Hunter, J. Appl. Pol. Sci. 1 (1), 3 (1959).

REPLICATION OF HIGH PRECISION ASPHERICAL LENSES USING UV-CURABLE COATINGS

R.J.M. Zwiers and G.C.M. Dortant

Philips Research Laboratories, P.O.Box 80 000, 5600 JA Eindhoven, The Netherlands

SYNOPSIS

The photopolymerization of polyfunctional methacrylate resins suitable for the production of aspherical replica lenses was investigated. The influence of molecular structure and polymerization conditions on the physical properties of the formed coating are described.

1. INTRODUCTION

In optical scanning systems such as the video or audio disk system, complicated lenses are required to focus the scanning lightspot. With diffraction-limited aspherical lenses the number of optical components, the weight of the optical path and the lens aberations are considerably reduced (1). Mass production of these high precision aspherical lenses is accomplished by a so-called lens replication process (2). In this process a methacrylate resin is applied to a spherical glass substrate and polymerized against an aspherical quartz glass mould using UV light, as is shown in Figure 1.

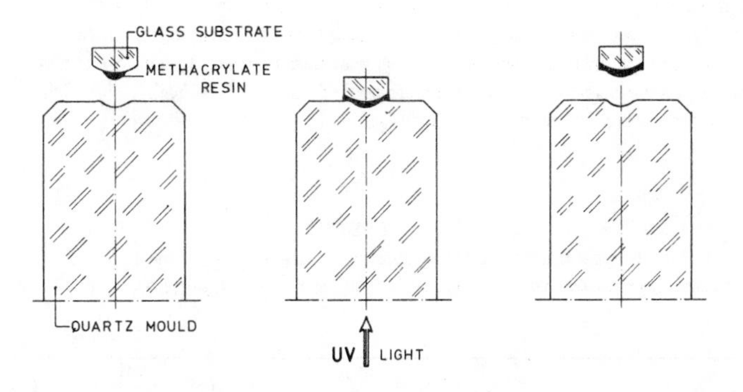

Figure 1.
Schematic drawing of the lens replication process.

Processes based on the photoinitiated radical polymerization are well known in the printing ink, printing plate, metal and wood coating industries (3,4,5). They are recognized for their high speed and the enormous variety of properties that they can provide, depending on the structure of the coating and the way the coating was produced. In order to be suitable for lens replication the coating has to possess additional properties like optical transparency, a relatively high refractive index and a low polymerization shrinkage.

For this purpose polymer networks obtained from ethoxylated bisphenol-A dimethacrylate are thought to be very attractive.
We have studied the influence of structure and polymerization conditions on the physical properties of these methacrylate resins, as well as the temperature dependence of these properties.

2. EFFECT OF MOLECULAR STRUCTURE ON ULTIMATE CONVERSION

The photopolymerization behaviour of a series of ethoxylated bisphenol-A dimethacrylate resins, with different chain lengths between the functional groups, as depicted in scheme 1, was evaluated by isothermal microcalorimetry with a specially equipped DSC apparatus (6). The amount of C = C unsaturation in the resins, from which the average value n could be estimated, was determined titrimetrically by thiol addition.

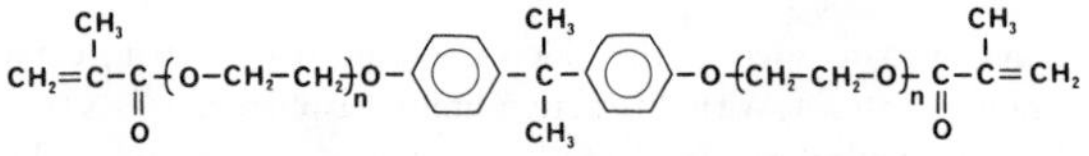

$$CH_2=C(CH_3)-C(=O)-(O-CH_2-CH_2)_n-O-C_6H_4-C(CH_3)_2-C_6H_4-O-(CH_2-CH_2-O)_n-C(=O)-C(CH_3)=CH_2$$

Scheme 1.

The photoinitiator used in all experiments was 2,2 dimethoxy-2-phenylacetophenone. The ultimate conversion (α), the maximum conversion rate (V_{max}) and the time to reach 80% of the ultimate conversion ($t_{0.8}$), were calculated from the heat of reaction at 80°C. In Table 1 the effect of chain length on the photopolymerization is nicely demonstrated. The difference is significant for the ultimate conversion and is even more pronounced for $t_{0.8}$.

TABLE 1.
The effect of chain length, n, on maximum conversion rate (V_{max}), time to reach 80% conversion ($t_{0.8}$) and the ultimate conversion (α). Conditions: I = 0.2 mW/cm^2; Lightsource TL-08, 4W; 80°C; Nitrogen atmosphere.

n	V_{max} (%/s)	$t_{0.8}$ (s)	(%)
0	4.2	120	54
1.0	7.6	55	78
1.6	7.9	30	84
2.8	7.9	12	95

A higher conversion of methacrylate groups is observed when n increases. Especially going from n = 0 to n = 1, thereby introducing ether linkages and a spacer between the aromatic and the methacrylate moiety, the difference is remarkable.

3. EFFECT OF POLYMERIZATION CONDITIONS ON ULTIMATE CONVERSION

Optimum mechanical and optical properties are required, and since they are combined in coatings at the ultimate conversion, we investigated the effect of polymerization temperature and irradiation intensity on the conversion.
It is widely-accepted that for highly crosslinked networks the ultimate conversion is limited by the approach of the Tg which in turn depends on the polymerization temperature.
This effect was also observed for a coating with n = 1.6 giving ultimate conversions of 51, 64 and 84% at 20, 40, and 80°C, respectively. The effect of light intensity on the ultimate conversion is shown in Table 2. The ultimate conversion increases at higher intensity.

TABLE 2.

The effect of light intensity (LI) on ultimate conversion for a coating with n = 1.6. Conditions: 25°C; Nitrogen atmosphere; a: TL-08, 4W; b: TL-09, 40W.

LI (mW/cm^2)	α (%)
0.02 (a)	47
0.20 (a)	51
2.45 (b)	59*

* Estimated by IR-spectroscopy

We explain this result by an increased local temperature at the reaction site due to the exothermic reaction, leading to a higher mobility and therefore a higher conversion. Also at higher light intensity the effect of volume relaxation is less predominant (7).

4. PHYSICAL PROPERTIES OF THE REPLICATION COATING

Among the great variety of commercially available (meth)acrylate systems, only a few are of potential interest for the replication process. This is mainly due to the additional optical demands, such as low reflection and scattering, high transmission at the laser wavelength (780 nm), relatively high refractive index and small polymerization shrinkage. The refractive index and the polymerization shrinkage are two important parameters for the lens design. They determine the aspherical contour for the desired optical performance. Next to a high optical quality the coating has to possess good mechanical properties.
The effect of irradiation time on the mechanical properties can be visualized by the hardness of the coating. The hardness was determined in a separate series of experiments on 50 μm films polymerized at a higher light intensity. The hardness was measured as the reciprocal indentation depth of a diamond sphere (radius 7 μm) under a standard load (25 mN). As can be seen from Table 3, the hardness increases rapidly for relative short irradiation times up to 4 minutes, after which it levels off at a conversion of about 60%, as determined by IR-spectrometry.

TABLE 3.

The effect of irradiation time on coating hardness for a coating with n = 1.6.
Conditions: I = 2.4 mW/cm^2 ; TL-09, 40W; 25°C; Nitrogen atmosphere.

Irradiation time (min)	Hardness (μm^{-1})	Hardness (μm^{-1}) after 7 days
0.3	0.167	0.169
0.5	0.192	0.256
1.0	0.263	0.400
1.5	0.357	0.500
2.0	0.417	0.526
4.0	0.588	0.769
10.0	0.625	0.833
20.0	0.667	0.833
30.0	0.769	0.833

Longer irradiation times give only a slight increase in hardness without a notable increase in conversion (60%). This slight increase in hardness is attributed to volume relaxation. The increase of hardness upon standing, e.g. volume relaxation becomes very plain when the samples are measured 7 days after irradiation. The hardness of the coating increases in all cases to a maximum value of 0.833 μm^{-1}. In this case almost no difference is observed between 4 and 30 minutes irradiation time. Under these conditions the ultimate hardness is reached after about 4 minutes.
Some other relevant physical properties for a coating with n = 1.6 are given in Table 4.

TABLE 4.

Physical properties of a lens replication coating with n = 1.6.

n_d^{25}	volume shrinkage	100 μm $T_{780\ nm}$	E-modulus	T-strength	ϵ
1.569	6.0%	99%	1500 MPa	39.5 MPa	2.75%

When the coating is subjected to annealing at high temperature for one hour an increase is found of E-modulus and tensile strength to a value of 2000 MPa and 41.5 MPa, respectively. The ultimate elongation (ϵ) decreased to 1.3%. We explain these phenomena by an additional conversion of methacrylate groups due to thermal mobilization of trapped radicals, encapsulated during the formation of the polymer network and some volume relaxation (7).
Since the lens has to operate even under drastic temperature changes ($-30 < T < 85$°C) the temperature coefficient of the refractive index and the linear expansion coefficient have to be such that constant optical properties are obtained in this temperature region. These conditions are reached by the annealing step at high temperature. Without annealing the desired properties are observed up to 65°C. Upon annealing we increased this level to at least 100°C. This effect is illustrated in the temperature coefficient of the refractive index and the linear expansion coefficient that became 8.7x10^{-5}/°C and 11.2x10^{-5}/°C, compared with initial values of 14.2x10^{-5}/°C and 16.5x10^{-5}/°C respectively.

5. CONCLUSIONS

Methacrylate resins based on ethoxylated bisphenol-A are attractive candidates as coatings in a lens replication process. Their optical and mechanical behaviour is strongly related to the ultimate conversion of functional groups, which is determined by the molecular structure and the polymerization conditions.

REFERENCES

1. J.J.M. Braat, S.P.I.E., Proc., 399, 294, (1983).
2. J.J.M. Braat, A. Smid and M.M.B. Wijnakker, Appl. Opt., accepted for publication.
3. A.J. Bean and R.W. Bassemir, in "UV Curing Science and Technology" (Ed. S.P. Pappas), 187, (1978).
4. E.D. Sallee, Rad. Curing, **3**, 13, (1976).
5. G.W. Gruber, in "UV Curing Science and Technology", (Ed. S.P. Pappas), 172, (1978).
6. J.G. Kloosterboer and G.J.M. Lippits, J. Rad. Curing, **11**, 1, (1984).
7. J.G. Kloosterboer, G.M.M. v.d. Hei, R.G. Gossink and G.C.M. Dortant, Polymer Comm., **25**, 322, (1985).